Symbol	Meaning
p	Sample proportion
H_0	Null hypothesis
H_a	Alternate hypothesis
σ_p	Sample standard deviation for proportions
$t_{0.025}$	Student's t distribution at 5% critical value
α	Probability of Type-I error
χ^2	Tests independence of 2 variables
E	Expected frequency
O	Observed frequency
ANOVA	Analysis of variance
SS	Sum of squares
MS	Mean square
F	F distribution
$\chi^2_{0.05}$	Chi square at 5% significance level
μ_R	Mean number of runs
σ_R	Standard deviation of the number of runs
$=$	Equal to
$<$	Less than
$\leq$	Less than or equal to
$>$	Greater than
$\geq$	Greater than or equal to
$\approx$	Approximately equal to
z	Standard score
df	Degree of freedom

Statistics and Probability In Modern Life

FOURTH EDITION

Joseph Newmark

The College of Staten Island
of the City University of New York

SAUNDERS COLLEGE PUBLISHING

New York   Chicago   San Francisco
Philadelphia   Montreal   Toronto
London   Sydney   Tokyo

—

Text Typeface: Caledonia
Compositor: General Graphic Services, Inc.
Acquisitions Editor: Robert B. Stern
Developmental Editor: Ellen Newman
Project Editor: Sally Kusch
Copy Editor: Robin Bonner
Art Director: Carol Bleistine
Art Assistant: Doris Roessner
Text Designer: Emily Harste
Cover Designer: Lawrence R. Didona
Text Artwork: ANCO/Boston, Inc.
Production Manager: Harry Dean, Merry Post

Cover Credit: THE IMAGE BANK/Michel Tcherevkoff

Printed in the United States of America

STATISTICS AND PROBABILITY IN
MODERN LIFE, 4th edition

0-03-008367-2

Library of Congress Catalog Card Number: 87-31430

89   039   987654

Preface

As in earlier editions, the fourth edition of this text reflects the increasing range of applications of statistics and probability. It has been designed to serve as a general introduction to modern statistics and probability for college students in all academic areas. This text presents to the college student the basic statistical ideas needed in such areas as sociology, business, ecology, economics, education, medicine, psychology, and mathematics. Students in such fields must frequently demonstrate a knowledge of the language and methods of statistics.

Since the concepts of statistics are assuming an ever-increasing role in the social sciences, a special effort has been made in this edition to make these ideas available to students who are not prepared for elaborate symbolisms or complex arithmetic. Although the mathematical content is complete and correct, the language is elementary and easy to understand. Mathematical rigor has not been sacrificed. This makes the text comprehensible to students of varying backgrounds, especially those with little mathematical knowledge. Introductory high school algebra is a sufficient prerequisite to use this text satisfactorily.

I have been careful to avoid expressions that students have difficulty comprehending. The introduction of new terminology has been held to a minimum. In frequent COMMENTS, various points that students often misunderstand or miss completely are carefully discussed. Each chapter begins with Objectives, followed by Statistics in Use, which features a newspaper or magazine clipping that sets the stage for the material to be covered. The ideas introduced are explained in terms of an example chosen from real-life situations with which the student can identify. Each chapter concludes with a Summary and a Study Guide to reinforce the concepts covered, as well as a Formulas to Remember section that serves as a quick reference for the student.

THE CHANGES IN THE FOURTH EDITION

The comments and suggestions of users of the previous editions of this text have proved invaluable in helping me prepare this revision. In this edition, all of the chapters have been reorganized, updated, or expanded to make them more streamlined and to allow the teacher greater flexibility in selecting topics. The following changes have been made to attain these objectives:

1. MINITAB computer output is now included for most chapters (Chapters 1–3 and 8–14). These enable the student to see the MINITAB output results when analyzing one of the procedures or topics discussed within the chapter.

2. A discussion of cumulative frequency histograms, cumulative frequency polygons, and stem-and-leaf diagrams has been added to Chapter 2 to enable the student to look at data in a slightly different way. Additionally, various techniques that some people often use "to lie with graphs" are discussed. The common pitfalls to avoid are presented.

3. A more detailed analysis of summation notation, an interpretation of the standard deviation via Chebyshev's theorem, and ogives are all presented in Chapter 3.

4. Chapter 6 on probability distributions has been reworked to emphasize the distinction between discrete and continuous random variables.

5. The discussion of sampling and sampling error has been expanded in Chapter 8.

6. The material on hypothesis testing and tests concerning means and differences between means for both small and large samples has been considerably expanded to enable the reader to identify the null hypothesis, the alternative hypothesis, and the appropriate test statistic.

7. Chapter 11 on linear correlation and regression has been thoroughly reorganized to allow for (a) obtaining the regression equation in a more natural way, (b) testing inferences, (c) the construction of prediction intervals, and (d) the analysis of multiple regression. There is also a discussion of the coefficient of determination and its interpretation.

8. Goodness of fit is discussed immediately after the chi-square distribution is presented, even before contingency tables are introduced.

9. Chapter 13 on analysis of variance now includes material on one-way ANOVA with unequal sample sizes, as well as an introduction to two-way ANOVA.

10. Detailed boxed procedures have been added throughout the text, especially in the hypothesis-testing chapters. These serve both to summarize and emphasize techniques as well as to help determine the most appropriate procedure that should be used in a given situation.

11. The one-sample sign test, the Wilcoxon signed-rank test, and the Kruskal Wallis H-test have been added to the nonparametric statistics chapter.

12. The number of exercises has been increased substantially. There are now well over 1200 exercises throughout the book.

To The Instructor

In order to write a statistics and probability book that is easily understood by students with a limited mathematical background and to help the student become actively involved in the learning process, I have incorporated the following teaching and learning aids:

1. *Chapter Objectives* Each chapter begins with a list of Objectives that highlight the ideas to be discussed in the chapter.
2. *Statistics in Use* Each chapter begins with a newspaper article taken from the daily press or a magazine clipping that presents the ideas discussed in the chapter in applied context. This feature is intended to motivate the student by showing how the concepts covered in the chapter are applied in real-life situations. Many of the exercises are also based on newspaper articles. This helps make statistics and probability more relevant.
3. *Introduction* Each chapter has an introductory section to set the stage for the topics discussed in the chapter.
4. *Student-Oriented Comments* Throughout the book comments are included to reinforce mathematical concepts that students often miss or find confusing.
5. *Open Format* The book is presented in a visually appealing two-color open format that highlights important rules and theorems in the text.
6. *Historical Notes* A number of chapters contain brief historical notes of interest on important mathematicians such as Carl Friedrich Gauss and Pierre Fermat.
7. *Numerous Illustrations* Many charts, graphs, and sketches illustrate the concepts discussed in each chapter.
8. *Student Aids* Each chapter concludes with a Summary, a Study Guide, and a Formulas to Remember section. The Study Guide and Formulas to Remember sections list the significant new terms and formulas introduced in the chapter and include page numbers for easy reference. In previous editions students have found these aids extremely useful in studying and preparing for exams.
9. *Examples and Exercises* All the examples and exercises have been selected to maintain student interest. They cover a wide range of areas, including the social sciences, psychology, education, medicine, ecology, and busi-

ness, and relate directly to the student's own experience. Some of the exercises are specifically designed to be worked out using a hand-held calculator and are marked with this symbol ⌨ ; however, they can be solved without a calculator.

10. *Mastery Tests*   Each chapter contains two sets of Mastery Tests designed to measure a student's comprehension of the material covered in the chapter, as well as to present a few challenging problems for the more sophisticated student. Form A tests require only short answers and essentially test the material covered in the chapter. Form B tests require longer answers. In general, they are more challenging and supplement the regular exercises at the end of each section. Thus, the Mastery Tests can be used by both the student and teacher.

11. *Suggested Course Outline*   An instructor should have no difficulty in selecting material for a one- or two-semester course. The following outlines indicate how this text can be used.

One-semester Course (meets 40 times per semester, 40 minutes per session)

Text Material	Approximate Amount of Time	Prerequisite Needed for Each Chapter
Chapter 1	1 lesson	none
Chapter 2	6 lessons	none
Chapter 3	6 lessons	Chapter 2, Section 2.2 or the equivalent
Chapter 4	5 lessons	none
Chapter 5 (skip 5.5)	4 lessons	Chapter 4
Chapter 6 (skip 6.7 & 6.8)	5 lessons	Chapter 4
Chapter 7	5 lessons	Chapter 2
Chapter 8	4 lessons	Chapter 7
	36*	

*The remaining meetings can be devoted to exams and review.

Two-semester Course
Semester 1

Text Material	Approximate Amount of Time	Prerequisite Needed for Each Chapter
Chapter 1	1 lesson	none
Chapter 2	6 lessons	none
Chapter 3	6 lessons	Chapter 2, Section 2.2 or the equivalent
Chapter 4	5 lessons	none
Chapter 5	5 lessons	Chapter 4
Chapter 6	7 lessons	Chapter 4
Chapter 7	5 lessons	Chapter 2
Chapter 8	4 lessons	Chapter 7
	39*	

*The remaining meetings can be devoted to exams and review.

Two-semester Course
Semester 2

Text Material	Approximate Amount of Time	Prerequisite Needed for Each Chapter
Chapter 9	5 lessons	Chapter 7
Chapter 10	7 lessons	Chapter 7
Chapter 11	8 lessons	Chapter 10
Chapter 12	4 lessons	Chapter 10
Chapter 13	7 lessons	Chapter 10
Chapter 11	7 lessons	Chapters 8, 10, 11 and 12
	38*	

*The remaining meetings can be devoted to exams and review.

Since comments and suggestions from instructors using previous editions of this test proved invaluable in completing this revision, I would be grateful to continue receiving them. Please address all correspondence to me directly at The College of Staten Island, 715 Ocean Terrace, Staten Island, NY 10301.

SUPPLEMENTS

The following supplements are available for use with this book:

1. A MINITAB supplement specifically keyed to this text has been prepared by Dr. Ernest Blaisdell of Elizabethtown College, teaching the reader how to use the popular statistical package MINITAB. No prior computer knowledge is assumed.

2. *20/20 Statistics* is a visual, interactive statistics courseware package that can be used both as a classroom teaching aid and as supplemental material for the students to work on independently. The package contains a tutorial workbook with software instructions and a module for each topic, a program disk that runs on the Apple PC, and an instructor's manual, which provides answers to exercises in the tutorial workbook and suggestions for use of the courseware by both students and instructors. The *20/20 Statistics* software and Instructor's Manual are written by Professor George W. Bergeman of Northern Virginia Community College, and the tutorial workbook is written by Professor Bergeman and Professor James P. Scott of Creighton University. No prior computer knowledge is required for use of this package.

3. An *Instructor's Manual*, written by Professor Gerald Hobbs of West Virginia University, gives detailed solutions to all of the exercises, as well as to the Mastery Test questions. Lecture suggestions are also included.

4. A *Student Solutions Manual,* written by Professor Karen Zak of the United States Naval Academy, includes worked-out solutions to every other odd-numbered problem in the exercises and Mastery Tests.

5. *Prepared Tests,* written by the author, contain four additional tests for each chapter, as well as two final examinations.

ACKNOWLEDGMENTS

I would like to thank the many instructors and students who received the earlier editions of this book so warmly. I am grateful to those who took the time to send comments, suggestions, and corrections. I am also grateful to the following people who reviewed this edition of the text and made valuable suggestions for its improvement:

Eleanor Allison, Sacred Heart College
William H. Beyer, University of Akron
James E. Holstein, University of Missouri
James Lang, Valencia Junior College
Hubert Lilliefors, George Washington University
Larry J. Ringer, Texas A & M University
John B. Rushton, Metropolitan State College
Charles W. Sinclair, Portland State University
W. Robert Stephenson, Iowa State University

Linn M. Stranak, Union University
Richard Uschold, Canisius College
Steve Vardeman, Iowa State University

Special thanks to the following people for their accuracy reviews of all examples, exercises, and Mastery Tests:

Gerald Hobbs, West Virginia State University
Karen E. Zak, United States Naval Academy

I would also like to express my gratitude to the following people for their excellent work on the various ancillary items that accompany this text:

George W. Bergeman, Northern Virginia Community College (20/20 Statistics)
Ernest Blaisdell, Elizabethtown College (MINITAB Supplement)
Gerald Hobbs, West Virginia University (Instructor's Manual)
James P. Scott, Creighton University (20/20 Statistics)
Karen E. Zak, United States Naval Academy (Student Solutions Manual)

My thanks also go to the authors and publishers who granted permission to use the statistical tables so necessary for this work. Among others, I am indebted to the Literary Executor of the late Sir Ronald A. Fisher, F.R.S., and to Oliver and Boyd, Ltd., Edinburgh, for their permission to reprint tables from *Statistical Methods for Biological, Agricultural, and Medical Research*.

Finally, and most importantly, I wish to thank my wife Trudy and our children Sharon, Rochelle, and Stephen for their understanding and patience as they endured the enormous strain associated with completing this project. Without their encouragement it could not have been undertaken or completed.

Joseph Newmark

To The Student

This is a nonrigorous mathematical text on elementary probability and statistics. If you are afraid that your mathematical background is rusty, don't worry. The only math background needed to use this book is introductory high school algebra or the equivalent; however, those of you who are fairly knowledgeable in math will see how statistics can be used in interesting and challenging problems.

The examples are plentiful and are chosen from everyday situations. Also, the ideas of statistics are applied to a variety of subject areas. This variety indicates the general applicability of statistical methods. In frequent COMMENTS, appropriate explanations of statistical theory are given and the "why's" of statistical methods answered.

Occasionally a section or exercise is starred. This means that it is slightly more difficult and may require some time and thought.

Some of the exercises are designated with the symbol ▦. These exercises have been specifically designed to be worked out using a hand-held calculator. However, they can also be easily worked out without such a calculator.

Each chapter concludes with a summary and a review of the formulas and major terms and concepts introduced in the chapter; page numbers are included for easy reference. In addition, there are Mastery Tests which will help in preparing for exams. Answers to selected exercises and all Mastery Test questions are given in the back of the book.

I hope that you will find reading and using this book an enjoyable and rewarding experience, since a basic knowledge of statistics is essential. Good luck.

Contents

CHAPTER 3
Numerical Methods For Analyzing Data 87

CHAPTER 4
Probability 149

CHAPTER 5
Rules of Probability 209

CHAPTER 6

Some Discrete Probability Distributions and Their Properties

255

CHAPTER 7

The Normal Distribution

313

CHAPTER 8
Sampling 351

CHAPTER 9
Estimation 383

CHAPTER 10
Hypothesis Testing 417

CHAPTER 11

Linear Correlation and Regression   *465*

CHAPTER 12

Analyzing Count Data: The Chi-Square Distribution   *527*

CHAPTER 13

Analysis of Variance **565**

CHAPTER 14

Nonparametric Statistics **599**

APPENDIX

Statistical Tables   *A.1*

The Nature of Statistics

What is Statistics?

OBJECTIVES

- *To discuss* the nature of statistics and numerous examples of how they are used.

- *To identify* the two major areas of statistics—descriptive statistics and inferential statistics. **Descriptive statistics** involves collecting data and tabulating it in a meaningful way. **Inferential statistics** involves making predictions based upon the sample data.

- *To distinguish* between part of a group, called a sample, and the whole group, called the population.

- *To present* a brief discussion of the historical development of statistics and probability.

A DECADE
OF CLEANING
THE AIR

The first Great American Smokeout took place on Nov. 20, 1976. At that time, 37 percent of the U.S. adult population were smokers. A decade later, 30 percent of the population still smokes. A total of 37 million Americans are ex-smokers. As the American Cancer Society reports, however, "The battle is far from over. Those who still smoke seem to be smoking more heavily, and smoking among women and young people is still a particular cause for concern. The harmful effects of involuntary, or passive smoking—the inhaling of cigarette smoke by non-smokers—have come to light. Cigarette companies are mounting an aggressive campaign against the anti-smoking movement."

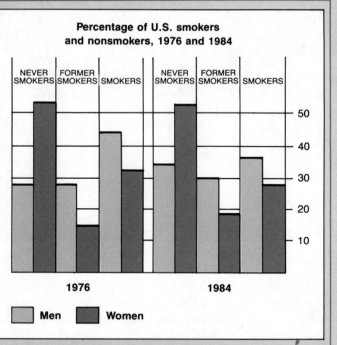

Percentage of U.S. smokers and nonsmokers, 1976 and 1984

Source: National Health Interview Survey (U.S. DHHS NCHS 1984). Reported in "Facts and Figures On Smoking, 1976-1986," American Cancer Society.

The article above indicates that new statistics have been gathered regarding the effects of even passive cigarette smoking. Obviously, the cigarette companies interpret such statistics in a different manner. It is apparent that such statistical data must be analyzed very carefully before arriving at any conclusions.

1.1 INTRODUCTION

What is **statistics?** Most of us tend to think of statistics as having something to do with charts or tables of numbers. While this idea is not wrong, mathematicians and statisticians use the word statistics in a more general sense. Roughly speaking, the term statistics, as used by the statistician, involves collecting numerical information called **data,** analyzing it, and making meaningful decisions based upon the data. Collected data, which represent observations or measurements of something of interest, can be classified into two general types: qualitative and quantitative. **Qualitative data** refer to observations that are nonnumerical or that are attributes. For example, a person's sex, eye color, or blood type, or the brand of whiskey preferred by an individual, etc. Each person can be placed in one and only one category depending upon the situation and how the categories are defined.

On the other hand, **quantitative data** represent observations or measurements that are numerical. For example, the weight of a student, the percentage of homes that are contaminated with radon gas in New Jersey, or the number of college students receiving financial aid at your college. In each of these cases, an individual observation measures some quantity. This gives us quantitative data.

The role played by statistics in our daily activities is constantly increasing. As a matter of fact, the nineteenth-century prophet H. G. Wells predicted that "statistical thinking will one day be as necessary for efficient citizenship as the ability to read and write."

The following examples on the uses of statistics indicate that a knowledge of statistics today is quickly becoming an important tool, even for the layman.

1. Department of Commerce statistics show that we experienced a double-digit rate of inflation throughout 1980.
2. Statistics show that there may be an oversupply of doctors in the 1990s in certain areas of the United States.
3. Statistics collected by the Social Security Administration show that people, on the average, are living longer today than did their parents. This has important implications for predicting the future fiscal soundness of our social security system. (See the article on the top of the next page.)
4. Statistics show that the number of hours in the American workweek is constantly shrinking and that, if the present trend continues, the average workweek will decline to 34 hours. (See Figure 1.1.)
5. Statistics show that if the air we breathe is excessively polluted, then undoubtedly some people will become ill.
6. Statistics show that the world's population is growing at a faster rate than the availability of food. (See Figure 1.2.)
7. Statistics show that the United States is facing a serious energy shortage and that alternative sources of energy must be found to meet the ever increasing demand.

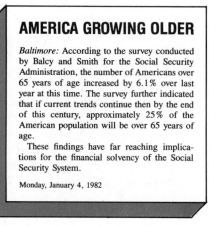

AMERICA GROWING OLDER

Baltimore: According to the survey conducted by Balcy and Smith for the Social Security Administration, the number of Americans over 65 years of age increased by 6.1% over last year at this time. The survey further indicated that if current trends continue then by the end of this century, approximately 25% of the American population will be over 65 years of age.

These findings have far reaching implications for the financial solvency of the Social Security System.

Monday, January 4, 1982

8. Statistics show that any student, no matter what his or her high school background, will succeed in college if properly motivated.

9. Both the Gallup and Harris Polls use statistics in determining public opinion on controversial issues.

10. The statistics given in Table 1.1 show the victimization rates for persons age 12 and over by race, annual family income, and type of crime.

11. Statistics show that it is often impossible to obtain accurate records on the prevalence of crime because many of the victims simply do not report the crime to police. This makes apprehension of the offenders impossible. (See Table 1.2.)

12. Federal Reserve Bank statistics indicate that the number of bank failures has been increasing over the past few years, with more than 100 such failures predicted in each of the next few years. Such statistics have important implications for the integrity of our banking and monetary system.

THE SHRINKING AMERICAN WORKWEEK
Average Hours Worked Per Week in Private Nonfarm Jobs

1910 — 51.3 hours
1930 — 42.1 hours
1950 — 39.8 hours
1960 — 38.6 hours
1970 — 37.1 hours
1975 — 36.1 hours
1980 — 35.1 hours
1985 — 34.8 hours

FIGURE 1.1

Note: Figures for 1910 and 1930 are for workers in manufacturing. *Source:* U.S. Department of Labor.

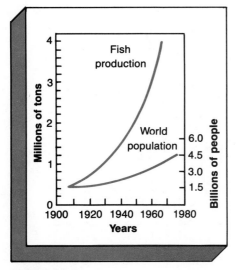

FIGURE 1.2

Over the years, both the world population and the production of fish have increased considerably. Will the world's fish production continue to increase indefinitely, or will we reach a maximum point beyond which production will decrease? *Source:* Fisheries of North America.

Very little formal mathematical knowledge is needed to collect and tabulate data. However, the interpretation of the data in a meaningful way requires careful analysis. If not done by a statistician or mathematician who has been trained to interpret data, statistics can be misused. The following example indicates how statistics can be misused.

In the city of Bushtown the occurrence of polio, thought to be nonexistent, increased by 100% from the year 1984 to 1985. Such a statistic would horrify any parent. However, upon careful analysis it was found that in 1984 there were 2 reported cases of polio out of a population of five million and in 1985 there were 4 cases.

1.2 CHOICE OF ACTIONS SUGGESTED BY STATISTICAL STUDIES

Since the word statistics is often used by many people in different ways, statisticians have divided the field of statistics into two major areas called **descriptive statistics** and **inferential statistics.**

Descriptive statistics involves collecting data and tabulating results using, for example, tables, charts, or graphs to make the data more manageable and meaningful. Very often, certain numerical computations are made that enable

TABLE 1.1
Personal Crimes: Victimization Rates for Persons Age 12 and Over, by Race, Annual Family Income of Victims, Type of Crime, 1984*

(Rate per 1000 population age 12 and over)

Race and Income	Crimes of Violence	Rape	Robbery			Assault			Crimes of Theft	Personal larceny	
			Total	With Injury	Without Injury	Total	Aggravated	Simple		With Contact	Without Contact
White											
Less than $3,000 (10,907,000)	54.4	3.1	10.7	4.4	6.3	40.5	15.9	24.6	83.9	4.3	79.6
$3,000–$7,499 (30,186,000)	35.3	1.0	7.2	2.8	4.4	27.1	11.3	15.8	80.3	3.6	76.7
$7,500–$9,999 (16,700,000)	33.8	[1]0.6	6.6	1.9	4.7	26.6	11.6	15.0	95.3	3.0	92.3
$10,000–$14,999 (38,650,000)	27.2	0.5	4.3	1.2	3.1	22.4	9.2	13.1	93.8	1.7	92.1
$15,000–$24,999 (29,168,000)	27.3	0.4	5.2	1.6	3.6	21.7	8.6	13.2	115.6	2.5	113.1
$25,000 or more (9,888,000)	25.6	[1]0.6	5.7	1.5	4.3	19.3	5.8	13.5	128.3	2.7	125.6
Black											
Less than $3,000 (3,377,000)	48.0	4.1	15.9	7.8	8.1	27.9	15.9	12.0	65.5	7.6	58.0
$3,000–$7,499 (6,469,000)	39.9	1.9	14.8	3.8	11.0	23.1	13.7	9.4	73.0	5.5	67.5
$7,500–$9,999 (1,926,000)	46.5	[1]1.3	17.2	8.8	8.4	28.0	15.4	12.6	91.2	8.2	83.0
$10,000–$14,999 (2,914,000)	32.9	[1]0.5	10.6	[1]2.5	8.2	21.8	11.7	10.1	93.3	4.8	88.4
$15,000–$24,999 (1,533,000)	42.0	[1]2.5	15.0	[1]5.1	9.9	24.5	10.8	13.6	129.5	[1]6.1	123.5
$25,000 or more (242,000)	10.7	[1]0.0	[1]10.7	[1]5.5	[1]5.1	[1]0.0	[1]0.0	[1]0.0	110.7	[1]4.5	106.1

Note: Detail may not add to total shown because of rounding. Numbers in parentheses refer to population in the group: excludes data on persons whose income level was not ascertained.

[1]Estimate, based on zero or on about 10 or fewer sample cases, is statistically unreliable.

*Source: *Criminal Victimization in the United States—A National Crime Survey Report.* U.S. Department of Justice, Law Enforcement Assistance Administration, National Criminal Justice Information and Statistics Service.

TABLE 1.2
Personal Crimes of Violence: Percent Distribution of Reasons for Not Reporting Victimizations to the Police, by Victim-Offender Relationship and Type of Crime, 1984*

Victim-Offender Relationship and Type of Crime	Total	Nothing Could Be Done; Lack of Proof	Not Important Enough	Police Would Not Want to Be Bothered	Too Inconvenient or Time Consuming	Private or Personal Matter	Fear of Reprisal	Reported to Someone Else	Other and Not Given
Involving strangers									
Crimes of violence	100.0	24.2	23.6	8.1	3.7	9.7	3.9	8.2	18.6
Rape	100.0	34.0	[1]1.8	[1]13.3	[1]0.0	[1]9.9	[1]6.9	[1]10.2	24.0
Robbery	100.0	31.0	17.1	10.3	4.6	7.5	5.5	7.6	16.2
Assault	100.0	21.4	26.7	7.1	3.6	10.4	3.2	8.3	19.2
Involving nonstrangers									
Crimes of violence	100.0	8.7	17.6	6.0	1.4	29.3	6.2	14.5	16.3
Rape	100.0	[1]2.1	[1]2.3	[1]9.3	[1]2.8	22.8	[1]0.7	[1]5.3	34.9
Robbery	100.0	12.5	17.8	[1]2.9	[1]0.0	21.6	10.4	17.3	17.5
Assault	100.0	8.2	18.2	6.2	1.5	30.4	5.5	14.6	15.4

Note: Detail may not add to total shown because of rounding.
[1]Estimate, based on zero or on about 10 or fewer sample cases, is statistically unreliable.
*Source: Criminal Victimization in the United States—A National Crime Survey Report. U.S. Department of Justice, Law Enforcement Assistance Administration, National Criminal Justice Information and Statistics Service.

us to analyze data more intelligently. Most of us are concerned with this branch of statistics. For example, even if we know the income of every family in California, we are still unable to analyze the figures because there are too many to consider. These figures, somehow, must be condensed so that meaningful statements can be made.

Inferential statistics or **statistical inference,** on the other hand, is much more involved. To understand why, imagine that we are interested in the average height of students at the University of California. Since there are so many students attending the university, it would require an enormous amount of work to interview each student and gather all the data. Furthermore, the procedure would undoubtedly be costly and could take too much time. Possibly we can obtain the necessary information from a sample of sufficient size that would be accurate for our needs. We could then use the data based upon this sample to make predictions about the entire student body, called the **population.** This is exactly what statistical inference involves. We have the following definitions.

DEFINITION 1.1

*A **sample** is any small group of individuals or objects selected to represent the entire group called the **population,** where the population is the set of all measurements of interest to the sample collector.*

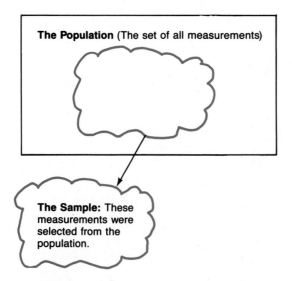

The Population (The set of all measurements)

The Sample: These measurements were selected from the population.

FIGURE 1.3
The relationship between a population and its sample.

> DEFINITION 1.2
>
> **Inferential statistics** *is the study of procedures by which we draw conclusions and make decisions or predictions about a population on the basis of a sample.*

Of course, we would like to make the best possible decisions about the population. To do this successfully we will need some ideas from the theory of probability. Statisticians must therefore be familiar with both statistics and probability. Exactly how such decisions are made will be discussed in later chapters.

1.3 STATISTICS IN MODERN LIFE

Statistics is so important to our way of living that many of us often use statistical analysis in making decisions without even realizing it. Today statistics are used not only by the mathematician but also by the nonmathematician in such areas as psychology, ecology, sports, insurance, education, biology, business, agriculture, music, and sociology, to name but a few. The fields of study to which statistics and probability are being applied is constantly increasing. Their usefulness in the fields of biology, economics, and psychology are so enormous that the subjects of biometrics, econometrics, and psychometrics have come into being.

As statistics developed, probability began to assume more importance because of its wide range of applications. Today the application of probability in gambling is but one of its minor uses. In recent years, statistics has even been applied to determine the total population of various species of living things. In particular, by using very simple procedures, statisticians have been able to predict the total population of such endangered species as the whooping crane and various fish. In each case a number of birds or fish are caught, tagged with a label or some other form of identification, and then released for breeding. When they are recaptured or sighted at a later date, the proportion of tagged fish or birds out of the total catch is calculated and used to predict the total population of the species.

Statistics and probability have recently been used to answer such questions as "Did Shakespeare author a newly discovered poem?" (*Science*, January 24, 1986, p. 335) or "What is the probability that a randomly selected person who has tested positive in a drug screening actually uses drugs?" (*Newsweek*, November 29, 1986, p. 18.)

In the future many new and interesting applications of statistics and probability are likely to be found.

1.4 HISTORICAL NOTES

Although statistics is one of the oldest branches of mathematics, it was not until the twentieth century that its use became widespread. Originally, it involved summarizing data by means of charts and tables. Historically, the use of statistics can be traced back to the ancient Egyptians and Chinese who used statistics for keeping state records. The Chinese under the Chou Dynasty, 2000 B.C., maintained extensive lists of revenue collection and government expenditures. They also maintained records on the availability of warriors.

The study of statistics was really begun by Englishman John Graunt (1620–1674). In 1662 he published his book, *Natural and Political Observations Upon the Bills of Mortality*. Graunt studied the causes of death in different cities and noticed that the percentage of deaths from different causes was about the same and did not change considerably from year to year. For example, deaths from suicide, accidents, and certain diseases not only occurred with surprising regularity but with approximately the same percentage from year to year. Furthermore, Graunt's statistical analysis led him to discover that there were more

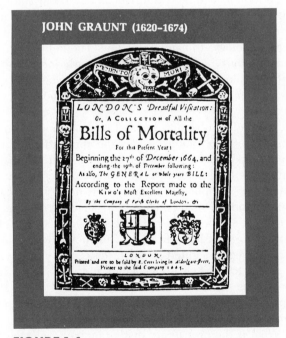

FIGURE 1.4
Illustration "Bills of Mortality" redrawn from *Devils, Drugs,* and *Doctors* by Howard W. Haggard, M.D. Copyright 1929 by Harper and Row, Publishers, Inc.; renewed 1957 by Howard W. Haggard. Reprinted by permission of the publisher.

male than female births. But, since men were more subject to death from occupational hazards, diseases, and war, it turned out that at marriageable age the number of men and women was about equal. Graunt believed that this was nature's way of assuring monogamy.

After Graunt published his *Bills of Mortality*, many other mathematicians became interested in statistics and made important contributions. Pierre-Simon Laplace (1749–1827), Abraham De Moivre (1667–1754), and Carl Friedrich Gauss (1777–1855) studied and applied the **normal distribution** (see page 315). Karl Pearson (1857–1936) and Sir Francis Galton (1822–1911) studied the **correlation coefficient** (see page 467). These are but a few of the many mathematicians who made valuable contributions to statistical theory. In later chapters we will further discuss their works.

Although a great deal of modern statistical theory was known before 1930, it was not commonly used, simply because the accumulation and analysis of statistical data involved time-consuming, complicated computations. However, things changed with the invention of the computer and its ability to perform long and difficult calculations in a relatively short period of time. Statistics soon began to be used for **inference,** that is, in making generalizations on the basis of samples. Also, probability theory was soon applied to the statistical analysis of data. The use of statistics for inference resulted in the discovery of new techniques for treating data.

Interestingly enough, the principles of the theory of probability were developed in a series of correspondences between Blaise Pascal (1623–1662) and Pierre Fermat (1602–1665). Pascal was asked by the Chevalier de Méré, a French mathematician and professional gambler, to solve the following problem: In what proportion should two players of equal skill divide the stakes remaining on the gambling table if they are forced to stop before finishing the game? Although Pascal and Fermat agreed on the answer, they both gave different proofs. It is in these correspondences during the year 1654 that they established the modern theory of probability.

A century earlier the Italian mathematician and gambler Girolomo Cardan (1501–1576) wrote *The Book On Games Of Chance*. This is really a complete textbook for gamblers since it contains many tips on how to cheat successfully. The origins of the study of probability are to be found in this book. Cardan was also an astrologer. According to legend, he predicted his own death astrologically and to guarantee its accuracy he committed suicide on that day. (Of course, that is the most convincing way to be right!) He also had a temper and is said to have cut off his son's ears in a fit of rage.

1.5 WHAT LIES AHEAD

In the preceding sections we mentioned several uses of statistics to convince you that the development of statistics and probability is not static. It is constantly changing. Who knows what a beginning course in statistics will be like by the year 2000? Undoubtedly, different things will be stressed and new applications for statistics and probability will be found. Yet certain basic ideas of probability and its uses in statistical studies will not be changed. Such ideas are too fundamental. It is with these ideas that we will concern ourselves in this text.

As mentioned earlier, with the advent and widespread use of both the microcomputer and the large mainframe computer in the 1980s, many time-consuming and tedious calculations can now be performed quickly and efficiently by the computer. Very little background in computer programming is actually required since the programs are usually incorporated as part of the routine statistical tasks. There are many such statistical computer packages currently on the market such as COMPSTAT, MINITAB, and SPSS. One of the most widely used of these is the SPSS (Statistical Package for the Social Sciences) Batch System.

Along with the widespread acceptance and use of statistical concepts has come the unfortunate use of statistics for deception. Statistics can often be manipulated so that they show what a person wants them to show. In the next chapter we will indicate how statistical graphs can distort information presented by data. Statistical mistakes can be honest also. In its January 2, 1967 issue, *Newsweek* reported on page 10 that Mao-Tse Tung cut the salaries of certain Chinese government officials by 300%. Is it possible for a salary to be cut by 300%?

Also, consider the advertisement by the Morgan Nursing Home chain, which claims that the medical care rendered by its medical staff is so superior that it has not had any stroke victim deaths in its 25 years of operations. What the advertisement fails to mention is that all seriously ill patients are transferred to a nearby hospital. The patient deaths are then recorded on the hospital records, where the deaths actually occurred, rather than on the nursing home records.

The above are but two examples of the misuses (intentionally or not) of statistics. Read the book *How to Lie with Statistics* by Darrell Huff for a number of interesting examples on the misuses of statistics. (See the list of suggested further reading at the end of this chapter.)

In the following chapters we will develop the techniques used in all applications of statistics and discuss the role played by probability in these applications. If you follow the well-defined statistical rules that are given, you will avoid falling into some of the fore-mentioned traps.

EXERCISES

1. After carefully studying 5000 pregnancies, a medical research team concludes that cigarette smoking or drug use by the mother is harmful to *any* unborn child. This conclusion involves
   a. quantitative data.
   b. qualitative data.
   c. descriptive statistics.
   d. inferential statistics.

    e. all of these.
    f. none of these.

2. A statistician made the following claim: Between 1960 and 1985 the salary of the average American rose by the same rate as did the cost for medical insurance. He concluded that the more money an American worker earns, the higher the cost for medical insurance. Do the facts support this claim? Explain your answer.

3. A television commercial claims that "our superior tiles are manufactured to such high standards that they will keep their shine 50% longer." If you are considering purchasing tiles, should you buy this brand of tiles because of its superior "shining" ability?

4. A national health food magazine claims that "95% of its subscribers who follow the magazine's recommendation and take a particular vitamin supplement are healthy and vigorous." Should you begin taking this vitamin supplement?

5. A student who is registered for a math course claims that she will not do well in the course since math has never been an easy subject for her. This conclusion involves
    a. descriptive statistics only.
    b. inferential statistics only.
    c. both descriptive and inferential statistics.
    d. no statistics at all.
    e. none of these.

6. Consider the article below. Do you agree that the new net helps to protect dolphins or is it possible that the number of dolphins in the sea has already been reduced from earlier catches?

NEW TUNA FISH NET PROVES SUCCESSFUL

Los Angeles (Sept 3): Responding to claims that tuna fish nets have been inadvertently capturing and killing dolphins also, the tuna fish industry has developed a new kind of net for catching tuna fish. The new net has been in use for one year. Statistics show that the percentage of dolphins out of the total tonnage of tuna fish caught has dropped radically when compared to the average percentage over the preceding twenty-five years.

Said a spokesperson for the industry "We think that we have the problem under control."

Tuesday—September 3, 1985

7. The Stapleton Medical Association claims that 8 out of every 11 doctors carry catastrophe malpractice insurance. If you select 11 Stapleton doctors at random, would you expect 8 of them to carry this type of insurance? Explain your answer.

8. The word "statistics" is often used by newscasters in radio and television newscasts. Classify the uses as descriptive or inferential.

9. In a recent study it was found that 99% of alcoholics were depressed before they drank.* Can you conclude that if you are depressed, then you are likely to drink and become an alcoholic? Explain.

10. In a study of the incidence of heart attacks in men, a researcher noted that men who had only one job had, on the average, somewhat fewer heart attacks than men who had two jobs. The researcher concluded that having more than one job increases a man's chance of a heart attack. Do the facts support the conclusion? Explain your answer.

11. Suppose we are interested in determining the percentage of men who live at least as long as their fathers did. Can we obtain this percentage by selecting all of our friends and relatives as a random sample? Explain your answer.

1.6 USING COMPUTER PACKAGES

We live in a computer age. The computer is a very important work machine. At home, as well as at the office or in school, personal computers are appearing in greater numbers and are being used for a variety of purposes such as word-processing, record-keeping, accounting, and other workaday tasks. The use of the computer in the field of statistics is no exception. The widespread availability of computers today (both mainframe and personal) and the supporting software, allow the user to perform many detailed statistical calculations.

As with learning to drive a car, operating a computer (mainframe or personal) is an easy task once you learn how to navigate the keyboard, how to negotiate the disk operating system (DOS), what each applications program can do, and which key gets it to do these things. Fortunately, it is rarely necessary for you to write your own computer program to perform statistical analysis. There is an ample supply of computer software currently on the market to perform most statistical tasks. These programs are ideally suited to perform the tedious and time-consuming statistical calculations. Many mainframe computers already have such programs stored in memory and can be called upon for use with a few simple commands.

Although there are many readily available statistical computer programs, the most popular of these are MINITAB, SAS (Statistical Analysis System), SPSS (Statistical Package for the Social Sciences), and BMDP.† Check with your local computer center to see which programs are available. (These centers have consultants who can show you how to use these programs if you need help.) Throughout this book, you will see examples of computer printouts illustrating particular ideas.

At this point we merely mention the existence of statistical packages. We

---

*Briggs and Allyn, 1985.

†For a detailed discussion of some of these packages, see J. Lefkowitz, *Introduction to Statistical Computer Packages* (Boston: Duxbury Press, 1985).

will present the details of the packages at the end of almost every chapter of this text. In this way, you can see how the packages can be used to do the calculations discussed within the chapter.

1.7 SUMMARY

In this chapter we discussed the nature of statistics and how they are used. We distinguished between descriptive statistics and statistical inference. A brief discussion on the origins of statistics and probability was given. Finally, we pointed out how statistics and probability are constantly gaining importance because of their ever-increasing wide range of applications.

STUDY GUIDE

You should now be able to demonstrate your knowledge of the following terms by giving definitions, descriptions, or specific examples. Page references are given in parentheses for each term so that you can easily check your answer.

Statistics (page 3)   Sample (page 8)
Data (page 3)   Population (page 8)
Quantitative data (page 3)   Inferential statistics (page 9)
Qualitative data (page 3)   Statistical inference (page 9)
Descriptive statistics (page 5)

Mastery tests appear at the end of each chapter. You will probably find these tests more useful if you take them after you have solved most of the exercises given in the chapter.

MASTERY TESTS

Form A

Some of the following questions involve statistical ideas to be discussed in greater detail in later chapters. Nevertheless, they are presented here to alert you to the type of thinking that is necessary in the statistical analysis of a problem. Based upon your own experiences, you should be able to discuss and answer most, if not all, of them.

1. After carefully analyzing the available food supply and the world population, several environmentalists have concluded that the world will experience a worldwide food shortage by the year 2010. This conclusion involves
   a. quantitative data.
   b. qualitative data.
   c. descriptive statistics.
   d. statistical inference.
   e. all of these.
   f. none of these.

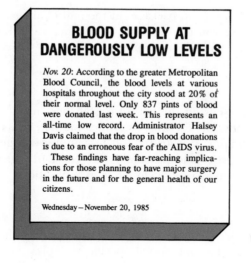

BLOOD SUPPLY AT DANGEROUSLY LOW LEVELS

Nov. 20: According to the greater Metropolitan Blood Council, the blood levels at various hospitals throughout the city stood at 20% of their normal level. Only 837 pints of blood were donated last week. This represents an all-time low record. Administrator Halsey Davis claimed that the drop in blood donations is due to an erroneous fear of the AIDS virus.

These findings have far-reaching implications for those planning to have major surgery in the future and for the general health of our citizens.

Wednesday—November 20, 1985

2. Consider the newspaper article shown above. What part of the article is an example of descriptive statistics?

3. In the previous exercise, what part of the article is an example of inferential statistics?

4. A researcher decided to investigate the reason for the drop in blood donations. He interviewed 5000 college students and determined that the reason was indeed the fear of contracting the AIDS virus. He concluded that this reason seems to be true for the entire population. This conclusion involves
   a. quantitative data only.
   b. qualitative data only.
   c. descriptive statistics.
   d. statistical inference.
   e. none of these.

5. After analyzing accident statistics from an insurance company, an analyst concludes that "since there are more automobile accidents when the roads are dry than when the roads are covered with ice, it is safer to drive when there is ice on the roads." Do you agree with this conclusion?

6. A researcher claims that "Now that women have entered the job market and assumed stress-ridden managerial jobs, then they also will be subject to the same type of illnesses which until now primarily afflicted men only." Do you agree?

For questions 7–9 use the following information: The following statistics are available from the Marlboro Transit Corporation.

Year	Commuter Fare Charged	Usual Number of Commuters Using Company Buses Daily
1982	$2.00	268,000
1983	2.20	250,000
1984	2.35	240,000
1985	2.50	210,000
1986	2.80	200,000

7. The conclusion "The usual number of commuters using company buses daily has been decreasing over the years" involves what type of statistics: statistical inference or descriptive statistics?

8. The conclusion "There is a considerable difference between the number of commuters using the buses when the fare charged is $2.80 instead of $2.00" involves what type of statistics: statistical inference or descriptive statistics?

9. The conclusion "Any future rise in the fare charged will result in a further decrease in the number of commuters using company buses" involves what kind of statistics: statistical inference or descriptive statistics?

10. The average income per person in Stanton is $10,000 per year. Can we conclude that the average income for a family of 5 is $50,000 per year? Explain your answer.

MASTERY TESTS

Form B

1. In 1985, only 25% of the registered Democrats actually voted in local elections. Can we conclude that 75% of the registered Democrats did not care who would win the election?

2. *Is Smoking Dangerous to Your Health?*   The American Cancer Society claims that the statistics overwhelmingly support this claim. Yet the tobacco industry claims that the statistics are not conclusive. Explain how both sides in this controversy interpret the results. (See the newspaper article on page 2.)

3. Due to economic conditions, the Gandy Company required all employees to take a temporary 5% cut in salary. A year later, the economic health of the company had improved drastically, and the company gave each employee a 5% raise in salary. Are the workers now earning as much as they were before? Explain your answer.

4. To decrease the number of accidents, the Highway Patrol issued 79,869 speeding tickets in 1985 to motorists on the Riverview Expressway. In 1986, only 41,602 speeding tickets were issued. Can one conclude that people are obeying the posted speed limits to decrease the number of accidents?

5. In 1920 there were 21 houses of worship in Dover. In 1986, there were 47 houses of worship. Can one conclude that the reason for the increase in the number of houses of worship is because people are more religious today?

6. Records indicate that graduates of a prestigious Ivy League college have an average annual salary of $63,000 ten years after graduation. On the other hand, graduates of a local college have an annual salary of $38,000 ten years after graduation. Can we conclude that the education received by the graduates of the prestigious Ivy League college is superior?

7. What part of the article below involves descriptive statistics?

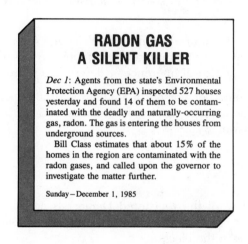

RADON GAS
A SILENT KILLER

Dec 1: Agents from the state's Environmental Protection Agency (EPA) inspected 527 houses yesterday and found 14 of them to be contaminated with the deadly and naturally-occurring gas, radon. The gas is entering the houses from underground sources.

Bill Class estimates that about 15% of the homes in the region are contaminated with the radon gases, and called upon the governor to investigate the matter further.

Sunday – December 1, 1985

8. What part of the article involves inferential statistics?

9. A cardiologist claimed: "Statistics show that the average blood serum cholesterol level of Americans is dropping because the foods we eat contain less fatty items." Based upon your own research, would you say that the facts support this claim? Explain your answer.

10. Over the years, numerous claims have been made that fluoridation of our water supply is associated with a higher incidence of cancer than occurs with use of nonfluoridated water. After a careful and detailed analysis of the water supply systems of 20 American cities, P. D. Oldham* concluded that there is no significant difference in the cancer rates after adjustments for the nonhomogeneity of the populations are made. Does the newspaper

*Oldham, P. D., "Fluoridation of Water Supplies and Cancer: A Possible Association?" *Applied Statistics* 26, (1977), pp. 125–135.

CANCER AND FLUORIDATION OF OUR WATER SUPPLY

June 1: Medical researchers reported yesterday that preliminary studies of the fluoridation of our water system showed an apparent increase in the death rate from cancer per 100,000 of population. Further studies will be undertaken before any definite conclusions can be drawn.

June 1, 1985

article above indicate that Oldham's analysis of the water supply systems and his conclusions are wrong? Explain your answer.

11. Read the article "Statistics-Watching: A Guide for the Perplexed" in the *New York Times* (November 21, 1975, p. 45). This article discusses how economic statistics can often be confusing.

SUGGESTED FURTHER READING

1. Beniger, J. R., and D. L. Robyn. "Quantitative graphs in statistics: A brief history," *The American Statistician* (February 1978), pp. 1–11.
2. Campbell, Stephen K. *Flaws and Fallacies in Statistical Thinking*. Englewood Cliffs, NJ: Prentice-Hall, 1974.
3. Dale, Edwin L. Jr. "Statistics-Watching: A Guide for the Perplexed" *The New York Times*, November 21, 1975, p. 45.
4. Galton, Francis. "Classification of Men According to Their Natural Gifts" in *The World of Mathematics* edited by James R. Newman. New York: Simon & Schuster, 1956. Vol. 2, part VI, chap. 2.
5. Graunt, John. "Foundations of Vital Statistics" in *The World of Mathematics* edited by James R. Newman. New York: Simon & Schuster, 1956. Vol. 3, part VIII, chap. 1.
6. Huff, Darrell. *How to Lie with Statistics*. New York: Norton, 1954.
7. Martin, Thomas L. Jr. *Malice in Blunderland*. New York: McGraw-Hill, 1973.
8. Newmark, Joseph, and Frances Lake. *Mathematics as a Second Language*, 4th ed. Reading, MA: Addison-Wesley, 1987.
9. Tanur, Judith M., F. Mosteller, W. H. Kruskal, R. F. Link, R. S. Pieters, and G. R. Rising. *Statistics: A Guide to the Unknown*. (E. L. Lehmann, Special Editor.) San Francisco, CA: Holden-Day, 1978.

CHAPTER 2

The Description of Sample Data

OBJECTIVES

- *To discuss* what is meant by a frequency distribution. This is a convenient way of grouping data so that meaningful patterns can be found.

- *To draw* a histogram, which is nothing more than a graphical representation of the data in a frequency distribution.

- *To analyze* circle graphs or pie charts, which are often used when discussing distributions of money.

- *To study* frequency polygons and bar graphs, which are used when we wish to emphasize changes in frequency, for example in business and economic situations.

- *To introduce* the graph of a frequency distribution, which resembles what we will call a normal curve.

- *To learn* how the area under a curve is related to relative frequency and, hence, probability.

- *To present* an alternate way of analyzing data graphically by means of stem-and-leaf diagrams, which are particularly well-suited for computer sorting techniques.

- *To show* how graphs can be misused.

- *To explain* the use of index numbers, which are used to show changes over a period of time.

- *To show* how we use such words as random sample, frequency, class mark, relative frequency, area under a curve, and pictograph. These words are used frequently when studying statistics.

DOLLAR'S BUYING POWER DECLINED 60% SINCE 1970

Despite slowing inflation, the purchasing power of the dollar has eroded nearly 60% since 1970, the Conference Board said. The business information firm said an average family of four that earned $10,000 in 1970 now needs $25,450 to maintain the same after-tax buying power. While inflation was cited as the major cause of the change, higher federal income and Social Security taxes were said to account for about one-third of the drop.

Wednesday—June 19, 1985

Many unions negotiate an escalator clause in their contract with management. When the Consumer Price Index (CPI) or some other index rises, then so will an employee's salary. Thus, it is hoped that the employee's purchasing power can keep up with inflation. How are such index numbers computed? In this chapter, we indicate how index numbers are calculated.

2.1 INTRODUCTION

Suppose that John, a student in a sociology class, is interested in determining the average age at which women in New York City marry for the first time (see the accompanying newspaper article). He could go to the Marriage License Bureau and obtain the necessary data. (We will assume that all marriages are reported to the Bureau.) Since there are literally thousands of numbers to analyze, he would want to organize and condense the data so that meaningful interpretations can be drawn from them. How should he proceed?

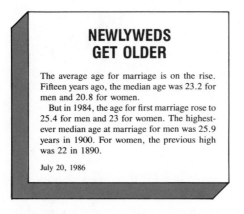

NEWLYWEDS GET OLDER

The average age for marriage is on the rise. Fifteen years ago, the median age was 23.2 for men and 20.8 for women.

But in 1984, the age for first marriage rose to 25.4 for men and 23 for women. The highest-ever median age at marriage for men was 25.9 years in 1900. For women, the previous high was 22 in 1890.

July 20, 1986

One of the first things to do when given a mass of data is to group it in some meaningful way and then construct a frequency distribution for the data. After this is done, one can use various forms or graphs of the distribution so that different distributions can be discussed and compared.

In this chapter we will discuss some of the common methods of describing data graphically. In the next chapter, we will discuss the numerical methods for analyzing data. In each case, we will use examples from everyday situations so that you can see how and where statistics is applied.

2.2 FREQUENCY DISTRIBUTIONS

Rather than analyze thousands of numbers, John decides to take a sample. He will examine the records of the Marriage Bureau on a given day and then use the information obtained from this sample to make inferences about the ages of *all* women who marry for the first time.

The most important thing in taking a sample of this sort is the requirement that it be a **random sample.** This means that each individual of the population, a woman who applies for a marriage license in our case, must have an equally likely chance of being selected. If this requirement is not satisfied, one cannot make meaningful inferences about the population based upon the sample. Since we will have more to say about random samples in Chapter 8, we will not analyze here whether or not John's sampling procedure is random.

John has obtained the ages of 150 women for a given day from the Marriage Bureau. Table 2.1 lists the ages in the order in which they were obtained from the Bureau records.

TABLE 2.1
Age at Which 150 Women in New York City Married for the First Time

20	17	24	18	16	23	21	17	21	27	23	17	16	19	14
23	26	38	33	20	33	26	22	26	33	26	35	28	35	24
21	18	21	23	22	21	19	23	19	18	19	15	23	21	18
27	40	27	26	22	25	27	25	25	25	30	19	26	32	22
22	19	22	19	24	15	20	22	23	26	23	18	21	20	21
34	41	35	20	29	20	29	27	29	32	29	29	32	28	31
19	22	23	25	23	23	18	19	24	24	21	20	24	22	20
30	31	39	43	38	37	30	37	33	30	36	34	36	32	26
25	23	17	24	18	24	24	21	16	20	18	22	25	24	17
28	29	34	31	25	34	25	36	28	27	31	27	28	30	28

The only thing we can say for sure is that most women were in their 20s and that one was as young as 14 while another woman was as old as 43 years when they married. However, since the ages are not arranged in any particular order, it is somewhat difficult to conclude anything else. Clearly, the data must be reorganized. We will use a frequency distribution to do so. A **frequency distribution** is a convenient way of grouping data so that meaningful patterns can be found.

DEFINITION 2.1

*The word **frequency** will mean how often some number occurs.*

A frequency distribution is easy to construct. We first make a list of numbers starting at 14, the youngest age, and going up to 43, the oldest age, to indicate the age of each women. This list will be the first column. In the second column we indicate tally marks for each age, that is, we go through the list of original numbers and put a mark in the appropriate space for each age. Finally, in the third column we enter the total number of tally marks for each particular age. The sum for each age gives us the frequency column. When applied to our example we get the distribution shown in Table 2.2.

A number of interesting things can be seen at a glance from Table 2.2. Since 23 occurred most often, this is the age at which most women married. Also, the frequency began to decrease as age increased beyond the age of 23.

The data in Table 2.2 can also be arranged into a more compact form as shown in Table 2.3. In this table we have arranged the data by age groups, also called **classes or intervals**. We select a group size of 3 years since this will

TABLE 2.2
Construction of a Frequency
Distribution for Age at Which 150
Women in New York City Married for
the First Time

Column 1	Column 2	Column 3
Age	Tally	Frequency
14	|	1
15	||	2
16	|||	3
17	卌	5
18	卌 |||	8
19	卌 ||||	9
20	卌 ||||	9
21	卌 卌	10
22	卌 卌	10
23	卌 卌 ||	12
24	卌 卌	10
25	卌 ||||	9
26	卌 |||	8
27	卌 ||	7
28	卌 |	6
29	卌 |	6
30	卌	5
31	||||	4
32	||||	4
33	||||	4
34	||||	4
35	|||	3
36	|||	3
37	||	2
38	||	2
39	|	1
40	|	1
41	|	1
42		0
43	|	1

result in 10 different groups. Although any number of intervals could be used, for this example, we have chosen to work with 10. With the possible exception of the first and last interval, classes are usually of equal size. Generally speaking, if we subtract the smallest age from the largest age and divide the results by 10, we will get a number, rounded off if necessary, which can be used as the

TABLE 2.3
Age at Which 150 Women in New York City Married for the First Time

Class Number	Ages	Class Mark	Tally	Class Frequency	Relative Frequency
1	14–16	15	|||||		6	6/150
2	17–19	18	||||| ||||| ||||| ||||| ||	22	22/150
3	20–22	21	||||| ||||| ||||| ||||| ||||| ||||	29	29/150
4	23–25	24	||||| ||||| ||||| ||||| ||||| |||||		31	31/150
5	26–28	27	||||| ||||| ||||| |||||		21	21/150
6	29–31	30	||||| ||||| |||||	15	15/150
7	32–34	33	||||| ||||| ||	12	12/150
8	35–37	36	||||| |||	8	8/150
9	38–40	39	||||	4	4/150
10	41–43	42	||	2	2/150

Total Frequency = 150

size of each group. In our case, we have

$$\frac{43 - 14}{10} = \frac{29}{10} = 2.9 \quad \text{or 3 when rounded}$$

COMMENT   Some people prefer to have each class begin with a multiple of 5 so as to make information easy to read.

We have labeled the first column of Table 2.3 **class number** because we will need to refer to the various age groups in our later discussions. Thus, we will refer to age group 26–28 as class 5. Notice also that each class has an *upper limit*, the oldest age, and a *lower limit*, the youngest age. A **class mark** represents the point which is midway between the limits of a class, that is, it is the midpoint of a class. Thus, 18 years is the class mark of class 2. We have indicated the class mark for each class in the third column of Table 2.3. This table, containing a series of intervals and the corresponding frequency for each interval, is an example of **grouped data.**

COMMENT   The class mark need not be a whole number.

COMMENT   Although we mentioned that equal class intervals are usually used, this does not necessarily include the first and last classes.

In the fifth column of Table 2.3 we have the **class frequency,** that is, the total tally for each class. Finally, the last column gives the relative frequency of each class. Formally stated, we have the following definition:

DEFINITION 2.2

*The **relative frequency** of a class is defined as the frequency of that class divided by the total number of measurements (the total frequency). Symbolically, if we let f_i denote the frequency of class i where i represents any of the classes, and we let n represent the total number of measurements, then*

$$Relative\ frequency = \frac{f_i}{n} \quad for\ class\ i$$

Since in our example the total frequency is 150, the relative frequency of class 6, for example, is 15 divided by 150, or $\frac{15}{150}$. Here $i = 6$ and $f_i = 15$. Similarly, the relative frequency of class 7 where $i = 7$ and $f_i = 12$ is $\frac{12}{150}$. The relative frequency of class 10 is $\frac{2}{150}$.

Once we have a frequency distribution, we can present the information it contains in the form of a graph called a **histogram.** To do this we first draw 2 lines, one horizontal (across) and one vertical (up-down). We mark the class boundaries along the horizontal line and indicate frequencies along the vertical line. We draw rectangles over each interval, with the height of each rectangle equal to the frequency of that class. All of our rectangles have equal widths. Generally, the areas of the rectangles should be proportional to the frequencies.

The histogram for the data of Table 2.3 is shown in Figure 2.1. The area under the histogram for any particular rectangle or combination of rectangles is proportional to the relative frequency. Thus the rectangle for class 4 will contain $\frac{31}{150}$ of the total area under the histogram. The rectangle for class 7 will contain $\frac{12}{150}$ of the total area under the histogram. The rectangles for classes 9 and 10 together will contain $\frac{6}{150}$ of the total area under the histogram since

$$\frac{4}{150} + \frac{2}{150} = \frac{4 + 2}{150} = \frac{6}{150}$$

The area under a histogram is important in statistical inference and we will discuss it in detail in later chapters.

COMMENT By changing the number of intervals used, we can change the appearance of a histogram and hence the information it gives. (See Exercises 15 and 16 at the end of this section.)

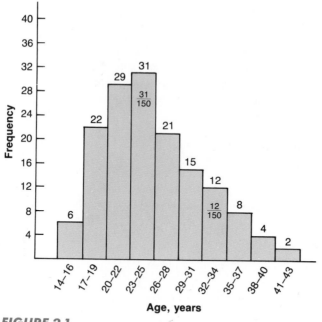

FIGURE 2.1

We can summarize the procedure to be used in the following rule:

RULES FOR GROUPING DATA AND FOR DRAWING HISTOGRAMS

When unorganized data are grouped into intervals and histograms are drawn, the following guidelines should be observed.

1. The intervals should be of equal size (with the possible exception of the end intervals).
2. The number of intervals should be between 5 and 15. (Using too many intervals or too few intervals will result in much of the data not effectively presented.)
3. The intervals can never overlap each other. If an interval ends with a counting number, then the following interval begins with the next counting number.
4. Any score to be tallied can fall into one and only one interval.
5. Draw the histogram by using rectangles placed next to each other. The rectangles are placed together to indicate that the next interval begins as soon as one interval ends. The height of each rectangle represents its frequency (assuming the widths of the rectangles are equal). There are no gaps between the rectangles drawn in a histogram (with the exception of an interval having a frequency of zero).

To further illustrate the concept of a frequency distribution and its histogram, let us consider the following example:

● **Example 1**   *Stock Transactions*

Manya, a stock broker, keeps a daily list of the number of transactions where her customers buy the stock on margin. The following number of stocks were bought on margin by her customers during the first 10 weeks of 1986.

```
30   28   16   23   22   10   8   15   16   23
29   30   22   15   24    9   5    4   15   24
21   24   16   14   21   13   6    2   23   22
24   21   27   13   20    8   4    3   24   20
22   18   23   23   17    7   6    1   16   17
```

Construct a frequency distribution for the preceding data and then draw its histogram.

Solution

We will use 10 classes. The largest number is 30 and the smallest number is 1. To determine the class size, we subtract 1 from 30 and divide the result by 10 and get:

$$\frac{30 - 1}{10} = \frac{29}{10} = 2.9 \quad \text{or 3 when rounded}$$

Thus, our group size will be 3. We now construct the frequency distribution. It will contain six columns as indicated.

Class Number	Number of Stocks	Class Mark	Tally	Class Frequency	Relative Frequency
1	1–3	2	|||	3	3/50
2	4–6	5	|||||	5	5/50
3	7–9	8	|||||	4	4/50
4	10–12	11			1	1/50
5	13–15	14	|||||		6	6/50
6	16–18	17	||||| ||	7	7/50
7	19–21	20	|||||	5	5/50
8	22–24	23	||||| ||||| ||||	14	14/50
9	25–27	26			1	1/50
10	28–30	29	||||	4	4/50

Total Frequency = 50

The relative frequency column is obtained by dividing each class frequency by the total frequency, which is 50. We draw the following histogram, which has the frequency on the vertical line and the number of stocks on the horizontal line.

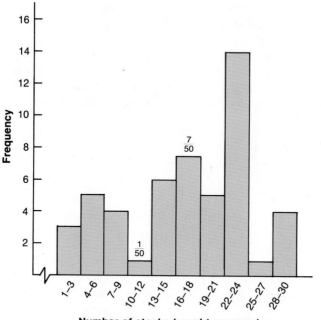

Number of stocks bought on margin

The rectangle for class 6 will contain $\dfrac{7}{50}$ of the total area under the histogram. Similarly, the rectangle for class 4 contains $\dfrac{1}{50}$ of that area. Finally, the rectangles for classes 1 and 2 contain

$$\frac{3}{50} + \frac{5}{50} = \frac{3 + 5}{50} = \frac{8}{50}$$

of the total area under the histogram. What part of the area is contained in *all* the rectangles under the histogram?

Often we may be interested in picturing data that have been arranged in relative frequency form. We can then draw a **relative frequency histogram** by constructing a histogram in which the rectangle heights are relative frequencies of each class. The procedure is illustrated in the following example.

● **Example 2**   *Young Business Owners*
According to Rick Hendricks, regional manager of First National Bank, the age at which business owners first open their businesses has been decreasing. In a random survey of 1000 business owners in the Hampton District, the following results were obtained:

Age at Which Owner Started Business	Frequency	Relative Frequency (In Decimal Form)
16–20	20	20/1000 = 0.02 or 2%
21–25	120	120/1000 = 0.12 or 12%
26–30	230	230/1000 = 0.23 or 23%
31–35	200	200/1000 = 0.20 or 20%
36–40	160	160/1000 = 0.16 or 16%
41–45	80	80/1000 = 0.08 or 8%
46–50	70	70/1000 = 0.07 or 7%
51–55	60	60/1000 = 0.06 or 6%
56–60	50	50/1000 = 0.05 or 5%
61–65	10	10/1000 = 0.01 or 1%
Total Frequency = 1000		*Total* = 100%

The frequency histogram as well as the relative-frequency histogram for the above data are shown below.

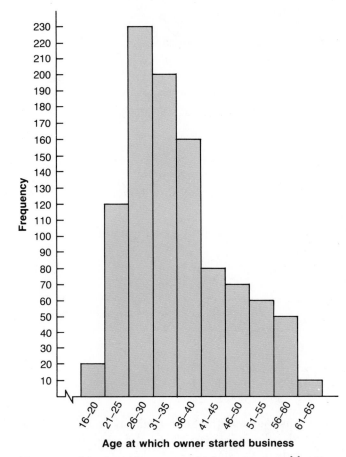

Frequency histogram for age at which owner started business.

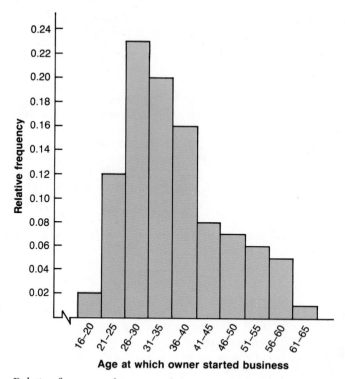

Relative-frequency histogram for age at which owner started business.

EXERCISES

For each of Exercises 1–6, construct a frequency distribution table and then draw its histogram. (Use ten classes in each.)

1. The ages of the 100 participants in a recent jogging marathon were as follows:

28	18	56	23	74	32	33	42	23	53
31	25	73	38	61	31	56	30	31	21
46	28	68	44	50	42	71	31	35	29
29	19	31	56	40	23	39	39	38	37
34	20	27	23	31	31	28	42	41	69
36	32	46	19	26	71	70	45	27	32
23	36	35	18	29	69	62	23	62	69
29	29	41	33	53	62	65	27	71	28
26	43	57	29	42	51	51	19	23	21
30	51	69	67	31	54	50	18	42	22

2. According to Sergeant Frank Pierce of the State Highway Patrol, the number of arrests daily for driving while under the influence of alcohol during the months of November and December 1986 throughout the state was as follows:

12	15	40	20	29	16	23	19	9	21
31	11	34	16	24	20	28	25	13	38
18	32	27	23	36	38	15	31	10	29
29	22	12	29	12	15	14	40	15	40
30	24	22	33	21	10	20	19	17	16
16	39	30	11	3	30	35	15	21	29
14									

3. Rhapsody Johnson is employed by the Environmental Protection Agency. Her job is to inspect the exhaust systems of automobiles to make sure that they do not pollute the atmosphere. The following list gives the scores of 50 cabs that were tested on the exhaust emission control machine on February 20, 1986.

15.2	40.2	33.5	10.0	28.6	75.8	60.6	53.9	45.3	20.1
18.8	7.7	19.1	62.2	55.5	12.5	16.7	59.4	36.7	72.2
29.6	60.1	48.3	20.8	39.9	37.3	19.8	75.5	74.8	69.7
38.1	52.0	69.8	71.7	29.2	16.5	23.4	61.1	38.2	51.6
71.4	26.3	28.6	53.9	18.8	40.2	18.3	69.0	42.9	31.2

4. In order to tighten security at airports, all passengers must pass through metal detectors. However, some of these detectors will often sound an alarm when a passenger is carrying not a dangerous item but rather a calculator, a pen, coins, or keys. Records indicate that one metal-detecting machine at a busy airport in Florida sounded its alarm the following number of times daily during the first three months of 1986.

0	5	27	1	6	3	18	21	9	12
21	18	16	12	15	13	16	25	14	18
29	7	9	18	19	28	0	13	26	16
18	13	16	29	4	10	12	16	8	5
16	21	25	4	6	16	15	0	14	11
20	16	28	2	18	21	10	3	17	15
9	28	0	16	9	13	17	2	3	19
28	12	2	0	11	28	21	8	5	23
17	29	11	17	7	0	1	6	1	29

5. Forty-two 2-family houses were sold in Petersville during December 1986. The following are the selling prices of these houses.

$120,000	$275,000	$129,000	$170,000	$254,000	$270,000	$172,000
180,000	160,000	180,000	270,000	189,000	198,000	185,000
200,000	180,000	265,000	268,000	194,000	215,000	251,000
250,000	129,000	248,000	220,000	142,000	120,000	208,000
175,000	181,000	125,000	138,000	165,000	158,000	198,000
225,000	200,000	176,000	145,000	215,000	143,000	137,000

6. Priscilla Yablonski placed an ad in the campus newspaper offering to sell her present computer system so that she could purchase a larger system. The following bids for the computer system were received.

$1200	$1375	$1440	$1500	$1230	$1200	$1475	$1200	$1550	$1600
2000	1550	1300	1600	1480	1600	1525	1300	1750	1800
1800	1400	1950	1430	1840	1950	1800	1500	1600	1925
1750	1625	1575	1250	1980	1700	1200	1950	1350	1200
1900	2000	1200	1700	1550	1375	1900	1700	1975	2000

7. Gordon Blaisley is Director of the Motor Vehicle Bureau in his state. He supervises the issuance of driver's licenses. During the first 60 working days of 1986, his agency issued the following number of driver's licenses (on a daily basis).

180	223	50	69	78	281	300	286	165	50
300	310	75	83	189	145	99	199	300	125
185	190	150	210	98	158	63	149	100	300
250	198	100	299	275	143	293	300	75	298
279	276	150	187	160	190	87	145	53	232
300	300	290	125	136	138	60	124	89	248

a. Construct a frequency distribution and its histogram using ten classes.
b. Construct a frequency distribution and its histogram using only five classes.
c. Construct a frequency distribution and its histogram using 15 classes.
d. Compare the histograms in parts (a), (b) and (c). What information, if any, is lost by using fewer classes? by using too many classes?

8. A dress company employs 40 workers who each complete similar products in its two factories. In each factory there are 20 employees. The number of products completed by each employee in each factory is as follows:

Factory A						Factory B				
17	24	16	28	11		7	26	11	9	24
25	5	7	10	30		17	14	32	5	28
9	4	31	21	13		19	25	28	31	12
35	26	21	19	28		31	27	25	37	19

a. Construct frequency distributions and histograms for each factory.
b. Combine the data for both factories, and construct a frequency distribution and its histogram.
c. Compare the histograms in parts (a) and (b). Comment.

9. Refer to Exercise 8. The personnel manager decides to change the working conditions in Factory A to determine what effect this will have on production. She installs new lighting facilities, new air conditioning, carpeting, piped music, and a new coffee machine. She now notices that the production of the employees in Factory A is

11	21	37	15
14	21	29	17
17	18	25	19
21	16	35	16
28	28	28	21

a. Construct the frequency distribution and its histogram for the new data.
b. Compare the new histogram with the original histogram for Factory A and the histogram for the combined data. Comment.

10. Seymour and Alfred both deliver packages for two competing delivery services. During December, the number of packages delivered daily by each is as follows:

Seymour							Alfred					
52	43	38	27	27	32		53	45	41	38	27	43
42	63	38	61	47	71		29	36	37	44	33	38
52	51	39	47	38	53		46	49	47	28	60	58
41	27	63	27	55	62		55	46	39	56	53	37
43	33	38	57	41	61		47	37	35	60	38	58
38							55					

a. Construct a histogram for each employee.
b. Who is a better employee (in terms of delivering packages)? Explain.

11. Why is it important that the classes be of equal size?

12. A cardiologist tested the blood serum cholesterol levels of 85 male patients in the "over 50 years" category. The results are given below.

180	320	300	250	200	225	280	250	185	198	230
189	200	185	189	234	320	200	236	310	260	187
210	290	311	180	225	300	250	275	234	189	295
300	260	201	239	293	259	235	198	189	200	310
315	234	215	261	280	220	190	315	305	320	250
190	190	256	299	320	250	320	300	195	234	
250	199	300	220	180	280	300	230	199	190	
305	220	310	195	235	291	195	320	320	300	

Draw the relative-frequency histogram for the data.

13. An IQ test was given to 100 fifth-grade students. The results of the test are given below:

IQ Score	Frequency
80 or less	1
81–85	4
86–90	9
91–95	12
96–100	17
101–105	24
106–110	14
111–115	11
116–120	6
Above 120	2

Draw the histogram for the data and answer the following:
  a. What part of the area is below, that is, to the left of, the 96–100 IQ category?
  b. What part of the area is in the 111 or above IQ category?
  c. What part of the area is in the 96–110 IQ category?
  d. What part of the area is in the 111–115 category?
  e. What part of the area is not in the 111–115 category?

14. Fifty owners of video cassette recorders (VCRs) were asked to indicate the number of movies each had rented during the previous 12 months. Their responses are as follows:

```
7    6   23   17   11   11   21   29   15   10
2   12    7   16   17   10   16   12   14   14
3   10    9    8   16    9   11   18    2    7
5    8    0   12   22   14   10    3    8    3
1   14    4   10   10   17   15    7   11   15
```

Draw the histogram for the data (using 10 intervals, each of length 3) and then answer the following:
  a. What part of the area is below, that is, to the left of 12?
  b. What part of the area is 18 or above?
  c. What part of the area is between 3 and 26 inclusive?
  d. What part of the area is either below 3 or above 26?
  e. What part of the area is *not* between 12 and 20?
  f. What part of the area is above 29?

15. Refer back to the data of Example 1 on page 29. Construct a frequency distribution and histogram using only
  a. 5 classes **b.** 15 classes

16. Compare the histograms obtained in Exercise 15 with the histogram given on page 30. Comment.

17. Draw a histogram and a relative frequency histogram for the following data:

```
39   28   35   49   28   27
43   37   27   29   33   39
52   63   43   38   47   42
61   27   51   61   43   53
52   38   53   60   48   41
28   44   41   40   53   27
37   53   57   30   41   36
39   41   47   29   38   61
```

2.3 OTHER GRAPHICAL TECHNIQUES

In the preceding section we saw how frequency distributions and histograms are often used to picture information graphically. In this section we will discuss some other forms of graphs which are often of great help in picturing information contained in data.

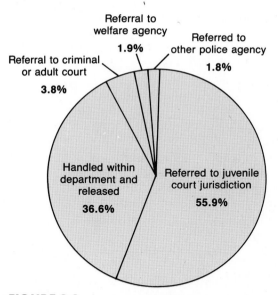

FIGURE 2.2

Percent distribution of juveniles taken into police custody by method of disposition, United States, 1984. *Source:* U.S. Department of Justice, Federal Bureau of Investigation *Uniform Crime Reports for the United States,* 1984 (Washington, D.C.) U.S. Government Printing Office, 1985.

Circle Charts or Pie Charts

A common method for graphically describing qualitative data is the **circle chart** or **pie chart.** A circle (which contains 360° or 360 degrees) is broken up into various categories of interest in the same way as one might slice a pie. Each category is assigned a certain percentage of the 360° of the total circle, depending upon the data.

The magazine clipping shown in Figure 2.2 is an example of a pie chart that reveals interesting statistics. Since crime committed by juveniles is on the increase, many people are interested in knowing how the courts are handling such cases. Are the courts too liberal? According to the chart, 36.6% of all juveniles taken into police custody in 1984 were handled within the department and released. Also, 55.9% of them were referred to juvenile court jurisdiction for appropriate action. Similarly, the pie chart given in Figure 2.3 indicates the reason that people gave for possessing a handgun or a pistol.

The information contained in both of these charts has important implications for our criminal justice system in particular and/or our society in general.

To illustrate the procedure of drawing pie charts, consider the data given in Table 2.4, which indicates the monthly living expenses of a doctoral student at a state university during 1985.

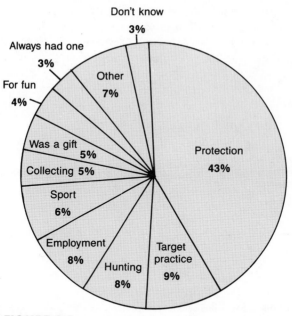

FIGURE 2.3

Gun-owners' reasons for possessing a handgun or pistol in the United States, 1985. *Source: Sourcebook of Criminal Justice Statistics.* U.S. Government Printing Office, Washington, D.C., p. 204.

TABLE 2.4
Monthly Living Expenses in 1985 for a Doctoral Student at a State University

Item	Amount (In Dollars)
Food	100
Apartment	75
Car and transportation	40
Entertainment	85
Laundry	20
Miscellaneous	40
	Total = 360

Since the total expenditure was \$360 and there are 360° in a circle, we can construct the pie chart directly (without any conversions). We draw a circle and partition it in such a way that each category will contain the appropriate number of degrees, as shown in Figure 2.4.

It is more convenient to work with percentages than with amounts of money. Thus, we may want to convert the amount of money spent in each category into percentages. Thus, the student's money was spent as follows:

Relative Frequency

$$\frac{100}{360} = 0.2778, \text{ or } 27.78\%, \text{ for food}$$

$$\frac{75}{360} = 0.2083, \text{ or } 20.83\%, \text{ for the apartment}$$

$$\frac{40}{360} = 0.1111, \text{ or } 11.11\%, \text{ for the car}$$

$$\frac{85}{360} = 0.2361, \text{ or } 23.61\%, \text{ for entertainment}$$

$$\frac{20}{360} = 0.0556, \text{ or } 5.56\%, \text{ for laundry}$$

$$\frac{40}{360} = 0.1111, \text{ or } 11.11\%, \text{ for miscellaneous items}$$

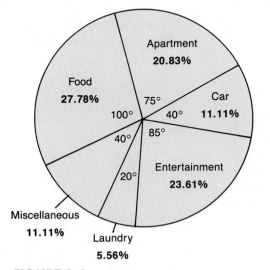

FIGURE 2.4
Pie chart showing the monthly living expenses in 1985 for a doctoral student at a state university.

We have indicated these percentages in the pie chart of Figure 2.4.

COMMENT   We convert a fraction into a decimal by dividing the denominator, that is, the bottom number, into the numerator, that is, the top number. Thus $\frac{100}{360}$ becomes

$$
\begin{array}{r}
0.27777 \\
360\overline{)100.00000} \\
\underline{720} \\
2800 \\
\underline{2520} \\
2800 \\
\underline{2520} \\
2800
\end{array}
$$

When rounded, this becomes 0.2778. This number is written in percentage form as 27.78%.

COMMENT   The sum of the percentages may not necessarily be 100%. This discrepancy is due to the rounding off of numbers. However, in the preceding example, the sum is 100%.

We will further illustrate the technique of drawing pie charts by working several examples.

● Example 1

During 1985, a nationwide auto-leasing company sold 15,000 cars to the individuals who had originally leased them. The types of cars involved are shown below:

Type of Car	Number Sold
Manufactured by Japanese companies	3100
Manufactured by General Motors	4800
Manufactured by Chrysler Corp.	2000
Manufactured by Ford Motor Co.	1150
Manufactured by American Motors	850
Manufactured by German companies	2330
Manufactured by other companies	770
	15,000 = *Total Sold*

Draw the pie chart for these data.

Solution

We first convert the numbers into percentages by dividing each by the total 15,000. Thus, we have

Type of Car	Number Sold	Percentage of Total
Japanese companies	3100	$\frac{3100}{15000} = 0.2067$, or 20.67%
General Motors	4800	$\frac{4800}{15000} = 0.32$, or 32%
Chrysler Corp.	2000	$\frac{2000}{15000} = 0.1333$, or 13.33%
Ford Motor Co.	1150	$\frac{1150}{15000} = 0.0767$, or 7.67%
American Motors	850	$\frac{850}{15000} = 0.0567$, or 5.67%
German companies	2330	$\frac{2330}{15000} = 0.1553$, or 15.53%
Other companies	770	$\frac{770}{15000} = 0.0513$, or 5.13%

Now we multiply each percentage by 360° (the number of degrees in a circle) to determine the number of degrees to assign to each part. We get

$0.2067 \times 360° = 74.41°$, or 74°, for Japanese companies

$0.32 \times 360° = 115.20°$, or 115°, for General Motors

$0.1333 \times 360° = 47.99°$, or 48°, for Chrysler Corp.

$0.0767 \times 360° = 27.61°$, or 28°, for Ford Motor Co.

$0.0567 \times 360° = 20.41°$, or 20°, for American Motors

$0.1553 \times 360° = 55.91°$, or 56°, for German companies

$0.0513 \times 360° = 18.47°$, or 18°, for other companies

Then, we use a protractor and compass to draw each part in order, using the appropriate number of degrees. In our case we obtain the pie chart represented in Figure 2.5.

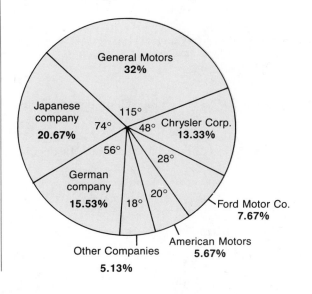

FIGURE 2.5

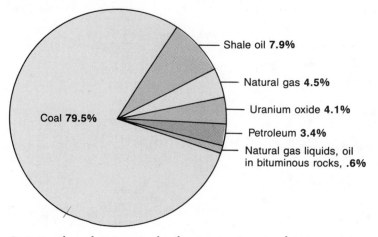

Statistics show that America has been using its natural resources at an increasing rate. As a result, government officials are looking for alternate sources of energy. The circle graph indicates some possible sources.

● **Example 2** *Oil Reserves*

It has been estimated that the number of billions of barrels of oil in the western hemisphere, excluding Alaska, is given by the pie chart shown in Figure 2.6. Assuming that there are 130 billion barrels of oil in reserve, answer the following:

a. How many barrels are in reserve in the United States?
b. How many barrels are there in Mexico?
c. How many barrels are there in Canada?

Solution

a. The United States has 29.7% of the 130 billion barrels. Thus, since

$$0.297 \times 130 = 38.61$$

the United States has 38.61 billion barrels of oil. (In decimal form 29.7% is written as 0.297.)

b. Mexico has 59.93 billion barrels of oil since

$$0.461 \times 130 = 59.93$$

c. Canada has 8.19 billion barrels of oil since

$$0.063 \times 130 = 8.19$$

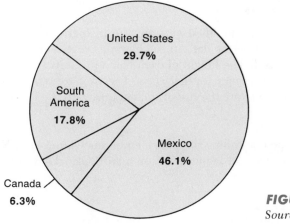

FIGURE 2.6
Source: Department of the Interior.

COMMENT Again we wish to point out that pie charts are commonly used to summarize qualitative (or categorical) data. Some people believe that pie charts are more difficult to read than other graphs; however, such charts are particularly useful when discussing distributions of money.

We summarize the procedure to be used in constructing pie charts:

HOW TO CONSTRUCT A PIE CHART OR CIRCLE GRAPH

1. Determine all the categories that are of interest from the data.
2. For each category determined in step 1, calculate its relative frequency.
3. Draw a circle and assign a slice of the circle to each category. The size of each slice should be proportional to the fraction of observations in that category. Also, the central angle should be an angle whose measure is 360° times the relative frequency for that category. The sum of the measures of all the central angles should be 360° (except for possible rounding errors).
4. Place an appropriate label in each category and indicate the percentage of the total number of observations in the category. This can be found by using the fact that for each category,

Percentage = relative frequency × 100

The sum of all the percentages must always be 100% (except for possible rounding errors).

Bar Graphs and Pictographs

The **bar graph** and a simplified version, the **pictograph,** are also commonly used to graphically describe data. In such graphs **vertical bars** are usually (but not necessarily) used. The height of each bar represents the number of members, that is, the frequency, of that class. The bars are often also drawn horizontally. We will illustrate the use of such graphs with the following examples:

● **Example 3**

The number of people calling the police emergency number in New York City for assistance during a 24-hour period on a particular day was as follows:

Time		Number of
Starting at	Ending at (Up to but not Including)	Calls Received
12 midnight– 2 A.M.		138
2 A.M.– 4 A.M.		127
4 A.M.– 6 A.M.		119
6 A.M.– 8 A.M.		120
8 A.M.–10 A.M.		122
10 A.M.–12 noon		124
12 noon– 2 P.M.		125
2 P.M.– 4 P.M.		128
4 P.M.– 6 P.M.		131
6 P.M.– 8 P.M.		139
8 P.M.–10 P.M.		141
10 P.M.–12 midnight		140

The bar graph for this data is shown in Figure 2.7.

From the bar graph we see that the least number of calls was received during the hours of 4 A.M.–6 A.M. The number of calls received after that period steadily increased until a maximum occurred during the 8 P.M.–10 P.M. period.

When such statistical information is presented in a bar graph, rather than in a table, it can be readily used by police officials to determine the number of police officers needed for each time period.

COMMENT Although the time periods in Example 3 are consecutive, the bars in the bar graph are drawn apart from each other for emphasis and ease in interpretation.

COMMENT Note that the beginning of the vertical line in Figure 2.7 is drawn with a jagged edge. We should always start a vertical scale with 0. However, when all of the frequencies are large, we insert this jagged edge in the beginning to indicate a break in the scale.

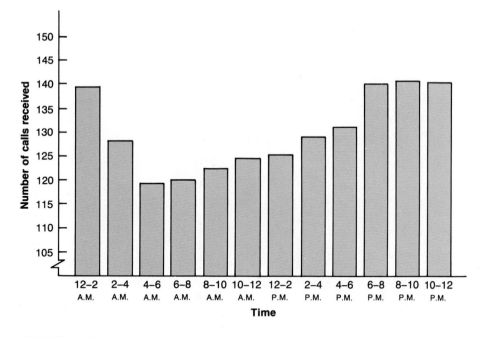

FIGURE 2.7

● **Example 4**

The number of Asians residing in California during 1970 and during 1980 is shown in Table 2.5. The bar graph for these data is given in Figure 2.8.

TABLE 2.5
Asian Population in California

	1980	1970	Percent Increase
Filipino	357,514	138,859	157
Chinese	322,340	170,131	89
Japanese	261,817	213,280	23
Korean	103,891	15,756	559

Source: U.S. Census Bureau

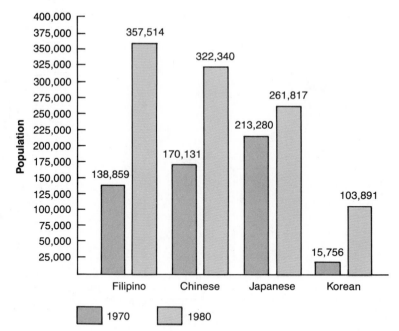

FIGURE 2.8

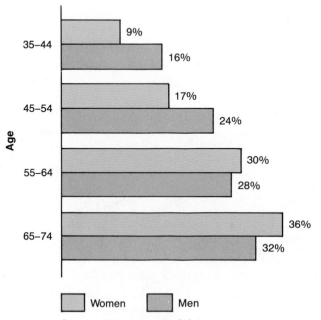

FIGURE 2.9   *Source:* U.S. Bureau of the Census.

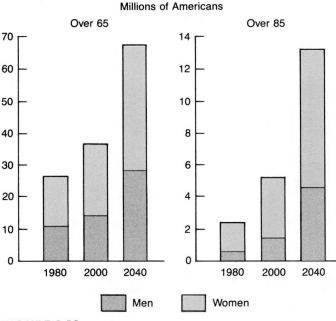

**Population Explosion
of Old People**

The rate of increase for those over 85
will be almost triple the rate of increase
for those over 65 by the year 2040.

Millions of Americans

FIGURE 2.10

Source: Social Security Administration.

When both bar graphs are drawn on the same scale, the U.S. Census Bureau can determine easily which segment of the Asian population increased the most between 1970 and 1980. Often the bars of the bar graph are placed side by side or superimposed one upon the other for easy comparison. This is shown in the bar graphs in Figures 2.9 through 2.11. Also note that the bars have been drawn horizontally.

Several variations of the bar graph are commonly used. In such modifications, columns of coins, pictures, or symbols are used in place of bars. When symbols and pictures are used, the bars are sometimes drawn horizontally. We call the resulting graph a **pictograph.** Pictographs do not necessarily have to be drawn in the form of a bar graph. Several pictographs are shown in Figures 2.12 through 2.14.

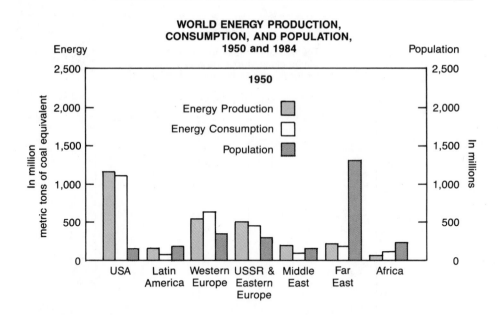

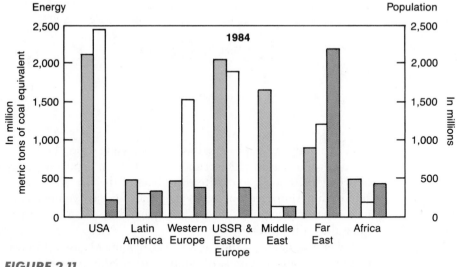

FIGURE 2.11

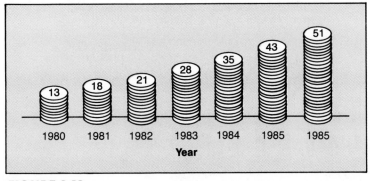

FIGURE 2.12
Expenditure for social services by a large northeastern city government. Each symbol represents six million dollars.

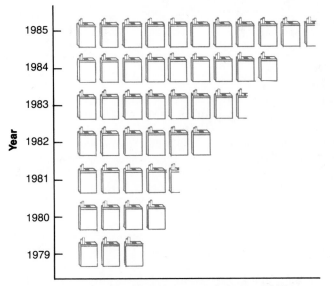

Number of packages of cigarettes sold

FIGURE 2.13
The number of packs of cigarettes sold weekly in the Magway Bus Terminal. Each symbol represents 2000 packages.

Public and Nonpublic School Enrollment, K–12
New York State, 1970, 1980 and 1990

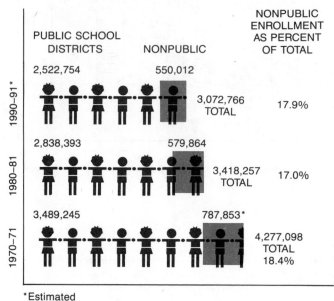

*Estimated

FIGURE 2.14

Enrollment in elementary and secondary schools increased annually during the 1960s. However, beginning in 1971, enrollment began to decline. This declining trend, expected to continue through the 1980s, is due both to reductions in the birth rate in recent years and out-migration from the State. Nonpublic school enrollment is expected to decline at a much slower rate than public school enrollment. Between 1983–84 and 1990–91, nonpublic school enrollment is expected to decline by 2.6 percent, while public school enrollment should decrease by 4.4 percent. *Source:* "Education Statistics New York State," January 1985, State Education Department.

Frequency Polygons

Another alternate graphical representation of the data of a frequency distribution is a **frequency polygon.** Here again the vertical line represents the frequency and the horizontal line represents the class boundaries.

DEFINITION 2.3

If the midpoints (class marks) of the tops of the bars in a bar graph or histogram are joined together by straight lines, then the resulting figure, without the bars, is a **frequency polygon.**

● **Example 5**

Draw the frequency polygon for the data of Example 1 on page 29.

Solution

Since the histogram has already been drawn (page 30), we place dots on the midpoints of the top of each interval and then join these dots. The result is the frequency polygon shown in Figure 2.15.

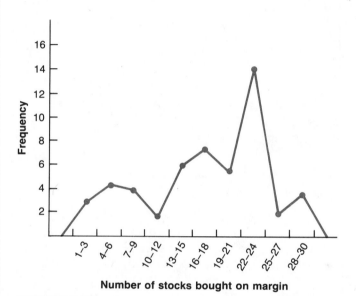

Number of stocks bought on margin

FIGURE 2.15

● **Example 6**

The Bookerville Red Cross Chapter compiles statistics on the number of pints of blood that it collects during the year. For the first six months of 1986, it has constructed the graph shown in Figure 2.16.

Answer the following:

a. During which months of the year were there at least 8000 pints of blood collected?

b. During which month(s) was the most blood collected?

COMMENT Since frequency polygons emphasize changes (rise and fall) in frequency more clearly than any other graphical representation, they are often used to display business and economic data.

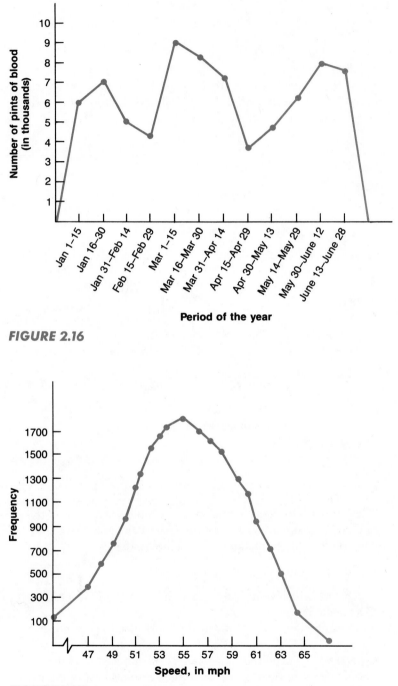

FIGURE 2.16

FIGURE 2.17

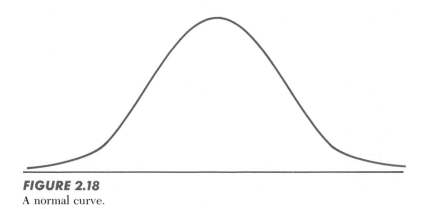

FIGURE 2.18
A normal curve.

● **Example 7**
Consider the frequency polygon showing the distribution of the speeds of cars as they passed through a particular speed enforcement station (radar trap). See Figure 2.17.

There are many frequency distributions whose graphs resemble a bell-shaped curve as shown in Figure 2.18. Such graphs are called **normal curves** and their distributions are known as **normal distributions.** Since many things that occur in nature are normally distributed, it is no surprise that they are studied in great detail by mathematicians. We will discuss this distribution in detail in a later chapter.

Cumulative Frequency Histograms and Cumulative Frequency Polygons

Suppose we analyze the data presented in the table below which represents the weights of 250 army recruits admitted to a particular training base.

Weight (rounded to the nearest lb)	Frequency (number)
131–140 lb	23
141–150	43
151–160	59
161–170	69
171–180	56
	250

A histogram representing the grouped data is shown below:

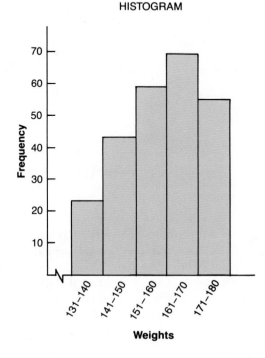

Based upon the frequency distribution and the histogram we can say that the weights of 23 of the recruits were in the interval 131–140 lb, that the weights of 43 of the recruits were in the interval 141–150 lb, and so forth.

Often we are interested in answering questions of the type "How many of the army recruits weighed less than or equal to a certain weight?" For example, suppose we wanted to know "How many of the recruits weighed less than or equal to 170 pounds?" We can answer this question by adding or "accumulating" the frequencies in the grouped data. Thus by adding the frequencies for the four lowest intervals, 23 + 43 + 59 + 69, we find that 194 of the recruits weighed 170 pounds or less. A histogram that displays these "accumulated" figures is called a **cumulative frequency histogram.** For our example, the cumulative frequency histogram is given below.

CUMULATIVE FREQUENCY HISTOGRAM

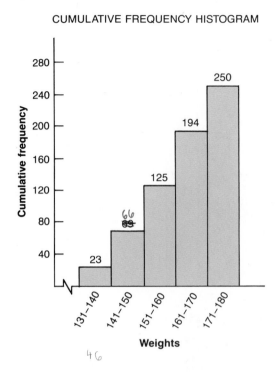

4 6

COMMENT The frequency scale for our cumulative frequency histogram will go from 0 to 250 which represents the total frequency for our data.

We can also draw a cumulative frequency polygon for the above data. A **cumulative frequency polygon** is simply a line graph connecting a series of points that answer the question mentioned earlier, namely, "How many of the army recruits weighed less than or equal to a certain weight?" Such a graph is shown at the top of the next page.

We construct the cumulative frequency polygon as follows: For the interval 171–180, a point is placed at the upper right of the bar to show that 250 of the recruits weighed 180 pounds or less. For the interval 161–170, a point is placed at the upper right of the bar to show that 194 of the recruits weighed 170 pounds or less.

We continue this process of placing dots for all the five intervals. You will notice that the last dot is placed at the bottom left of the interval 131–140 since "0 of the recruits weighed 130 pounds or less. The line graph connecting these six points represents the cumulative frequency polygon.

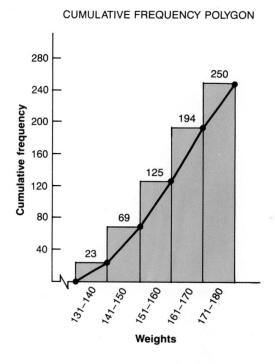

CUMULATIVE FREQUENCY POLYGON

COMMENT   It is also possible to label the vertical scale of a cumulative frequency histogram (or polygon) in a slightly different manner, namely, one involving percents. This will be done in Chapter 3 when we discuss percentiles.

EXERCISES

1. During a 1986 blood donor drive, 6000 pints of blood were collected by a local chapter of the American Red Cross. The blood was classified as follows:

Number of Pints Collected	Blood Type
480	Type B, Rh positive
120	Type B, Rh negative
240	Type AB, Rh positive
60	Type AB, Rh negative
2100	Type A, Rh positive
300	Type A, Rh negative
2340	Type O, Rh positive
360	Type O, Rh negative

Draw a pie chart to picture this information.

2. In a recent nationwide survey, 4000 elementary schools were asked to indicate what type of microcomputer the school used for instructional and administrative purposes. The results were as follows:

Number of Schools	Type of Computer Used
203	Atari
1296	Apple
501	Commodore
291	TRS-80
1239	IBM-PC
345	Radio Shack
125	All others

Draw a pie chart to picture this information.

3. After the breakup of AT&T on January 1, 1984, many manufacturers began selling phones. A survey of 4320 families on January 1, 1986, produced the following pie chart:

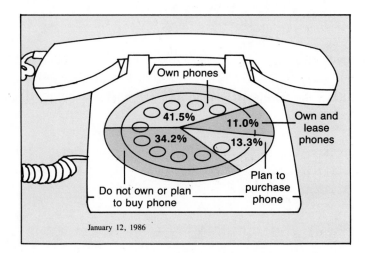

a. How many families owned their own phone?
b. How many families own and lease phones?
c. How many families plan to purchase phones?
d. How many families do not own nor plan to buy a phone?

4. On February 20, 1986, there were 569 inmates at the Charlesville correctional facility. The crimes committed by the inmates are shown in the following pie chart.

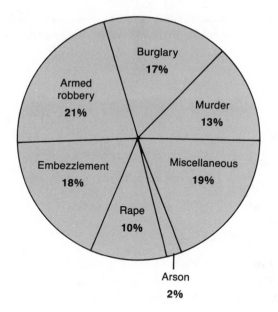

How many inmates were in jail as a result of
a. Armed robbery?
b. Burglary or embezzlement?
c. Murder, rape, or arson?

5. The distribution of IQ scores in the Marleville School system is given in the following table:

IQ Range	Frequency
71–80	67
81–90	170
91–100	370
101–110	562
111–120	308
121–130	189
Above 130	42

Draw a bar graph to picture the above data.

6. Each of the world's six largest suspension bridges is longer than 1000 meters. Use the information given in the graph on the next page to answer the questions:
a. Find the length of each bridge to the nearest 100 meters.
b. What is the approximate difference in length between the Humber bridge and the Verrazano Narrows bridge?
c. What is the approximate difference in length between the Golden Gate bridge and the Bosporus bridge?

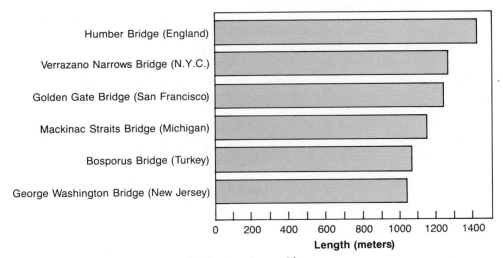

The six longest suspension bridges in the world.

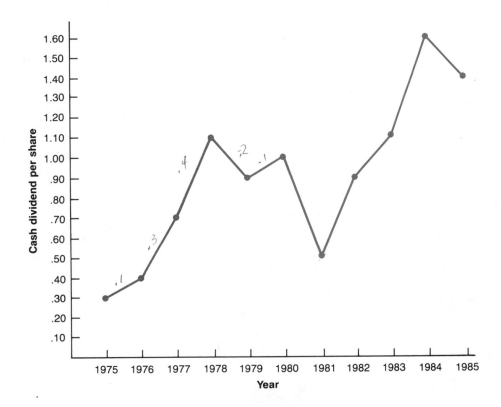

7. The graph above (although not a frequency polygon) shows the annual cash dividend per share paid by a large communications company to its shareholders during the years 1975–1985.

a. In what year was the dividend the lowest paid?

b. Between what two years did the dividend increase the most?

c. Between what two years did the dividend decrease most sharply?

d. What is the range of the dividends paid by this company over the ten-year period from 1975 to 1985?

8. In a survey conducted by Hadley and Kolb, people were asked whether they favored building additional nuclear energy-generating facilities or were opposed.* The results were then compared to earlier such surveys. The findings are given below:

Year	Opposed	In Favor	Not Sure
1980	940	760	300
1981	760	960	280
1982	840	940	220
1983	620	1180	200
1984	500	1340	160
1985	360	1580	60

Draw a bar graph to picture this information.

9. Consider the newspaper article below. The following list gives the total monthly rainfall (in inches) for the region over the past year. Draw the graph for the data.

Oct.	11.3	Feb.	5.8	June	5.1
Nov.	7.6	Mar.	10.3	July	4.2
Dec.	8.4	Apr.	7.9	Aug.	3.7
Jan.	6.9	May	6.4	Sept.	3.6

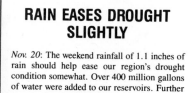

RAIN EASES DROUGHT SLIGHTLY

Nov. 20: The weekend rainfall of 1.1 inches of rain should help ease our region's drought condition somewhat. Over 400 million gallons of water were added to our reservoirs. Further conservation is still needed.

Wednesday – Nov. 20, 1985

*Hadley and Kolb, New York, 1986.

10. Seven hundred parking permits were issued by Spoduk College to authorize students to park their cars on campus grounds. The ages of those cars are shown in the following pie chart.

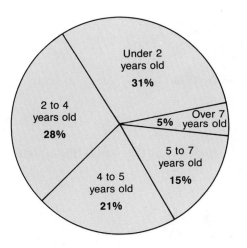

a. How many cars were over 5 years old?
b. How many cars were less than 7 years old?

11. Each student in the entering class of a university was asked to select a major or area of specialization. The students selected the following:

Major or Area of Specialization	Number of Students
Nursing	169
Business	187
Computer Science	223
Engineering	128
Pre-Medical	86
Pre-Law	43
Other	204

a. Draw a bar graph to picture this information.
b. Draw a pie chart to picture this information.
c. Which graph (bar graph or pie chart) presents the information in a more useful form?

12. The following pictograph indicates the number of foreign cars sold by a used car dealership over the past seven years. (Each symbol represents 200 cars.)
a. How many foreign cars were sold in 1984?
b. By approximately how many cars did the number of foreign cars sold increase from 1979 to 1985?
c. What is the total number of foreign cars sold by this dealership during the past seven years?

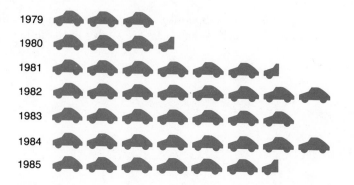

13. The following pictograph indicates the 1980 population of the world's most populated countries. (Each symbol represents 100 million people.)

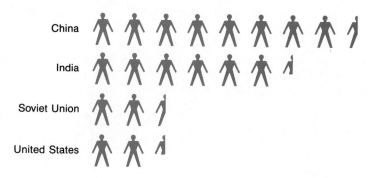

a. Approximately how many people did each of these countries have in 1980?
b. Approximately how many more people did China have than did the Soviet Union in 1980?

14. Draw a cumulative frequency histogram and a cumulative frequency polygon for the data given in each of the following:
a. The heights of 250 army recruits at a particular training base.

Height (rounded to the nearest inch)	Frequency (number)
62–64	21
65–67	67
68–70	82
71–73	47
74–76	33
	250

b. The test scores of 250 army recruits on a physical fitness exam.

Test Score	Frequency
61–70	39
71–80	48
81–90	84
91–100	62
101–110	17
	250

15. Based upon your own experience, which of the following do you think is (are) likely to be normally distributed?
a. Weight of individuals in an aerobics class
b. Length of newborn babies
c. Cost of auto insurance
d. The amount of energy consumption by individuals in the United States
e. The deductions for charity claimed on Federal 1040 tax returns

2.4 STEM-AND-LEAF DIAGRAMS

The graphical techniques discussed to this point are well-suited to handle most situations. In recent years, however, a new technique known as **stem-and-leaf diagrams** has become very popular. It represents a combination of sorting techniques often used by computers and a graphical technique.

To see how this new method works let us analyze some information on the number of people using the cash machines daily at an automated banking facility. The following data are available for 30 business days.

162	146	110	219	174	165
128	159	197	205	153	166
151	142	188	212	162	123
203	137	167	178	183	152
178	198	143	179	189	138

Using stem-and-leaf diagrams we can group the data and at the same time obtain a display that looks like a histogram. This is done as follows: The first number on the list is 162. We designate the first 2 leading digits (16) as its **stem.** We call the last (or trailing) digit its **leaf** as illustrated here.

Stem (First or Leading Digits)	Leaf (Last or Trailing Digit)
16	2

The stem and leaf of the number 128 are 12 and 8, respectively. Also, the stem and leaf of the number 151 are 15 and 1, respectively.

To form a stem-and-leaf display for the above data, we first list all stem possibilities in a column starting with the smallest stem (11, which corresponds

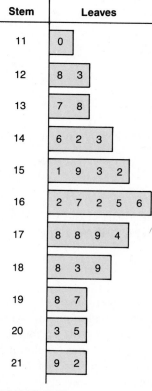

Stem	Leaves
11	0
12	8 3
13	7 8
14	6 2 3
15	1 9 3 2
16	2 7 2 5 6
17	8 8 9 4
18	8 3 9
19	8 7
20	3 5
21	9 2

FIGURE 2.19

to the number 110) and ending with the largest stem (21, which corresponds to the number 219). Then we place the leaf of each number from the original data in the row of the display corresponding to the number's stem. This is accomplished by placing the last (or trailing) digit on the right side of the vertical line opposite its corresponding leading digit or stem. For example, our first data value is 162. The leaf 2 is placed in the stem row 16. Similarly, for the number 128, the leaf 8 is placed in the stem row 12. We continue in this manner until each of the leaves is placed in the appropriate stem rows. The completed stem-and-leaf display will appear as shown in Figure 2.19.

The stem-and-leaf diagram arranges the data in a convenient form since we can now count the number of leaves for each stem. We then obtain the frequency distribution. From this it is very easy to draw the histogram or bar graph. If we turn the above stem-and-leaf display on its side we obtain the same type of bar graph provided by the frequency distribution. This is shown in Figure 2.20:

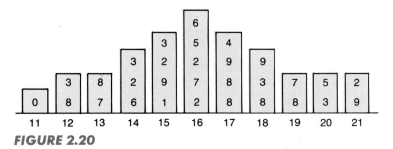

FIGURE 2.20

COMMENT One major advantage of a stem-and-leaf diagram over a frequency distribution is that the original data are preserved. The stem-and-leaf diagram displays the value of each individual score as well as the size of each data class. This is not possible with a bar graph.

COMMENT A stem-and-leaf diagram presents the data in a more convenient form. This will make it easy to perform various arithmetic calculations to be studied in the next chapter.

We summarize the procedure to be used in constructing stem-and-leaf diagrams in the following rule.

> *RULE*
>
> To construct a stem-and-leaf diagram, proceed as follows:
>
> 1. Determine how the stems and leaves will be identified.
> 2. Arrange the stems in order in a vertical column, starting with the smallest stem and ending with the largest.
> 3. Go through the original data and place a leaf for each observation in the appropriate stem row.
> 4. If the display looks too cramped and narrow, we can stretch the display by using two lines (or more) per stem so that we can place leaf digits 0, 1, 2, 3, and 4 on one line of the stem and leaf digits 5, 6, 7, 8, and 9 on the other line of the stem.

Let us illustrate the preceding rule with another example.

● **Example 1**

A new drug treatment clinic recently opened in the East and the number of addicts treated with methadone per day during the first month was as follows:

37	95	34	26	45
88	89	24	61	28
42	78	67	32	79
67	29	28	24	35
68	72	91	86	78

Draw a stem-and-leaf diagram for the above data.

Solution

Let us use the first digit of each of the numbers as the stem and the second digit as the leaf. Then we arrange the stems in order in a vertical column. Although they can be arranged horizontally, the stems are usually arranged vertically. We get

Stem

Now we go through the original data and place a leaf for each observation in the appropriate stem row. The stem-and-leaf diagram is shown in Figure 2.21:

Stem	Leaves
2	9 4 8 6 4 8
3	7 4 2 5
4	2 5
5	
6	7 8 7 1
7	8 2 9 8
8	8 9 6
9	5 1

FIGURE 2.21

EXERCISES

Construct a stem-and-leaf diagram for each of the following:

1. Sam is interested in determining the average height of students in his math class. The following list gives the height, in inches, of each student in the class.

68	69	60	76	68	69
67	65	65	77	65	65
74	71	59	69	71	67
73	70	72	70	73	70
71	69	75	69	75	72
74	70	69	65	72	68

2. Are drivers observing the 50 miles-per-hour speed limit on the state speedway? To answer this question several legislators set up radar equipment and clocked the speed of the first 45 cars that passed. The following speeds were recorded.

45	54	51	57	51	59	56	51	49
60	59	50	49	54	46	56	48	57
62	61	62	62	47	46	48	58	58
63	48	61	53	61	49	51	56	51
60	56	59	55	60	53	57	53	55

* 3. Thirty students in a statistics class received the following scores on the mathematics section of the Scholastic Aptitude Test (SAT). Group the numbers so that 20 values could fall on each line. Thus, the stems will be 4 and 5 where each stem will have five lines. Values falling between 400 and 419 (inclusive) on the first leaf, values between 420 and 439 (inclusive) falling on the second leaf, etc.

586	512	478	531	532
483	576	493	523	428
512	419	501	513	488
510	576	526	486	449
433	528	512	503	573
463	593	572	517	586

* 4. Randy is interested in buying an elaborate and complete computer system for his child. Numerous dealers have quoted him the following prices, in dollars, for the same system. Group the numbers so that 40 values fall on each line. Thus, the stems will be 1900–1939, 1940–1979, and so on.

$2000	$2149	$2050	$1875	$2000
2198	2249	1945	2249	1998
1945	1901	1989	1976	1987
1998	2150	2149	2004	2049
2000	1945	2200	1984	2145

5. The Environmental Protection Agency performs mileage tests on all new car models to determine their miles-per-gallon rating. On a certain new car model, the following ratings were obtained.

39.7	33.9	34.4	33.1	39.3	34.8	38.8	35.6	31.8	32.8	36.3	40.2
36.4	38.6	37.6	37.8	31.7	35.7	35.3	38.1	35.7	39.1	37.8	39.5
38.3	40.1	38.8	33.9	30.8	37.6	39.1	35.8	38.4	34.5	37.9	38.2
36.7	30.8	37.8	35.5	39.8	36.9	31.8	32.4	35.9	36.1	38.1	37.6
31.4	31.7	32.9	30.7	37.5	31.8	40.2	33.8	34.7	39.0	36.0	37.3

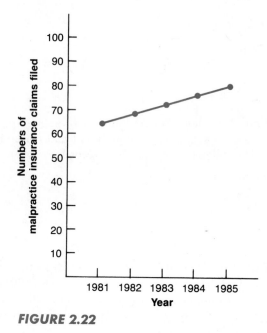

FIGURE 2.22

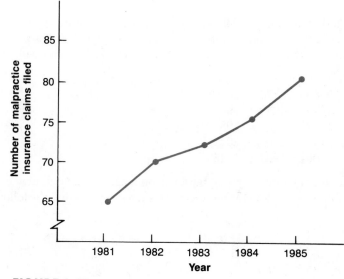

FIGURE 2.23

2.5 HOW TO LIE WITH GRAPHS

In the previous sections, we indicated how to analyze a list of numbers by graphical techniques. Nevertheless, it is possible to use these techniques to present the data in a misleading way. The following examples indicate some misuses of statistics and the reason for the incorrect use.

Since frequency polygons emphasize changes, (rise and fall) in frequency more clearly than any other graphical representation, they are used often to display business and economic data. However, this must be done with great care. Figures 2.22 and 2.23 both represent the same idea, that is, the number of cases of malpractice insurance filed against neurosurgeons in a large northeastern city in the United States during the years 1981–1985. Figure 2.23 seems to indicate that the number of malpractice insurance cases filed increased significantly over this five-year period. Figure 2.22 also indicates that the number of such cases increased, although not as dramatically. How can we have two *different* graphs representing the same situation? Which graph better displays the situation?

Notice that the vertical scales in the two graphs are not the same. The vertical scale of Figure 2.23 has been truncated or cut off. **Truncated graphs often tend to distort the information presented.** Consider the graphs on a drought situation shown in Figures 2.24 and 2.25. Again notice the truncated graph. How bad is the drought situation?

Another misuse of statistics involves **scaling.** Consider the graphs given in Figures 2.26 and 2.27 to show how the circulation of a computer magazine has

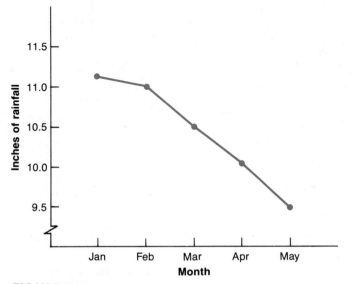

FIGURE 2.24

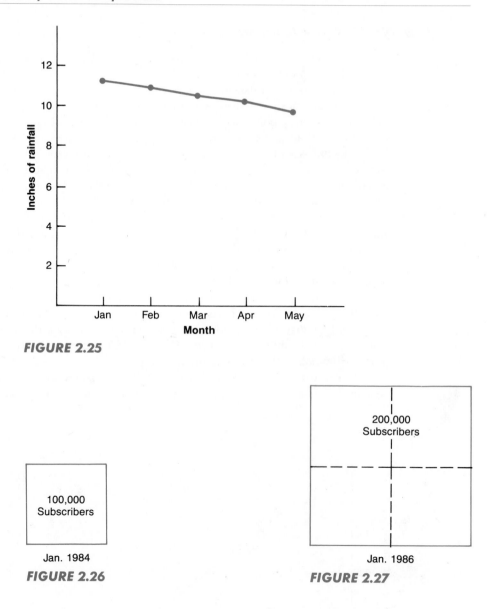

FIGURE 2.25

FIGURE 2.26

FIGURE 2.27

doubled. Anything wrong? The graph in Figure 2.27 is twice as tall as the graph in Figure 2.26. But it is also twice as wide, so that it is four times as large. In Figure 2.27, each of the four rectangles is exactly the same size as the one in Figure 2.26. Do people draw graphs in such a misleading way? Unfortunately, the answer is yes.

Up to this point, we have merely indicated two ways that statistical graphs can be misused. Many other ways are discussed in Darrell Huff's book *How to Lie with Statistics* (New York: W. W. Norton, 1954).

COMMENT By now the point should be obvious. Read and construct statistical graphs carefully to avoid the common pitfalls.

EXERCISES

1. The number of arrests for drug abuse on a particular college campus over a five-year period was presented graphically in the college student newspaper and in the local city newspaper. Which graph is misleading? Explain your answer.

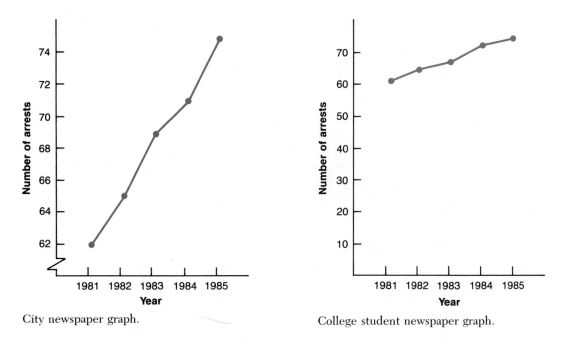

City newspaper graph.

College student newspaper graph.

2. Consider the following advertisement promoted by a swimming pool construction company. Anything wrong with the claim?

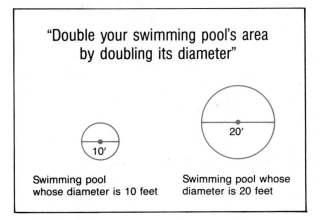

"Double your swimming pool's area by doubling its diameter"

Swimming pool whose diameter is 10 feet

Swimming pool whose diameter is 20 feet

2.6 INDEX NUMBERS

Often when analyzing data, we are interested in obtaining a clear picture of trend. We could then use index numbers which give us a good analysis of changes over a period of time. Index numbers allow us to compare such things as prices, production figures, sales figures, and so on, for a given period of time with corresponding values in some earlier period of time. The earlier period is usually referred to as the **base period.** We have:

DEFINITION 2.2

An **index number** is a special form of ratio that is used to show percentage changes over a period of time.

The best known index number is the **Consumer Price Index (CPI)** published by the U.S. Bureau of Labor Statistics. In the media, this is often called the cost-of-living index. It reflects the changes in prices of goods and services purchased by a typical wage earner in a large city. Since the labor contracts negotiated by some unions often tie increases in salary to changes in the CPI through cost of living escalation clauses, a knowledge of how to compute index numbers is important.

The CPI is published each month and includes the results of surveys for the prices of over 400 goods and services including among other things food, clothing, medical costs, automobile repair, college tuition, and so on. Thus it can be used as a measure of inflation. When it goes up, wages, pensions, and social security payments go up.

COMMENT Great care must be exercised when using the CPI. The prices of the items included in the computation must be weighted according to their importance.

To calculate an index number, we use the following formula:

FORMULA 2.1

$$\textit{Index number} = \frac{\textit{Given year's values}}{\textit{Base year's value}} \times \textit{100}$$

We illustrate the use of this formula with several examples.

● **Example 1**

The Musicktone Corporation sells various stereo components. The annual sales for these components over the years 1980–1985 is as follows:

Year	Sales
1980	$400,000
1981	425,000
1982	491,000
1983	350,000
1984	410,000
1985	380,000

Using 1980 as a base year, calculate the sales index for each year.

Solution

Since we are using 1980 as our base year, we must divide each year's value by the base year's value and multiply the result by 100. We have

Year	Ratio	Ratio $\times$ 100 = Index Number
1980	$\dfrac{400,000}{400,000} = 1$	$1 \times 100 = 100$
1981	$\dfrac{425,000}{400,000} = 1.06$	$1.06 \times 100 = 106$
1982	$\dfrac{491,000}{400,000} = 1.23$	$1.23 \times 100 = 123$
1983	$\dfrac{350,000}{400,000} = 0.875$	$0.875 \times 100 = 87.5$
1984	$\dfrac{410,000}{400,000} = 1.025$	$1.025 \times 100 = 102.5$
1985	$\dfrac{380,000}{400,000} = 0.95$	$0.95 \times 100 = 95$

The index numbers calculated in the previous example show us the percent change from the base year. This percent change is the difference from 100. We have

a. The sales index for 1981 indicates an increase of 6% when compared to 1980 since $106 - 100 = 6$.

b. The sales index for 1982 indicates an increase of 23% when compared to 1980 since $123 - 100 = 23$.

c. The sales index for 1983 indicates a decrease of 12.5% when compared to 1980 since $87.5 - 100 = -12.5$.

Similar conclusions can be arrived at for the other years.

● **Example 2** *Oil Dependence of the U.S.*

The number of thousands of barrels of oil imported by the United States from OPEC sources for the years 1981–1984 is as follows:

Year	Oil Imported
1981	1,849,017
1982	2,260,482
1983	2,057,468
1984	2,023,341

Source: American Petroleum Institute

Using 1981 as a base year, compute the import index for each of the years given and interpret the results.

Solution

We use the same tabular arrangement as we did in the previous example. We have

Year	Ratio	Ratio × 100 = Index Number
1981	$\dfrac{1,849,017}{1,849,017} = 1$	$1 \times 100 = 100$
1982	$\dfrac{2,260,482}{1,849,017} = 1.22$	$1.22 \times 100 = 122$
1983	$\dfrac{2,057,468}{1,849,017} = 1.11$	$1.11 \times 100 = 111$
1984	$\dfrac{2,023,341}{1,849,017} = 1.09$	$1.09 \times 100 = 109$

The import indexes for these years indicate

a. for 1982 an increase of 22% when compared to 1981.
b. for 1983 an increase of 11% when compared to 1981.
c. for 1984 an increase of 9% when compared to 1981.

COMMENT   In the previous example we can use a different base year and arrive at different conclusions. Thus if 1983 is used as a base year we get index numbers of 90, 110, 100, and 98, respectively, for the years 1981, 1982, 1983, and 1984. These index numbers give us different interpretations than when we used 1981 as a base year.

COMMENT   Often when using index numbers, an important decision involves deciding which base year is best to use.

EXERCISES

For Exercises 1–4 use the following information. The price of a gallon of unleaded premium gas at a particular gas station on U.S. Highway No. 1 on Labor Day over a period of seven years was as follows:

Year	Price per Gallon
1979	$1.09
1980	1.17
1981	1.19
1982	1.29
1983	1.35
1984	1.51
1985	1.41

1. Using 1979 as a base year, what was the price index for 1985?

2. Using 1979 as a base year, what does the price index for 1982 tell us?

3. Using 1985 as a base year, what does the price index for 1982 tell us?

4. Using 1985 as a base year, what does the price index for 1981 tell us?

5. Juan Piedrahita is analyzing the tuition costs of his children's education. He has accumulated the following data.

Year	Total Tuition Costs
1981	$2400
1982	2750
1983	2975
1984	3500
1985	4360

Using 1981 as a base year, compute cost indexes for each of the years given and interpret the results.

6. The average charge by an obstetrician for delivering a baby (except by caesarean section) in Petersburg was as follows:

Year	Average Charge
1981	$1600
1982	1710
1983	1725
1984	1760
1985	1790

Using 1984 as a base year, compute cost indexes for each of the years given and interpret the results.

7. Refer back to Exercise 6. Using 1981 as a base year, compute cost indexes for each of the years given. Compare the results to those obtained in Exercise 6. Comment.

8. The Budget Director of the Beck Corporation has computed price indexes for several items directly related to employees. The following is available.

Index of Costs (January 1982 = 100)

Month	Labor Costs	Health Insurance Costs	Pension Costs
Jan. 1983	106	115	103
Jan. 1984	108	111	91
Jan. 1985	109	112	101

a. Which cost showed the smallest change between January 1983 and January 1985?
b. Which cost showed the greatest change between January 1983 and January 1985?

9. Refer back to Exercise 8. Pension costs had an index of 91 for the month of January 1984. Explain this price index.

10. Refer back to the newspaper article given at the beginning of this chapter, on page 22. Yolanda Stuart works for the Beck Corporation mentioned in Exercise 8. Official government records indicate that the CPI for Yolanda's region was 105.9 in 1984 and 107.3 for 1985. Yolanda earned $53,000 in 1984 and $54,000 in 1985. Did her purchasing power increase or decrease over the year?

2.7 USING COMPUTER PACKAGES

To illustrate how we can use the MINITAB computer package to draw a histogram, let us consider the following data representing the number of credit card sales reported by a large department store chain at all of its branches over a 50-day period.

Day	Number of Sales	Day	Number of Sales	Day	Number of Sales
Mon.	52	Mon.	60	Mon.	53
Tues.	68	Tues.	62	Tues.	58
Wed.	51	Wed.	58	Wed.	62
Thurs.	62	Thurs.	48	Thurs.	61
Fri.	49	Fri.	61	Fri.	60
Mon.	37	Mon.	45	Mon.	45
Tues.	74	Tues.	49	Tues.	61
Wed.	64	Wed.	53	Wed.	53
Thurs.	48	Thurs.	67	Thurs.	42
Fri.	63	Fri.	49	Fri.	37

Day	Number of Sales	Day	Number of Sales
Mon.	34	Mon.	47
Tues.	68	Tues.	59
Wed.	51	Wed.	38
Thurs.	47	Thurs.	32
Fri.	56	Fri.	76
Mon.	59	Mon.	57
Tues.	69	Tues.	41
Wed.	42	Wed.	39
Thurs.	38	Thurs.	31
Fri.	36	Fri.	37

After logging in on the computer, we can construct the histogram for the above data by typing in the following instructions:

```
MTB   >  SET THE FOLLOWING DATA INTO C1
DATA  >  52  68  51  62  49  37  74  64  48  63
DATA  >  60  62  58  48  61  45  49  53  67  49
DATA  >  53  58  62  61  60  45  61  53  42  37
DATA  >  34  68  51  47  56  59  69  42  38  36
DATA  >  47  59  38  32  76  57  41  39  31  37

MTB   >  HISTOGRAM OF C1
```

The first line tells the computer that the data should be placed in column C1 of the worksheet that **MINITAB** maintains in the computer. The next few lines contain the data themselves. The last line instructs the computer to construct the histogram.

After typing the above information, the computer will automatically print out the frequency distribution and histogram shown below. Notice that the histogram is printed sideways with the bars (actually they are asterisks) drawn horizontally rather than vertically.

53,000 = 105.9

54,000   107.3

```
C1
  MIDDLE OF      NUMBER OF
  INTERVAL       OBSERVATIONS
       30            2      **
       35            5      *****
       40            6      ******
       45            4      ****
       50            8      *******
       55            5      *****
       60           12      ***********
       65            3      ***
       70            3      ***
       75            2      **
MTB > STOP
```

COMMENT   Our objective here is not to teach you how to use the MINITAB statistical package, but merely to familiarize you with its general nature and usefulness. The details can be obtained from manuals. For a manual specifically dealing with MINITAB, see T. Ryan, B. Joiner, and B. Ryan, *MINITAB STUDENT HANDBOOK* (Boston: Duxbury Press, 1985).

2.8 SUMMARY

In this chapter we discussed the different graphical methods that one can use to picture a mass of data so that meaningful statements can be made. When given a large quantity of numbers to analyze, it is recommended that frequency tables be constructed. Some forms of graphical representation should then be used to serve as visual aids for thinking about and discussing statistical problems in a clear and easily understood manner. We discussed the different graphical techniques that one can use and also pointed out in Figures 2.22 and 2.24 on pages 68 and 69 how they can be misused. We studied an alternate and relatively new way of analyzing data. This is by means of stem-and-leaf diagrams. We also analyzed index numbers, which are special ratios that can be used to study percentage changes over a period of time. We pointed out that great care must be exercised when interpreting such numbers. The best known of these index numbers is the CPI published by the U.S. Bureau of Labor Statistics. There are many other indexes that can be computed. Some of these measure changes in regional and national retail and wholesale prices, industrial and agricultural production, and so on. Each of these is computed in a manner similar to the way we computed index numbers.

STUDY GUIDE

You should now be able to demonstrate your knowledge of the following ideas presented in this chapter by giving definitions, descriptions, or specific examples. Page references are given for each term so that you can check your answer.

Random sample (page 23)
Frequency distribution (page 24)
Frequency (page 24)
Classes (intervals) (page 24)
Grouped data (page 26)
Class number (page 26)
Upper and lower limit (page 26)
Class mark (page 26)
Class frequency (page 26)
Relative frequency (page 27)
Histogram (page 27)
Relative-frequency histogram (page 30)
Circle graph or pie chart (page 37)
Bar graph (page 44)
Pictograph (page 44)
Frequency polygon (page 50)
Normal curve (page 53)
Normal distribution (page 53)
Stem-and-leaf diagrams (page 63)
Misuse of statistics (page 69)
Truncated graphs (page 69)
Scaling (page 69)
Base year (page 72)
Index number (page 72)
CPI, Consumer's Price Index (page 72)

FORMULAS TO REMEMBER

The following list summarizes the formulas given in this chapter.

1. Rules for grouping data (page 28)

2. Rules for drawing histograms (page 28)

3. Rules for drawing pie charts, bar graphs, and frequency polygons (pages 43–47)

4. Rules for constructing stem-and-leaf diagrams (page 65)

5. Index number $= \dfrac{\text{Given year's values}}{\text{Base year's values}} \times 100$ (page 72)

The tests of the following section will be more helpful if you take them after you have studied the examples given in this chapter and solved the exercises at the end of each section.

MASTERY TESTS

Form A

1. The graph below shows the distribution of scores on a computer skills test. How many students took the test?   *15*

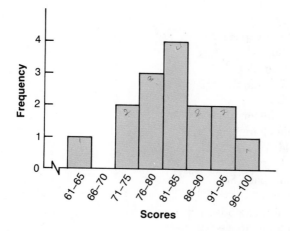

For questions 2–4, use the following information. The test scores of 16 students were 96, 83, 91, 77, 58, 88, 80, 62, 89, 100, 87, 93, 64, 99, 88, and 86.

2. Complete the following table:

Interval	Frequency		Relative Frequency
91–100	⊤⊦⊦	*5*	*5 / 16*
81–90	⊦⊦⊦⊦		*6 / 16*
71–80	II		*2 / 16*
61–70	II		*2 / 16*
51–60	I		*1 / 16*

3. Draw the frequency histogram and relative-frequency histogram based on the data.

4. Draw a stem-and-leaf diagram for the data.

Use the following information to answer questions 5 and 6. The graph on the next page shows how the life expectancy for U.S. men has changed over the period 1940–1980.

**Life Expectancy in Years
for U.S. Males**

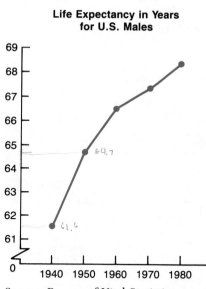

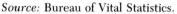

Source: Bureau of Vital Statistics.

5. In what decade did the life expectancy increase the most?
a. 1940–1950   **b.** 1950–1960   **c.** 1960–1970
d. 1970–1980   **e.** none of these

6. In what decade did the life expectancy increase the least?
a. 1940–1950   **b.** 1950–1960   **c.** 1960–1970
d. 1970–1980   **e.** none of these

7. In a certain tennis club, the marital status of the women players is indicated by the following pie chart: If there are 1000 women in the club, how many are *not* married?
a. 643   **b.** 64.3   **c.** 357   **d.** 35.7   **e.** none of these

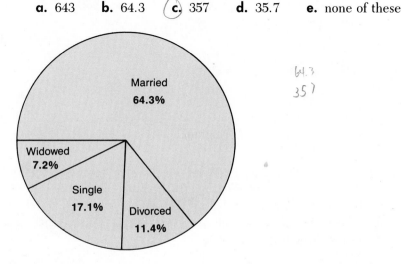

8. The number of calls from residents of Calhoun to the AAA for assistance in starting stalled cars for each of the days of January was as follows:

```
38  34  19  33  22  27  38
19  42  29  16  19  39
16  17  32  44  20  ·31
29  29  40  39  35  25
53  16  22  27  46  18
```

Construct a frequency distribution for this data and then draw its histogram.

9. The number of residential burglar alarms installed by the Gant Corporation and the prices charged for each installation during the years 1980–1985 are as follows:

Year	Number of Alarms Installed	Price per Installation
1980	840	$400
1981	970	440
1982	1120	420
1983	1050	430
1984	820	460
1985	710	490

Find index numbers for the number of alarms installed. Use 1980 as a base year.

10. Refer to Exercise 9. Find index numbers for the price charged per installation. Use 1985 as a base year. Compare both sets of index numbers. Comment.

1. At a certain college there are 2729 freshmen, 2468 sophomores, 1976 juniors, and 2163 seniors. Construct a pie chart to indicate the make-up of the student body at this school.

For questions 2–4, use the following information. During 1984, the circulation of a new magazine was as follows:

Months	Circulation
Jan.–Feb.	9,000
Mar.–Apr.	10,000
May–June	12,265
July–Aug.	13,480
Sept.–Oct.	14,790
Nov.–Dec.	15,050

2. Draw a bar graph to picture the above data.

3. Draw a pie chart to picture the above data.

4. Two graphs have been prepared, one by a statistician and one by an advertising agency. Parts of these are presented below.

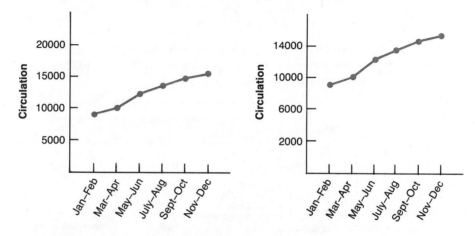

Which is the correct graph? Explain your answer.

5. The pictogram below is intended to illustrate the fact that the total real estate tax paid by families in Greentown had doubled from 1980 to 1985. How should it be modified so that it will convey a fair impression of the actual change?

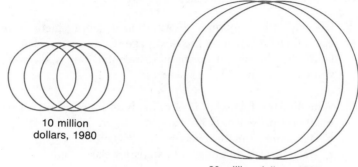

10 million dollars, 1980

20 million dollars, 1985

6. The price of a share of stock of the Med Corp. on the first day of each month for the previous year was as follows:

Jan.	38¢	Apr.	45¢	July	33¢	Oct.	46¢
Feb.	47¢	May	49¢	Aug.	38¢	Nov.	54¢
Mar.	53¢	June	41¢	Sept.	42¢	Dec.	63¢

Draw a graph to picture the above information.

7. Construct both a frequency histogram and a stem-and-leaf diagram for the following data, which represent the electric energy sales (in billions of kilowatt hours) for several states:

7.8	4.3	13.7	13.1	12.2
8.3	4.8	12.3	12.8	6.5
9.9	7.1	10.7	10.2	8.5
4.9	11.8	9.1	6.7	9.4

8. Consider the bar graph below, which presents data from the FBI analysis of the weapons used in committing murders in 1982. Draw a circle graph to picture the information given.

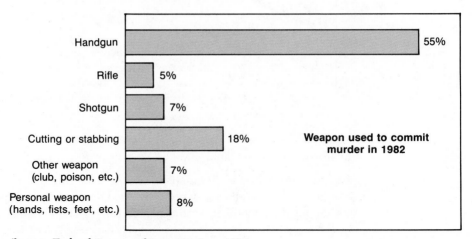

Source: Federal Bureau of Investigation, 1984.

9. Consider the frequency distribution of family income for a large northeastern city, shown on the next page. Contrary to what you might expect, family income is not normally distributed but is skewed to the left. Can you explain why?

10. Which of the following is (are) likely to be normally distributed?
  a. Blood pressure of 50-year-old women
  b. IQ score of math teachers
  c. Height of individuals in a statistics class

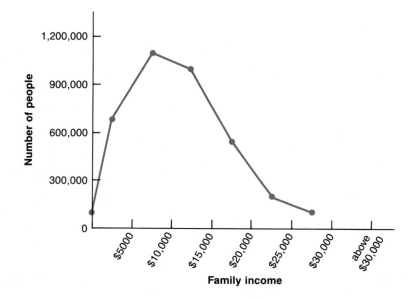

SUGGESTED FURTHER READING

1. Book, S. *Statistics: Basic Techniques for Solving Applied Problems.* New York: McGraw-Hill, 1977.
2. Haber, A., and R. Runyan. *General Statistics*, 3rd ed. Reading, MA: Addison-Wesley, 1977.
3. Huff, D. *How to Lie with Statistics.* New York: W. W. Norton, 1954.
4. McClare, J. T., and P. G. Benson. *Statistics for Business and Economics*, 2nd ed. San Francisco: Dellen, 1982.
5. Newmark, Joseph, and Frances Lake. *Mathematics as a Second Language*, 4th ed. Reading, MA: Addison-Wesley, 1987.
6. O'Toole, A. L. *Elementary Practical Statistics.* New York: Macmillan, 1965.
7. Tanur, J. M., F. Mosteller, W. H. Kruskul, R. F. Link, R. S. Pieters, and G. R. Rising. *Statistics: A Guide to the Unknown.* San Francisco: Holden-Day, 1978.
8. Tukey, J. W. *Exploratory Data Analysis.* Reading, MA: Addison-Wesley, 1977.

CHAPTER 3

Numerical Methods for Analyzing Data

OBJECTIVES

- *To introduce* summation notation, which is used to denote the addition of numbers.

- *To discuss* the mean, median, and mode as three different ways of measuring some general trend or location of the data. These will be called measures of central tendency.

- *To study* the range, standard deviation, variance, and average deviation. Even when we know the mean, median, or mode, we might still want to know whether the numbers are close to each other or whether they are spread out. These are called measures of dispersion or measures of variation.

- *To present* several shortcut formulas for calculating the mean, variance, and standard deviation for grouped or ungrouped data.

- *To find* an interpretation for the standard deviation.

- *To mention* Chebyshev's theorem, which gives us some information about where many terms of a distribution will lie.

- *To draw* the graphs of cumulative frequency distributions and the graphs of cumulative relative frequency distributions. These are called ogives.

- *To analyze* the idea of percentiles, which give us the relative standing of one score when compared to the rest. Specifically, percentiles tell us what percent of the scores are above or below a given score.

- *To learn* how to compute z-scores. These give us the relative position of a score with respect to the mean, and are expressed in terms of standard deviations.

STATISTICS IN USE

MUST WE SPEED?

Nov. 10 – A new study conducted by the Department of Transportation found that more and more drivers are ignoring the states' legal 55 mph speed limit on the highways and freeways. A survey of 500 cars found them travelling at an average speed of 63 mph. Stricter enforcement of existing speed limits was urged.

November 10, 1985

CREDIT CARD INTEREST STILL AN AVERAGE 18%

New York – (Dec. 2) Despite the recent sharp drop in interest rates, many of our country's major lending institutions have still not lowered the annual interest rates charged to their credit card customers. A survey of 100 banks nationwide disclosed that the average annual interest rate charged was 18%. Several U.S. senators have called this legal usury.

December 2, 1985

Note the use of the word "average" in both of the newspaper articles. Is the word "average" being used in the same manner in both instances? How do we calculate such averages?

3.1 INTRODUCTION

In the preceding chapter we learned how to summarize data and present it graphically. Although the techniques discussed there are quite useful in describing the features of a distribution, statistical inference usually requires more precise analysis of the data. In particular, we will discuss many measures that locate the *center* of a distribution of a set of data and analyze how dispersed or spread out the distribution is. In this chapter we will examine measures of central tendency and measures of variation (dispersion).

Most analyses involve various arithmetic computations that must be performed on the data. In each case the operation of addition plays a key role in these calculations. It is for this reason that we introduce a special shorthand notation called **summation notation.**

Now consider the card from the Emerson Medical Laboratory shown below.

Emerson Medical Laboratory
Staten Island, N.Y.

Smith, Mary Ellen	F	32	Nov 1, 1986
patient's name	sex	age	date

Results	109	256
Percentile rank	79	68
	Triglycerides	Cholesterol

The card gives Mary Ellen Smith her percentile rank for triglycerides and cholesterol. How should she interpret this result? Is it better to have a higher or lower percentile rank? It should be apparent that we cannot answer this unless we understand the meaning of percentiles and learn how to calculate them. This will be done in this chapter.

After the data have been analyzed numerically, the techniques of statistical inference can be applied. These will be discussed in later chapters.

3.2 SUMMATION NOTATION

Often when working with a distribution of many numbers, we use letters with subscripts, that is, small numbers attached to them. Thus, we write x_1, x_2, x_3, . . . , x_n, which is read "x sub one, x sub two, x sub three, . . . , x sub n." To be specific, consider the following set of numbers:

7, 15, 5, 3, 9, 8, 14, 21, 10

Here we let x_1 denote the first number, that is $x_1 = 7$, and we let x_2 denote the second number, that is $x_2 = 15$. Similarly, $x_3 = 5$, $x_4 = 3$, $x_5 = 9$, $x_6 = 8$, $x_7 = 14$, $x_8 = 21$ and $x_9 = 10$.

To indicate the operation of taking the sum of a sequence of numbers, we use the Greek symbol Σ, which is read as sigma. To add all of the above x's we would write

$$\sum_{i=1}^{9} x_i$$

This tells us to add all the consecutive values of x starting with x_1 and proceeding to x_9. Thus

$$\sum_{i=1}^{9} x_i = x_1 + x_2 + x_3 + x_4 + x_5 + x_6 + x_7 + x_8 + x_9$$
$$= 7 + 15 + 5 + 3 + 9 + 8 + 14 + 21 + 10$$
$$= 92$$

If we only wanted to add $x_1 + x_2 + x_3 + x_4 + x_5$, we would write $\sum_{i=1}^{5} x_i$. The i (or j) in the summation symbol $\sum_{i=1}^{n}$ is referred to as the **index.**

Throughout our study of statistics, we shall have need to work with various applications of the summation Σ symbol. Great care must be exercised when using this symbol. For example, using the Σ notation, we have the following:

$$\sum_{i=1}^{n} x_i^2 = x_1^2 + x_2^2 + x_3^2 + \cdots + x_n^2$$

$$\sum_{i=1}^{n} x_i y_i = x_1 y_1 + x_2 y_2 + \cdots + x_n y_n$$

$$\sum_{i=1}^{n} x_i^2 f_i = x_1^2 f_1 + x_2^2 f_2 + \cdots + x_n^2 f_n$$

COMMENT   The symbols $\sum_{i=1}^{n} x_i^2$ and $\left(\sum_{i=1}^{n} x_i \right)^2$ are quite different. The symbol $\sum_{i=1}^{n} x_i^2$ means that we first square the numbers and add them together, whereas the symbol $\left(\sum_{i=1}^{n} x_i \right)^2$ means that we add all the x_i's together to obtain a sum and then square the sum.

Let us illustrate the use of the Σ symbols with several examples.

● **Example 1**

Find $\sum_{i=1}^{5} x_i^2$ and $\left(\sum_{i=1}^{5} x_i\right)^2$ for the following data:

$x_1 = 3,\ x_2 = 7,\ x_3 = 6,\ x_4 = 3,\ x_5 = 9$

Solution

$\sum_{i=1}^{5} x_i^2$ means $x_1^2 + x_2^2 + x_3^2 + x_4^2 + x_5^2$

Thus

$$\sum_{i=1}^{5} x_i^2 = 3^2 + 7^2 + 6^2 + 3^2 + 9^2$$
$$= 9 + 49 + 36 + 9 + 81$$
$$= 184$$

Also $\left(\sum_{i=1}^{5} x_i\right)^2$ means $(x_1 + x_2 + x_3 + x_4 + x_5)^2$.

Thus

$$\left(\sum_{i=1}^{5} x_i\right)^2 = (3 + 7 + 6 + 3 + 9)^2$$
$$= 28^2$$
$$= 784$$

Therefore, $\sum_{i=1}^{5} x_i^2$ and $\left(\sum_{i=1}^{n} x_i\right)^2$ are quite different.

NOTATION   Throughout this text, whenever we use the Σ notation, we will want to add *all* the available data. Therefore, to simplify the formulas, we will sometimes write the summation symbol without any index. *So that when no index is indicated, it is understood that all of the data are being used.*

COMMENT   In Example 1 we can write $\sum_{i=1}^{5} x_i^2$ and $\left(\sum_{i=1}^{5} x_i\right)^2$ as $\sum x^2$ and $\left(\sum x\right)^2$ since we are using all of the available data.

● **Example 2**

Using the data given below, find (a) Σx, (b) Σy, (c) Σxy, (d) $\Sigma x \cdot \Sigma y$.

x	3	5	6	8	11
y	8	7	9	10	12

a. Σx means $3 + 5 + 6 + 8 + 11 = 33$. Thus $\Sigma x = 33$.
b. Σy means $8 + 7 + 9 + 10 + 12 = 46$. Thus $\Sigma y = 46$.
c. To find Σxy, we must first find all the products of the corresponding x and y values, and then add these products together. Thus, we have

$$\begin{aligned} \Sigma xy &= 3(8) + 5(7) + 6(9) + 8(10) + 11(12) \\ &= 24 + 35 + 54 + 80 + 132 \\ &= 325 \end{aligned}$$

Therefore, $\Sigma xy = 325$
d. The symbol $\Sigma x \cdot \Sigma y$ means the product of the two summations Σx and Σy. From parts (a) and (b) we already know that $\Sigma x = 33$ and $\Sigma y = 46$ so that

$$\Sigma x \cdot \Sigma y = 33(46) = 1518$$

● **Example 3**

If $x_1 = 7$, $x_2 = 3$, $x_3 = 9$, $x_4 = 4$, $f_1 = 5$, $f_2 = 1$, $f_3 = 8$, and $f_4 = 2$, find

a. $\displaystyle\sum_{i=1}^{4} x_i$ b. $\displaystyle\sum_{i=1}^{4} f_i$ c. $\displaystyle\sum_{i=1}^{4} x_i f_i$.

Solution

a. $\displaystyle\sum_{i=1}^{4} x_i$ means $x_1 + x_2 + x_3 + x_4$, so that

$$\sum_{i=1}^{4} x_i = 7 + 3 + 9 + 4 = 23$$

b. $\displaystyle\sum_{i=1}^{4} f_i = f_1 + f_2 + f_3 + f_4 = 5 + 1 + 8 + 2 = 16$

c. $\displaystyle\sum_{i=1}^{4} x_i f_i = x_1 f_1 + x_2 f_2 + x_3 f_3 + x_4 f_4$

$$= 7(5) + 3(1) + 9(8) + 4(2) = 118$$

When working with summations, there are certain rules that we will use. These rules are easily verified by simply writing out in full what each of the summations represents.

RULE 1

$$\sum_{i=1}^{n} (x_i + y_i) = \sum_{i=1}^{n} x_i + \sum_{i=1}^{n} y_i$$

RULE 2

$$\sum_{i=1}^{n} kx_i = k \cdot \sum_{i=1}^{n} x_i$$

RULE 3

$$\sum_{i=1}^{n} k = n \cdot k$$

In words, the first rule says that the summation of a sum of two terms equals the sum of the individual summations. The second rule says that we can "factor" a constant out from under the operation of a summation. Thus

$$\sum_{i=1}^{n} kx_i = kx_1 + kx_2 + kx_3 + \cdots + kx_n$$

$$= k(x_1 + x_2 + \cdots + x_n)$$

$$= k \cdot \sum_{i=1}^{n} x_i$$

The third rule says that the summation of a constant is simply n times that constant or the constant times the number of indicated terms in the summation. These rules are easy to use, and will be applied throughout the book.

● **Example 4**

If $x_1 = 4$, $x_2 = 5$, $x_3 = 7$ and $x_4 = 9$, find $\sum_{i=1}^{4} (3x_i - 1)$.

Solution

$$\sum_{i=1}^{4} (3x_i - 1) = (3x_1 - 1) + (3x_2 - 1) + (3x_3 - 1) + (3x_4 - 1)$$
$$= (3 \cdot 4 - 1) + (3 \cdot 5 - 1) + (3 \cdot 7 - 1) + (3 \cdot 9 - 1)$$
$$= 11 + 14 + 20 + 26$$
$$= 71$$

Thus, $\sum_{i=1}^{4} (3x_i - 1) = 71$ for the above values of x_i.

COMMENT Example 4 can also be evaluated by using Rules 1–3. Can you see how?

EXERCISES

1. Rewrite each of the following using summation notation.
   a. $y_1 + y_2 + y_3 + y_4 + y_5 + y_6 + y_7$
   b. $y_1^2 + y_2^2 + y_3^2 + y_4^2 + y_5^2 + y_6^2 + y_7^2$
   c. $y_1 f_1 + y_2 f_2 + y_3 f_3 + y_4 f_4 + y_5 f_5 + y_6 f_6 + y_7 f_7$
   d. $17y_1 + 17y_2 + 17y_3 + 17y_4 + 17y_5 + 17y_6 + 17y_7$
   e. $(2y_1 + x_1) + (2y_2 + x_2) + (2y_3 + x_3) + \cdots + (2y_n + x_n)$

2. Rewrite each of the following without summation notation.
   a. $\displaystyle\sum_{i=1}^{5} y_i$ **b.** $\displaystyle\sum_{i=5}^{10} x_i y_i$ **c.** $\displaystyle\sum_{i=3}^{10} (5x_i + 2)$

   d. $\displaystyle\sum_{i=1}^{10} 2x_i^2$ **e.** $\displaystyle\sum_{i=1}^{6} y_i^2 f_i$ **f.** $\displaystyle\sum_{i=3}^{7} y_i^2$

3. If $x_1 = 4$, $x_2 = 17$, $x_3 = 7$, $x_4 = 19$, and $x_5 = 21$, find each of the following:
   a. Σx **b.** $(\Sigma x)^2$ **c.** Σx^2
   d. $\Sigma(x + 1)$ **e.** $\Sigma(x + 1)^2$ **f.** $\Sigma(2x + 3)$
   g. $\Sigma(2x + 3)^2$

4. If $x_1 = 4$, $x_2 = 7$, $x_3 = 13$, $x_4 = 31$, $f_1 = 2$, $f_2 = 6$, $f_3 = 8$, $f_4 = 10$, $y_1 = 0$,
   $y_2 = 3$, $y_3 = -2$ and $y_4 = -5$, find each of the following:
   a. Σx^2 **b.** $\Sigma x \cdot f$ **c.** $\Sigma x \cdot \Sigma y$
   d. Σxy **e.** $\Sigma x \cdot y \cdot f$ **f.** $\Sigma(x - y)f$

5. If $\displaystyle\sum_{i=1}^{10} x_i = 15$ and $\displaystyle\sum_{i=1}^{10} x_i^2 = 40$, find each of the following:

   a. $\displaystyle\sum_{i=1}^{10} (x_i + 9)$ **b.** $\displaystyle\sum_{i=1}^{10} (x_i + 9)^2$

6. Show that each of the following are true.
   a. $\displaystyle\sum_{i=1}^{n} (x_i - y_i) = \sum_{i=1}^{n} x_i - \sum_{i=1}^{n} y_i$

   b. $\displaystyle\sum_{i=1}^{n} (x_i + k) = \sum_{i=1}^{n} x_i + nk$

3.3 MEASURES OF CENTRAL TENDENCY

To help us understand what we mean by measures of central tendency, consider the Metropolis Police Department, which recently purchased tires from two different manufacturers. The police department is interested in determining which tire is superior and has compiled a list on the number of miles each set of tires lasted before replacement was needed for fourteen of its identical police cars. Seven cars were fitted with Brand X tires and seven with Brand Y. The number of miles each set lasted before replacement is indicated in the following chart:

Brand X	Brand Y
14,000	10,000
12,000	8,000
12,000	14,000
14,000	10,000
14,000	8,000
11,000	40,000
14,000	8,000
Total = 91,000	98,000

It would appear that Brand Y is the better tire since the seven police cars were driven a combined total of 98,000 miles using Brand Y tires but only 91,000 miles with Brand X tires. Let us, however, analyze the data by computing the average number of miles driven with each brand of tires. We will divide each total by the number of police cars used for each brand. We have the following:

Average

Brand X	Brand Y
$\dfrac{91,000}{7} = 13,000$	$\dfrac{98,000}{7} = 14,000$

Since Brand Y tires lasted on the average 14,000 miles and Brand X on the average 13,000 miles, it again appears that Brand Y is superior. If we look at the data more carefully, however, we find that Brand X tires consistently lasted around 13,000 miles. As a matter of fact, they lasted 14,000 miles most often, 4 times. Brand Y, on the other hand, lasted 14,000 miles only once. They lasted 8,000 miles most often. Thus in terms of consistency of performance one might say that Brand X is more consistent. Let us arrange the data for each brand in order from smallest to largest. We have

Brand X	Brand Y
11,000	8,000
12,000	8,000
12,000	8,000
(14,000)	(10,000)
14,000	10,000
14,000	14,000
14,000	40,000

In this chart we have circled two numbers. These are the numbers that are in the middle for each brand: 14,000 for Brand X and 10,000 for Brand Y. All the preceding ideas are summarized in the following definitions:

DEFINITION 3.1

*The **mean**, or **average**, of a set of numbers is obtained by adding the numbers together and dividing the sum by the number of numbers added. We will denote the mean by the symbol $\bar{x}$, which is read x bar, or by the Greek letter μ, which is read mu, depending upon the situation. The use of each symbol will be explained shortly.*

DEFINITION 3.2

*The **mode** of a set of numbers is the number (or numbers) that occurs most often. If no number occurs more than once, there is no mode.*

DEFINITION 3.3

*The **median** of a set of numbers is the number that is in the middle when the numbers are arranged in order from smallest to largest. The median is easy to calculate when we have an odd number of numbers. If we have an even number of them, the median is defined as the average of the middle two numbers when arranged in increasing order of size. We denote the median by the symbol $\tilde{x}$, read "x tilde."*

COMMENT A set of numbers may have more than one mode. (See Example 1 of this section.)

When the preceding definitions are applied to our example, we get

	Brand X	Brand Y
Mean	13,000	14,000
Median	14,000	10,000
Mode	14,000	8,000

Which number should the police department use to determine the superior tire: the mean, median, or mode? You might say that the mean is not particularly helpful since one set of Brand Y tires lasted 40,000 miles and this instance had the effect of increasing the average for Brand Y considerably. In terms of consistency Brand X appears to be superior to Brand Y.

Notice that in calculating the mean we had to add numbers. Since the operation of addition plays a key role in our calculations, we use the summation notation discussed in the last section.

Consider the following set of numbers:

7, 15, 5, 3, 9, 8, 14, 21, 10

We have:

$$\Sigma x = x_1 + x_2 + x_3 + x_4 + x_5 + x_6 + x_7 + x_8 + x_9$$

$$\Sigma x = 7 + 15 + 5 + 3 + 9 + 8 + 14 + 21 + 10$$

$$\Sigma x = 92$$

If we divide this sum by n, the number of terms added, the result will be the mean for these numbers.

COMMENT   Remember that usually no indexes will be shown in the summation formulas in this book: the summations are understood to be over all the available data.

FORMULA 3.1   Sample Mean

The mean of a set of sample values x_1, x_2, x_3, . . . , x_n is given by

$$\textbf{\textit{Mean}} = \bar{x} = \frac{\Sigma x}{n} = \frac{x_1 + x_2 + x_3 + \cdots + x_n}{n}$$

For our set of numbers the mean is

$$\bar{x} = \frac{\Sigma x}{n} = \frac{92}{9} = 10.22$$

● **Example 1**   *The Better Worker*
The district office of a state unemployment insurance department recently hired two new employees, Rochelle and Sharon, to interview prospective aid recipients. Their supervisor is interested in determining who is the better worker. The following chart indicates the number of clients interviewed daily by each on seven randomly selected days.

Rochelle	Sharon
54	38
67	51
46	46
52	49
45	46
39	38
41	44
344	312

Calculate the sample mean, median, and mode for each employee.

Solution

Let x represent the number of clients interviewed daily by Rochelle and let y represent the number of clients interviewed daily by Sharon. For Rochelle the sample mean is

$$\bar{x} = \frac{\Sigma x}{n} = \frac{x_1 + x_2 + x_3 + x_4 + x_5 + x_6 + x_7}{7}$$

$$= \frac{54 + 67 + 46 + 52 + 45 + 39 + 41}{7}$$

$$= \frac{344}{7} = 49.14$$

For Sharon the sample mean is

$$\bar{y} = \frac{\Sigma y}{n} = \frac{y_1 + y_2 + y_3 + y_4 + y_5 + y_6 + y_7}{7}$$

$$= \frac{38 + 51 + 46 + 49 + 46 + 38 + 44}{7}$$

$$= \frac{312}{7} = 44.57$$

Let us now arrange the numbers for each employee in order from the lowest to the highest. We get the following:

Rochelle	39	41	45	(46)	52	54	67
Sharon	38	38	44	(46)	46	49	51

Notice that one number has been circled for each worker. This number is in the middle and it represents the median. For both workers the median is 46.

The mode is the number that occurs most often. For Rochelle there is no mode since no number occurs more than once. For Sharon there are two modes, 38 and 46.

Which statistic, the mean, mode, or median, should the supervisor use in determining who is the better worker?

● **Example 2**

The personnel manager of the Manhattan Detective Agency has compiled the following list on the number of years several former employees worked before retiring:

Worker	Number of Years of Service Before Retiring
Fay	43
Renée	38
Trudy	47
Hilda	35
Jack	42
Pedro	41
José	39
Beebabats	31

Find the mean, median, and mode.

Solution

We first arrange the numbers in order from lowest to highest. The median is the number that is in the middle. Since there are an even number of terms, the median is between the two circled numbers, 39 and 41.

31,   35,   38,   ㊴,   ㊶,   42,   43,   47

Definition 3.3 tells us that it is the average of 39 and 41. Thus we have

$$\tilde{x} = \text{Median} = \frac{39 + 41}{2} = 40$$

Since no number occurred more than once, there is no mode. The mean is obtained by dividing the total sum, which is 316, by the number of terms, 8:

$$\text{Mean} = \frac{316}{8} = 39.5$$

Summarizing we have

Mean: 39.5 years
Median: 40 years
Mode: none

Which is more useful to the personnel manager?

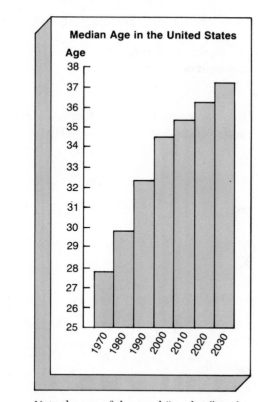

Note the use of the word "median" in this newspaper clipping. The information contained in it has important implications for our future.

● **Example 3**

Calculate the sample mean of the following numbers:

28	19	25	17	28	19	26	17	28	25
17	28	31	22	31	17	31	28	31	14
31	14	17	28	24	31	17	14	26	24
24	24	19	24	14	28	22	31	17	22
25	12	26	19	26	12	19	26	19	28

Solution

One way of calculating the mean would be to add the numbers and divide the result by 50, which is the number of terms. This takes a considerable amount of time but nevertheless we have

$$\bar{x} = \frac{\Sigma x}{n} = \frac{x_1 + x_2 + \cdots + x_{50}}{n}$$

$$= \frac{28 + 19 + 25 + \cdots + 28}{50}$$

$$= \frac{1145}{50} = 22.9$$

Thus the sample mean is 22.9.

We can also group the data (not necessarily in the usual interval format) as follows:

Column 1 Number (x)	Column 2 Tally	Column 3 Frequency (f)	Column 4 x · f
12	||	2	24
14	||||	4	56
17	ЖГ ||	7	119
19	ЖГ |	6	114
22	|||	3	66
24	ЖГ	5	120
25	|||	3	75
26	ЖГ	5	130
28	ЖГ |||	8	224
31	ЖГ ||	7	217
		Total = 50	1145

Column 4 was obtained by multiplying each number by its frequency. If we now sum column 3 and column 4 individually and divide the column 4 total by the column 3 total, our answer will be 22.9, which is the mean. Although we get the same result as we did before, it was considerably easier to obtain it by grouping the data as we did.

FORMULA 3.2

*The **sample mean of a distribution of grouped data** is given by*

$$\bar{x} = \frac{\Sigma xf}{\Sigma f}$$

where xf represents the product of each class mean and its frequency and Σf represents the total number of items in the distribution.

COMMENT It may seem that Formula 3.2 is considerably different from Formula 3.1 for calculating the sample mean. In reality it is not. In ungrouped data the frequency of each observation is 1 and the total number of terms, Σf, is n. So, Formulas 3.1 and 3.2 are really the same.

To further illustrate the use of Formula 3.2, we consider the following example (where we have grouped the data in interval form).

● **Example 4**

George is a maintenance person in the Auburn Shopping Center. He keeps accurate records on the life of the special security light bulbs that he services. He has recorded the life of 60 light bulbs in the following chart. (*Note:* A bulb that lasted exactly 50 hours is included in the 40–50 hours category. Similarly, a bulb that lasted exactly 80 hours is included in the 70–80 hours category.):

Life of Bulb (Hours)	Class Mark x	Frequency f	Product $x \cdot f$
40–50	45	5	225
50–60	55	7	385
60–70	65	8	520
70–80	75	9	675
80–90	85	12	1020
90–100	95	9	855
100–110	105	6	630
110–120	115	4	460
		Total = 60	4770

Calculate the sample mean life of a light bulb.

Solution

In using Formula 3.2 we use the class mark (see page 26) for each interval. Thus we assume that any bulb that lasted between 40 and 50 hours actually lasted 45 hours. Although this may introduce a slight error, the error will be minimal when the number of bulbs is large. Thus, we use the class mark. Applying Formula 3.2 we have

$$\bar{x} = \frac{\Sigma xf}{\Sigma f} = \frac{225 + 385 + \cdots + 460}{5 + 7 + \cdots + 4} = \frac{4770}{60} = 79.5$$

In the previous example, we assumed that any bulb that lasted between 40 and 50 hours actually lasted 45 hours, which is the class mark. Some statisticians prefer to compute medians (and percentiles, to be discussed later) for grouped data by assuming that the entries in a class are evenly distributed in that class. For example, suppose we were interested in calculating the median of a distribution for grouped data. These statisticians would then use the formula

$$\tilde{x} = L + \frac{j}{f} \cdot c$$

where L is the lower boundary into which the median must fall, f is its frequency, c is its interval length, and j is the number of items still to be counted after reaching L. To illustrate, suppose we were interested in computing the median for the grouped data given below:

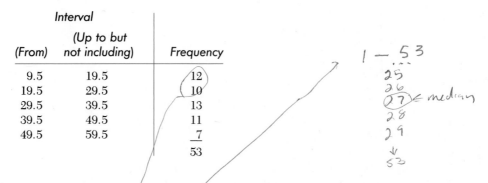

Interval		Frequency
(From)	(Up to but not including)	
9.5	19.5	12
19.5	29.5	10
29.5	39.5	13
39.5	49.5	11
49.5	59.5	7
		53

When the numbers are arranged in order of size (from smallest to largest), the median will be the 27th term. This falls in the 29.5–39.5 interval. Since there are 12 + 10 or 22 below this category, we need another 5 items in addition to the 22 which fall below this class. We add $\dfrac{5}{13}$ of the class interval to the lower bound of that class. Here we have $L = 29.5$, $f = 13$, $c = 10$, and $j = 5$, so that

$$\tilde{x} = 29.5 + \frac{5}{13} \cdot 10$$

$$= 33.35 \quad \text{when rounded}$$

Now let us consider the following examples in which all values are not of equal importance. To solve such examples we must use Formula 3.3.

FORMULA 3.3

*If $w_1, w_2, \ldots , w_n$ are the weights assigned to the numbers $x_1, x_2, \ldots , x_n$, then the **weighted sample mean** denoted by the symbol $\bar{x}_w$ is given by*

$$\bar{x}_w = \frac{\Sigma xw}{\Sigma w} = \frac{x_1w_1 + x_2w_2 + \cdots + x_nw_n}{w_1 + w_2 + \cdots + w_n}$$

This formula indicates that we multiply each number by its weight and divide the sum of these products by the sum of the weights.

● **Example 5** *Calculating Term Grades*

The grades that Liz received on her exams in a statistics course and the weight assigned to each are as follows:

	Grade	Weight Assigned
Test 1	84	1
Test 2	73	2
Test 3	62	5
Test 4	91	4
Final Exam	96	3

Find Liz's average term grade.

Solution

Since each test did not have the same weight, that is, count as much, Formula 3.1 or 3.2 has to be modified. This change is necessary because Formula 3.1 assumes that all numbers are of equal importance, which is not the case in this example. To calculate a weighted mean we use Formula 3.3. When this formula is applied to our example we have

$$\bar{x}_w = \frac{x_1 w_1 + x_2 w_2 + \cdots + x_5 w_5}{w_1 + w_2 + \cdots + w_5}$$

$$= \frac{(84 \cdot 1) + (73 \cdot 2) + (62 \cdot 5) + (91 \cdot 4) + (96 \cdot 3)}{1 + 2 + 5 + 4 + 3}$$

$$= \frac{1192}{15} = 79.47 \text{ when rounded off.}$$

Thus the weighted mean is 79.47, not 81.2, which is obtained by adding the numbers together and dividing the sum by n, which is 5.

● **Example 6** *Average Cost of Gasoline*

On a recent vacation trip Bill Hunt kept the following record of his gasoline purchases:

	Price per Gallon	Number of Gallons Purchased
Town 1	$1.53	17
Town 2	1.46	21
Town 3	1.49	16
Town 4	1.35	11
Town 5	1.51	19

What is the average cost per gallon of gasoline for the entire trip?

Solution
Since Bill did not purchase an identical amount of gasoline in each town, we cannot use Formula 3.1. We must use instead Formula 3.3. We have

$$\bar{x}_w = \frac{(1.53)(17) + (1.46)(21) + (1.49)(16) + (1.35)(11) + (1.51)(19)}{17 + 21 + 16 + 11 + 19}$$

$$= \frac{124.05}{84} = \$1.48 \text{ (rounded off)}.$$

Thus the weighted average cost per gallon of gasoline for the entire trip was $1.48.

There are other measures that describe in some way the middle or center of a set of numbers. One of these is known as the **midrange,** which is found by taking the average of the lowest value L and the highest value H. Thus,

$$\text{Midrange} = \frac{L + H}{2}$$

For the sample 3, 7, 11, 15, 16, 18, 21, 22, and 23, the lowest value L is 3 and the highest value H is 23. Therefore, the midrange is

$$\frac{3 + 23}{2} = 13.$$

Other measures that are sometimes used are the **geometric mean** and the **harmonic mean.** (See Exercise 15 at the end of this section.)

We mentioned earlier that both $\bar{x}$ and μ will be used to represent the mean. The symbol $\bar{x}$ is used to represent the sample mean or the mean of a sample. Thus, the mean $\bar{x}$ of the sample values $x_1, x_2, x_3, \ldots, x_n$ is given by the formula

$$\text{Sample mean } \bar{x} = \frac{\Sigma x}{n}$$

The symbol μ is used to represent the mean of the entire population. Thus the mean of a population of N values, $x_1, x_2, x_3, \ldots, x_N$ is given by the formula

$$\text{Population mean } \mu = \frac{\Sigma x}{N}$$

where the N values of x constitute the entire population.

COMMENT Generally speaking, Greek letters are used when we are referring to a description of the population as opposed to English letters, which are used when we are referring to a description of a sample.

COMMENT   Throughout this section we used $\bar{x}$ since we were calculating sample means. From a given problem, we can almost always tell whether we are referring to only part of the population or to the entire population.

EXERCISES

1. *Business failures*   According to government records, the total number of businesses that failed and filed for bankruptcy during 1986 in Greensville is as follows:

Month	Number of Business Failures
Jan.	23
Feb.	17
Mar.	16
Apr.	8
May	31
June	14
July	12
Aug.	17
Sept.	15
Oct.	17
Nov.	15
Dec.	13

Find the mean, median, and mode for the number of business failures. Which is more useful?

2. *Salaries*   There are 19 employees of the GAP Corporation. During 1986 their annual salaries were as follows:

$21,000 $16,500 $18,600 $200,000 $19,300
 17,000   18,600   21,700   20,500   20,700
 19,000   22,400   18,600   17,800   18,000
 27,000   18,400   19,100   16,400

Calculate the mean, median, and modal salary in 1986 for the GAP Corporation.

3. Refer back to Exercise 2. Neglecting the Executive Officer's salary of $200,000, calculate the mean, median, and mode for the remaining data. Compare the answers obtained here with the answers obtained in Exercise 2. Comment.

4. Refer back to Exercise 2. Because of unexpected profits, the Board of Directors has decided to increase each employee's salary for 1987 by $1000. Find the mean, median, and modal salary under these new conditions.

5. Professor Alex Bailey, a prominent member of the local chapter of the American Cancer Society, believes that there is too much smoking on his college campus. He petitions the president of the college to have the cigarette vending machines removed from campus. In support of his claim, he has obtained the following statistics on the number of packages of cigarettes sold by the machines on campus during 1986.

Jan.	198	Apr.	195	July	159	Oct.	188
Feb.	201	May	220	Aug.	148	Nov.	195
Mar.	167	June	165	Sept.	167	Dec.	217

Calculate the mean, median, and mode for the above data.

6. Mary Aquilla is the telephone switchboard operator for the AMTEX Company. The number of "wrong number" telephone calls that she received over the past 70 business days is as follows:

```
 3    4  13  31  33  16  23   9  17  13
12    7   8  23  16  14  12  16  16  19
 5   14  24  17  14  12  14  12  14  11
 6   12  21  19  22  11  13  13  10  16
 4   19  17  18  19  19   2  10   8  10
10   21  16  23  12  17   1  11   6  11
 8   11  19  25  13  16   8  14   1   3
```

By grouping the data, find the mean, median, and mode. Which is more important to the management of the company?

7. An analysis of the records of three little league baseball teams reveals the following information about the performance of their pitchers.

Pitcher	Earned Run Average (ERA)	Number of Innings Pitched
A	3.8	10
B	2.7	8
C	3.1	11

Find the overall earned run average for the three pitchers.

8. *Cost averaging*   Bill purchased 80 shares of ABC stock on January 2 at a price of $50 per share. On February 2, he purchased 100 shares of the same stock at a price of $45 per share. On March 2, he purchased 150 shares of the same stock at a price of $48 per share. What is Bill's average cost of a share of stock of this company?

9. Consider the newspaper article on the next page. Find the average price of an admission ticket paid by these 100 people.

STATE TO INVESTIGATE TICKET SCALPING

Dover: The Attorney General announced yesterday that he would launch an immediate investigation into the widespread practice of ticket scalping for tickets to rock concerts. A survey of 100 ticket holders for last night's rock concert indicated that these people paid exorbitant prices for the tickets. The prices paid for the same general admission ticket were as follows:

Number of people	Price paid
36 people	$30
28 people	35
19 people	40
17 people	45

The consumer is being victimized by this unlawful practice.

Friday—December 1, 1985

---

*10. The speed of 50 cars as they passed a posted "Maximum 40 mph speed limit" sign was measured. The results are as follows:

Speed (mph)	Frequency
31–35	3
36–40	6
41–45	12
46–50	14
51–55	8
56–60	5
61–65	2

Find the mean, median, and mode for the data.

*11. The number of illegally parked cars that were towed away daily over the first 100 days of 1986 in Meadowview was as follows:

Number of Cars Towed Away	Frequency
1–5	17
6–10	13
11–15	18
16–20	19
21–25	11
26–30	12
31–35	8
36–40	2

Find the mean, median, and mode for the data.

12. During December, the Able Finance Company approved 20 loan applications. If the combined amount loaned was $280,000, find the average amount of a loan.

13. Three manufacturers of exercise equipment claim that the average life of their exercise equipment under normal use is five years. A consumer's group decides to test each manufacturer's claim. It compiles the following list on the life, in years, of the exercise equipment manufactured by each:

Manufacturer A:	0.5	1.6	2	3.5	4	4.5	6	7	7.9	8	10
Manufacturer B:	4	4	5	5	5	6	11	13	14	15	16
Manufacturer C:	2	3	4	4	6	13	14	15			

a. Which measurement of average is each manufacturer using to support the claim made?

b. From which manufacturer would you buy exercise equipment? Why?

14. The average family income of all the female students at Boduk College is $25,000. The average family income of all the male students at the college is $35,000. Is it safe to assume that the average family income for *all* college students at Boduk College is $30,000? Explain your answer.

*15. The **harmonic mean** of n numbers $x_1, x_2, \ldots, x_n$ is defined as n divided by the sum of the reciprocals of the numbers, that is,

$$\text{Harmonic mean} = \frac{n}{\sum\limits_{i=1}^{n} 1/x_i}$$

Also, the **geometric mean** of a set of n positive numbers $x_1, x_2, \ldots, x_n$ is defined as the nth root of their product, that is,

$$\text{Geometric mean} = \sqrt[n]{x_1 \cdot x_2 \cdots x_n}$$

Find the harmonic mean and geometric mean of the numbers 2, 4, 2, and 1. (Both the harmonic mean and the geometric mean are used in certain applications.)

3.4 MEASURES OF VARIATION

Although the mean, median, or mode are very useful in analyzing a distribution, there are some disadvantages in using them alone. These measures only locate the center of the distribution. In certain situations location of the center may not be adequate. We need some method of analyzing variation, that is, the difference among the terms of a distribution. In this section we will discuss some of the most commonly used methods for analyzing variation.

First let us consider Christina, who is interested in determining the best route to drive to work. During one week she drove to work on the Brooks Expressway and during a second week she drove on the Kingston Expressway. The number of minutes needed to drive to work each day was:

Brooks Expressway:	15	26	30	39	45
Kingston Expressway:	29	30	31	32	33

In each case the average time that it took her to drive to work was 31 minutes. Which way is better?

When she used the Brooks Expressway, the time varied from 15 to 45 minutes. We then say that for the Brooks Expressway the **range** is 45 − 15 = 30 minutes.

On the Kingston Expressway the time varied from 29 to 33 minutes. Thus, the range is 33 − 29 = 4 minutes.

DEFINITION 3.4

*The **range** of a set of numbers is the difference between the largest number in the distribution and the smallest number in the distribution.*

The range is frequently used by manufacturers as a measure of dispersion (spread) in specifying the variation in the quality of a product. So, although the average diameter of a drill bit may be $\frac{15}{32}$ inches, in reality the range in size may be enormous. The manufacturer usually specifies the range to prospective customers.

The range is also used frequently by stock brokers to describe the prices of certain stock. One often hears such statements as "Stock X had a price range of 15 to 75 dollars, or 60 dollars, during the year."

The range is by far the simplest measure of variation to calculate since only two numbers are needed to calculate it; however, it does not tell us anything about how the other terms vary. Furthermore, if there is one extreme value in a distribution, the dispersion or the range will appear very large. If we remove the extreme term, the dispersion may become quite small. Because of this, other measures of variation such as variance, standard deviation, or average deviation are used.

To calculate the sample variance of a set of numbers, we first calculate the sample mean of the numbers. We then subtract the sample mean from each number and square the result. Finally we divide the result by $n - 1$ where n is the number of items in the sample. The result is called the **sample variance** of the numbers. If we now take the square root of the sample variance, we get the **sample standard deviation** for the numbers.* If instead of squaring the differences from the mean we take the absolute value (that is, we neglect any negative signs) of these differences and find the average of these absolute values,

---

*A knowledge of how to compute square roots is not needed. Such values can be obtained by using a calculator or a square root table.

the resulting number is called the **average deviation.** The symbol for absolute value is two vertical lines. Thus $|+8|$ is read as "the absolute value of $+8$."

Let us illustrate the preceding ideas by calculating the sample variance, sample standard deviation, and average deviation for the two routes that Christina uses to drive to work. Since the sample mean, $\bar{x}$, is 31 we can arrange our calculations as shown in the following chart.

	Time (x)	Difference from Mean $(x - \bar{x})$	Square of Difference $(x - \bar{x})^2$	Absolute Value of Difference $\|x - \bar{x}\|$
	15	$15 - 31 = -16$	$(-16)^2 = 256$	16
By Way of	26	$26 - 31 = -5$	$(-5)^2 = 25$	5
Brooks	30	$30 - 31 = -1$	$(-1)^2 = 1$	1
Expressway	39	$39 - 31 = 8$	$8^2 = 64$	8
	45	$45 - 31 = 14$	$14^2 = 196$	14
		Sum = 0	Sum = 542	Sum = 44

Therefore, if Christina travels to work by way of the Brooks Expressway, the sample variance is $\dfrac{542}{5 - 1}$, or 135.5, the sample standard deviation is $\sqrt{135.5}$ or 11.64, and the average deviation is $\dfrac{44}{5}$, or 8.8.

Notice that in computing these measures of variation, we used symbols. Thus $\bar{x}$ represents the sample mean, $x - \bar{x}$ represents the difference of any number from the mean, $(x - \bar{x})^2$ represents the square of the difference, and $|x - \bar{x}|$ represents the absolute value of the difference from the sample mean. Furthermore, the sum of the differences from the sample mean, $\Sigma(x - \bar{x})$, is 0. Can you see why?

Let us now compute the sample variance, sample standard deviation, and average deviation for traveling to work by way of the Kingston Expressway. Again we will use symbols.

	Time (x)	Difference from Mean $(x - \bar{x})$	Square of Difference $(x - \bar{x})^2$	Absolute Value of Difference $\|x - \bar{x}\|$
	29	$29 - 31 = -2$	$(-2)^2 = 4$	2
By Way of	30	$30 - 31 = -1$	$(-1)^2 = 1$	1
Kingston	31	$31 - 31 = 0$	$0^2 = 0$	0
Expressway	32	$32 - 31 = 1$	$1^2 = 1$	1
	33	$33 - 31 = 2$	$2^2 = 4$	2
		Sum = 0	Sum = 10	Sum = 6

In this case the sample variance is $\dfrac{10}{5-1}$, or 2.5; the sample standard deviation is $\sqrt{2.5}$, or 1.58; and the average deviation is $\dfrac{6}{5}$, or 1.2. Here again the sum of the differences from the mean, $\Sigma(x - \bar{x})$, is 0. This is always the case.

We now formally define sample variance, sample standard deviation, and average deviation.

DEFINITION 3.5   Sample Variance

The sample variance of a sample of n numbers is a measure of the spread of the numbers about the sample mean and is given by

$$\text{Sample variance} = s^2 = \frac{\Sigma(x - \bar{x})^2}{n-1}$$

DEFINITION 3.6   Sample Standard Deviation

The sample standard deviation of a sample of n numbers is the positive square root of the variance. Thus,

$$\text{Sample standard deviation} = s = \sqrt{s^2} = \sqrt{\frac{\Sigma(x - \bar{x})^2}{n-1}}$$

DEFINITION 3.7

*The **average deviation** of a sample of numbers is the average of the absolute value of the differences from the sample mean. Symbolically,*

$$\text{Average deviation} = \frac{\Sigma|(x - \bar{x})|}{n}$$

If we are working with the entire population rather than with a sample, then we have the following:

$$\text{Population variance} = \sigma^2 = \frac{\Sigma(x - \mu)^2}{N}$$

$$\text{Population standard deviation} = \sigma = \sqrt{\frac{\Sigma(x - \mu)^2}{N}}$$

COMMENT   When calculating the sample variance (or sample standard deviation) we use $n - 1$ instead of n in the denominators of the two formulas given in Definitions 3.5 and 3.6. However, when we calculate the population variance

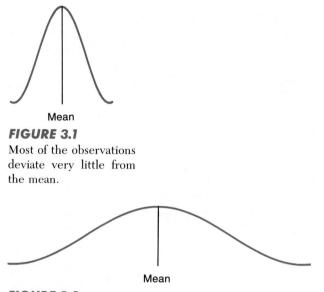

Mean

FIGURE 3.1

Most of the observations
deviate very little from
the mean.

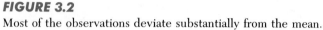

Mean

FIGURE 3.2

Most of the observations deviate substantially from the mean.

(or standard deviation) we use N instead of $N - 1$ in the denominators of the formulas. There is a sound statistical reason for doing this. We will not concern ourselves with the reason at this point. To summarize: The difference between σ and s is whether we divide by N or by $n - 1$ and whether we use the population mean, μ or the sample mean, $\bar{x}$. When calculating the population standard deviation we use μ and divide by N and denote our result by σ, whereas when calculating a sample standard deviation we use $\bar{x}$ and divide by $n - 1$ and denote our result by s. Thus s is really an estimate of σ, the population standard deviation. Very often statisticians will refer to s as *the* standard deviation, even though it is only an estimate.

COMMENT It may seem that the standard deviation is a complicated and useless number to calculate. At the moment, let us say that it is a useful number to the statistician. Just as the measures of central tendency help us locate the "center" of a relative frequency distribution, the standard deviation helps us measure its "spread." It tells us how much the observations differ from the mean. Notice that most of the observations in Figure 3.1 deviate very little from the mean of the distribution. As opposed to this, most of the observations in Figure 3.2 deviate substantially from the mean of the distribution. When we discuss the normal distribution in later chapters, you will understand the significance and usefulness of the standard deviation.

In most statistical problems we do not have all the data for the population. Instead, we have only a small part, that is, a sample, of the population. It is for this reason that we use the sample variance and sample standard deviation, which we restate below using a formula number for later reference.

FORMULA 3.4

Sample Variance   *Sample Standard Deviation*

$$s^2 = \frac{\Sigma(x - \bar{x})^2}{n - 1}$$ $$s = \sqrt{\frac{\Sigma(x - \bar{x})^2}{n - 1}}$$

We will illustrate the preceding ideas with another example.

● **Example 1**

The number of hours per day that several technicians spent adjusting the timing mechanism on CAT scan machines during the past week is 5, 3, 2, 6, 4, 2, and 6. Find the sample variance, sample standard deviation, and average deviation.

Solution

We arrange the data in order as shown in the following chart and perform the indicated calculations:

The sample mean is

$$\bar{x} = \frac{\Sigma x}{n} = \frac{2 + 2 + 3 + 4 + 5 + 6 + 6}{7}$$

$$= \frac{28}{7} = 4$$

Number of Hours x	Difference from Mean $(x - \bar{x})$	Square of Difference $(x - \bar{x})^2$	Absolute Value of Difference $\lvert x - \bar{x} \rvert$
2	$2 - 4 = -2$	$(-2)^2 = 4$	2
2	$2 - 4 = -2$	$(-2)^2 = 4$	2
3	$3 - 4 = -1$	$(-1)^2 = 1$	1
4	$4 - 4 = 0$	$0^2 = 0$	0
5	$5 - 4 = 1$	$1^2 = 1$	1
6	$6 - 4 = 2$	$2^2 = 4$	2
6	$6 - 4 = 2$	$2^2 = 4$	2
	Sum = 0	Sum = 18	Sum = 10

Our answers then are

$$\text{Sample variance} = \frac{\Sigma(x - \bar{x})^2}{n - 1} = \frac{18}{7 - 1} = 3$$

$$\text{Sample standard deviation} = \sqrt{\text{variance}} = \sqrt{3}, \text{ or } 1.73$$

$$\text{Average deviation} = \frac{\Sigma \lvert x - \bar{x} \rvert}{n} = \frac{10}{7} = 1.43$$

3.5 COMPUTATIONAL FORMULA FOR CALCULATING THE VARIANCE

Although the formulas in Definitions 3.5 and 3.6 of the preceding section can always be used for calculating the sample variance, it turns out that, in practice, the calculations become quite tedious. For this reason we use more convenient formulas, which follow:

FORMULA 3.5 *Computational Formulas*

$$\text{Population variance} = \sigma^2 = \frac{\Sigma x^2}{N} - \frac{(\Sigma x)^2}{N^2} \quad \text{or} \quad \frac{N\,\Sigma x^2 - (\Sigma x)^2}{N^2}$$

$$\text{Sample variance} = s^2 = \frac{n\Sigma x^2 - (\Sigma x)^2}{n(n-1)}$$

$$\text{Population standard deviation} = \sigma = \sqrt{\frac{\Sigma x^2}{N} - \frac{(\Sigma x)^2}{N^2}}$$

$$\text{Sample standard deviation} = s = \sqrt{\frac{n\Sigma x^2 - (\Sigma x)^2}{n(n-1)}}$$

In Formula 3.5, Σx^2 means that we square each number and add the squares together, whereas $(\Sigma x)^2$ means we first sum the numbers and then square the sum.

Using summation notation for the data in Table 3.1, we have

$$\Sigma x = 1 + 2 + 3 + 4 + 5 = 15$$
$$\Sigma x^2 = 1^2 + 2^2 + 3^2 + 4^2 + 5^2$$
$$= 1 + 4 + 9 + 16 + 25 = 55$$

TABLE 3.1
Squares of Integers
(x^2 Means x Times x)

x	x^2
1	1
2	4
3	9
4	16
5	25
15	55
$\Sigma x = 15$	$\Sigma x^2 = 55$

We now use Formula 3.5 to calculate the sample variance for the data of Table 3.1:

$$s^2 = \frac{n\Sigma x^2 - (\Sigma x)^2}{n(n-1)}$$

$$= \frac{5(55) - (15)^2}{5(5-1)}$$

$$= \frac{275 - 225}{5(4)}$$

$$= \frac{50}{20} = 2.5$$

If you now calculate the sample variance by using the formula in Definition 3.5 of the preceding section and compare the results, your answer will be the same. It is considerably simpler, however, to get the answer by using Formula 3.5. The sample standard deviation is obtained by taking the square root of the variance. Thus,

$$\text{Standard deviation} = \sqrt{\text{variance}}$$

$$= \sqrt{2.5} \approx 1.58$$

(The symbol $\approx$ stands for approximately.)

COMMENT   The advantage of computing the sample variance by Formula 3.5 is that we do not have to subtract the sample mean from each term of the distribution.

BEWARE   Do not confuse the symbols Σx^2 and $(\Sigma x)^2$. The symbol Σx^2 represents the sum of the squares of each number, whereas the symbol $(\Sigma x)^2$ represents the square of the sum of the numbers. If your calculation of the variance results in a negative number, you probably have confused the two symbols.

How are the mean, variance, and standard deviation affected if we *multiply* each term of a distribution by some number? To see what happens, multiply each number of the distribution given in Table 3.1 above by, say 10, and compute the mean, variance and standard deviation of the new distribution. (You will be asked to do this in one of the exercises.)

How are the mean, variance, and standard deviation affected if we *add* the same constant to each term of a distribution? (Again, you will be asked to do this in one of the exercises.)

To calculate the variance and standard deviation of sample data that is presented in frequency distribution form, as is often the case with published data, we need a computational formula for determining the sample variance and sample standard deviation for grouped data. We can always use the definition.

Thus, we have

Sample Variance for Grouped Data

$$s^2 = \frac{\Sigma(x - \bar{x})^2 \cdot f}{n - 1}$$

Sample Standard Deviation for Grouped Data

$$s = \sqrt{\frac{\Sigma(x - \bar{x})^2 \cdot f}{n - 1}}$$

However, the computations involved in using these formulas can be very tedious. Formula 3.6 is a short-cut formula that gives the same results as if we had used the definition.

FORMULA 3.6 *Computational Formulas for Grouped Data*

Sample Variance for Grouped Data

$$s^2 = \frac{n(\Sigma x^2 \cdot f) - (\Sigma x \cdot f)^2}{n(n - 1)}$$

Sample Standard Deviation for Grouped Data

$$s = \sqrt{\frac{n(\Sigma x^2 \cdot f) - (\Sigma x \cdot f)^2}{n(n - 1)}}$$

● **Example 1**
The IQ of 50 students in the fifth grade of a special school was measured. The results are as follows:

IQ Score	Frequency
71–79	3
80–88	6
89–97	12
98–106	14
107–115	8
116–124	5
125–133	2

Compute the sample variance and sample standard deviation.

Solution
We rearrange the data as shown in the following table. Note that we use the class marks for each category. In grouped data, we assume that all the numbers in each group are at the class mark for that group.

IQ Scores	Class Marks x	Frequencies f	$x \cdot f$	$x^2 \cdot f$
71–79	75	3	225	16,875
80–88	84	6	504	42,336
89–97	93	12	1116	103,788
98–106	102	14	1428	145,656
107–115	111	8	888	98,568
116–124	120	5	600	72,000
125–133	129	2	258	33,282
		50	5019	512,505

Here $n = \Sigma f = 50$ since the sample consisted of 50 students. Then, using Formula 3.6, we get

$$s^2 = \frac{n(\Sigma x^2 \cdot f) - (\Sigma x \cdot f)^2}{n(n-1)}$$

$$= \frac{50(512,505) - (5019)^2}{50(50-1)}$$

$$= 177.5057$$

and

$$s = \sqrt{s^2} = \sqrt{177.5057} \approx 13.32.$$

Thus, the sample variance is 177.51 and the sample standard deviation is 13.32. (Do not use a rounded variance to get a standard deviation. Round all answers at the end of all calculations.)

EXERCISES

1. The weights of ten randomly selected new members who joined a health club are as follows:

158 pounds 181 pounds
123 pounds 211 pounds
187 pounds 137 pounds
169 pounds 147 pounds
142 pounds 161 pounds

a. Calculate the range of weights for these ten new members.
b. Find the sample variance and sample standard deviation.

2. The annual salaries of the 20 workers of the Bevy Printing Company are as follows:

$25,000	$27,950	$19,000	$38,950
34,700	26,280	28,000	44,761
41,795	29,640	26,500	23,280
28,200	23,500	27,000	24,512
26,400	17,000	24,000	31,710

Find the range of salaries for the 20 workers.

3. The number of chapters in ten math books published by a company is as follows: 12, 8, 11, 10, 7, 10, 15, 13, 14, and 9. Find the sample mean, sample variance, sample standard deviation, and average deviation for the number of chapters in a math book.

4. *Drugs and the senior citizen* Are senior citizens in nursing homes taking too many medicines? A sample of eight senior citizens indicated that they were taking 4, 2, 0, 3, 8, 2, 6, and 1 different kinds of medications. Find the range, sample mean, sample variance, sample standard deviation, and average deviation for the number of medications taken by these senior citizens.

5. Joyce has been comparison shopping at every store in town. The prices quoted for a particular model of video cassette recorder are $269, $318, $288, $305, $276, $302, $299, and $293. Find the population variance, population standard deviation, and average deviation.

6. The number of express mail packages received on February 12 by five of the offices located in the Johnson Building was 12, 8, 17, 11, and 19. Calculate the sample mean, sample variance, sample standard deviation, and the average deviation for this day.

7. Multiply each number in the preceding exercise by 3 and then compute the sample mean, sample variance, sample standard deviation, and average deviation for the new distribution. How do the results compare with those of the preceding exercise? Can you generalize?

8. The ages of 162 prospective jurors considered for a controversial case are shown below:

Ages (years)	Frequency
20–28	25
29–37	38
38–46	27
47–55	36
56–64	21
65–73	15

Find the sample variance and sample standard deviation.

*9. There are 125 convicts at a state correctional facility. The number of years remaining in the convicts' sentences is shown below:

Number of Years Remaining	Frequency
0–2	32
3–7	21
8–10	28
11–15	21
16–20	14
21–40	9

Find the sample variance and sample standard deviation.

10. The average cost of malpractice medical insurance last year in a certain city was $8700 for an orthopedist, with a standard deviation of $1275. This year the cost of malpractice insurance for every orthopedist in the city will be increased by $2000. What will be the new mean and new standard deviation?

11. The price of a 2-liter bottle of a popular soft drink charged by the seven dealers in Bolton during 1985 was $1.29, $1.15, $1.09, $.99, $1.11, $1.19, and $1.07.
   a. Calculate the population mean and population standard deviation by subtracting $1.07 from each price.
   b. Calculate the population mean and population standard deviation by subtracting $1.14 from each price.
   c. How are the (population) mean and (population) standard deviation affected if we subtract a different number from each price?

12. A trucking company tested two different brands of tires on its fleet of trucks to determine the life of each brand of tire. The following results are available:

	Brand A	Brand B
Average life	12,000 miles	13,000 miles
Standard deviation	500 miles	800 miles

Which brand of tire would you buy? Why?

13. *Discrimination*   The attorney general of a particular state believes that two supermarket chains doing business within the state discriminate against poor people by charging more for a quart of milk in poor neighborhoods than in middle-class neighborhoods. The supermarket chains deny the charge. Both sides agree to sample five company stores in each area and to calculate the average cost per quart of milk. When this is done the following data are obtained:

Cost Per Container of Milk

Poorer Neighborhoods	Middle-class Neighborhoods
56¢	79¢
61¢	47¢
60¢	57¢
59¢	58¢
64¢	59¢

Average price is 60¢ per quart of milk

Thus, the supermarket chains claim that the average price for a quart of milk in both areas is the same. Is there anything wrong with the above reasoning?

3.6 INTERPRETATION OF THE STANDARD DEVIATION: CHEBYSHEV'S THEOREM

Up to this point, we have been discussing formulas for calculating the standard deviation of a set of numbers. If the standard deviation turns out to be small, then we can conclude that all the data values are concentrated around the mean. On the other hand, if the standard deviation is large, then the data values will be widely scattered about the mean.

In a later chapter, when we discuss the normal distribution, we will notice that a substantial number of the data is bunched within 1, 2, or 3 standard deviations above or below the mean. Nevertheless, a more general result, which is true for any set of measurements—population or sample—and regardless of the shape of the frequency distribution is known as **Chebyshev's Theorem.** We state it now:

> ### CHEBYSHEV'S THEOREM:
>
> Let k be any number equal to or greater than 1. Then the proportion of any distribution that lies within k standard deviations of the mean is at least $1 - \dfrac{1}{k^2}$.

To see what Chebyshev's Theorem means let us compute some values of $1 - \dfrac{1}{k^2}$ as shown below.

Value of k	Value of $1 - \dfrac{1}{k^2}$
1	0
2	$\dfrac{3}{4}$ or 75%
3	$\dfrac{8}{9}$ or 89%

When $k = 2$, then if μ is the population mean and σ the population standard deviation, Chebyshev's theorem says that you will always find at least $\dfrac{3}{4}$, that is, 75% or more, of the measurements will lie within the interval $\mu - 2\sigma$ and $\mu + 2\sigma$ that is within two standard deviations of the mean on either side. This can be seen in Figure 3.3. Similarly, when $k = 3$, Chebyshev's theorem says that at least $\dfrac{8}{9}$ of the measurements, or 89% of the data, will lie within the interval $\mu - 3\sigma$ and $\mu + 3\sigma$, that is within 3 standard deviations of the mean on either side. This can be seen in Figure 3.4.

● **Example 1**

A statistician is analyzing the claims filed with an auto insurance company. A sample of 100 claims discloses that the average claim filed was $831, with a standard deviation of $150. If we let $k = 2$, then at least $1 - \dfrac{1}{2^2}$ or $\dfrac{3}{4}$, that is,

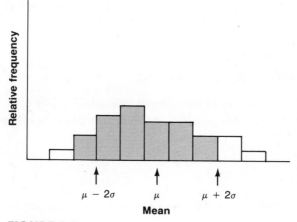

FIGURE 3.3

At least $\frac{3}{4}$ of the measurements will fall within the shaded portion.

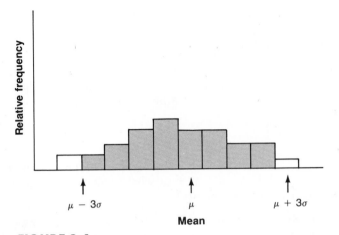

FIGURE 3.4
At least $\frac{8}{9}$ of the measurements will fall within the shaded portion.

75% or more, of the claims will be within

$$\$831 - 2(150) \text{ and } 831 + 2(150)$$

or between $531 and $1131

Similarly, at least $1 - \dfrac{1}{3^2}$ or $\dfrac{8}{9}$, or 89% or more, of the claims will be within

$$\$831 - 3(150) \text{ and } \$831 + 3(150)$$

or between $381 and $1281.

● **Example 2**

If all the light bulbs manufactured by a certain company have a mean life of 1600 hours with a standard deviation of 100 hours, at least what percentage of the bulbs will have a mean life of between 1450 and 1750 hours?

Solution
We first find $1750 - 1600 = 1600 - 1450 = 150$. Using Chebyshev's theorem, we know that k standard deviations or $k(100)$ will equal 150. That is,

$$100k = 150$$

so that

$$k = 1.5$$

Thus, at least $1 - \dfrac{1}{(1.5)^2} = 1 - \dfrac{1}{2.25} = 0.5556$, or at least 55.56% of the light bulbs will have a mean life between 1450 and 1750 hours.

EXERCISES

1. Verify Chebyshev's theorem with $k = 2$ and $k = 3$ for the following data:

   38   69   41   53   68   76   39   49   37   58.

2. Verify Chebyshev's theorem with $k = 2$ and $k = 3$ for the following data, which gives the number of inches of rainfall recorded monthly for a certain city:

   2.8   4.1   3.7   2.7   6.9   5.1   8.3
   1.2   1.9   2.5   3.3   4.5   7.1.

3. If all the bags of mixed nuts produced by a company contain an average of 14 walnuts with a standard deviation of 1.1 walnuts, at least what percentage of the bags of mixed nuts will contain between 12 and 16 walnuts?

4. If all the drivers of the Kent Bus Company have been driving an average of 10 years with a standard deviation of 2.2 years, at least what percentage of drivers will have been driving for the company between 7 and 13 years?

3.7 PERCENTILES AND PERCENTILE RANK

Consider the newspaper article in Figure 3.5. The article indicates that percentile ranks are used by many colleges in determining which students will be admitted. Also, if we analyze the card from the Emerson Medical Laboratory shown in Figure 3.6, we again notice the use of percentiles. How are percentiles calculated? How do we interpret them?

Now consider the table on the opposite page, which shows the distribution of the ages of 30 runners in a Florida jogging marathon. The histogram for the data is shown in Figure 3.7.

NEW ADMISSIONS GUIDELINES ADOPTED

Trudy Hoffman

March 20: As a result of further budget cutbacks the Chancellor's office announced new admission guidelines. Effective this fall, no students will be admitted to a senior college unless his or her high school average is 80% or higher or is in at least the 75th percentile in his or her graduating class. To be admitted to a community college, a student must have a 70% high school average or have a percentile rank of 40 or higher.

Sunday, March 20, 1981

FIGURE 3.5

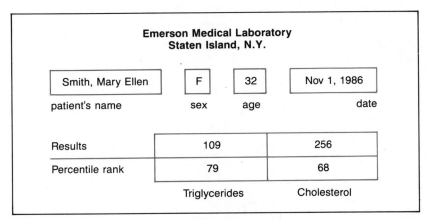

FIGURE 3.6

Age (Years)	Frequency
21–30	1
31–40	1
41–50	6
51–60	8
61–70	11
71–80	3

Suppose we were interested in determining how many runners were 70 years old or younger. We would add the frequencies in the five lowest categories. In our case, we would add $1 + 1 + 6 + 8 + 11 = 27$. We can obtain the same answers by drawing a **cumulative frequency histogram.** In essence, we accumulate the frequencies by adding the frequencies in each interval of the grouped data. The resulting cumulative frequency histogram is shown below in Figure 3.8. Such cumulative frequency or relative frequency graphs are called **ogives.**

Instead of using the cumulative frequency as the vertical scale, we can use percents where 100% corresponds to 30 runners, and 0% corresponds to 0 runners. Fifty percent would correspond to 15 runners, and so on. By having the percent on the vertical scale we can answer such questions as "What percent of the runners were 70 years old or younger?" We can read our answer from the graph, called a **cumulative relative frequency histogram,** given in Figure 3.9. The answer is 90%. More generally, a score that tells us what percent of the total population scored at or below that measure is called the **percentile.** How do we determine such percentiles?

To get us started, let us consider Lorraine, who received a 76 on her midterm psychology examination. There are 150 students, including her, in the class. She knows that 60% of the class got below 76, 10% of the class got 76, and the remaining 30% got above 76.

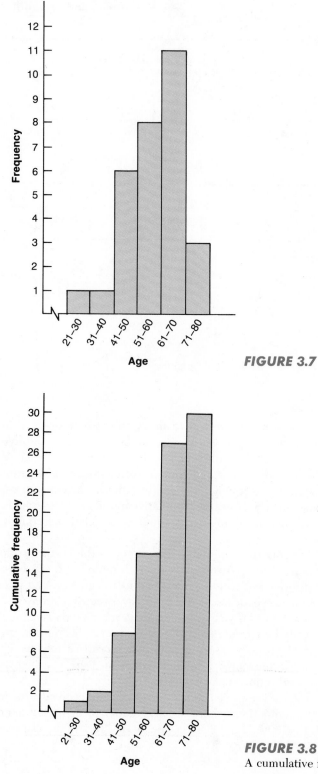

FIGURE 3.7

FIGURE 3.8
A cumulative frequency histogram.

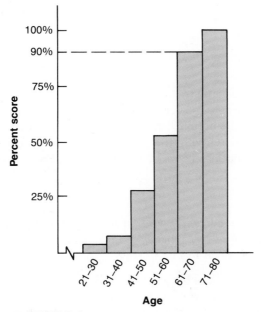

FIGURE 3.9
A cumulative relative frequency histogram.

Since 60% of the class got below her grade of 76 and 30% got above her grade, her percentile rank should be between 60 and 70. We will use 65, which is midway between 60 and 70. What we do is find the percent of scores that are below the given score and add one half of the percent of the scores that are the same as the given score. In our case 60% of the class grades were below

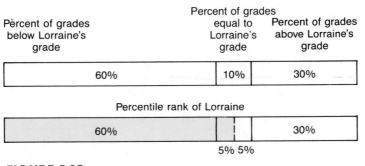

FIGURE 3.10

Lorraine's and 10% were the same as Lorraine's. Thus, the percentile rank of Lorraine's grade is

$$60 + \left(\frac{1}{2}\right)(10) = 65$$

Figure 3.10 illustrates the situation.

We now say that Lorraine's percentile rank is 65. This means that approximately 35% of the class did better than her on the exam and that she did better than 65% of the class. Essentially the percentile rank of a score tells us the percentage of the distribution that is below that score. Formally we have

---

DEFINITION 3.8

*The **percentile rank** or **percentile** of a term in a distribution is found by adding the percentage of terms below it with one half of the percentage of terms equal to the given term.*

Let X be a given score, let B represent the *number* of terms below the given score X, and let E represent the *number* of terms equal to the given score X. If there are N terms altogether (that is, if the entire population consists of N terms together), then the percentile rank of X is given by Formula 3.6.

FORMULA 3.6

$$\textit{Percentile rank of } X = \frac{B + \frac{1}{2}E}{N} \cdot 100$$

We will now illustrate the use of Formula 3.6.

● **Example 1**
Bill and Jill are twins, but both are in different classes. Recently they both got 80 on a math test. The grades of the other students in their class were as follows:

Jill's Class
64 67 73 73 73 74 77 77 78 78 79 80 80 82 91 94 100

Bill's Class
43 65 68 73 75 76 76 77 79 80 80 80 80 85 86 87
88 90 92 96

Find the percentile rank of each student.

Solution

We will use Formula 3.6. Jill's grade is 80. There were two 80s (including Jill's) in the class, so $E = 2$. There were 11 grades below 80, so $B = 11$. Since there are 17 students in the class altogether, so $N = 17$. Thus,

$$\text{Jill's percentile rank} = \frac{11 + \frac{1}{2}(2)}{17} \cdot 100$$

$$= \frac{11 + 1}{17} \cdot 100 = \frac{12}{17} \cdot 100$$

$$= \frac{1200}{17} = 70.59$$

Jill's percentile rank is 70.59. Using a similar procedure for Bill's class, we find that $B = 9$, $E = 4$, and $N = 20$. Thus,

$$\text{Bill's percentile rank} = \frac{9 + \frac{1}{2}(4)}{20} \cdot 100$$

$$= \frac{9 + 2}{20} \cdot 100 = \frac{11}{20} \cdot 100$$

$$= \frac{1100}{20} = 55$$

Bill's percentile rank is 55.

COMMENT The percentile rank of an individual score is often more helpful than the particular score value. Although both Bill and Jill had grades of 80, Jill's percentile rank is considerably higher. If we assume that the levels of competition are equivalent in both classes, this may indicate that Jill's performance is superior to Bill's performance when compared to the rest of their respective classes.

We often use the word **percentile** to refer directly to a score in a distribution. So, instead of saying that the percentile rank of Lorraine's grade is 65, we would say that her grade is in the 65th percentile. Similarly, if a term has a percentile rank of 40, we would say that it is in the 40th percentile.

Percentiles are used quite frequently to describe the results of achievement tests and the subsequent ranking of people taking those tests. This is especially true when applying for many civil service jobs. If there are more applicants than available jobs, candidates are often ranked according to percentiles. Many colleges use only percentile ranks, rather than the numerical high school average, to determine which candidates to admit. The reason is that percentile ranks of a student's high school average reflect how they did with respect to their classmates, whereas numerical averages only indicate an individual student's performance.

Since percentiles are numbers that divide the set of data into 100 equal parts, we can easily compare percentiles. Thus, in Example 1 we were able to find the percentile rank of Jill and Bill, even though they both were in different classes.

During World War II the United States Army administered the Army General Classification Tests (AGCT) to thousands of enlisted men. The results showed important differences in the average IQ of men in various jobs, ranging from 93 for miners and farmhands to around 120 for accountants, lawyers, and engineers. Figure 3.11 shows the IQ range between the 10th and 90th percentiles for workers in various occupations. Furthermore, the vertical bars represent the 50th percentile or median scores. Very often, when such tests are administered to large groups of people, the results are given in terms of

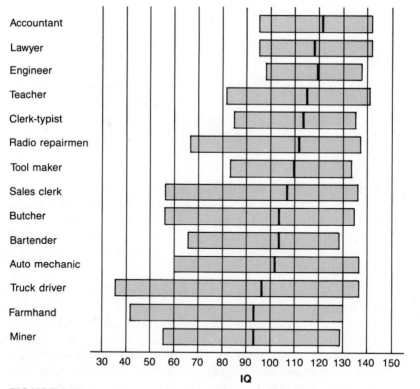

FIGURE 3.11

Each bar shows the IQ range between the 10th and 90th percentiles for men in that occupation. The vertical bars represent the 50th percentiles. Note that although the average IQ score of accountants was 121 and that of miners 93, some miners had higher IQ scores than some accountants.

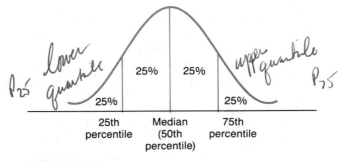

FIGURE 3.12
Some of the frequently used percentiles.

percentile bands, as shown in Figure 3.12. Since percentiles are used quite often, special names are given to the 25th and 75th percentiles of a distribution. Of course, the *50th percentile is the median of the distribution.* See Figure 3.12.

DEFINITION 3.9

*The 25th percentile of a distribution is called the **lower quartile.** It is denoted by P_{25}. Thus, 25% of the terms are below the lower quartile and 75% of the terms are above it.*

DEFINITION 3.10

*The 75th percentile of a distribution is called the **upper quartile.** It is denoted by P_{75}. Thus, 75% of the terms are below the upper quartile and 25% of the terms are above it. These percentiles are pictured in Figure 3.12 for a normal distribution.*

NOTATION   The various percentiles are denoted by the letter P with the appropriate subscript. Hence, P_{37} denotes the thirty-seventh percentile, P_{99} denotes the 99th percentile, and so on.

When calculating percentiles involving grouped data, we use the class mark. This is illustrated in the following example.

● **Example 2**
One hundred candidates have applied for a high-paying acting job. The candidates were tested and then rated according to their dancing ability, poise, and overall acting skill. The following ratings were obtained:

WANTS US TO UNDERSTAND P₇₅
what if you wanted to find P₇₅
it would be in 50-59 range
but if you don't want to use class marks

Rating	Frequency	Cumm Freq
10–19	17	17
20–29	8	25
30–39	6	31
40–49	18	49
50–59	28	77
60–69	16	93
70–79	4	97
80–89	3	100
Total = 100		

$49 + \underline{\quad} = 75$
need 26 more, there are 28 in whole thing
80
$50 + \left(\frac{26}{28} \cdot 10\right)$

Calculate the percentile rank of Heather McAllister, who scored 44.5 in the ratings.

Solution

Since Heather scored 44.5 in the ratings, she is in the 40–49 category. We know that there are $17 + 8 + 6$ or 31 below the 40–49 category. We now assume that all of the 18 candidates who scored in the 40–49 category scored 44.5, which is the class mark. Thus there are 18 people including Heather who scored 44.5. Using Formula 3.6 with $N = 100$, $B = 31$, and $E = 18$, we have:

$$\text{Heather's percentile rank} = \frac{31 + \dfrac{1}{2}(18)}{100} \times 100$$

$$= \frac{31 + 9}{100} \times 100$$

$$= 40$$

Heather is in the 40th percentile.

COMMENT   In the previous example, we assumed that all of the candidates who scored in the 40–49 category scored 44.5, which is the class mark. Some statisticians prefer to compute percentiles for grouped data by assuming that the entries in a class are evenly distributed in that class. These statisticians would then use the formula $\tilde{x} = L + \dfrac{j}{f} \cdot c$. Refer back to the discussion of this approach given on page 102.

3.8 z-SCORES

As we saw in the preceding section, one way of measuring the performance of an individual score in a population is by determining its percentile rank. Using percentile rank alone, however, can sometimes be misleading. For example, two students in different classes may have the same percentile rank. Yet one student may be far superior to his or her competitors, whereas the second may only slightly surpass the others in his or her class.

Statisticians have another very important way of measuring the performance of an individual score in a population. This measure is called the **z-score**. The z-score measures how many standard deviations an individual score is away from the mean. We define it formally as follows:

DEFINITION 3.11—FORMULA 3.7

The z-score of any number X in a distribution whose mean is μ and whose standard deviation is σ is given by

$$z = \frac{X - \mu}{\sigma} \quad or \quad z = \frac{X - \bar{x}}{s}$$

where X = *value of number in original units*
μ = *population mean*
σ = *population standard deviation*
$\bar{x}$ = *sample mean*
s = *sample standard deviation*

COMMENT   The z-score of a number in a population is sometimes called the **z-value** or **measurement in standard units**.

COMMENT   Since σ is always a positive number, z will be a negative number whenever X is less than μ, as $X - \mu$ is then a negative number. A z-score of 0 implies that the term has the same value as the mean.

We now illustrate how to calculate z-scores with several examples.

● **Example 1**
A certain brand of flashlight battery has a mean life, μ, of 40 hours and a standard deviation of 5 hours. Find the z-score of a battery which lasts

a. 50 hours   b. 35 hours   c. 40 hours.

Solution
Since $\mu = 40$ and $\sigma = 5$, we will use Formula 3.7.

a. The z-score of 50 is

$$\frac{50 - 40}{5} = \frac{10}{5} = 2$$

b. The z-score of 35 is

$$\frac{35 - 40}{5} = \frac{-5}{5} = -1$$

c. The z-score of 40 is

$$\frac{40 - 40}{5} = 0$$

- **Example 2**   *Testing Tuna Fish*

Two consumer's groups, one in New York and one in California, recently tested, at numerous local colleges, a number of different brands of canned tuna fish for taste appeal. Each consumer group used a different rating system. The following results were obtained:

Brand	New York Rating	Brand	California Rating
A	1	M	25
B	10	N	35
C	15	P	45
D	21	Q	50
E	28	R	70

Which brand has the greatest taste appeal?

Solution

At first glance it would appear that Brand R is superior since its California rating was 70. However, we see from the rating and from the given information that the two consumer's groups awarded their points differently, so that the point value alone is not enough of a basis for deciding among the different brands. We will therefore convert each of the ratings into standard scores. These calculations are shown in Tables 3.2 and 3.3.

   We can now use the z-scores as a basis for comparison of the different brands. Clearly, Brand R for which $z = 1.65$ is superior to Brand E for which $z = 1.41$.

   Notice that the sum of the z-scores for the New York ratings of the brands is 0:

$$(-1.52) + (-0.54) + (0) + (0.65) + (1.41) = 0$$

This means that the average of the z-scores is 0, since 0 divided by 5, the number of z-scores, is 0. Also the sum of the z-scores for the California rating of the brands is 0:

$$(-1.32) + (-0.66) + (0) + (0.33) + (1.65) = 0$$

Therefore the mean is 0. If you now compute the standard deviations of the z-scores in Tables 3.2 and 3.3, you will find that the standard deviation in each case is 1. We summarize these facts in the following rule.

TABLE 3.2
New York Rating of Tuna Fish ($\mu = 15$, $\sigma = 9.23$)

	Rating	Mean	Difference From Mean	z-Score $z = \dfrac{X - \mu}{\sigma}$
A	1	15	$1 - 15 = -14$	$\dfrac{-14}{9.23} = -1.52$
B	10	15	$10 - 15 = -5$	$\dfrac{-5}{9.23} = -0.54$
C	15	15	$15 - 15 = 0$	$\dfrac{0}{9.23} = 0$
D	21	15	$21 - 15 = 6$	$\dfrac{6}{9.23} = 0.65$
E	28	15	$28 - 15 = 13$	$\dfrac{13}{9.23} = 1.41$

TABLE 3.3
California Rating of Tuna Fish ($\mu = 45$, $\sigma = 15.17$)

Brand	Rating X	Mean μ	Difference From Mean $(X - \mu)$	z-Score $z = \dfrac{X - \mu}{\sigma}$
M	25	45	$25 - 45 = -20$	$\dfrac{-20}{15.17} = -1.32$
N	35	45	$35 - 45 = -10$	$\dfrac{-10}{15.17} = -0.66$
P	45	45	$45 - 45 = 0$	$\dfrac{0}{15.17} = 0$
Q	50	45	$50 - 45 = 5$	$\dfrac{5}{15.17} = 0.33$
R	70	45	$70 - 45 = 25$	$\dfrac{25}{15.17} = 1.65$

> **RULE**
>
> In any distribution the mean of the z-scores is 0 and the standard deviation of the z-scores is 1.

Formula 3.7 can be changed so that if we are given a particular z-score, we can calculate the corresponding original score. The changed formula is as follows:

> **FORMULA 3.8**
>
> $$X = \mu + z\sigma$$

● **Example 3**

In a recent swimming contest the mean score was 40 and the standard deviation was 4. If Carlos had a z-score of -1.2, how many points did he score?

Solution

Since $\mu = 40$, $\sigma = 4$, and $z = -1.2$, we can use Formula 3.8. Thus we have

$$X = 40 + (-1.2)(4)$$
$$= 40 - 4.8$$
$$= 35.2$$

Carlos's score was 35.2.

EXERCISES

1. *Ecology* Twenty students in an ecology class have been collecting aluminum soda cans so that they can be used for recycling. The number of cans collected by each of these students in one month is shown in the following chart:

Jennifer	407	John	406	Jimmy	348
Heather	378	Michele	382	Lisa	425
Ike	401	Christine	347	Ricardo	407
Moses	361	Rufus	312	Eddie	376
Cassandra	287	Michael	361	Terrie	391
Vincent	277	Robert	382	Carol	410
Patrick	345	Gregios	328		

Find the percentile rank of Michele and the percentile rank of Ricardo.

2. A study of the blood serum cholesterol levels of 100 people was undertaken by a medical research team to determine the relationship between high cholesterol levels and the incidence of heart attacks.* The following results were obtained:

*Baxter and Johnson, New York, 1984.

Cholesterol Levels (mg)	Frequency
240–249	5
250–259	8
260–269	12
270–279	28
280–289	26
290–299	11
300–309	8
310–319	2

a. Find the percentile rank of Melissa, whose cholesterol level was 264.5.

b. Find the percentile rank of Maurice, whose cholesterol level was 294.5.

3. Joanne Torpey has joined a weight-reduction class. The class consists of 50 people, where 18 people weigh less than Joanne and 6 other people weigh the same as Joanne. Find her percentile rank.

4. An analysis of the family income of the 600 members of the Bergenfield Golf and Country Club reveals the following:

P_{25} = $23,000
P_{50} = $36,000
P_{75} = $45,000
P_{90} = $51,000

What percent of the members have family incomes that are

a. less than $23,000?

b. more than $51,000?

c. less than $45,000?

d. more than $36,000?

e. between $23,000 and $51,000?

5. The table below shows the distribution of scores of 30 volunteers on an obstacle-avoidance driving test.

Scores	Frequency
11–20	3
21–30	4
31–40	9
41–50	7
51–60	5
61–70	2

a. Draw the cumulative frequency histogram for the above data.

b. Draw the cumulative relative frequency histogram (in percent form) for the above data.

6. A computer literacy test was given to the 100 secretaries at a particular college. The average score was 128, with a standard deviation of 13. Find the z-score of

a. Arlene, who scored 143 on the exam.

b. Fran, who scored 128 on the exam.

c. George, who scored 119 on the exam.

d. Lena, who scored 167 on the exam.

7. Refer to the previous exercise.

a. If Alice had a z-score of 1.79, what was her actual score?

b. If Janice had a z-score of -0.93, what was her actual score?

c. If Carolyn had a z-score of -3.76, what was her actual score?

8. Banks often survey their customers to determine which employees are courteous and efficient. The z-scores of six employees of a particular bank rated by the customers were as follows:

Joe	0	Anthony	-3.9	Caesar	$+1.2$
Cathy	-0.6	Aida	$+1.3$	Greta	$+2.7$

a. Rank these employees from lowest to highest.

b. Which of these employees were above the average?

c. Which of these employees were below the average?

9. Verify that the standard deviation of the z-scores of Tables 3.2 and 3.3 is 1.

10. Abellard and Erik have both applied for a high-paying mechanical engineering job. Abellard scored 74 on a special state engineering exam where the mean was 70 and the standard deviation was 2.13. Erik scored 123 on the company exam where the mean was 111 with a standard deviation of 9.67. Assuming that the company uses these results as the only criterion for hiring new engineers and that both tests are considered equal by company officials, who will get the job? Explain your answer.

11. A nurse in the geriatric division of a university hospital is analyzing the following information concerning the systolic blood pressure of the hospital's senior citizens.

Systolic Blood Pressure	z-Score	Percentile Rank
120	-2	4
130	-1	18
140	0	38
150	1	60
160	2	87

(Note: $\sigma = 10$)

a. What is the average blood pressure reading for these patients?

b. What percent of these patients at the hospital have blood pressure readings between 130 and 160?

c. On the basis of the information given in the above chart, what is the blood pressure reading of a patient whose z-score is -1.57?

ACME CLOTHING TO CLOSE

Dec. 5: Acme Clothing Company officials announced yesterday that they would close the Patchogue plant on Feb. 1, 1987, with the resulting loss of 1400 jobs. The closing is a direct result of rising foreign imports with which the company cannot compete.

Tuesday – Dec. 2, 1986

12. Refer to the newspaper article above. The company agreed to retrain some of the workers so that they could find other jobs. Each of the workers was given an aptitude test to determine suitability for particular jobs. Laura Snyder's results as well as the results of the other workers who took the test are shown below:

Skill	Average Test Score	Standard Deviation	Laura's Score	
Marketing/sales	43	6.8	43	$-43/6.8 = 0$
Plant maintenance	32	4.1	27	$-32/4.1 = -1.22$
Inventory/shipping	78	3.7	79	$-78/3.7 = 0.27$
Personnel	63	5.2	67	$-63/5.2 = 0.769$

(*a.*)

a. Transform each of Laura's scores into a *z*-score.
b. In which skill does she have the most talent? ~~Inventory~~ Personnel
c. In which skill does she have the least talent? plant maint.

13. The average running time in a certain marathon is 95 minutes, with a standard deviation of 7.62 minutes. Find the running times of Pete, Willie, and Tom if their *z*-scores are 3.1, −0.1, and 1.28, respectively.

14. Joe has a high school average of 89. The college that he wishes to attend will not accept any applicant with a percentile rank below 85. Is Joe sure he will be accepted by this college, or is it possible he will be denied admission? Explain your answer.

15. The distribution shown on p. 140 indicates the range of grades that can be expected on many intelligence tests. Notice that the scores are normally distributed.
a. If someone scores 400 on the ETS exam, what is the corresponding percentile rank?
b. What percent of the children taking the WISC exam will score higher than 115?
c. Find the percentile rank of a score which has a *z*-value of +2.

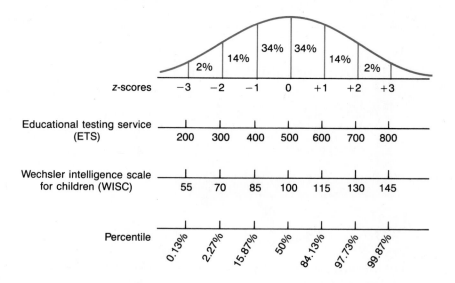

z-scores		−3	−2	−1	0	+1	+2	+3

2% 14% 34% 34% 14% 2%

Educational testing service (ETS): 200 300 400 500 600 700 800

Wechsler intelligence scale for children (WISC): 55 70 85 100 115 130 145

Percentile: 0.13% 2.27% 15.87% 50% 84.13% 97.73% 99.87%

16. The following relative frequencies were obtained by entering freshmen college students from a certain school on the Scholastic Aptitude Test administered by the Educational Testing Service.

Test Score	Relative Frequency	Relative Cumulative Frequency
200–249	0.01	.03
250–299	0.02	9
300–349	0.06	19
350–399	0.10	
400–449	0.17	36
450–499	0.19	55
500–549	0.19	74
550–599	0.14	88
600–649	0.08	96
650–699	0.03	99
700–749	0.01	100
750–799	0	

Approximately what score must a student have in order to be in the
a. 75th percentile?   **b.** 90th percentile?

550 – 599   600 – 649

3.9 USING COMPUTER PACKAGES

To illustrate how we can use the **MINITAB** computer package to compute the various statistical measures discussed in this chapter let us consider the following data on the ages of 10 patients who were admitted to the 1986 Rainville Drug Detoxification Program because of their addiction to "Crack." Their ages were 16, 24, 32, 19, 21, 14, 18, 19, 17, and 25. The following MINITAB program gets the data into column C1.

```
MTB  >  SET THE FOLLOWING DATA INTO C1
DATA  >  16   24   32   19   21
DATA  >  14   18   19   17   25
```

The DESCRIBE statement in MINITAB yields the population size (*N*), the mean, median, trimmed mean (TMEAN, which we will disregard), standard deviation, standard error of the mean, the maximum and minimum values, the upper (75th) percentile Q_3, and the lower (25th) percentile Q_1. The DESCRIBE statement for the above data produces the following computer printout:

```
MTB  >  DESCRIBE DATA IN C1
                     C1
N                    10
MEAN              20.50
MEDIAN            19.00
TMEAN             19.87
STDEV              5.28
SEMEAN             1.67
MAX               32.00
MIN               14.00
Q3                24.25
Q1                16.75
MTB  >  STOP
```

3.10 SUMMARY

In this chapter we discussed various numerical methods for analyzing data. In particular, we calculated and compared three measures of central tendency: the mean, median, and mode. We pointed out that each has its advantages and disadvantages. In addition, various properties of each measure were discussed. Thus we mentioned that the mean is affected by extreme values and that the sum of the differences from the mean is zero. The mean is the most frequently used measure of central tendency. We also demonstrated how to calculate a weighted mean when the terms of a distribution are not of equal weight. In the process we introduced summation notation.

We then discussed four measures of variation that tell us how dispersed, that is, how spread out, the terms of the distribution are around the *center* of the distribution. These were the range, variance, standard deviation, and average deviation. Various short cuts for computing the standard deviation and variance were introduced.

We also saw that an individual score is sometimes meaningless unless it is accompanied by a percentile rank or *z*-score. When scores are converted into percentile ranks or *z*-values, we can then make meaningful statements about them and compare them with other scores.

We discussed and demonstrated how to calculate percentile ranks as well as z-scores. The latter play an important role in the normal distribution to be discussed in a later chapter.

STUDY GUIDE

You should now be able to demonstrate your knowledge of the following terms by giving definitions, descriptions, or specific examples. Page references are given in parentheses for each term so that you can easily check your answer.

Summation notation (page 89)
Index (page 90)
Measures of central tendency (page 94)
Mean or average (page 96)
Mode (page 96)
Median (page 96)
Sample mean (page 97)
Weighted mean (page 103)
Midrange (page 105)
Population mean (page 107)
Geometric mean (page 107)
Harmonic mean (page 107)
Range (page 110)
Sample variance (page 112)
Sample standard deviation (page 112)
Average deviation (page 112)
Population variance (page 115)
Population standard deviation (page 115)
Chebyshev's Theorem (page 121)
Cumulative-frequency histogram (page 125)
Ogives (page 125)
Percentile (page 128)
Percentile rank (page 128)
Lower quartile (page 131)
Upper quartile (page 131)
z-score (page 133)
z-value (page 133)
Standard units (page 133)

FORMULAS TO REMEMBER

At this point you have learned some of the common terms used in statistical analysis and some of the graphic techniques. The formulas are important too.

The following list is a summary of all formulas given in the chapter. You should be able to identify each symbol, understand the relationships among the symbols expressed in each formula, understand the significance of each formula, and use the formulas in solving problems.

1. When using summation notation the following are true:

$$\sum_{i=1}^{n} (x_i + y_i) = \sum_{i=1}^{n} x_i + \sum_{i=1}^{n} y_i$$

$$\sum_{i=1}^{n} kx_i = k \cdot \sum_{i=1}^{n} x_i$$

$$\sum_{i=1}^{n} k = n \cdot k$$

2. Sample mean: $\bar{x} = \dfrac{\Sigma x}{n}$

   Population mean: $\mu = \dfrac{\Sigma x}{N}$

3. Sample mean for grouped data: $\bar{x} = \dfrac{\Sigma xf}{\Sigma f}$

4. Weighted sample mean: $\bar{x}_w = \dfrac{\Sigma xw}{\Sigma w}$

5. Sample variance: $s^2 = \dfrac{\Sigma(x - \bar{x})^2}{n - 1}$

6. Sample standard deviation: $s = \sqrt{\dfrac{\Sigma(x - \bar{x})^2}{n - 1}}$

7. Average deviation: $\dfrac{\Sigma|x - \bar{x}|}{n}$

8. Population variance: $\sigma^2 = \dfrac{\Sigma(x - \mu)^2}{N}$

9. Population standard deviation: $\sigma = \sqrt{\dfrac{\Sigma(x - \mu)^2}{N}}$

10. Computational formula for population variance: $\sigma^2 = \dfrac{\Sigma x^2}{N} - \dfrac{(\Sigma x)^2}{N^2}$

11. Computational formula for population standard deviation:

$$\sigma = \sqrt{\dfrac{\Sigma x^2}{N} - \dfrac{(\Sigma x)^2}{N^2}}$$

12. Computational formula for sample variance: $s^2 = \dfrac{n(\Sigma x^2) - (\Sigma x)^2}{n(n - 1)}$

13. Computational formula for sample standard deviation:

$$s = \sqrt{\frac{n(\Sigma x^2) - (\Sigma x)^2}{n(n - 1)}}$$

14. Computational formulas for grouped data:

Sample variance: $s^2 = \dfrac{n(\Sigma x^2 \cdot f) - (\Sigma x \cdot f)^2}{n(n - 1)}$

Sample standard deviation: $s = \sqrt{\dfrac{n(\Sigma x^2 \cdot f) - (\Sigma x \cdot f)^2}{n(n - 1)}}$

15. *Chebyshev's Theorem:* At least $1 - \dfrac{1}{k^2}$ of a set of measurements will lie within k standard deviation units of the mean.

16. Percentile rank of $X = \dfrac{B + \dfrac{1}{2}E}{N} \cdot 100$

17. z-score: $z = \dfrac{x - \mu}{\sigma}$ or $z = \dfrac{x - \bar{x}}{s}$

18. Original score: $x = \mu + z\sigma$

The tests of the following section will be more helpful if you take them after you have studied the examples given in this chapter and solved the exercises at the end of each section.

MASTERY TESTS

Form A

For questions 1–3 use the following information. The table below gives the distribution of test scores for a class of 20 students.

Test Score Interval	Number of Students (Frequency)
91–100	1
81–90	3
71–80	3
61–70	7
51–60	6

(handwritten notes: 95, 255, 225, 455, 330, 05, .15, .15)

1. Which interval contains the mean?
   a. 91–100   **b.** 81–90   **c.** 71–80   **d.** 61–70   **e.** 51–60

2. Which interval contains the lower quartile?

    a. 91–100   **b.** 81–90   **c.** 71–80   **d.** 61–70   **e.** 51–60

3. Which interval contains the median?

    a. 91–100   **b.** 81–90   **c.** 71–80   **d.** 61–70   **e.** 51–60

4. The ages of six senior citizens who attended the recent Thanksgiving Day parade were 80, 67, 72, 74, 75, and 80. Find the median of this distribution.

    a. 74   **b.** 80   **c.** 74.5   **d.** 75   **e.** none of these

5. A secretary typed a highly technical six-page report for the boss. The number of errors per page was 5, 3, 4, 3, 5, and 4. What is the (population) variance of this distribution?

    a. 1.5   **b.** 0.67   **c.** 0.5   **d.** 18   **e.** none of these

6. If $x_1 = 12$, $x_2 = 13$, $x_3 = 7$, $x_4 = 11$, and $x_5 = 17$, find Σx^2.

    a. 3600   **b.** 289   **c.** 772   **d.** 60   **e.** none of these

7. Ten babies were born on New Year's Day at Richmond Memorial Hospital. The weight (in pounds) of these babies was as follows:

Heather	6.8	Marie	9.1	Arlene	8.2	Priscilla	8.8
Jason	7.3	Curt	6.4	Pete	6.7		
Ann	8.9	Maurice	5.1	George	9.4		

The percentile rank of Arlene's weight (8.2 pounds) is

    a. P_{65} **b.** P_{45} **c.** P_{55} **d.** P_{60} **e.** none of these

8. Refer back to the previous question. The average length of a newborn child was 20.1 inches with a standard deviation of 1.3 inches. Arlene's z-score was -1.64. What was her actual length (in inches) at birth?

    a. 17.968   **b.** 22.232   **c.** 18.46   **d.** 18.80   **e.** none of these

9. A Peace Corps volunteer reported that six starving Ethiopian people that he first met had an average weight of 102.6 lb. with a standard deviation of 3.8 lb. Three months later, after receiving appropriate medical attention and sufficient food, each of the starving people had gained seven pounds. The new standard deviation is

    a. 10.8   **b.** 3.8   **c.** 102.6   **d.** 112.6   **e.** none of these

10. If the mean of four numbers is exactly 10, that is, if $\mu = 10$ when $N = 4$, find the sum of the four numbers, Σx.

    a. 10   **b.** 14   **c.** 4   **d.** 40   **e.** none of these

MASTERY TESTS

Form B

1. Ted's z-score on a test was 3.01. His actual score was 95 and the standard deviation on the test was 8.31. What was the mean score on the test?

2. During the first week of August, 15 women and 25 men joined a weight control club. The average age of the women was 32 and the average age of the men was 28. What was the average age for the entire group?
a. 30   **b.** 31   **c.** 29.5   **d.** 30.5   **e.** none of these

3. If the variance of a distribution is equal to 0.49, what is the standard deviation?
a. 0.7   **b.** 0.07   **c.** 7   **d.** 0.2339   **e.** none of these

4. The grades that George received on his statistics exams and the weight assigned to each exam are as follows:

	Grade	Weight Assigned
Test 1	81	1
Test 2	84	2
Test 3	75	3
Test 4	90	2
Final	88	4

Find George's average term grade.
a. 83.6   **b.** 83.83   **c.** 84   **d.** 83.5   **e.** none of these

5. On a recent qualifying test for the New York City Marathon, Augustus' z-score was -2.08. The average score was 58 and the standard deviation was 8.72. What was his actual score?
a. 55.92   **b.** 76.14   **c.** 66.72   **d.** 39.86   **e.** none of these

6. If $x_1 = 17$, $x_2 = 14$, $x_3 = 31$, $x_4 = 12$, and $x_5 = 31$, find $\Sigma(x - \bar{x})$.
a. 21   **b.** 0   **c.** 5   **d.** 105   **e.** none of these

7. A Nielsen pollster recently asked all eight families in a village to indicate the number of movies that each had seen in the last three months. Their answers were as follows: 7, 4, 8, 2, 0, 5, 9, and 5. Find the (population) standard deviation for this distribution.
a. 2.828   **b.** 8   **c.** 11.66   **d.** 4.12   **e.** none of these

8. In a certain distribution, $\mu = 25$ and $\sigma = 0$. What is the z-score of any term in the distribution?
a. 0   **b.** 1   **c.** there is none   **d.** 25   **e.** none of these

9. Which of the following is always true?
a. The mean has an effect on extreme scores.
b. The median has an effect on extreme scores.
c. Extreme scores have an effect on the mean.
d. Extreme scores have an effect on the median.
e. none of these

10. Verify that the following two formulas for calculating the sample standard deviation are the same:

$$s = \sqrt{\frac{\Sigma(x - \bar{x})^2}{n - 1}} = \sqrt{\frac{n(\Sigma x^2) - (\Sigma x)^2}{n(n - 1)}}$$

11. *Pollution* Six polluted lakes in New York and five polluted lakes in neighboring New Jersey were analyzed by environmentalists of both states to determine the level of cancer-causing PCBs. (Each state has its own method for measuring the pollution level of its waterways.) The following results on the amount of pollution were obtained by these two environmental groups:

New York Lake	Rating	New Jersey Lake	Rating
A	21	M	58
B	16	N	63
C	29	O	54
D	18	P	71
E	14	Q	62
F	12		

a. Which of these 11 lakes tested has the greatest amount of pollution?
b. Which of these 11 lakes tested has the least amount of pollution?

12. The dispatcher of an express package delivery service is analyzing the number of packages received daily with the incorrect zip code on them. The following data are available:

	Number of Packages With Incorrect Zip Code
Mon.	7
Tues.	14
Wed.	12
Thurs.	11
Fri.	10
Sat.	14
Mon.	21
Tues.	18
Wed.	9
Thurs.	15
Fri.	18
Sat.	19

Find the (sample) mean and variance for the above data.

13. *Case Study* Despite governmental supervision, many companies often dispose of industrial wastes by dumping them into nearby rivers, lakes, or streams, thereby polluting the water supply for the animals and fish that feed in these lakes. In one such study, an environmental group documented the devastating effect of such activities for a particular river.* They collected data on the number of fish and deer that died (based upon pathological studies) as a result of industrial dumping. The data below are for a six-month period.

Month	Number of Dead Deer Found
April	12
May	14
June	6
July	9
August	8
September	11

Find the (sample) mean and standard deviation for the above data.

SUGGESTED FURTHER READING

1. Harrell, T. W., and M. S. Harrell. "Army General Classification Test Scores for Civilian Occupations" in *Educational and Psychological Measurements* (1945), pp. 222–239.
2. Mendelhall, W. *Introduction to Probability and Statistics*, 6th ed. Belmont, CA: Duxbury Press, 1983.
3. Minum, E. W. *Statistical Reasoning in Psychology and Education*. New York: Wiley, 1970.
4. Mode, E. B. *Elements of Probability and Statistics*. Englewood Cliffs, NJ: Prentice-Hall, 1966.
5. Newmark, J., and F. Lake. *Mathematics as a Second Language*, 4th ed. Reading, MA: Addison-Wesley, 1987.
6. Newmark, J. *Using Finite Mathematics*. New York: Harper and Row, 1982.
7. O'Toole, A. L. *Elementary Practical Statistics*. New York: Macmillan, 1965.
8. Senter, R. J. *Analysis of Data: Introductory Statistics for the Behavioral Sciences*. Glenview, IL: Scott Foresman, 1969.
9. Tanur, J. M., F. Mosteller, W. H. Kruskal, R. G. Link, R. S. Pieters, and G. R. Rising. *Statistics: A Guide to the Unknown*. (E. L. Lehman, Special editor.) San Francisco: Holden-Day, 1978.

*Boggs and Jones, New York, 1986.

Allentown College Continuing Education and Summer Sessions

ACCESS
MID-TERM GRADE REPORT

Academic Year ___1989-90___

Name of Student: ___HAROLD YOHN___

Course Name: ___(PROB) STAT___

Course Number: ___MA 111 A___

Session or Semester: ___A___

Mid-Term Grade: ___A___

Special problems or difficulties: _____

_____

_____

[ ] I would like an ACCESS staff person to contact this
student to discuss the above.

Date ___1/26/90___   ___Ann Arterbrenlerk___

Instructor's Signature

CHAPTER 4

Probability

OBJECTIVES

- *To define* what is meant by the word "probability."

- *To explain* the use of the word "probability" in such statements as "the *probability* of a particular thing happening is 0.78."

- *To discuss* a formula for determining the number of possible outcomes when an experiment is performed. This is done so that we can find the total number of possible outcomes, which we use for probability calculations.

- *To analyze* the number of different ways of arranging things, depending upon whether or not order counts. Thus we will analyze permutations and combinations.

- *To present* a convenient notation used to represent a special type of multiplication. This is the factorial notation.

- *To study* a computational device for calculating the number of possible combinations. This is Pascal's triangle.

- *To learn* about "odds" and "mathematical expectation." These words, often used by gamblers, represent the payoff for a situation and the likelihood of obtaining it.

GOVERNORS ASK FOR MORE STATE AID

May 10: It was decided yesterday at a governor's meeting that a committee of four governors would be selected to go to Washington to plead for more state aid. Many governors expressed the view that the current level of Federal aid to the states was appalling, and that it was the obligation of the Federal government to share the burden.

May 10, 1986

The newspaper article indicates that a committee of four governors would be selected to go to Washington to plead for state aid. How do we select such committees? How many four-governor committees can possibly be formed from among all the governors?

4.1 INTRODUCTION

Although the word "probability" may sound strange to you, it is not as unfamiliar as you may think. In everyday situations we frequently make decisions and take action as a result of the probability of certain events. Thus, if the weather forecaster predicts rain with a probability of 80%, we undoubtedly would prepare ourselves accordingly.

Let us, however, analyze the weather forecaster's prediction. What the forecaster really means is that based upon past records, 80% of the time when the weather conditions have been as they are today, rain has followed. Thus, the probability calculations and resultant forecasts are based upon past records. They are based upon the assumption that since in the *past* rain has occurred a certain percentage of the time, it will occur the same percentage of times in the *future*. This is but one usage of probability. It is based on **relative frequency.** We will explain this idea in greater detail shortly.

Probability is also used in statements that express a personal judgment or conviction. This can be best illustrated by the following statements: "If the United States had not dropped the atomic bomb on Japan, World War II would *probably* have lasted several more years" or "If all the New York Mets players had been healthy the entire season, they *probably* would have won the pennant last year."

Probability can also be used in other situations. For example, if a fair coin* is tossed, we would all agree that the probability is $\frac{1}{2}$ that heads comes up.

This is because there are only two possible outcomes when we flip a coin, heads or tails.

Now consider the following conversation overheard in a student cafeteria.

Bill: I am going to cut math today.
Eric: Why?
Bill: I didn't do my homework.
Eric: So what? I didn't either.
Bill: So, since the teacher calls on at least half of the class each day for answers, he will *probably* call on me today and find out that I am not prepared.
Eric: The teacher called on me yesterday, so he *probably* won't get me today. I am going to class.

In the preceding situation each student is making a decision based upon probability.

Basically, the theory of probability deals with the study of uncertainties. Thus, it has been found to have wide applications in the following situations:

1. It is used by insurance companies when they calculate insurance premiums and the *probable* life expectancies of their policyholders.

*A fair coin is a coin that has the same chance of landing on heads as on tails. Throughout this book we will always assume that we have fair coins unless told otherwise.

2. It is used (formally or informally) by a gambler who decides to bet 10 to 1 on a particular horse.
3. It is used by industry officials in determining the reliability of certain equipment.
4. It is used by medical researchers who claim that smoking increases your *chance* of getting lung cancer.
5. It is used by biologists in their study of genetics.
6. It is used by pollsters in such polls as the Harris Poll, the Gallup Poll, and the Nielsen ratings to determine the reliability of their polls.
7. It is used by an investor who decides that a particular stock has a greater chance for future growth than any other stock.
8. It is used by business managers in determining which products to manufacture, which products to advertise, and through which media: television, radio, magazines, newspapers, subway and bus advertisements, and so on.
9. It is used by psychologists in predicting reactions or behavioral patterns under certain stimuli.
10. It is used by government economists in predicting that the inflation rate will increase or decrease in the future.

Since probability has so many possible meanings and uses, we will first analyze the nature of probability and how to calculate it. This will be done in this chapter. In the next chapter we will discuss various rules that allow us to calculate probabilities for many different situations.

HISTORICAL NOTE

Historically, probability had its origin in the gambling room. The Chevalier de Méré, a professional French gambler, had asked his friend Blaise Pascal (1623–1662) to solve the following problem: In what proportion should two players of equal skill divide the stakes remaining on the gambling table if they are forced to stop before finishing the game? Pascal wrote to and began an active correspondence with Pierre Fermat (1601–1665) concerning the problem. Although Pascal and Fermat agreed on the answer, they gave different proofs. It was in this series of correspondences during the year 1652 that they developed the modern theory of probability.

A century earlier the Italian mathematician and gambler Girolomo Cardan (1501–1576) wrote *The Book on Games of Chance*. For a further discussion of Cardan, see p. 11.

Another famous mathematician who contributed to the theory of probability was Abraham De Moivre (1667–1754). Like de Méré, De Moivre spent many hours with London gamblers and wrote a manual for gamblers entitled *Doctrine of Chances*. Like Cardan, De Moivre predicted the day of his death. A rather interesting story is told of De Moivre's death. De Moivre was ill and each day he noticed that he was sleeping 15 minutes longer than he did on the preceding day. Using progressions, he computed that he would die in his sleep on the day after he slept 23 hours and 45 minutes. On the day following a sleep of 24 hours, De Moivre died.

The French mathematician Pierre Simon de Laplace (1749–1827) also contributed much to the historical development of probability. In 1812 he published *Théorie Analytique des Probabilités*, in which he referred to probability as a science that began with games but that had wide-ranging applications. In particular, he applied probability theory not to gambling situations but as an aid in astronomy.

Over the course of many years probability theory has left the gambling rooms and has grown to be an important and ever-expanding branch of mathematics.

4.2 DEFINITION OF PROBABILITY

Probability theory can be thought of as that branch of mathematics that is concerned with calculating the probability of outcomes of experiments.

Since many ideas of probability were derived from gambling situations, let us consider the following experiment. An honest die (the plural is dice) was rolled many times and the number of 1's that came up was recorded. The results are

Number of 1's that came up	1	11	18	99	1001	10,001
Number of rolls of the die	6	60	120	600	6000	60,000

Notice that in each case the number of 1's that appeared is approximately $\frac{1}{6}$ of the total number of tosses of the die. It would then be reasonable to conclude that the probability of a 1 appearing is $\frac{1}{6}$.

Although when a die is rolled there are six equally likely possible outcomes if it is an honest die (see Figure 4.1), we are concerned with the number of 1's

FIGURE 4.1
Possible outcomes when a die is rolled once.

appearing. Each time a 1 appears, we call it a **favorable outcome.** There are six possible outcomes of which only one is favorable. The probability is thus the number of favorable outcomes divided by the total number of possible outcomes, which is 1 divided by 6, or $\frac{1}{6}$. The preceding chart indicates that our guess, that the probability is approximately $\frac{1}{6}$, is correct.

Possible outcomes when a coin is flipped.

Similarly if a coin is tossed once, we would say that the probability of getting a head is $\frac{1}{2}$, since there are two possible outcomes, heads and tails, and only one is favorable. These are the only two possible outcomes in this case. All the possible outcomes of an experiment are referred to as the **sample space** of the experiment. We will usually be interested in only some of the outcomes of the experiment. The outcomes that are of interest to us will be referred to as an **event.** Thus, in flipping a coin once, the sample space is Heads or Tails, abbreviated as H,T. The event of interest is H.

In the rolling of a die the sample space consists of six possible outcomes: 1, 2, 3, 4, 5, and 6. We may be interested in the event "getting a 1."

If we toss a coin twice, the sample space is HH, HT, TH, and TT. There are four possibilities. In this abbreviated notation, HT means heads on the first toss and tails on the second toss, whereas TH means tails on the first toss and heads on the second toss. The event "getting a head on both tosses" is denoted by HH. The event "no head" is TT.

To further illustrate the idea of sample space and event, consider the following examples.

● **Example 1**
Two dice are rolled at the same time. Find the sample space.

Solution
There are 36 possible outcomes as pictured in Figure 4.2. The possible outcomes are summarized as follows:

FIGURE 4.2
Thirty-six possible outcomes when two dice are rolled at the same time.

Die 1	Die 2	Die 1	Die 2	Die 1	Die 2
1	1	3	1	5	1
1	2	3	2	5	2
1	3	3	3	5	3
1	4	3	4	5	4
1	5	3	5	5	5
1	6	3	6	5	6
2	1	4	1	6	1
2	2	4	2	6	2
2	3	4	3	6	3
2	4	4	4	6	4
2	5	4	5	6	5
2	6	4	6	6	6

The event "sum of 7 on both dice together" can happen in six ways. These are

Die 1	Die 2	Die 1	Die 2
1	6	3	4
6	1	5	2
4	3	2	5

Similarly the event "sum of 9 on both dice together" can happen in four ways. The event "sum of 13 together" can happen in zero ways.

● **Example 2**

Two contestants, Jack and Jill, are on a quiz show. Each is in a soundproof booth and cannot hear what the other is saying. They are each asked to select a number from 1 to 3. They each win $10,000 if they select the same number. In how many different ways can they win the prize?

Solution

They will win the prize if they both say 1, 2 or 3. The sample space for this experiment is

Jill Guesses	Jack Guesses	Jill Guesses	Jack Guesses	Jill Guesses	Jack Guesses
1	1	2	1	3	1
1	2	2	2	3	2
1	3	2	3	3	3

Thus the event "winning the prize" can occur in three possible ways.

We now define formally what we mean by probability.

DEFINITION 4.1

*If an event can occur in n equally likely ways and if f of these ways are considered favorable, then the **probability** of getting a favorable outcome is*

$$\frac{\textbf{Number of favorable outcomes}}{\textbf{Total number of outcomes}} = \frac{f}{n}$$

Thus, the probability of any event equals the number of favorable outcomes divided by the total number of possible outcomes.
We use the symbol p(A) to stand for "the probability of event A."

Definition 4.1 is often called the **classical interpretation of probability.** It assumes that all outcomes of an experiment are equally likely. Thus, if we say that the probability of getting heads when flipping an honest coin is $\frac{1}{2}$, we are basing this on the following facts: When a coin is flipped once there are 2 possible outcomes, heads and tails. Therefore, the probability of getting heads is $\frac{1}{2}$ (1 out of a possible 2 outcomes.) Similarly, the probability of getting a 1 when an honest die is rolled is $\frac{1}{6}$ since there are 6 possible outcomes, only 1 of which is favorable.

An alternate interpretation of probability is called the **relative frequency concept of probability.** Suppose we tossed a coin 100 times and it landed heads 50 of the times. It would seem reasonable to claim that the probability of landing heads is approximately $\frac{50}{100}$ or $\frac{1}{2}$. We can think of the probability as the relative frequency of the event. Of course, to convince ourselves that our answer is reasonable, we could toss the coin many, many additional times. Since it is much easier to determine relative frequencies, this interpretation is very easy to understand and is commonly used. The relative-frequency interpretation represents the percentage of times that the event will happen in repeated experiments.

Probability can also be defined from a strictly mathematical, that is, axiomatic point of view; however, this is beyond the scope of this text. Let us now illustrate the concept of probability with several examples.

● **Example 3**
A family plans to have three children. What is the probability that all three children will be girls? (Assume that the probability of a girl being born in a given instance is $\frac{1}{2}$).

Solution
We will use Definition 4.1. We first find the total number of ways of having three children, that is, the sample space. There are eight possibilities as shown in the following table:

Child 1	Child 2	Child 3
Boy	Boy	Boy
Boy	Boy	Girl
Boy	Girl	Boy
Boy	Girl	Girl
Girl	Boy	Boy
Girl	Girl	Boy
Girl	Boy	Girl
Girl	Girl	Girl

Of these only one is favorable, namely the outcome Girl, Girl, Girl. Thus,

$$p(3 \text{ girls}) = \frac{1}{8}$$

1970 Draft Lottery

Birth-day	January Draft-priority number	Birth-day	February Draft-priority number	Birth-day	March Draft-priority number	Birth-day	April Draft-priority number	Birth-day	May Draft-priority number	Birth-day	June Draft-priority number
1	305	1	86	1	108	1	32	1	330	1	249
2	159	2	144	2	29	2	271	2	298	2	228
3	251	3	297	3	267	3	83	3	40	3	301
4	215	4	210	4	275	4	81	4	276	4	20
5	101	5	214	5	293	5	269	5	364	5	28
6	224	6	347	6	139	6	253	6	155	6	110
7	306	7	91	7	122	7	147	7	35	7	85
8	199	8	181	8	213	8	312	8	321	8	366
9	194	9	338	9	317	9	219	9	197	9	335
10	325	10	216	10	323	10	218	10	65	10	206
11	329	11	150	11	136	11	14	11	37	11	134
12	221	12	68	12	300	12	346	12	133	12	272
13	318	13	152	13	259	13	124	13	295	13	69
14	238	14	4	14	354	14	231	14	178	14	356
15	17	15	89	15	169	15	273	15	130	15	180
16	121	16	212	16	166	16	148	16	55	16	274
17	235	17	189	17	33	17	260	17	112	17	73
18	140	18	292	18	332	18	90	18	278	18	341
19	58	19	25	19	200	19	336	19	75	19	104
20	280	20	302	20	239	20	345	20	183	20	360
21	186	21	363	21	334	21	62	21	250	21	60
22	337	22	290	22	265	22	316	22	326	22	247
23	118	23	57	23	256	23	252	23	319	23	109
24	59	24	236	24	258	24	2	24	31	24	358
25	52	25	179	25	343	25	351	25	361	25	137
26	92	26	365	26	170	26	340	26	357	26	22
27	355	27	205	27	268	27	74	27	296	27	64
28	77	28	299	28	223	28	262	28	308	28	222
29	349	29	285	29	362	29	191	29	226	29	353
30	164			30	217	30	208	30	103	30	209
31	211			31	30			31	313		

Birth-day	July Draft-priority number	Birth-day	August Draft-priority number	Birth-day	September Draft-priority number	Birth-day	October Draft-priority number	Birth-day	November Draft-priority number	Birth-day	December Draft-priority number
1	93	1	111	1	225	1	359	1	19	1	129
2	350	2	45	2	161	2	125	2	34	2	328
3	115	3	261	3	49	3	244	3	348	3	157
4	279	4	145	4	322	4	202	4	266	4	165
5	188	5	54	5	82	5	24	5	310	5	56
6	327	6	114	6	6	6	87	6	76	6	10
7	50	7	168	7	8	7	234	7	51	7	12
8	13	8	48	8	184	8	283	8	97	8	105
9	277	9	106	9	263	9	342	9	80	9	43
10	284	10	21	10	71	10	220	10	282	10	41
11	248	11	324	11	158	11	237	11	46	11	39
12	15	12	142	12	242	12	72	12	66	12	314
13	42	13	307	13	175	13	138	13	126	13	163
14	331	14	198	14	1	14	294	14	127	14	26
15	322	15	102	15	113	15	171	15	131	15	320
16	120	16	44	16	207	16	254	16	107	16	96
17	98	17	154	17	255	17	288	17	143	17	304
18	190	18	141	18	246	18	5	18	146	18	128
19	227	19	311	19	177	19	241	19	203	19	240
20	187	20	344	20	63	20	192	20	185	20	135
21	27	21	291	21	204	21	243	21	156	21	70
22	153	22	339	22	160	22	117	22	9	22	53
23	172	23	116	23	119	23	201	23	182	23	162
24	23	24	36	24	195	24	196	24	230	24	95
25	67	25	286	25	149	25	176	25	132	25	84
26	303	26	245	26	18	26	7	26	309	26	173
27	289	27	352	27	233	27	264	27	47	27	78
28	88	28	167	28	257	28	94	28	281	28	123
29	270	29	61	29	151	29	229	29	99	29	16
30	287	30	333	30	315	30	38	30	174	30	3
31	193	31	11			31	79			31	100

In December 1969 the United States Selective Service established a priority system for determining which young men would be drafted into the army (see chart above). Capsules representing each birth date were placed in a drum and selected at random. Those men whose numbers were selected first were almost certain that they would be drafted. The different birthdays were given draft priority numbers as indicated in the clipping on the draft lottery. Supposedly, each birthday had an equally likely probability of being selected. However, by analyzing the numbers carefully, we find that the majority of those birthdays that occurred later in the year had a higher priority number than those that occurred earlier in the year. Did each birthday have an equal probability of being selected?

● **Example 4** *Playing Cards*

A card is selected from an ordinary deck of 52 cards. What is the probability of getting

a. a queen?
b. a diamond?
c. a black card?
d. a picture card?
e. the king of clubs?

Solution

Since there are 52 cards in the deck, the total number of outcomes is 52. This is shown in Figure 4.3.

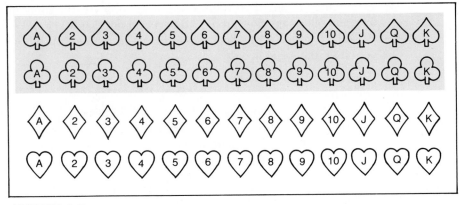

FIGURE 4.3
The sample space when a card is drawn from an ordinary deck of 52 cards.

a. As shown in Figure 4.3, there are 4 queens in the deck, so there are 4 favorable outcomes. Definition 4.1 tells us that

$$p(\text{queen}) = \frac{4}{52} = \frac{1}{13}$$

b. There are 13 diamonds in the deck, so there are 13 favorable outcomes. Therefore,

$$p(\text{diamonds}) = \frac{13}{52} = \frac{1}{4}$$

c. Since a black card can be either a spade or a club, there are 26 black cards in the deck as shown in Figure 4.3. Therefore,

$$p(\text{black card}) = \frac{26}{52} = \frac{1}{2}$$

d. There are 12 picture cards (4 jacks, 4 queens, and 4 kings), so there are 12 favorable outcomes. Therefore,

$$p(\text{picture card}) = \frac{12}{52} = \frac{3}{13}$$

e. There is only one king of clubs in a deck of 52 cards. Thus,

$$p(\text{king of clubs}) = \frac{1}{52}$$

● **Example 5**

Mary and her friend Gwendolyn are visiting the Sears' Tower in Chicago. They each enter a different elevator in the main lobby that is going up and that can let them off on any floor from 1 to 6. Assuming that Mary and Gwendolyn are as likely to get off at one floor as another, what is the probability that they both get off on the same floor?

Solution

Since both Mary and Gwendolyn can get off at any floor between 1 and 6, there are 36 possible outcomes as shown in the table below. A favorable outcome occurs if both get off on floor 1, floor 2, and so on as shown. Thus, there are 6 favorable outcomes out of 36 possible outcomes, so that

$$\text{Prob (both get off on same floor)} = \frac{6}{36} = \frac{1}{6}$$

Floor Where Mary Gets Off	Floor Where Gwendolyn Gets Off	Floor Where Mary Gets Off	Floor Where Gwendolyn Gets Off
1	1 ← Favorable outcome	4	1
1	2	4	2
1	3	4	3
1	4	4	4 ← Favorable outcome
1	5	4	5
1	6	4	6
2	1	5	1
2	2 ← Favorable outcome	5	2
2	3	5	3
2	4	5	4
2	5	5	5 ← Favorable outcome
2	6	5	6
3	1	6	1
3	2	6	2
3	3 ← Favorable outcome	6	3
3	4	6	4
3	5	6	5
3	6	6	6 ← Favorable outcome

● **Example 6**

A wheel of fortune has the numbers 1 through 50 painted on it (see the next page). Tickets numbered 1 through 50 have been sold. The wheel will be rotated and the number on which the pointer lands will be the winning number. If the pointer stops on the line between two numbers, the wheel must be turned again. A prize of a new car will be awarded to the person whose ticket number matches the winning number. What is the probability that

a. ticket number 42 wins?
b. a ticket between the numbers 26 and 39 wins?
c. ticket number 51 wins?

Solution

Since 50 tickets were sold, the total number of possible outcomes, that is, the sample space, is 50.

a. Only 1 ticket numbered 42 was sold. There is only one favorable outcome. Thus,

$$p(\text{ticket number 42 wins}) = \frac{1}{50}$$

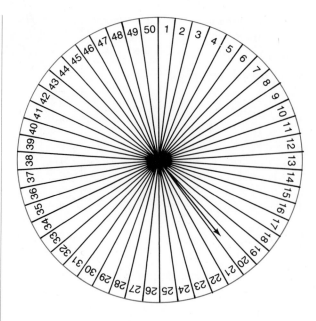

b. There are 12 ticket numbers between 26 and 39, not including 26 and 39. Thus,

$$p(\text{a ticket between the numbers 26 and 39 wins}) = \frac{12}{50} = \frac{6}{25}$$

c. Since tickets numbered up to 50 were sold, ticket number 51 can never be a winning number. Thus,

$$p(\text{ticket number 51 wins}) = \frac{0}{50} = 0$$

An event that can never happen is called a **null event** and its probability is 0.

● **Example 7**

Two fair dice are rolled at the same time and the number of dots appearing on both dice is counted. Find the probability that this sum

a. is 7.
b. is an odd number larger than 6.
c. is less than 2.
d. is more than 12.
e. is between 2 and 12, including these two numbers.

Solution

When two dice are rolled, there are 36 possible outcomes, that is, the sample space has 36 possibilities. These were listed in Example 1 on page 155.

a. The sum of 7 on both dice together can happen in 6 ways, so that

$$p(\text{sum of 7}) \ = \ \frac{6}{36} \ \text{or} \ \frac{1}{6}$$

b. The statement "a sum that is an odd number larger than 6" means a sum of 7, a sum of 9, or a sum of 11. A sum of 7 on both dice together can happen in 6 ways. Similarly, a sum of 9 on both dice together can happen in 4 ways, and a sum of 11 can happen in 2 ways. There are then 12 favorable outcomes out of 36 possibilities. Thus,

$$p(\text{a sum that is an odd number larger than 6}) \ = \ \frac{12}{36} \ = \ \frac{1}{3}$$

c. When two dice are rolled, the minimum sum is 2, and we cannot obtain a sum less than 2. There are *no* favorable events. This is the null event. Hence,

$$p(\text{a sum that is less than 2}) \ = \ 0$$

d. When two dice are rolled, the maximum sum is 12. We cannot obtain a sum that is more than 12. This is the null event. Thus,

$$p(\text{a sum that is more than 12}) \ = \ 0$$

e. When two dice are rolled we *must* obtain a sum that is between 2 and 12, including the numbers 2 and 12. There are 36 possible outcomes and *all* these are favorable. Thus,

$$p(\text{a sum between 2 and 12, including 2 and 12}) \ = \ \frac{36}{36} \ = \ 1$$

Therefore, a favorable outcome *must* occur in this case.

An event that is certain to occur is called the **certain event** or **definite event** and its probability is 1.

COMMENT   Any event, call it A, may or may not occur. If it is sure to occur, we have the certain event and its probability is 1. If it will never occur, we have the null event and its probability is 0. Thus, if we are given any event A, then we know that its probability *must* be between 0 and 1 and possibly equal to 0 or 1. This is because the event may or may not occur. Probability can *never* be a negative number.

COMMENT   We mentioned earlier that probability can be thought of as the fraction of times that an outcome will occur in a long series of repetitions of an experiment. However, there may be certain experiments that cannot be repeated. For example, if Gary's kidney has to be removed surgically, we cannot think of this as an experiment that can be repeated over and over again, at least as far as Gary is concerned. How do we assign probabilities in this case? This is not an easy task. Calculating the probability in such situations requires the judgment and experience of a doctor familiar with *many* experiments of a similar type. Thus, if doctors tell you that you have an 80% chance of surviving the operation, they mean that based upon their previous experiences with such situations, 80% of the patients with similar operations have survived. Usually an experienced surgeon can assign a fairly reasonable probability to the success of a nonrepeatable operation.

EXERCISES

1. The following is a breakdown of the 200 members of a health club.

Categories	Number of Men	Number of Women
Single	38	26
Married	51	43
Divorced	21	11
Widowed	7	3

What is the probability that a club member selected at random will be
   a. a single woman?
   b. a divorced man?
   c. a widowed person?
   d. a man?

2. A charity box is known to contain 24 nickels, 13 dimes, 17 quarters, and 6 half-dollars. When the box is shaken a coin falls out. Find the probability that the coin is
   a. a half-dollar.
   b. a penny.

3. Angela and José both board a train that will make four stops before going out of service. Assuming that Angela and José are as likely to get off at one stop as another, find the probability that they both get off at the same stop.

4. Which of the following cannot be the probability of some event?
   a. 0.97   **b.** 1.36   **c.** −0.01   **d.** 0   **e.** $\dfrac{4007}{4009}$

*5. At a recent grab-bag party, each of the six guests brought a small gift. These gifts were then placed in a sack. The sack contained no other gifts. Each guest was blindfolded and asked to select a gift from the sack. Find the probability that each guest selects the gift that he or she brought. $\frac{1}{6} \times \frac{1}{5} \times \frac{1}{4} \times \frac{1}{3} \times \frac{1}{2} \times \frac{1}{1} = \frac{1}{720}$

*6. Refer to Exercise 5. Find the probability that half of the guests select the same gift that they brought.

7. A supermarket receives a shipment of 70 cases of fresh milk. The driver forgets to rotate the stock and places them in the refrigerator where there are 6 cases of spoiled milk. Later, the stock clerk selects a case of milk at random from the refrigerator. Find the probability that the case of milk selected is from the spoiled group. $6|76 = 3|38$

8. The Remco Delivery Service operates a fleet of 200 vehicles daily as follows:

Type of Vehicle	Number in Fleet
Trucks	47
Trailers	23
Vans	63
Cars	39
Pick-up trucks	28
	200

$200 - 63 = 127$ $127/200$

The dispatcher receives a call that one of the company's vehicles has been involved in a minor accident. Find the probability that the vehicle involved in the accident is *not* a van. (Assume that one type of vehicle is as likely to be in an accident as another. Is this a reasonable assumption?)

9. The following is an ethnic and minority group breakdown of the 300 new employees that the Remco Delivery Service has hired since January 1, 1985:

Classification	Number
Puerto Rican	59
Black	61
Oriental	37
Mexican	83
White woman	60
	300

Find the probability that a randomly selected new employee is
a. a white woman. $60|300 - 20|100 = 2|10 = 1|5$
b. a Black or Mexican. $61 + 83 = 144 /300 = 72|150 = 36|75 = 12|25$
c. an American Indian. 0

10. All of the 300 newly hired employees mentioned in Exercise 9 are eligible for different company pension plans according to their age. The ages of the new employees are as follows:

Number of Employees

Classification	Under 30 Years	Between 30–45 Years	Over 45 Years
Puerto Rican	21	27	11
Black	23	37	1
Oriental	12	15	10
Mexican	39	24	(20)
White woman	25	26	9

Find the probability that an employee selected at random from this group will be
a. a Mexican over 45 years of age. $20 | 300 > 2/30 > 1/15$
b. a Puerto Rican not between 30–45 years of age. $32 | 300$ $16/150 = 8/75$
c. an Oriental or a Black under 30 years of age. $35/300 = 7/60$

11. *Affirmative Action*   Several feminist groups have accused the Bellow Savings Bank of discrimination by not promoting a sufficient number of women to high-level executive positions. The following statistics are available on the bank's promotions for several years:

Year	Number of Women Promoted	Total Number of Promotions
1981	12	21
1982	16	39
1983	14	27
1984	19	39
1985	22	64
1986	15	38

a. If an employee who was promoted during these years is selected at random, find the probability that the employee is a woman. $98/228 = 49/114$
b. Is the feminists group's claim justified? Explain. 43%

12. A pharmaceutical company is experimenting with a new process that produces 4 different kinds of viruses, A, B, C, and D, in the ratio 8:5:3:2, respectively. A medical researcher randomly selects a virus that has been produced by the new process. Find the probability that it is a virus
a. of type A.
b. of type B or C.
c. of type B, C, or D.
d. not of type A or D.

13. *Gambling*   In a gambling casino there is a slot machine that consists of three wheels that rotate when $5 is deposited in the machine. Each of the wheels will display a picture of a lemon, a cherry, or an apple independently of the picture displayed by another wheel. If all the wheels show the same item, the player wins $100. If Luciano deposits $5 and plays the game, what is the probability that he wins $100?

14. *Birthdays*   The following is a list of the birthdays of the presidents of the United States. Find the probability that a randomly selected president was born in June. (Base your answer on the data given in the table.)

President	Birthday	President	Birthday
1. George Washington	Feb. 22, 1732	21. Chester Arthur	Oct. 5, 1830
2. John Adams	Oct. 30, 1735	22. Grover Cleveland	Mar. 18, 1837
3. Thomas Jefferson	Apr. 13, 1743	23. Benjamin Harrison	Aug. 20, 1833
4. James Madison	Mar. 16, 1751	24. William McKinley	Jan. 29, 1843
5. James Monroe	Apr. 28, 1758	25. Theodore Roosevelt	Oct. 27, 1858
6. John Quincy Adams	July 11, 1767	26. William Taft	Sept. 15, 1857
7. Andrew Jackson	Mar. 15, 1767	27. Woodrow Wilson	Dec. 28, 1857
8. Martin Van Buren	Dec. 5, 1782	28. Warren Harding	Nov. 2, 1865
9. William H. Harrison	Feb. 9, 1773	29. Calvin Coolidge	July 4, 1872
10. John Tyler	Mar. 29, 1790	30. Herbert Hoover	Aug. 10, 1874
11. James Knox Polk	Nov. 2, 1795	31. Franklin D. Roosevelt	Jan. 30, 1882
12. Zachary Taylor	Nov. 24, 1784	32. Harry S Truman	May 8, 1884
13. Millard Fillmore	Jan. 7, 1800	33. Dwight D. Eisenhower	Oct. 14, 1890
14. Franklin Pierce	Nov. 23, 1804	34. John F. Kennedy	May 29, 1917
15. James Buchanan	Apr. 23, 1791	35. Lyndon B. Johnson	Aug. 27, 1908
16. Abraham Lincoln	Feb. 12, 1809	36. Richard M. Nixon	Jan. 9, 1913
17. Andrew Johnson	Dec. 29, 1808	37. Gerald Ford	July 14, 1913
18. Ulysses S. Grant	Apr. 27, 1822	38. Jimmy Carter	Oct. 1, 1924
19. Rutherford Hayes	Oct. 4, 1822	39. Ronald Reagan	Feb. 6, 1911
20. James Garfield	Nov. 19, 1831		

*15. *Presidents*   Consider the 39 presidents of the United States. If a president is selected at random, find the probability that the president did not complete his term in office (either because of death, assassination, or resignation).†

*16. What is the probability that a president elected in a year divisible by 20 (since 1840) died while in office?†

17. A secretary types three letters and addresses three envelopes. Before inserting each letter into the appropriate envelope, she drops them all on the floor. If she picks up all the letters and envelopes and inserts a letter at random (without looking) into an envelope, what is the probability that each letter will be inserted in its correct envelope? *Hint:* Find the sample space.

18. At a recent New Year's party, four of the guests became drunk and were asked to leave. As each of the drunk guests left the party, he or she picked up a coat that definitely belonged to one of these four people. Since these guests were drunk, however, they did not know whose coat they were taking. Find the probability that each person got his or her own coat.

19. In Example 6 and again in Exercise 3 we assumed that people are as likely to get off at one floor or station as another. Is this assumption of equally likely outcomes reasonable?

†If necessary, consult an encyclopedia to determine the number of presidents in the specified categories.

4.3 COUNTING PROBLEMS

In determining the probability of an event, we must first know the total number of possible outcomes. In many situations it is a rather simple task to list all the possible outcomes and then to determine how many of these are favorable. In other situations there may be so many possible outcomes that it would be too time consuming to list all of them. Thus when two dice are rolled, there are 36 possible outcomes. These are listed on page 155. Exercise 13 of Section 4.2 has 27 possible outcomes. When there are too many possibilities to list, we can use rules, which will be given shortly, to determine the actual number.

One technique that is sometimes used to determine the number of possible outcomes is to construct a **tree diagram.** The following examples will illustrate how this is done.

● **Example 1**

By means of a tree diagram we can determine the number of possible outcomes when a coin is repeatedly tossed. If one coin is tossed, there are two possible outcomes, heads or tails, as shown in the tree diagram in Figure 4.4. If two

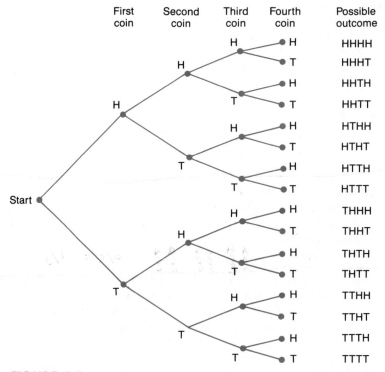

FIGURE 4.4

Tree diagram of the number of possible outcomes when a coin is repeatedly tossed.

coins are tossed, there are now four possible outcomes. These are HH, HT, TH, and TT, since each of the possible outcomes on the first toss can occur with each of the two possibilities on the second toss. So, if heads appeared on the first toss, we may get heads or tails on the second toss. The same is true if tails appeared on the first toss. The tree diagram shows that there are four possibilities. When three coins are tossed, the diagram shows that there are eight possible outcomes. Also, there are sixteen possible outcomes when four coins are tossed.

● Example 2

Hazel Brown is about to order dinner in a restaurant. She can choose any one of three main courses and any one of four desserts:

Main Course	Dessert
Hamburger	Hot-fudge Sundae
Steak	Jello
Southern-fried Chicken	Cake
	Fruit

Using a tree diagram, find all the possible dinners that Hazel can order.

Solution

Hazel may order any one of the three dishes as a main course. With each main course she may order any one of four desserts. These possibilities are pictured in Figure 4.5 on the next page. The diagram shows that there are twelve possible meals that Hazel can order.

COMMENT   Although counting the number of possible outcomes by using a tree diagram is not difficult when there are only several possibilities, it becomes very impractical to construct a tree when there are many possibilities. For example, if 10 coins are tossed, there are 1024 different outcomes. Similarly, if a die is rolled 4 times, there are 1296 different outcomes. For situations such as these we need a rule to help us determine the number of possible outcomes.

Before stating the rule, however, let us analyze the following situations. When one die is rolled, there are six possible outcomes, 1, 2, 3, 4, 5, 6. When a second die is rolled, there are 6×6, or 36, possible outcomes for both dice together. These are listed on page 155.

If again we analyze Exercise 13 of Section 4.2, we find that the first wheel can show a lemon, cherry, or apple. This gives us three possibilities. Similarly, the second wheel can also show a lemon, cherry, or apple. The same is true for the third wheel. This gives us 3 possibilities for wheel 1, 3 possibilities for

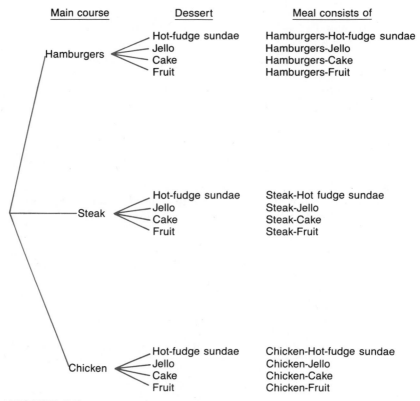

Main course	Dessert	Meal consists of
Hamburgers	Hot-fudge sundae Jello Cake Fruit	Hamburgers-Hot-fudge sundae Hamburgers-Jello Hamburgers-Cake Hamburgers-Fruit
Steak	Hot-fudge sundae Jello Cake Fruit	Steak-Hot fudge sundae Steak-Jello Steak-Cake Steak-Fruit
Chicken	Hot-fudge sundae Jello Cake Fruit	Chicken-Hot-fudge sundae Chicken-Jello Chicken-Cake Chicken-Fruit

FIGURE 4.5
Tree diagram of the possible dinners that can be ordered from a choice of any one of three main courses and any one of four desserts.

wheel 2, and 3 possibilities for wheel 3. We then have a total of 27 possible outcomes since

$$3 \times 3 \times 3 = 27$$

This leads us to the following rule.

COUNTING RULE

If one thing can be done in m ways, and if after this is done, something else can be done in n ways, then both things can be done in a total of $m \cdot n$ different ways in the stated order. (The same rule extends to more than two things done in sequence.)

● **Example 3**

A geology teacher plans to travel from New York to Florida and then on to Mexico to collect rock specimens for her class. From New York to Florida she can travel by train, airplane, boat, or car. However, from Florida to Mexico she cannot travel by train. In how many different ways can she make the trip?

$$4 + 3 = 12$$

Solution

Since she can travel from New York to Florida in 4 different ways and from Florida to Mexico in 3 different ways, she can make the trip in 4 × 3, or 12, different ways.

● **Example 4**

$$LLL \ DD$$
$$\underset{26 \ \ 26 \ \ 26}{\rule{0pt}{0pt}} \times \underset{9}{\rule{0pt}{0pt}} \times \underset{10}{\rule{0pt}{0pt}}$$

In a certain state, license plates have 3 letters followed by 2 digits. If the first digit cannot be 0, how many different license plates can be made if

a. repetitions of letters or numbers are allowed?
b. repetitions of letters are not allowed?

Solution

There are 26 letters and 10 possible digits (0, 1, 2, . . . , 9).

a. If repetitions are allowed, the same letter can be used again. Since 0 cannot be used as the first digit, the total number of different license plates is

26 × 26 × 26 × ⑨ × 10 = 1,581,840

Thus, 1,581,840 different license plates are possible. Note that the circled position has only 9 possibilities. Why?

b. If repetition of letters is not allowed, there are 26 possibilities for the first letter, but only 25 possibilities for the second letter since once a letter is used it may not be used again. For the third letter there are only 24 possibilities. There are then a total of 1,404,000 different license plates since

26 × 25 × 24 × 9 × 10 = 1,404,000

● **Example 5**

In Example 1 (see page 168) there are two possible outcomes for the first toss, two possibilities for the second toss, and two possibilities for the third toss. Thus, there are a total of eight possible outcomes when three coins are tossed or when one coin is tossed three times, since

2 × 2 × 2 = 8

This is the same result we obtained using tree diagrams. It is considerably easier to do it this way.

EXERCISES

1. Marjorie Valentine is the keynote speaker at tomorrow's Board of Directors meeting. She can select any one of eight skirts, five blouses, four blazers, and six pairs of shoes to wear for the meeting. How many different outfits are possible? $8 \cdot 5 \cdot 4 \cdot 6 = 960$

2. There are seven pitchers and four catchers on a baseball team. In how many different ways can the manager select a starting battery (pitcher and catcher) for a game? $7 \cdot 4 = 28$

3. There are five different roads that lead to a stadium. In how many different ways can a person drive to the stadium by one road and leave by a different road? $5 \cdot 4 = 20$

4. **a.** How many different seven-digit telephone numbers are possible if the first digit cannot be a zero or a one? (Assume no other restrictions and that repetitions are allowed.)

   b. If there are 20 different area codes that can be used, how many telephone numbers are possible using the same restrictions as in part (a)?

5. The *New York Daily News* has as a daily feature, a five- or six-letter nonsensical word that the reader must unscramble to make a meaningful word. In how many different ways can the letters of the word "MAGIC" be arranged? (*Note:* Each arrangement does not necessarily have to form a meaningful word.) $5 \cdot 4 \cdot 3 \cdot 2 \cdot 1 = 120$

6. How many different license plates can be formed if each is to consist of three letters followed by three numbers and if:
   a. repetition of numbers or letters is allowed?
   b. repetition of numbers or letters is not allowed?
   c. only letters cannot be repeated?

7. Matthew Piefsky is in a gambling casino and observes a card dealer selecting 3 cards (without replacement) from a deck of 52 cards. In how many different ways can the 3 cards be selected? (Assume order counts.)

8. Each calculator manufactured by the Texas Calculator Company is inscribed with a six-digit serial number preceded *and* followed by a letter. Using this serial number scheme, how many different codes are possible?

9. A postal employee is scheduling this morning's seven special deliveries. How many different schedules are possible?

*10. In the baseball playoffs, two teams play against each other in a series of up to seven games. The first team to win four games is the winner. Assume that the two teams are labelled A and B. Make a tree diagram showing all the possible ways in which the playoffs can end.

11. Construct a tree diagram showing all the possibilities, male and female, for a couple that has five children.

12. There are seven members on the Board of Directors of the Ross Chemical Corp. as follows: Alice Johnson, Steve Bachtel, George Klangman, Jeremy Billingsley, Fern Wash-

ington, Wallace McKenzie, and Eugene Brooks. At any stockholder's meeting, these members are seated at a (straight) head table.

a. In how many different ways can the Board members be seated?

*b. If Jeremy Billingsley and Wallace McKenzie are bitter enemies and cannot be seated together, how many different seating arrangements are possible?

*c. Find the probability that the two female board members (Alice Johnson and Fern Washington) are seated together.

13. *Environmental Protection*   In an effort to avoid polluting the environment, the health department of a certain state requires that all chemical wastes produced within that state be placed in special drums and that each of these drums be stamped with numbers or letters as follows:

a. The company number as assigned by the department; there are nine companies within the state.

b. The plant location within the state; the state is divided into six geographic areas labeled A, B, C, D, E, and F.

c. The date of sealing: month and last two digits of the year.

For example, a drum with a code 3 C 0380 means that it was sealed by company 3 located in area C during March 1980. Using this scheme, how many different codes are possible for drums sealed during the years 1980–1986?

14. How many different three-digit numbers larger than 700 can be formed using the digits 4, 6, 8, and 9 if

a. repetition is allowed?

b. repetition is not allowed?

15. During a recent ice storm, a falling tree severed three electrical power lines as shown in the accompanying diagram. A repair crew is sent to the scene and must reconnect the wires exactly as they were originally.

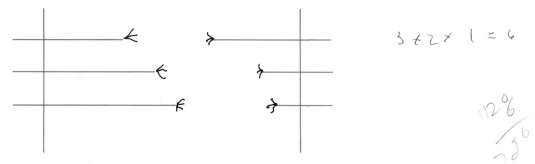

In how many different ways can the three wires be connected?

16. *Smoke Signals*   The Felquois Indians sent messages from one tribe to another by using the following different types of smoke signals:

Long, black smoke cloud
Short, black smoke cloud
Long, white smoke cloud
Short, white smoke cloud

Each of these clouds was followed by a two-minute pause, after which a second smoke cloud was created. Again a two-minute pause followed, after which a third smoke cloud was created. Each message consisted of three smoke clouds.

a. Using this scheme, how many different messages could the Felquois Indians transmit from one tribe to another?

b. Assuming that all messages are equally likely to be transmitted, what is the probability that a transmitted message will consist of three long, black smoke clouds one after another?

17. Heather has been advised by her counselor that in order to graduate in June she must complete one course from each of the three categories shown in the table. All these courses will be offered in the spring semester.

Categories

A		B		C	
Math	231	Educ	117	Soc	114
Bio	100	Eng	113	Eco	412
Chem	041	Hist	112		

Neglecting any other considerations, how many *different* programs can Heather select? Use tree diagrams.

18. Jennifer is a clerk in a department store. A customer purchases an item for $86 and pays for it with one $50 bill, one $20 bill, one $10 bill, one $5 bill, and one $1 bill. Jennifer's cash register drawer has five different compartments, one for each of the bills mentioned. If Jennifer picks up the bills and, without looking, randomly places the bills in the drawer (one per compartment), find the probability that she inserts each bill in its proper compartment.

4.4 PERMUTATIONS

Consider the following situation: Three vacationing students, Mel, Carl, and Rhoda, have purchased standby tickets on a transatlantic flight that will take them from London back to New York. There are three aisle seats remaining vacant in the plane; one in the forward section, one in the mid section, and one in the tail section of the airplane. In how many different ways can the three travelers line up to board the plane and pick the best of the remaining seats?

There are six ways as shown in the accompanying table. In this case you will notice that, in each arrangement, order is important.

First Person to Board Plane	Second Person to Board Plane	Last Person to Board Plane
Mel	Carl	Rhoda
Mel	Rhoda	Carl
Carl	Mel	Rhoda
Carl	Rhoda	Mel
Rhoda	Carl	Mel
Rhoda	Mel	Carl

Thus, the arrangement Mel, Carl, Rhoda means that Mel boards first, then Carl, and finally Rhoda. If the number of seats is limited, then the person who boards first has first choice of selecting the best of the remaining seats as compared with the person who boards last.

This important idea in mathematics, in which a number of objects can be arranged in a particular order, is known as a permutation.

DEFINITION 4.2

*A **permutation** is any arrangement of distinct objects in a particular order.*

Let us now examine another such problem:

● **Example 1**

Four young ladies, Stephanie Marie Gallagher, Liz Armstrong, Ann Sullivan, and Patricia Beth O'Connell, are finalists in a state beauty contest. The judges must select a winner and an alternate from among these four contestants. In how many different ways can this be done?

Solution
The winner and the alternate can be selected in 12 different ways. We list these possibilities in a table.

Winner	Alternate	Winner	Alternate
Stephanie	Liz	Ann	Stephanie
Stephanie	Ann	Ann	Liz
Stephanie	Patricia	Ann	Patricia
Liz	Stephanie	Patricia	Stephanie
Liz	Ann	Patricia	Liz
Liz	Patricia	Patricia	Ann

COMMENT Perhaps you are wondering whether there are any formulas that can be used to determine the number of possible permutations. The answer is yes and this will be done shortly.

● **Example 2**

A doctor has five examination rooms. There are five patients in the waiting room. In how many different ways can the patients be assigned to the examination rooms?

Solution

We can solve this problem by numbering the waiting rooms 1 through 5 as shown below, and then considering them in sequence.

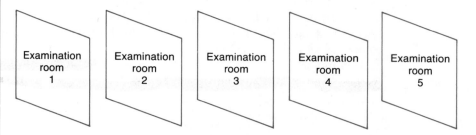

Examination room 1 can be used by any one of the five patients. Once a patient has been assigned to examination room 1, there are four patients who can use examination room 2, so there are (using the counting rule)

$5 \cdot 4 = 20$ ways

of using examination room 1 and examination room 2. Similarly, once examination rooms 1 and 2 are occupied, there are three patients remaining who can use examination room 3, so there are

$5 \cdot 4 \cdot 3 = 60$ ways

of using examination rooms 1, 2, and 3. Continuing in this manner we find that there are

$5 \cdot 4 \cdot 3 \cdot 2 \cdot 1 = 120$ ways

of using examination rooms 1, 2, 3, 4, and 5. Thus, there are 120 different ways in which the five patients can be assigned to the 5 examination rooms.

Notice in Example 2 that we had to multiply $5 \cdot 4 \cdot 3 \cdot 2 \cdot 1$. Often in mathematics we have to multiply a series of numbers together starting from a given whole number and multiplying this by the number that is 1 less than that, and so on until we get to the number 1, which is where we stop. For this type of multiplication, we introduce **factorial notation** and use the symbol $n!$ read as "n factorial." In this case we would write the product as $5!$. Thus,

$5! = 5 \cdot 4 \cdot 3 \cdot 2 \cdot 1 = 120.$

Accordingly,

$1! = 1$
$2! = 2 \cdot 1 = 2$
$3! = 3 \cdot 2 \cdot 1 = 6$
$4! = 4 \cdot 3 \cdot 2 \cdot 1 = 24$
$5! = 5 \cdot 4 \cdot 3 \cdot 2 \cdot 1 = 120$
$6! = 6 \cdot 5 \cdot 4 \cdot 3 \cdot 2 \cdot 1 = 720$
$7! = 7 \cdot 6 \cdot 5 \cdot 4 \cdot 3 \cdot 2 \cdot 1 = 5{,}040$
$8! = 8 \cdot 7 \cdot 6 \cdot 5 \cdot 4 \cdot 3 \cdot 2 \cdot 1 = 40{,}320$
$9! = 9 \cdot 8 \cdot 7 \cdot 6 \cdot 5 \cdot 4 \cdot 3 \cdot 2 \cdot 1 = 362{,}880$
$10! = 10 \cdot 9 \cdot 8 \cdot 7 \cdot 6 \cdot 5 \cdot 4 \cdot 3 \cdot 2 \cdot 1 = 3{,}628{,}800$

$0! = 1$

Also, we define $0!$ to be 1. This makes formulas involving $0!$ meaningful.

Let us now return to the beauty contest example (Example 1) discussed earlier. In that example the judges were interested in selecting two winners from among the four finalists. This actually represents the number of permutations of four things taken two at a time, or the number of possible permutations of two things that can be formed out of a possible four things. We abbreviate this by writing $_4P_2$. This is read as the number of permutations of four things taken two at a time.

NOTATION $_nP_r$ represents the number of permutations of n things taken r at a time, and $_nP_n$ represents the number of permutations of n things taken n at a time.

To determine the number of possible permutations, we use the following formula.

FORMULA 4.1

$$_nP_r = \frac{n!}{(n-r)!}$$

● **Example 3**
  a. Find $_6P_4$.
  b. Find $_7P_5$.
  c. Find $_5P_5$.

Solution

a. $_6P_4$ means the number of possible permutations of six things taken four at a time. Using Formula 4.1 we have $n = 6$, $r = 4$, and $n - r = 6 - 4 = 2$. Thus,

$$_6P_4 = \frac{6!}{(6-4)!} = \frac{6!}{2!} = \frac{6 \cdot 5 \cdot 4 \cdot 3 \cdot 2 \cdot 1}{2 \cdot 1} = \frac{6 \cdot 5 \cdot 4 \cdot 3 \cdot \cancel{2} \cdot \cancel{1}}{\cancel{2} \cdot \cancel{1}}$$

$$= 6 \cdot 5 \cdot 4 \cdot 3 = 360$$

Thus, $_6P_4 = 360$.

b. $_7P_5$ means the number of permutations of seven things taken five at a time, so that $n = 7$ and $r = 5$. Using Formula 4.1 we have

$$_7P_5 = \frac{7!}{(7-5)!} = \frac{7!}{2!} = \frac{7 \cdot 6 \cdot 5 \cdot 4 \cdot 3 \cdot 2 \cdot 1}{2 \cdot 1}$$

$$= \frac{7 \cdot 6 \cdot 5 \cdot 4 \cdot 3 \cdot \cancel{2} \cdot \cancel{1}}{\cancel{2} \cdot \cancel{1}}$$

$$= 7 \cdot 6 \cdot 5 \cdot 4 \cdot 3 = 2520$$

Thus, $_7P_5 = 2520$.

c. $_5P_5$ means the number of permutations of five things taken five at a time, so that $n = 5$ and $r = 5$. Using Formula 4.1 we have

$$_5P_5 = \frac{5!}{(5-5)!} = \frac{5!}{0!}$$

$$= \frac{5!}{1} \quad \text{(Since 0! is 1)}$$

$$= 5! = 5 \cdot 4 \cdot 3 \cdot 2 \cdot 1$$

$$= 120$$

Thus, $_5P_5 = 5! = 120$.

● **Example 4**

A baseball scout has received a list of 15 promising prospects for the team's consideration. He is asked to list, in order of preference, the 5 most outstanding of these prospects. In how many different ways can he select the 5 best players?

Solution

As there are 15 promising prospects, $n = 15$. Also only the top 5 are to be considered, so that $r = 5$ and $n - r = 15 - 5 = 10$. Using Formula 4.1, we have

$$_nP_r = \frac{n!}{(n-r)!}$$

$$= \frac{15!}{10!}$$

$$= \frac{15 \cdot 14 \cdot 13 \cdot 12 \cdot 11 \cdot \cancel{10} \cdot \cancel{9} \cdot \cancel{8} \cdot \cancel{7} \cdot \cancel{6} \cdot \cancel{5} \cdot \cancel{4} \cdot \cancel{3} \cdot \cancel{2} \cdot \cancel{1}}{\cancel{10} \cdot \cancel{9} \cdot \cancel{8} \cdot \cancel{7} \cdot \cancel{6} \cdot \cancel{5} \cdot \cancel{4} \cdot \cancel{3} \cdot \cancel{2} \cdot \cancel{1}}$$

$$= 15 \cdot 14 \cdot 13 \cdot 12 \cdot 11 = 360{,}360$$

Thus, the scout can select the top 5 players in 360,360 different ways.

COMMENT In Example 3 we calculated $_5P_5$, which represents the number of permutations of five things taken five at a time. This really means the number of different ways of arranging five things, where order counts. Our answer turned out to be 5! This leads us to Formula 4.2.

FORMULA 4.2

The number of possible permutations of n things taken n at a time, denoted as $_nP_n$, is

$$_nP_n = n!$$

● **Example 5**

How many different permutations can be formed from the letters of the word CAT?

Solution

There are six permutations. These are

CAT CTA TAC TCA ACT ATC

This actually represents the number of possible permutations of three letters taken three at a time. Formula 4.2 tells us that there are 3!, or 6, possible permutations since

$$3! = 3 \cdot 2 \cdot 1 = 6.$$

These are listed above.

Suppose a librarian has two identical algebra books and three identical geometry books to be shelved. In how many different ways can this be done? Since there are five books altogether and all have to be shelved, we are tempted to use Formula 4.2 or Formula 4.1. Unfortunately, since the two algebra books are identical, we cannot tell them apart. There are actually 10 possible permutations. These are:

algebra	algebra	geometry	geometry	geometry
algebra	geometry	geometry	geometry	algebra
algebra	geometry	algebra	geometry	geometry
algebra	geometry	geometry	algebra	geometry
geometry	geometry	geometry	algebra	algebra
geometry	geometry	algebra	algebra	geometry
geometry	geometry	algebra	geometry	algebra
geometry	algebra	algebra	geometry	geometry
geometry	algebra	geometry	algebra	geometry
geometry	algebra	geometry	geometry	algebra

Of course, if we label the algebra books as copy 1 and copy 2, which is sometimes done by some libraries, we can use Formula 4.1. Thus,

| Algebra copy 1 | Algebra copy 2 | Geometry copy 1 | Geometry copy 2 | Geometry copy 3 |

would be a different permutation than

| Algebra copy 2 | Algebra copy 1 | Geometry copy 1 | Geometry copy 2 | Geometry copy 3 |

However, this is usually not done. Thus, these two permutations have to be counted as the same or as only one permutation. Formulas 4.1 and 4.2 have to be revised somewhat to allow for the possibility of repetitions. This leads us to the following formula:

FORMULA 4.3

The number of different permutations of n things of which p are alike, q are alike, or r are alike, and so on, is

$$\frac{n!}{p!\,q!\,r!}\,\cdots\cdots$$

It is understood that $p + q + r + \cdots = n$

In our example we have five books to be shelved, so $n = 5$. Since the two algebra books are identical and the three geometry books are also identical, $p = 2$ and $q = 3$. Formula 4.3 tells us that the number of permutations is

$$\frac{5!}{2! \cdot 3!}$$

$$= \frac{5 \cdot 4 \cdot 3 \cdot 2 \cdot 1}{2 \cdot 1 \cdot 3 \cdot 2 \cdot 1}$$

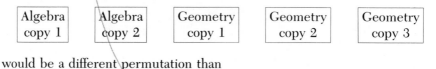

not on test

$$= \frac{\overset{2}{5 \cdot \cancel{4} \cdot \cancel{3} \cdot \cancel{2} \cdot \cancel{1}}}{\cancel{2} \cdot \cancel{1} \cdot \cancel{3} \cdot \cancel{2} \cdot \cancel{1}}$$

$$= 5 \cdot 2 = 10$$

Thus, there are 10 permutations, which were listed previously.

● **Example 6**

How many different permutations are there of the letters in the word

a. "IDIOT?"
b. "STATISTICS?"

Solution

a. The word "IDIOT" has five letters, so $n = 5$. There are 2 I's, so $p = 2$. Formula 4.3 tells us that the number of permutations is

$$\frac{5!}{2!1!1!1!}$$

$$= \frac{5 \cdot 4 \cdot 3 \cdot \cancel{2} \cdot \cancel{1}}{\cancel{2} \cdot \cancel{1} \cdot 1 \cdot 1 \cdot 1}$$

$$= 5 \cdot 4 \cdot 3 = 60$$

There are 60 permutations.

b. Since the word "STATISTICS" has ten letters, $n = 10$. The letter "S" is repeated three times, "T" is repeated three times, and "I" is repeated twice, so $p = 3$, $q = 3$, and $r = 2$. Formula 4.3 tells us that the number of permutations is

$$\frac{10!}{3!3!2!1!1!}$$

$$= \frac{10 \cdot 9 \cdot 8 \cdot 7 \cdot \cancel{6} \cdot 5 \cdot \cancel{4} \cdot \cancel{3} \cdot \overset{2}{\cancel{2}} \cdot \cancel{1}}{\cancel{3} \cdot \cancel{2} \cdot \cancel{1} \cdot \cancel{3} \cdot \cancel{2} \cdot \cancel{1} \cdot \cancel{2} \cdot \cancel{1} \cdot 1 \cdot 1}$$

$$= 10 \cdot 9 \cdot 8 \cdot 7 \cdot 5 \cdot 2 = 50{,}400$$

not on test

There are 50,400 permutations.

EXERCISES

1. Evaluate each of the following symbols:

a. 7!   **b.** 6!   **c.** 2!   **d.** $\dfrac{5!}{5!}$

e. $\dfrac{0!}{3}$ **f.** $\dfrac{4!}{2!2!}$ **g.** $\dfrac{7!}{3!4!}$ **h.** $\dfrac{6!}{3!3!}$

i. $_7P_6$ **j.** $_5P_4$ **k.** $_6P_2$ **l.** $_3P_2$

m. $_7P_7$ **n.** $_2P_2$ **o.** $_0P_0$ **p.** $_4P_0$

2. Medical researchers are experimenting with a new drug (in pill form) to be given to patients to cure the common cold. Ten people have volunteered to take the pill. However, the researchers have decided that they will use only three of these volunteers. The first will be given the new pill, the second will be given a placebo, and the third will be given an aspirin. In how many different ways can these three subjects be selected for the experiment?

3. Each year, readers of a certain magazine are asked to rank the top 3 best-dressed men from among a list of 12 candidates. In how many different ways can this be done?

4. The ten-member Board of Directors of a company wishes to elect a chairperson, a secretary, an executive officer, and a comptroller from among the ten members. In how many different ways can this be done?

5. There are eight rooms available for use by the statistics department. This summer the statistics department is offering eight courses during a given hour. In how many different ways can these courses be assigned to a room? (Assume that order counts.)

6. Priscilla asks her kid brother (who is too young to read) to place her four-volume dictionary on a shelf.
   a. In how many different ways can the books be placed on the shelf?
   b. Find the probability that the little boy places the books on the shelf in sequential order, volume I first, then volume II, and so on.

7. A baseball manager has selected his nine starting players.
   a. How many different batting orders are possible?
   b. Assuming that all the different batting orders are equally likely, what is the probability that the pitcher bats last?
   c. Is the assumption given in part (b) reasonable?

8. A child is playing with a telephone and dials a long-distance (ten-digit) number.
   a. How many possible long-distance numbers are there? (Assume that there are no restrictions.)
   b. What is the probability that the child dials his or her own number?

9. Stephanie Wilson is planning a trip to see her aunt and uncle who live in Washington D.C. There are ten different places of interest that Stephanie wishes to visit. However, Stephanie's schedule will permit her the time to see only six of them. If the order of visits matters, in how many different ways can Stephanie plan her trip?

10. A chemist mixed six chemicals together in a solution and created a new synthetic drug. Unfortunately, he does not recall the order in which the chemicals were introduced into the solution. He decides to repeat the experiment. How many possibilities are there?

11. There are ten cars entered in a particular race. In how many different ways can four of them finish first, second, third, and fourth? $_{10}P_4 = \frac{10!}{(10-4)!} = \frac{10!}{6!} \frac{10 \cdot 9 \cdot 8 \cdot 7 \cdot 6!}{6!} = 5040$

12. In how many different ways can six crane operators be assigned to operate six cranes? (Assume that order counts.)

13. In how many different ways can the letters of the word "SIMILAR" be arranged if the letter "I" must be the first letter of each arrangement? $6! = 720$

14. How many different permutations are there of the letters in the words
   a. "SCHEDULE"   **b.** "CONSTRUCTION"
   c. "DISTRIBUTION"   **d.** "MATHEMATICS"

15. In how many different ways can the letters of the word "TRIGONOMETRY" be arranged if a vowel must be the first letter of each permutation?

16. A librarian is arranging four algebra books, three geometry books, two calculus books, and three statistics books on a bookshelf. None of the books are identical.
   a. How many different permutations of these books are there?
   b. How many different permutations of these books are there if books on the same subject are to be grouped together?

17. Six couples are to be honored for their outstanding humanitarian and philanthropic efforts. Each husband will sit with this wife at a head (straight) table. In how many different ways can these 12 people be seated?

18. The number of different ways in which n distinct objects can be arranged in a circle is $(n - 1)!$
   a. Explain why this formula is valid.
   b. In how many distinct ways can a baker display six different cakes in a circular arrangement in a showcase window?

19. Six people are seated around a circular table playing cards. One of the players is annoyed by the smoking at an adjoining table. It is decided that the players should change seats. In how many different ways can the players rearrange themselves?

4.5 COMBINATIONS

Imagine that a six-person rescue party is climbing a mountain in search of survivors of an airplane crash. They suddenly spot the plane on a ledge but the passageway to the ledge is very narrow and only four people can proceed. The remaining two people will have to return. How many different four-person rescue groups can be formed to reach the ledge?

In this situation we are obviously interested in selecting four out of six people. However, since the order in which the selection is to be made is not important, Formula 4.1 of Section 4.4 (see page 177) has to be changed.

Any selection of things in which the order is not important is called a **combination.** Let us determine how many possible combinations there actually are. If the names of the people of the six-person rescue party are Alice, Betty, Calvin, Drew, Ellen, and Frank, denoted as A, B, C, D, E, and F, then Formula 4.1 of Section 4.4 tells that there are $_6P_4$ possible ways of selecting the four people out of a possible six. This yields

$$_6P_4 = \frac{6!}{(6-4)!}$$

$$= \frac{6!}{2!}$$

$$= 360 \text{ possibilities}$$

However, we know that permutations take order into account. Since we are not interested in order, there cannot be 360 possible combinations. Thus, if the four-person rescue party consists of A, B, C, and D, there would be the following 24 different permutations:

$$
\begin{array}{llll}
A\,B\,C\,D & B\,A\,C\,D & C\,A\,B\,D & D\,A\,B\,C \\
A\,B\,D\,C & B\,A\,D\,C & C\,A\,D\,B & D\,A\,C\,B \\
A\,D\,C\,B & B\,D\,A\,C & C\,B\,A\,D & D\,B\,A\,C \\
A\,D\,B\,C & B\,D\,C\,A & C\,B\,D\,A & D\,B\,C\,A \\
A\,C\,B\,D & B\,C\,A\,D & C\,D\,A\,B & D\,C\,B\,A \\
A\,C\,D\,B & B\,C\,D\,A & C\,D\,B\,A & D\,C\,A\,B
\end{array}
$$

Since these permutations consist of the same four people, A, B, C, and D, we consider them as only one combination of these people.

Similarly for any other combination of four people there are 24 permutations of these people. Thus, it would seem reasonable to divide the 360 by 24 getting 15 and concluding that there are only 15 different combinations.

Notice that $24 = 4!$. Therefore, the number of combinations of six things taken four at a time is

$$\frac{_6P_4}{4!} = \frac{6!}{4!2!}$$

$$= 15$$

In general, consider the problem of selecting r objects from a possible n objects. We have the following definitions and formula.

DEFINITION 4.3

A **combination** is a selection from a collection of distinct objects where order is not important.

DEFINITION 4.4

The number of different ways of selecting r objects from a possible n distinct objects, where the order is not important, is called the **number of combinations of n things taken r at a time** and is denoted as $_nC_r$. Some books use the symbol $\binom{n}{r}$ instead of $_nC_r$. $\binom{n}{r}$ is called a **binomial coefficient.**

FORMULA 4.4

The number of combinations of n things taken r at a time is

$$\binom{n}{r} = {}_nC_r = \frac{n!}{r!(n-r)!}$$

Formula 4.4 is especially useful when calculating the probability of certain events. We illustrate the use of this formula with several examples.

● **Example 1**

Consider any five people, whom we shall name A, B, C, D, and E.

a. In how many ways can a committee of three be selected from among them?
b. What is the probability of selecting the three-person committee consisting of A, B, and C? (Assume that each committee is equally likely to be selected.)
c. In how many ways can a committee of five be selected from among them?

Solution

a. We must select any three people from a possible five, and order does not matter. Since this is the number of combinations of five things taken three at a time, we want $_5C_3$. Using Formula 4.4 with $n = 5$ and $r = 3$, we have

$$\binom{5}{3} = {}_5C_3 = \frac{5!}{3!(5-3)!} = \frac{5!}{3!2!} = \frac{\overset{2}{5} \cdot \cancel{4} \cdot \cancel{3} \cdot \cancel{2} \cdot \cancel{1}}{\cancel{3} \cdot \cancel{2} \cdot \cancel{1} \cdot \cancel{2} \cdot \cancel{1}} = 10$$

There are ten possible three-person committees that can be formed. We can verify this answer by listing them:

A B C   A B E   A C E   B C D   B D E
A B D   A C D   A D E   B C E   C D E

b. There are ten possible three-person committees that can be formed, as listed in part (a). Of these only one consists of A, B, and C. Thus,

$$p(\text{committee consists of A, B, and C}) = \frac{1}{10}$$

c. We are interested in selecting any five people from a possible five, and order does not matter. This is $_5C_5$. Thus,

$$\binom{5}{5} = {}_5C_5 = \frac{5!}{5!(5 - 5)!}$$

$$= \frac{5!}{5!0!} \quad (\text{Remember } 0! = 1)$$

$$= \frac{5 \cdot 4 \cdot 3 \cdot 2 \cdot 1}{5 \cdot 4 \cdot 3 \cdot 2 \cdot 1 \cdot 1}$$

$$= 1$$

So there is only one combination containing all five people.

● **Example 2** *Nuclear Accident*

On Wednesday, March 27, 1979, the nuclear generating facility on Three Mile Island near Middletown, Pennsylvania, malfunctioned and began discharging radiation into the air. In an attempt to prevent a nuclear "meltdown," ten nuclear physicists were contacted to shed some light on the problem. (Later, additional physicists were called in.) Because the ten scientists were busily engaged on another project, they agreed that only seven of them would come to the crippled nuclear plant. (Any three of them could remain to oversee their other existing project.) In how many different ways could the seven scientists have been selected?

Solution

Order was not important, so the seven scientists had to be selected out of a possible 10. This could be done in $_{10}C_7$ ways. Using Formula 4.4 we have

$$\binom{10}{7} = {}_{10}C_7 = \frac{10!}{7!(10 - 7)!} = \frac{10!}{7!3!} = 120$$

Thus, the 7 scientists could have been selected in 120 different ways.

The four cooling towers at the Three Mile Island Nuclear Power Plant. (*Photo:* UPI/ Bettmann Newsphotos)

● **Example 3** *Football Teams*

How many different 11-member football teams can be formed from a possible 20 players if any player can play any position?

Solution

We are interested in the number of combinations of 20 players taken 11 at a time. So $n = 20$ and $r = 11$. Thus,

$$\binom{20}{11} = {}_{20}C_{11} = \frac{20!}{11!(20 - 11)!}$$

$$= \frac{20!}{11!9!}$$

$$= 167,960$$

There are 167,960 possible 11-member football teams.

● **Example 4**

a. How many different poker hands consisting of 5 cards can be dealt from a deck of 52 cards?

b. What is the probability of being dealt a royal flush in five-card poker? (A royal flush consists of the ten, jack, queen, king, and ace of the same suit.)

Solution

a. Since order is not important, we are interested in the number of combinations of 5 things out of a possible 52. So $n = 52$ and $r = 5$. Thus,

$$\binom{52}{5} = {}_{52}C_5 = \frac{52!}{5!(52-5)!}$$

$$= \frac{52!}{5!47!}$$

$$= \frac{52 \cdot 51 \cdot 50 \cdot 49 \cdot 48 \cdot 47!}{5 \cdot 4 \cdot 3 \cdot 2 \cdot 1 \cdot 47!}$$

$$= 2{,}598{,}960$$

There are 2,598,960 possible poker hands.

b. Out of the possible 2,598,960 different poker hands only four are favorable. These are the ten, jack, queen, king, and ace of hearts; the ten, jack, queen, king, and ace of clubs; the ten, jack, queen, king, and ace of diamonds; and the ten, jack, queen, king, and ace of spades. Thus,

$$p(\text{royal flush}) = \frac{4}{2{,}598{,}960} = \frac{1}{649{,}740}$$

COMMENT For those familiar with poker, we present certain probabilities in Table 4.1:

TABLE 4.1
Different Poker Hands and Their Probabilities

Type of Hand	Probability of It Being Dealt to You
Royal flush (ace, king, queen, jack, 10 in the same suit)	0.0000015
Four of a kind (four of a kind, like four 4's or four queens)	0.00024
Flush (five cards in a single suit but not straight)	0.0020
Two pairs	0.0475
Nothing of interest	0.5012

● **Example 5**

John has ten single dollar bills of which three are counterfeit. If he selects four of them at random, what is the probability of getting two good bills and two counterfeit bills?

(handwritten: 10 total / 7 good / 3 counter)

$$\left(\begin{smallmatrix}7\\2\end{smallmatrix}\right)\left(\begin{smallmatrix}3\\2\end{smallmatrix}\right)\Big/\left(\begin{smallmatrix}10\\4\end{smallmatrix}\right)$$

Solution

We are interested in selecting four bills from a possible ten. Thus the number of possible outcomes is $_{10}C_4$. The two good bills to be drawn must be drawn from the seven good ones. This can happen in $_7C_2$ ways. Also, the two counterfeit bills to be drawn must be drawn from the three counterfeit ones. This can happen in $_3C_2$ ways. Thus,

$$p(\text{two good bills and two counterfeit bills}) = \frac{_7C_2 \cdot {_3C_2}}{_{10}C_4}$$

Now

$$_7C_2 = \frac{7!}{2!(7-2)!}$$

$$= \frac{7!}{2!5!} = 21$$

and

$$_3C_2 = \frac{3!}{2!(3-2)!}$$

$$= \frac{3!}{2!1!} = 3$$

Also,

$$_{10}C_4 = \frac{10!}{4!(10-4)!}$$

$$= \frac{10!}{4!6!} = 210$$

(handwritten: $\binom{7}{4}$ / $\binom{10}{4}$)

Therefore,

$$p(\text{two good bills and two counterfeit bills}) = \frac{21 \cdot 3}{210}$$

$$= \frac{3}{10}$$

● **Example 6**

Find the probability of selecting all good bills in the preceding example.

Solution

Since we are interested in selecting only good bills, we will select four good bills from a possible seven and no counterfeit bills. Thus

$$p(\text{all good bills}) = \frac{_7C_4 \cdot \ _3C_0}{_{10}C_4}$$

$$= \frac{35 \cdot 1}{210} = \frac{1}{6}$$

The probability of selecting all good bills is $\frac{1}{6}$.

There is an alternate method for computing the number of possible combinations of n things taken r at a time, $_nC_r$, which completely avoids the factorial notation. This can be accomplished by using **Pascal's triangle.** Such a triangle is shown in Figure 4.6. How do we construct such a triangle?

Each row has a 1 on either end. All the in-between entries are obtained by adding the numbers immediately above and directly to the right and left of them as shown by the arrows in the diagram of Figure 4.7. For example, to obtain the entries for the eighth row we first put a 1 on each end (moving over slightly). Then we add the 1 and 7 from row 7, getting 8 as shown. We then add 7 and 21, getting 28; then we add 21 to 35, getting 56, and so on. Remember to place the 1s on each end. The numbers must be lined up as shown in the diagram. Such a triangle of numbers is known as **Pascal's triangle** in honor of

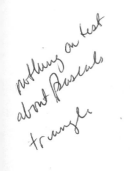
nothing on test about Pascals triangle

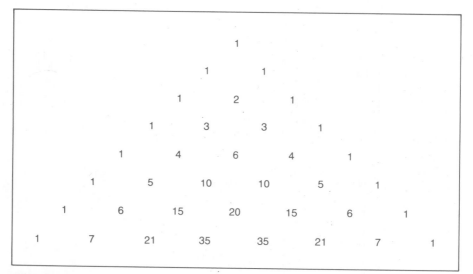

FIGURE 4.6
Pascal's triangle.

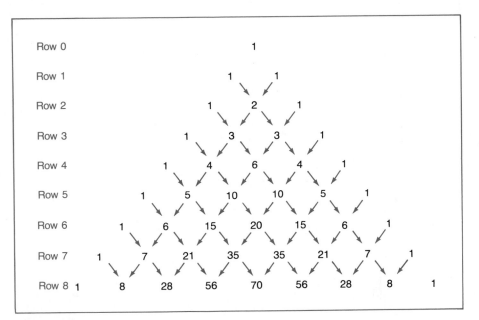

FIGURE 4.7
Construction of Pascal's triangle.

the French mathematician who found many applications for it. Although the Chinese knew of this triangle several centuries earlier (see Figure 4.8), it is named for Pascal.

Let us now apply Pascal's triangle to solve problems in combinations.

● **Example 7**

Professor Matthews tells four jokes each semester. His policy is never to repeat any combination of four jokes once they are used. How many semesters will eight different jokes last the professor?

Solution

Since order is not important, we are interested in the number of combinations of eight things taken four at a time. So $n = 8$ and $r = 4$. Therefore, we must evaluate $_8C_4$. We will find $_8C_4$ first using Formula 4.4 and then using Pascal's triangle. Using Formula 4.4 we have

$$_8C_4 = \frac{8!}{4!(8-4)!} = \frac{8!}{4!4!} = 70$$

Thus, the eight jokes will last him for 70 semesters.

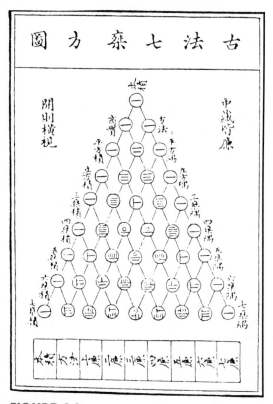

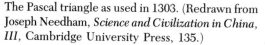

FIGURE 4.8

The Pascal triangle as used in 1303. (Redrawn from
Joseph Needham, *Science and Civilization in China,
III*, Cambridge University Press, 135.)

Let us now evaluate $_8C_4$ using Pascal's triangle. Since $n = 8$, we must write
down the first nine rows of Pascal's triangle as shown in Figure 4.9.

Now we look at row 8. Since we are interested in selecting four things from
a possible eight things, we skip the 1 and move to the fourth entry appearing
after the 1. It is 70. This number represents the value of $_8C_4$. Thus $_8C_4 = 70$.
Similarly, to find $_8C_6$ we move to the sixth entry after the end 1 on row 8. It
is 28. Thus $_8C_6 = 28$. Also the first entry after the end 1 is 8, so that $_8C_1 =$
8. If we wanted $_8C_8$, we would move to the eighth entry after the end 1. It
is 1, so that $_8C_8 = 1$. To find $_8C_0$ we must move to the zeroth entry after
the end 1. This means we do not go anywhere; we just stay at the 1. Thus,
$_8C_0 = 1$.

COMMENT When you use Pascal's triangle, remember that the rows are numbered from 0 rather than from 1, so that the rows are labeled row 0, row 1,
row 2, etc.

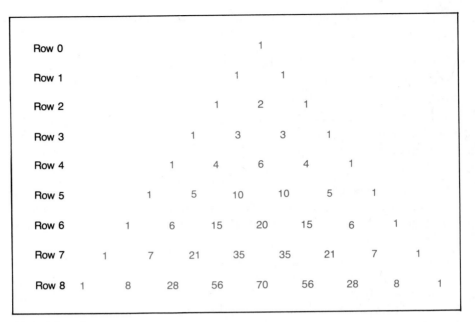

Row 0						1				
Row 1					1		1			
Row 2				1		2		1		
Row 3			1		3		3		1	
Row 4		1		4		6		4		1
Row 5	1		5		10		10		5		1
Row 6	1	6		15		20		15		6	1
Row 7	1	7	21		35		35		21	7	1
Row 8	1	8	28	56		70		56	28	8	1

FIGURE 4.9
First nine rows of Pascal's triangle.

● **Example 8**
Using Pascal's triangle find the following:

a. $_5C_2$ b. $_5C_3$ c. $_6C_6$ d. $_7C_0$

Solution
We will use the Pascal triangle shown in Figure 4.9.

a. To find $_5C_2$, go to row 5 and move across to the second number from the left after the end 1. The entry is 10. Thus $_5C_2 = 10$.
b. To find $_5C_3$, go to row 5 and move across to the third number from the left after the end 1. The entry is 10. Thus $_5C_3 = 10$.
c. To find $_6C_6$, go to row 6 and move across to the sixth number from the left after the end 1. The entry is 1. Thus $_6C_6 = 1$.
d. To find $_7C_0$, go to row 7 and do not move anywhere after the end 1. Remain there. Thus $_7C_0 = 1$.

EXERCISES

1. Evaluate each of the following:

a. $_7C_6$ b. $_6C_4$ c. $_8C_3$ d. $_{10}C_0$

e. $_5C_1$ f. $\begin{pmatrix} 9 \\ 6 \end{pmatrix}$ g. $\begin{pmatrix} 7 \\ 2 \end{pmatrix}$ h. $\begin{pmatrix} 8 \\ 4 \end{pmatrix}$

i. $\begin{pmatrix} 7 \\ 8 \end{pmatrix}$ i. $\begin{pmatrix} 7 \\ 7 \end{pmatrix}$

2. Eight police officers arrive at Memorial Hospital to give blood for a wounded fellow cop. However, the hospital needs only five volunteers. In how many different ways can these volunteers be selected?

3. The Metro Insurance Company received nine claims for fire damage during one week. Management decides to select three of these claims randomly and to investigate them thoroughly for the possibility of fraud or arson. In how many different ways can this be done?

4. The Calco Food Processing Company randomly selects three loads of every day's production and thoroughly checks them for food contamination. On January 7, the company produced 18 loads of food. In how many different ways can the three loads to be inspected be selected?

5. The Marvo Computer Company is expanding its West Coast operations and expects to hire five new male and five new female employees. After interviewing numerous candidates, it is determined that 16 men and 11 women qualify for the job. In how many different ways can the jobs be filled from among these candidates?

6. The Argavon Corp. believes that it is in the company's best interest to maintain the physical fitness of its 35 employees. It recently purchased 6 exercise machines to be used by employees during their lunch break or after work. In how many different ways can 6 of the 35 employees be assigned to the different machines?

7. Matthew Priofsky is in a gambling casino and observes a card dealer selecting 7 cards (without replacement) from a deck of 52 cards. In how many different ways can the 7 cards be selected?

*8. A committee of four is to be set up to investigate charges of medical incompetence at Kings Hospital. The members of the committee are to be selected from the seven cardiologists, four surgeons, three anesthesiologists, and two obstetricians on the hospital staff. In how many different ways can the committee be selected, if
a. one member must be from each of the above mentioned specialties?
b. at least one cardiologist and one surgeon must be on the committee?

9. How many different worker-management committees can be formed from a total of 14 workers and 8 management personnel if each committee is to consist of 4 workers and 3 management personnel?

10. The Able Nursing Home employs 13 nurses on its morning shift and 11 nurses on its afternoon shift. As a cost-reducing technique, management decides to fire 2 nurses. What is the probability that
a. both fired nurses will be from the morning shift?
b. both fired nurses will be from the afternoon shift?
c. one of the fired nurses will be from the morning shift and one will be from the afternoon shift?

11. There are 20 faculty members in the School of Business at a large university. According to student evaluations, 9 are rated excellent, 7 are rated good, and 4 are rated poor. A student registers for two courses taught by these teachers by randomly selecting them. Find the probability that the student gets

   a. two excellent teachers.

   b. one good and one poor teacher.

   c. two poor teachers.

12. The police department of a small city plans to hire 7 new police officers for the coming fiscal year. After interviewing prospective employees, it is determined that 17 men and 11 women qualify for the job. In how many different ways can the new police officer job openings be filled if

   a. the new police recruits must be 4 men and 3 women?

   b. the new police recruits must be more women than men?

*13. Verify that the probabilities given in Table 4.1 are accurate.

*14. Explain how the identity $\binom{n+1}{r} = \binom{n}{r} + \binom{n}{r-1}$ enables us to construct Pascal's triangle. (*Hint:* First verify the identity.)

15. A tour operator is arranging a winter travel package from the East Coast to the West Coast. Although there are 11 airlines that could be used for the travel package, the operator wishes to consider only 5 of the airlines. This will cut down on the paperwork involved.

   a. In how many different ways can the 5 airlines to be considered be selected?

   b. In how many different ways can the 6 airlines *not* to be considered be selected?

   c. Compare the answers obtained in parts (a) and (b). They should be the same. Can you explain why?

16. It is known that two microwave ovens in today's production of 20 ovens by a company are defective. The quality control engineer randomly selects 3 ovens from each day's production and carefully checks each one. In how many different ways can the quality control engineer select 3 ovens for inspection

   a. so that none of the defective ovens are selected?

   b. so that one of the defective ovens is selected?

   c. so that both of the defective ovens are selected?

17. Refer to Exercise 16. Find the probability that the quality control engineer selects both defective ovens (so that no defective item is shipped).

*18. Using Pascal's triangle, show that $\binom{n}{r} = \binom{n}{n-r}$, that is, numbers in a particular row read from left to right are the same as read from right to left.

*19. Using Pascal's triangle, show that the sum of the numbers in any row is 2^n where n is the row number.

4.6 ODDS AND MATHEMATICAL EXPECTATION

Gamblers are frequently interested in determining which games are profitable to them. What they would really like to know is how much money can be earned in the long run from a particular game. If, in the long run nothing can be won, that is, if as much money that is won will be lost, then why play at all? Similarly, if in the long run no money can be won but money can be lost, then the gambler will not play such a game. He or she will play any game only if money can be won. When applied to gambling or business situations, the **mathematical expectation** of an event is the amount of money to be won or lost in the long run.

COMMENT In Chapter 6, when we discuss random variables and their probability distributions in detail, we will present a more general definition of mathematical expectation.

Since many applied examples of mathematical expectations involve gambling or business situations, where money can be won or lost in the long run, we have the following convenient formula for such cases.

FORMULA 4.5

Consider an event that has probability p_1 of occurring and that has a payoff, that is, the amount won, m_1. Consider also a second event with probability p_2 and payoff m_2, a third event with probability p_3 and payoff m_3, and so on. The mathematical expectation of the events is

$$m_1p_1 + m_2p_2 + m_3p_3 + \cdots + m_np_n$$

where the event has n different payoffs.
   (The numbers attached to the letters are called subscripts. They have no special significance. We use them only to avoid using too many different letters.)

We illustrate these ideas with several examples.

● **Example 1**

A large company is considering opening two new factories in different towns. If it opens in town A, it can expect to make $63,000 profit per year with a probability of $\frac{4}{7}$. However, if it opens in town B, it can expect to make a profit of $77,000 with a probability of only $\frac{3}{7}$. What is the company's mathematical expectation?

Solution
We use Formula 4.5. We have

$$(63,000)\left(\frac{4}{7}\right) + (77,000)\left(\frac{3}{7}\right) = 36,000 + 33,000 = 69,000.$$

Thus, the company's mathematical expectation is $69,000.

● **Example 2**
A contractor is bidding on a road construction job that promises a profit of $200,000 with a probability of $\frac{7}{10}$ and a loss, due to strikes, weather conditions, late arrival of building materials, and so on, of $40,000 with a probability of $\frac{3}{10}$. What is the contractor's mathematical expectation?

CAN YOUR CHANCES OF WINNING IN A LOTTERY BE IMPROVED?

Sept. 17: Because winning lottery numbers are generated randomly, no set of numbers has a higher probability of winning than any other, according to Dr. Jim Maxwell of the American Mathematics Society. "It is my understanding that lotteries are designed so that no one will have an advantage over anyone else," he said. "You could play the same number over and over again and have the same chance of winning as you would if you changed the number each time." Nevertheless, lottery players can decrease the number of players with whom they share their winnings by decreasing the odds that other players will choose the same numbers. "This is done by staying away from common number combinations such as months of the year, birthdays, holidays, et cetera," Dr. Maxwell said, since numbers drawn from the calendar are usually 31 or less. Because many lotteries call for numbers up to 48, players may increase their winnings, though not their chances of winning, by selecting some numbers above 31.

September 17, 1986

The above newspaper article indicates that we can often make decisions (and select numbers accordingly) that affect our mathematical expectations.

Solution

A "+" sign will denote a gain and a "−" sign will denote a loss. Using Formula 4.5, we find that the contractor's mathematical expectation is

$$(+\$200,000)\left(\frac{7}{10}\right) + (-\$40,000)\left(\frac{3}{10}\right)$$

$$= \$140,000 - \$12,000 = \$128,000$$

The contractor's mathematical expectation is $128,000.

● **Example 3**

Peter selects one card from a deck of 52 cards. If it is an ace, he wins $5. If it is a club, he wins only $1. However, if it is the ace of clubs, then he wins an extra $10. What is his mathematical expectation?

Solution

When one card is selected from a deck of cards, we have the following probabilities:

$$p(\text{ace}) = \frac{4}{52}$$

$$p(\text{clubs}) = \frac{13}{52}$$

$$p(\text{ace of clubs}) = \frac{1}{52}$$

Thus, Peter's mathematical expectation is

$$5\left(\frac{4}{52}\right) + 1\left(\frac{13}{52}\right) + 10\left(\frac{1}{52}\right) = \frac{20}{52} + \frac{13}{52} + \frac{10}{52}$$

$$= \frac{20 + 13 + 10}{52} = \frac{43}{52} = 0.83$$

His mathematical expectation is 83 cents.

COMMENT 83¢ is the fair price to pay to play the game, or is the break-even point. If he pays more than 83¢ per game to play, then in the long run he will lose money. If he pays less, then in the long run he will win.

Another interesting application of probability is concerned with betting. Gamblers frequently speak of the odds of a game. To best understand this idea, consider George, who believes that whenever he washes his car, it usually rains the following day. If George has just washed his car and if the probability of it raining tomorrow is $\frac{3}{10}$, gamblers would say that the odds in favor of it raining are 3 to 7 and the odds against it raining are 7 to 3. The 7 represents the 7 chances of it not raining. Formally we have the following definitions:

DEFINITION 4.5

*The **odds in favor** of an event occurring are p to q, where p is the number of favorable outcomes and q is the number of unfavorable outcomes.*

DEFINITION 4.6

*If p and q are the same as in Definition 4.5, the **odds against** an event happening are q to p.*

We now illustrate these definitions with several examples.

● **Example 4**

What are the odds in favor of the New York Mets winning the world series if the probability of their winning is $\frac{4}{7}$ and the probability of their losing is $\frac{3}{7}$?

Solution

Since the probability of their winning is $\frac{4}{7}$, this means that out of 7 possibilities 4 are favorable and 3 are unfavorable. Thus, the odds in favor of the New York Mets winning the series are 4 to 3.

YANKEES ARE FAVORED TO WIN SERIES

New York (Sept. 20) – The New York Yankees are favored to win the upcoming baseball world series. Local oddsmakers were betting 7-to-4 odds that the Yankees would win the series in six games.

● **Example 5**

Leon is in a restaurant. He decides to give the waiter a tip consisting of only 1 coin selected randomly from among the 6 that he has in his pocket. What are the odds against him giving the waiter a penny tip, if he has a penny, a nickel, a dime, a quarter, a half-dollar, or a dollar piece?

Solution

There are 6 coins; 5 are favorable and 1 is unfavorable. Thus the odds against giving the waiter a penny tip are 5 to 1.

● **Example 6**

What are the odds in favor of getting a face card when selecting a card at random from a deck of 52 cards?

There are 52 possibilities; 12 are favorable and 40 are unfavorable. Thus, the odds in favor of getting a face card are 12 to 40.

EXERCISES

1. A baker has just baked several pies. There is a $\frac{5}{11}$ chance that they will be sold today, in which case the profit will be $28. There is a $\frac{6}{11}$ chance that they will not be sold today, in which case the baker will lose $15 since the baker does not have refrigerator space to store the pies. Find the baker's mathematical expectation.

2. Margaret Billingsley has just sent some construction plans to Europe via one of the delivery services. There is a 90% chance that they will arrive at their destination by 9:00 A.M. the next morning, in which case her company will be awarded a $50,000 contract. If the plans do not arrive on time, then the contract will be awarded to a competing company. Find her company's mathematical expectation.

3. Ronald Johnson has just invested $125,000 to open a new drive-up fast food store. If successful, he can expect an annual income of $60,000. If unsuccessful, he will lose $55,000. (The remaining $70,000 can be recovered by selling the equipment.) If the probability of success is 0.85, find his mathematical expectation for the first year.

4. Consider the newspaper clipping given below. Based upon the information contained in it,
   a. find the oil companies' mathematical expectation.
   b. what are the odds in favor of finding oil in the area?

EXPLORATION FOR OIL TO BEGIN TODAY

Prudhoe Bay (March 17): A conglomeration of oil companies will begin drilling for oil in an area which is 70 miles north of area A. Geologists estimate that the probability of finding oil in the area is 0.85. If oil is discovered in large enough quantities to make it commercially feasible to drill for, the oil companies are expected to make a profit of $7 million. On the other hand, if the quantity of oil discovered is not large enough, then the oil companies will lose 2.5 million dollars in exploration costs.

Thursday, March 17, 1981

5. George Salington is a salesperson for a medical supplies company. He is analyzing his five best accounts and comes up with the following information:

Account	Weekly Sales	Estimated Probability That Sales Will Be Realized
I	$1698	0.61
II	1203	0.82
III	1506	0.59
IV	1475	0.51
V	1388	0.46

Find the expected sales from each account.

6. Refer to Exercise 5. George's car has broken down and as a result George finds that he has enough time to visit only two of the accounts. Using mathematical expectation, help George decide which accounts to visit.

7. Beverly Gray has just written a letter to the president of a cosmetics company complaining about the quality of a product. If there is a $\frac{7}{9}$ chance that the president will respond to her letter, what are the odds in favor of the president responding to her letter?

8. Each year the state Health Department checks the sanitary conditions at about 80% of the camps within the state. What are the odds against the state checking Camp Morgan for sanitary condition violations?

9. There is a 75% chance that Gloria will contract mononucleosis. What are the odds that she contracts the disease?

10. According to a survey, 65% of the state's residents are in favor of the state's proposed anti-abortion referendum. What are the odds against the proposed anti-abortion referendum?

11. Along with seven other guests, Bob places his hat on the table. All the hats look alike. The hostess puts the hats in a closet. After the party, Bob returns and selects a hat at random. What are the odds against his getting his own hat?

12. In a special lottery, 50,000 $1 tickets are sold. The first prize is $10,000 and the five second-prize winners will share $1000 equally. Maria buys one ticket for the lottery.
   a. What is her mathematical expectation?
   b. What are the odds in favor of her being a first-prize winner?
   c. What are the odds against her being a winner?

4.7 SUMMARY

In this chapter we discussed various aspects of probability. We noticed that probability is concerned with the total number of possible outcomes and experiments. Among the different ways of defining probability, we mentioned both the classical and the relative-frequency interpretation of probability. In the classical definition, we have

$$p = \frac{\text{number of outcomes favoring an event}}{\text{total number of equally possible outcomes}}$$

We noticed that the probability of an event was between 0 and 1, the null event and the definite event, respectively.

To enable us to determine the total number of possible outcomes, we analyzed various counting techniques. Tree diagrams, permutations, and combinations were introduced and discussed in detail. Permutations represent arrangements of objects where order *is* important, whereas combinations represent selections of objects where order is *not* important. Applications of permutations and combinations to many different situations were given, in addition to the usual gambling problems.

Finally, probability was applied to determine the amount of money to be won in the long run in various situations. This was called the mathematical expectation of the event. We also discussed what is meant by statements such as odds in favor of an event and odds against an event. Definitions were given that allow us to calculate these odds.

STUDY GUIDE

You should now be able to demonstrate your knowledge of the following ideas presented in this chapter by giving definitions, descriptions, or specific examples. Page references are given for each term so that you can check your answer.

Favorable outcome (page 154)
Sample space (page 154)
Event (page 154)
Probability (page 156)
Relative frequency (page 157)
Null event (page 162)
Certain event or definite event (page 163)
Tree diagrams (page 168)
Permutation (page 175)
Factorial notation (page 177)
Combination (page 185)
Binomial coefficient (page 185)
Pascal's triangle (page 190)
Mathematical expectation (page 196)
Odds in favor of an event (page 199)
Odds against an event (page 199)

FORMULAS TO REMEMBER

You should be able to identify each symbol in the following formulas, understand the relationships among the symbols expressed in each formula, understand the significance of each formula, and use the formulas in solving problems.

1. $p(A) = \dfrac{\text{number of favorable outcomes}}{\text{total number of equally possible outcomes}} = \dfrac{f}{n}$

2. $_nP_r = \dfrac{n!}{(n - r)!}$ The number of permutations of n things taken r at a time.

3. $_nP_n = n!$ The number of permutations of n things taken n at a time.

4. $\dfrac{n!}{p!q!r!} \cdots$ The number of permutations of n things where p are alike, q are alike, and so on.

5. $_nC_r = \dfrac{n!}{r!(n - r)!}$ The number of combinations of n things taken r at a time.

6. $m_1p_1 + m_2p_2 + m_3p_3 + \cdots$ Mathematical expectation of an event.

7. Odds in favor of an event are p to q, and odds against an event are q to p,

where $\begin{cases} p = \text{the number of favorable outcomes} \\ q = \text{the number of unfavorable outcomes.} \end{cases}$

The following tests will be more useful if you take them after you have studied the examples and solved the exercises given in this chapter.

MASTERY TESTS

Form A

1. The letters A, E, N, and T are written on four individual cards (one letter per card) and placed in a container. Each has an equal likelihood of being drawn. One card is drawn from the container, the letter noted, and the card is returned to the container. A second card is drawn and the letter noted. Find the probability that both letters drawn are the same.
   a. $\dfrac{1}{4}$ **b.** $\dfrac{1}{2}$ **c.** $\dfrac{1}{16}$ **d.** $\dfrac{3}{16}$ **e.** none of these

2. The probability that a 1040 federal income tax return will be audited by the IRS is $\dfrac{3}{100}$ in New York. What are the odds against a tax return being audited in New York?
   a. 3 to 100   **b.** 100 to 3   **c.** 3 to 97
   d. 97 to 3   **e.** none of these

3. Cassandra is planning her intersession vacation. She will go either to Puerto Rico, Hawaii, or Jamaica, stay at a hotel, a motel, or a condominium, and go by a charter flight or by a regularly scheduled flight. In how many different ways can Cassandra complete her intersession vacation plans?
   a. 12   b. 18   c. 8   d. 27   e. none of these

4. Ten protestors have complained to the Defense Department about the Navy's decision to build a deep-sea homeport in their city. The Secretary has agreed to meet with three of the group. In how many different ways can the three people be selected?
   a. 120   b. 720   c. 30   d. 5040   e. none of these

5. Eight students are waiting for a professor to advise them for the coming semester and sign their registration forms. The professor has time to see only five students. In how many different ways can the five students to be counseled be selected? (Assume order counts.)
   a. 40,320   b. 6720   c. 56   d. 120   e. none of these

6. In how many different ways can the letters of the word "RIDICULOUS" be arranged?
   a. 3,628,800   b. 1,814,400   c. 907,200
   d. 453,600   e. none of these

7. Bill and Joan have just arrived at Kennedy Airport to board a plane bound for Australia. The plane will make four additional stops before arriving at its final destination. Assuming that Bill and Joan are equally likely to disembark at any of these five stops, what is the probability that they both will get off at the same airport?
   a. $\dfrac{1}{5}$ b. $\dfrac{1}{2}$ c. $\dfrac{4}{5}$ d. $\dfrac{2}{5}$ e. none of these

8. In Question 7, the probability that Joan took her suntan lotion with her is 0.94. What is the probability that she forgot to take her suntan lotion with her?
   a. 0.06   b. 0   c. 0.6   d. 0.94   e. none of these

9. How many different 7-card rummy hands can be dealt from a 52-card deck of cards?
   a. $\dfrac{52!}{45!}$ b. $\dfrac{52!}{7!45!}$ c. $\dfrac{52!}{7!}$ d. $\dfrac{45!}{7!}$ e. none of these

10. Richie is in an OTB office, where he has just placed bets on some of the horses entered in this afternoon's race. In how many different ways can the 10 horses entered come in first, second, and third place?
    a. 720   b. 120   c. 30   d. 1000   e. none of these

11. Refer to the newspaper article given at the beginning of this chapter, on page 150. Assume that there are 38 governors that are Democrats, and 12 governors that are Republicans. Find the probability that the committee

will consist of
a. all Democrats.
b. at least two Democrats.
c. no Democrats.

MASTERY TESTS
Form B

1. Nine tickets numbered 1 to 9 are in a box. If one ticket is drawn at random, what is the probability of drawing an even-numbered ticket?
  a. $\frac{5}{9}$ **b.** $\frac{1}{2}$ **c.** $\frac{2}{9}$ **d.** $\frac{4}{9}$ **e.** none of these

2. Eleven books consisting of five engineering books, four math books, and two chemistry books are to be placed on a shelf. In how many different ways can the books be arranged if books of the same subject are to stay together?
  a. 11!   **b.** $_{11}C_{11}$ **c.** 4!5!2!3!
  d. 4!5!2!   **e.** none of these

3. How many different three-digit numbers can be formed from the digits 4, 5, 7, 8, and 9 if the digits *can* be repeated?
  a. 60   **b.** 210   **c.** 280   **d.** 125   **e.** none of these

4. Given the numbers 4, 6, 7, 8, and 9, how many three-digit numbers larger than 700 can be formed if repetition is *not* allowed?
  a. 18   **b.** 36   **c.** 125   **d.** 60   **e.** none of these

5. According to the latest U.S. Census Bureau statistics, 23% of all families in the United States are one-parent families. If a family in the United States is randomly selected, what are the odds against it being a one-parent family?

6. A wheel of fortune is equally divided into five colored areas: blue, green, yellow, white, and red. If the wheel stops on red after one spin, then the prize is $8. If it stops on yellow, the prize is $6, and if it stops on white, the prize is $3. There is no prize when the wheel stops on the other colors.
  a. What are the odds of winning a prize in any one spin of the wheel?
  b. What is the mathematical expectation for someone who plays this game?

7. The Acme Maintenance Company employs eight people who repair washing machines. Two of these people are Felippe Johnson and Richie Vestar. Mary calls the Acme Maintenance Company to request service for her washing machine. However, it is company policy never to honor requests for specific repairmen. Mary does not like Felippe or Richie's work. If repairmen are randomly assigned, find the probability that she does not get Felippe or Richie to repair her washing machine.

8. *Statement Savings*   The Valley Savings Bank no longer issues bankbooks. Instead the bank uses the statement savings system whereby each depositor is assigned a 7-digit code and all transactions are recorded by the bank's computers according to the account number furnished. Matilda Lichten-stein recalls the first six digits of her code but has completely forgotten the last one. She guesses that it is a 6 or an 8. Find the probability that she guesses correctly.

*9. Construct a tree diagram to determine the number of possible ways that a coin can be flipped five times so that throughout the flips there are always at least as many heads as tails.

10. Each of the 13,878 students at a state university attends class by bicycle.
 a. Will a three-digit number carved on each bike to discourage theft be sufficient for all the bikes? Explain your answer.
 b. Will a three-letter code carved on each bike to discourage theft be sufficient for all the bikes? Explain your answer.

11. A food-processing company is considering using one of two identification codes to place on each can of processed food: three letters followed by two numbers or two letters followed by three numbers. Which of these schemes will result in more identification codes?

12. *Stock Investment*   Joe McDonald invests $360,000 in the stock market for one week. The price of the particular stock that he bought can go up with a probability of $\dfrac{5}{9}$, in which case the man will make a profit of $7000, or the price of the stock can go down with a probability of $\dfrac{1}{3}$, in which case he will lose $15,000. Of course, the price of the stock can remain the same, in which case he will lose $450 (commission charges). What is Joe's mathematical expectation?

13. A secretary copies three computer disks and types three labels with information that identifies each disk. However, before placing each label on the appropriate disk, she drops everything on the floor. If she picks up all the identifying labels and computer disks and attaches a label at random onto a disk, what are the odds in favor of each computer disk being identified with its correct label?

14. In a certain state, license plates consist of five letters, with repetition of letters allowed. What is the probability that the letters of your license plates will be "IDIOT"?

15. Ten people board a crowded commuter bus that has four vacant seats. In how many different ways can the four vacant seats be occupied?

16. A large book company intends to publish eleven math books and seven science books this coming year. Furthermore, the editor has decided to

$$\binom{11}{5}\binom{7}{3}$$

462

heavily promote five of the math books and three of the science books. In how many different ways can this be done?

17. An airline company is sponsoring a contest and has selected ten semifinalists from which it will pick four winners. Each winner will win an all-expense-paid vacation trip to Europe. In how many different ways can the four winners be selected?

$$\binom{10}{4} \quad \frac{10!}{4!\,6!} = \frac{10 \cdot 9 \cdot 8 \cdot 7 \cdot 6!}{4 \cdot 3 \cdot 2 \cdot 1 \cdot 6!} \quad 30 \cdot 7 = 210$$

18. There are eight cars in an auto garage waiting to be serviced. However, since several of the mechanics are out on sick leave, the manager believes that there is sufficient help available to service only five cars. Assuming order counts, in how many different ways can the cars to be serviced be selected?

SUGGESTED FURTHER READING

1. Adler, I. *Probability and Statistics for Everyman.* New York: New American Library, 1966.
2. Bell, E. T. *Men of Mathematics.* New York: Simon & Schuster, 1961. Chapter 5 contains a bibliography of Pascal.
3. Bergamini, D. and eds. *Life. Mathematics* (Life-Science Library). New York: Time Life Books, 1970. Pages 126 to 147 discuss figuring the odds in an uncertain world.
4. Epstein, R. A. *Theory of Gambling and Statistical Logic.* New York: Academic Press, 1967. Contains an interesting discussion on the fairness of coins.
5. Havermann, E. "Wonderful Wizard of Odds" in *Life* 51, no. 14 (Oct. 6, 1961), p. 30 ff. contain a discussion of odds.
6. Huff, Darrell. *How to Take a Chance.* New York: Norton, 1959.
7. Kasner, E., and J. Newman. *Mathematics and the Imagination.* New York: Simon & Schuster, 1940. See the chapter on chance and probability.
8. *Mathematics in the Modern World* (Readings from *Scientific American*) San Francisco: W. H. Freeman, 1968. Article 22 discusses chance, Articles 23 and 24 discuss probability.
9. Newmark, Joseph. *Using Finite Mathematics.* New York: Harper and Row, 1982.
10. Newmark, Joseph, and Frances, Lake. *Mathematics as a Second Language,* 4th ed. Reading, MA: Addison-Wesley, 1987. Chapter 11 discusses probability and its applications.
11. Ore, Øystern. *Cardano, the Gambling Scholar.* New York: Dover, 1965.
12. Polya, G. *Mathematics and Plausible Reasoning.* Princeton, NJ: Princeton University Press, 1957.
13. Weaver, W. *Lady Luck.* New York: Anchor Books, also Doubleday, 1963.

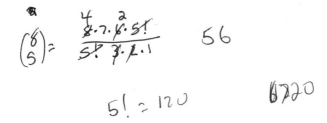

$$\binom{8}{5} = \frac{8 \cdot 7 \cdot 6 \cdot 5!}{5! \cdot 3 \cdot 2 \cdot 1} \qquad 56$$

$$5! = 120 \qquad 6720$$

CHAPTER 5

Rules of Probability

OBJECTIVES

- *To introduce* two addition rules that will allow us to determine the probability of either event A, event B, or both happening.

- *To discuss* whether particular events are or are not mutually exclusive and how this determines which addition rule to use.

- *To determine* when one event is affected by the occurrence or nonoccurrence of another and how this affects probability calculations. This is known as conditional probability.

- *To understand* that when one event is not affected at all by the occurrence of another, we have independent events and a simplified multiplication rule.

- *To learn* about Bayes' rule. Thus, if we know the outcome of some experiment, we might be interested in determining the probability that it occurred because of some specific event.

BEWARE (A LITTLE) FALLING METEORITES

A falling meteorite may not be one of the major hazards of modern life, but a person can be struck by one. Scientists in Canada have even calculated the magnitude of the risk.

The scientists, at Herzberg Institute of Astrophysics in Ottawa, used a network of 60 cameras in Western Canada for the past nine years to study meteorite falls. From the observations, they said in a recent letter to Nature, it can be calculated that one human should be hit in North America every 180 years.

The researchers, I. Halliday, A.T. Blackwell and A. A. Griffin, based their calculations on the number of meteorite falls of size large enough to be detected, the number of humans in the total Canadian and United States populations, the average human size, the time a person could be expected to be out of doors and other factors. On these assumptions they calculated an annual rate of human meteorite hits at 0.0055 per year, or one every 180 years.

Rare as these events should be, they noted, one such case actually occurred only 31 years ago. It is believed to be the only well-documented case of a collision between a meteorite and a human.

On Nov. 30, 1954, a nine-pound stony meteorite plunged though the roof of a home in Sylacauga, Ala., bounced off a large radio and struck Mrs. E.H. Hodges, inflicting painful bruises but causing her no serious injury.

"At first glance, it would appear unlikely that there would be even one known event only 31 years ago," the Canadian scientists said, "but the fact that there are no other verified cases elsewhere in the world indicates that the impacts on people are extremely rare."

Meteorite impacts on buildings are much less rare, they said, noting that there have been seven documented reports in North America during the past 20 years.

Worldwide, the Canadian scientists said, one could expect a human to be struck by a meteorite once in every nine years, while 16 buildings a year would be expected to sustain some meteorite damage.

November 14, 1986

What is the likelihood (probability) of a meteorite falling on a human being? The newspaper article indicates that given the average human size, the time that a person can be expected to be outdoors, and other factors, the probability of a human hit is still 0.0055 per year. How are such conditional probabilities calculated?

5.1 INTRODUCTION

In Chapter 4 we discussed the nature of probability and how to calculate its value. However, in order to determine the probability of an event in many situations, we must often first calculate the probability of other related events and then combine these probabilities. In this chapter we will discuss several rules for combining probabilities, including rules for addition, multiplication, conditional probability, and Bayes' rule. Depending upon the situation, these rules enable us to combine probabilities so that we may determine the probability of some event of interest.

5.2 ADDITION RULES

Mutually Exclusive Events

Let us look in on Charlie, who is playing cards. He is about to select one card from an ordinary deck of 52 playing cards. His opponent, Dick, will pay him $50 if the card selected is a face card (that is, a jack, queen, or king) *or* an ace. What is the probability that Charlie wins the $50?

To answer the question we first notice that a card selected cannot be a face card and an ace at the same time. Mathematically we say that the events of "drawing a face card" and of "drawing an ace" are **mutually exclusive.**

Since there are 12 face cards in a deck (4 jacks, 4 queens, and 4 kings), the probability of getting a face card is $\frac{12}{52}$, or $\frac{3}{13}$. Similarly, the probability of getting an ace is $\frac{4}{52}$, or $\frac{1}{13}$, since there are 4 aces in the deck. There are then 12 face cards and 4 aces. Thus there are 16 favorable outcomes out of a possible 52 cards in the deck. Applying the definition of probability we get

$$p(\text{face card or ace}) = \frac{16}{52} = \frac{4}{13}$$

Let us now add the probability of getting a face card with the probability of getting an ace. We have

$$p(\text{face card}) + p(\text{ace}) = \frac{12}{52} + \frac{4}{52}$$

$$= \frac{12 + 4}{52} \text{ (since the denominators are the same)}$$

$$= \frac{16}{52} = \frac{4}{13}$$

This indicates that

$$p(\text{face card or ace}) = p(\text{face card}) + p(\text{ace})$$

The same reasoning can be applied for any mutually exclusive events. First we have the following definition and formula.

DEFINITION 5.1

Consider any two events A and B. If both events cannot occur at the same time, we say that the events A and B are **mutually exclusive.**

FORMULA 5.1

If A and B are mutually exclusive, then

$$p(A \text{ or } B) = p(A) + p(B)$$
$$p(A \cup B)$$

We illustrate the use of Formula 5.1 with several examples.

● **Example 1**

Louis has been shopping around for a calculator and has decided to buy the scientific model calculator. The probability that he will buy the Hewlett Packard model is $\frac{1}{9}$ and the probability that he will buy the Texas Instruments model is $\frac{4}{9}$. What is the probability that he will buy either of these two models?

Solution

Since Louis will buy only one calculator, the events "buys the Texas Instruments model" and "buys the Hewlett Packard model" are mutually exclusive. Thus Formula 5.1 can be used. We have

$$p(\text{buys either model}) = p(\text{buys Texas Instruments model})$$
$$+ \; p(\text{buys Hewlett Packard model})$$

$$= \frac{4}{9} + \frac{1}{9}$$

$$= \frac{4 + 1}{9} = \frac{5}{9}$$

Therefore, $p(\text{Louis buys either model}) = \frac{5}{9}$.

● **Example 2**
Mary turns on the television set. The probability that the television is on Channel 2 is $\dfrac{1}{3}$, and the probability that it is on Channel 7 is $\dfrac{4}{13}$. What is the probability that it is either on Channel 2 or Channel 7?

Solution
Since the same television set cannot be on Channel 2 and on Channel 7 at the same time, the events "television set on Channel 2" and "television set on Channel 7" are mutually exclusive; thus Formula 5.1 can be used. We have

p(television set on Channel 2 or 7) = p(television set on Channel 2)

$$+ \; p\text{(television set on Channel 7)}$$

$$= \frac{1}{3} + \frac{4}{13}$$

$$= \frac{13}{39} + \frac{12}{39} = \frac{25}{39}$$

Therefore,

$$p(\text{TV on Channel 2 or 7}) = \frac{25}{39}$$

$\Bigg($ *Note:* We cannot add the fractions $\dfrac{1}{3}$ and $\dfrac{4}{13}$ together as they are, since they do not change the denominators. We must change both fractions to fractions with the same denominators. Thus $\dfrac{1}{3}$ becomes $\dfrac{13}{39}$ and $\dfrac{4}{13}$ becomes $\dfrac{12}{39}$. Hence, the probability that the television is on Channel 2 or 7 is $\dfrac{25}{39}$. $\Bigg)$

● **Example 3**
Two dice are rolled. What is the probability that the sum of the dots appearing on both dice together is 9 or 11?

Solution
Since the events "getting a sum of 9" and "getting a sum of 11" are mutually exclusive, Formula 5.1 can be used. When 2 dice are rolled there are 36 possible outcomes. These were listed on page 155. There are 4 possible ways of getting a sum of 9. Thus

$$p(\text{sum of 9}) = \frac{4}{36}$$

Also, there are only two possible ways of getting a sum of 11. Thus

$$p(\text{sum of } 11) = \frac{2}{36}$$

Therefore,

$$p(\text{sum of 9 or 11}) = p(\text{sum of 9}) + p(\text{sum of 11})$$

$$= \frac{4}{36} + \frac{2}{36}$$

$$= \frac{6}{36} = \frac{1}{6}$$

Hence the probability that the sum is 9 or 11 is $\frac{1}{6}$.

● **Example 4**

Doris and her friends plan to travel to Florida during the winter intersession period. The probability that they go by car is $\frac{2}{3}$, and the probability that they go by plane is $\frac{1}{5}$. What is the probability that they travel to Florida by car or plane only?

Solution

Since they plan to travel to Florida either by car or by plane, not by both, we are dealing with mutually exclusive events. Formula 5.1 can be used. Therefore,

$$p(\text{go by car or plane}) = p(\text{go by car}) + p(\text{go by plane})$$

$$= \frac{2}{3} + \frac{1}{5}$$

$$= \frac{10}{15} + \frac{3}{15}$$

$$= \frac{13}{15}$$

Complementary Events

● **Example 5**

Rosemary buys a ticket in the state lottery. The probability that she will win the grand prize of one million dollars is $\frac{1}{50,000}$. What is the probability that she does not win the one million dollars?

Solution

Since the events "Rosemary wins the million dollars" and "Rosemary does not win the million dollars" are mutually exclusive, we can use Formula 5.1. One of those events must occur so that the event "Rosemary wins the million dollars or does not win the million dollars" is the definite event. We know that the definite event has probability 1 (see page 163).* Thus

$$p(\text{Rosemary wins million dollars or does not win})$$
$$= \begin{array}{l} p(\text{wins million dollars}) \\ + \ p(\text{does not win million dollars}) \end{array}$$

$$1 = \frac{1}{50{,}000} + p(\text{does not win million dollars})$$

$$1 - \frac{1}{50{,}000} = p(\text{does not win million dollars})$$

$$\frac{50{,}000}{50{,}000} - \frac{1}{50{,}000} = p(\text{does not win million dollars})$$

$$\frac{49{,}999}{50{,}000} = p(\text{does not win million dollars})$$

Therefore, the probability that Rosemary does not win the million dollars is $\frac{49{,}999}{50{,}000}$.

More generally, consider any event A. Let $p(A)$ be the probability that A happens and let $p(A')$, read as the probability of A prime, be the probability that A does not happen. Since either A happens or does not happen, we can use Formula 5.1. Thus,

$$p(A \text{ happens or does not happen}) = \begin{array}{l} p(A \text{ happens}) \\ + \ p(A \text{ does not happen}) \end{array}$$

$$1 = p(A) + p(A')$$

$$1 - p(A) = p(A'). \text{ (We subtract } p(A) \text{ from both sides.)}$$

Therefore, the probability of A not happening is $1 - p(A)$. (*Note:* Some books refer to the event A' as the **complement** of event A.)

Addition Rule—General Case

Now consider the following problem. One card is drawn from a deck of cards. What is the probability of getting a king or a red card? At first thought we might say that since there are 4 kings and 26 red cards, then

* An event that is certain to occur is called the certain event or a definite event and its probability is 1

$$p(\text{king or red card}) = p(\text{king}) + p(\text{red card})$$

$$= \frac{4}{52} + \frac{26}{52}$$

$$= \frac{30}{52} = \frac{15}{26}$$

Thus, we would say that the probability of getting a king or a red card is $\frac{15}{26}$.

Notice, however, that in arriving at this answer we have counted some cards twice. The 2 red kings have been counted as both kings and red cards. Obviously we must count them only once in probability calculations. The events "getting a king" and "getting a red card" are not mutually exclusive. We therefore have to revise our original estimate of the total number of favorable outcomes by deducting the number of cards that have been counted twice. We will subtract 2. When this is done, we get

$$p(\text{king or red card}) = p(\text{king}) + p(\text{red card}) - p(\text{king also a red card})$$

$$= \frac{4}{52} + \frac{26}{52} - \frac{2}{52}$$

$$= \frac{4 + 26 - 2}{52}$$

$$= \frac{28}{52} = \frac{7}{13}$$

Thus, the probability of getting a king or a red card is $\frac{7}{13}$. This leads us to a more general formula.

ADDITION RULE (GENERAL CASE)

If A and B are events, the probability of obtaining either of them is equal to the probability of A plus the probability of B minus the probability of both occurring at the same time.

Symbolically, the addition rule is as follows:

FORMULA 5.2

If A and B are any events, then

$$p(A \text{ or } B) = p(A) + p(B) - p(A \text{ and } B)$$

We now apply Formula 5.2 in several examples.

● **Example 6**

Sylvester is a member of The Hamilton Bay Ensemble. The probability that Sylvester plays a guitar is $\dfrac{1}{4}$, and the probability that he plays a clarinet is $\dfrac{5}{8}$.

If the probability that he plays both of these instruments is $\dfrac{5}{24}$, what is the probability that he plays the guitar or that he plays the clarinet?

Solution

Since it is possible that Sylvester plays both these instruments, these events are not mutually exclusive. Thus we must use Formula 5.2. We have

$$p(\text{plays guitar or clarinet}) = p(\text{plays guitar}) + p(\text{plays clarinet})$$
$$- p(\text{plays guitar and clarinet})$$

$$= \frac{1}{4} + \frac{5}{8} - \frac{5}{24}$$

$$= \frac{6}{24} + \frac{15}{24} - \frac{5}{24}$$

$$= \frac{6 + 15 - 5}{24}$$

$$= \frac{16}{24} = \frac{2}{3}$$

Thus, the probability that he plays either instrument is $\dfrac{2}{3}$.

● **Example 7** *Voting*

Consider the accompanying newspaper article. What is the probability that either a husband will vote or his wife will vote in the coming mayoral election?

Greensburgh (Oct. 20) — According to a poll released yesterday, voter turnout for the coming mayoral election is expected to be at an all-time low. One out of eleven married men said that they would vote. For the women, the figure was one out of nine. Only one out of 28 couples said that they would both vote. Both candidates indicated that they were mobilizing their forces to get out more votes.

November 1, 1986

Reprinted by permission.

Solution

The events husband votes and wife votes are not mutually exclusive events, since both events can occur. Thus, we use Formula 5.2. We have:

Prob(husband or wife votes) =
$$\text{Prob(husband votes)} + \text{Prob(wife votes)} - \text{Prob(both vote)}$$

Based on the newspaper article

$$\text{Prob(husband votes)} = \frac{1}{11}$$

$$\text{Prob(wife votes)} = \frac{1}{9}$$

$$\text{Prob(both vote)} = \frac{1}{28}$$

Therefore,

$$\text{Prob(husband or wife votes)} = \frac{1}{11} + \frac{1}{9} - \frac{1}{28}$$

$$= \frac{252}{2772} + \frac{308}{2772} - \frac{99}{2772} = \frac{461}{2772}$$

Thus, Prob(husband or wife votes) $= \frac{461}{2772}$.

● **Example 8** *Controlling Pollution*

Environmentalists have accused a large company in the eastern United States of dumping nuclear waste material into a local river. The probability that *either* the fish in the river or the animals that drink from the river will die is $\frac{11}{21}$. The probability that only the fish will die is $\frac{1}{3}$ and the probability that only the animals that drink from the river will die is $\frac{2}{7}$. What is the probability that both the fish *and* the animals that drink from the river will die?

Solution

Since both the fish and animals may die, the events are not mutually exclusive. We then use Formula 5.2. We have

$$p(\text{fish or animals die}) = p(\text{fish die}) + p(\text{animals die})$$
$$- p(\text{both fish and animals die})$$

$$\frac{11}{21} = \frac{1}{3} + \frac{2}{7} - p(\text{both die})$$

$$\frac{11}{21} = \frac{7}{21} + \frac{6}{21} - p(\text{both die})$$

$$\frac{11}{21} = \frac{13}{21} - p(\text{both die})$$

$$\frac{11}{21} + p(\text{both die}) = \frac{13}{21}$$

$$p(\text{both die}) = \frac{13}{21} - \frac{11}{21} = \frac{2}{21}$$

Thus, the probability that *both* the fish and the animals that drink from the river will die is $\frac{2}{21}$.

COMMENT Although you may think that Formulas 5.1 and 5.2 are different, this is not the case. Formula 5.1 is just a special case of Formula 5.2. Formula 5.2 can always be used, since if the events A and B are mutually exclusive, the probability of them happening together is 0. In this case Formula 5.2 becomes

$$p(A \text{ or } B) = p(A) + p(B) - 0,$$

which is exactly Formula 5.1.

COMMENT Formula 5.2 can be used when we have only two events, A and B. For any three events A, B, and C, the probability of A or B or C is given by the formula

$$p(A \text{ or } B \text{ or } C) = p(A) + p(B) + p(C) - p(A \text{ and } B) - p(A \text{ and } C)$$
$$- p(B \text{ and } C) + p(A \text{ and } B \text{ and } C)$$

The use of this formula will be illustrated in the exercises.

EXERCISES

1. Determine which of the following events are mutually exclusive.
  a. Getting audited by the IRS and by the state income tax department
  b. Being overweight and having high blood pressure
  c. The N.Y. Yankees and N.Y. Mets both winning the same world series
  d. Being a U.S. Senator and being a member of the U.S. House of Representatives
  e. Taking out a life insurance policy from company A and taking out a life insurance policy from company B

2. Rosalie is conducting a survey of health club members' use of the various facilities. According to the responses received, 40% of the members use the sauna, 55% use the jogging track, and 11% use both the sauna and the jogging track. If a club member is

$$p(\text{sauna or track}) = \frac{40}{100} + \frac{55}{100} - \frac{11}{100} = 84\%$$

randomly selected, find the probability that the member uses the sauna *or* the jogging track.

3. Larry's car will not start. Based upon past experience, Larry knows that the probability that it is the battery is 0.65, the probability that it is the automatic choke on the carburator is 0.45 and the probability that it is both is 0.24. Find the probability that it is either of these two problems. $p(\text{either}) \, \frac{65}{100} + \frac{45}{100} - \frac{24}{100} = .86$

4. The Gable Corp. has placed bids for two different public works projects. The probability that it will be awarded the job for project A is 0.85, the probability that it will be awarded the job for project B is 0.62, and the probability that it will be awarded the jobs for both projects is 0.53. Find the probability that it will be awarded the job for either project.

5. According to a survey conducted by Kobb and Gerry, the probability that a woman in Batesville uses roll-on deodorant is 0.72 and the probability that she uses spray deodorant is 0.59. If the probability that she uses either is 0.93, find the probability that she uses both. $\frac{93}{100} = \frac{72}{100} + \frac{59}{100} - P(\text{both}) = .38$

6. According to an environmental study, the probability that the Croton Lake contains coliform bacteria at any given time is 0.82 and the probability that it contains toxic chemical wastes at any time is 0.58. If the probability that it contains both coliform bacteria and toxic chemical wastes is 0.53, find the probability that it contains either.

7. Regional transportation officials claim that the probability that a commuter train will break down is 0.12 and the probability that a commuter bus will break down is 0.08. If the probability that both a bus and a train will break down is 0.03, find the probability that there will be *no* breakdowns on either system.

8. In a certain city, the probability that a doctor has office hours in the evening is 0.37, the probability that the doctor has office hours on the weekend is 0.22, and the probability that the doctor has office hours both in the evening and on the weekend is 0.02. If a doctor practicing within the city is randomly selected, find the probability that the doctor has office hours either in the evening or on weekends.

9. In a certain community, the probability that a married man is a college graduate is $\frac{5}{14}$. The probability that his wife is a college graduate is $\frac{3}{7}$, and the probability that both the husband and wife are college graduates is $\frac{1}{7}$. What is the probability that either a man or his wife are college graduates in this community? $\frac{5}{14} + \frac{6}{14} - \frac{2}{14} \quad \frac{9}{14}$

10. Ed is planning to buy his son either a home computer system or a component stereo system (but not both) for his birthday. The probability that he buys him the home computer system is 0.34. If the probability that he buys him either the computer system or the stereo system is 0.78, find the probability that Ed buys his son the stereo system.

11. A college guidance counselor is analyzing the undergraduate transcripts of students who have submitted applications for admission to Princeton, Yale, and Harvard graduate schools. Her analysis is given in the accompanying table.

Schools Applied to	Probability
Princeton $p(A)$	$\dfrac{48}{60}$
Yale $p(B)$	$\dfrac{33}{60}$
Harvard $p(C)$	$\dfrac{39}{60}$
Princeton and Yale $p(AB)$	$\dfrac{27}{60}$
Princeton and Harvard $p(AC)$	$\dfrac{15}{60}$
Yale and Harvard $p(BC)$	$\dfrac{29}{60}$
$p(ABC)$ Princeton, Yale, and Harvard	$\dfrac{4}{60}$

Handwritten:

$p(A\ or\ B\ or\ C) \sim p(A) + p(B) + p(C)$
$- p(A\ and\ B) - p(A\ and\ C)$
$- p(B\ and\ C) + p(A\ and\ B\ and\ C)$

$= \dfrac{48}{60} + \dfrac{33}{60} + \dfrac{39}{60} - \dfrac{27}{60} - \dfrac{15}{60} - \dfrac{29}{60} + \dfrac{4}{60}$

$\dfrac{53}{60}$

If a graduating student is randomly selected, find the probability that the student has submitted an application to either of these three schools.

12. A flight attendant for a large airline company has been studying the reading habits of passengers who fly with the company. The accompanying table gives the probability that a passenger will ask for some of the various magazines available.

Magazine	Probability
Newsweek	$\dfrac{22}{47}$
Time	$\dfrac{16}{47}$
U.S. News and World Report	$\dfrac{31}{47}$
Newsweek and Time	$\dfrac{17}{47}$
Newsweek and U.S. News and World Report	$\dfrac{8}{47}$
Time and U.S. News and World Report	$\dfrac{7}{47}$
Any of these three magazines	$\dfrac{43}{47}$

Handwritten:

$\dfrac{43}{47} = \dfrac{22}{47} + \dfrac{16}{47} + \dfrac{31}{47} - \dfrac{17}{47} - \dfrac{8}{47} - \dfrac{7}{47} + p(\ )$

$\dfrac{6}{47}$

Assuming that a passenger will definitely ask for a magazine, find the probability that a passenger will ask for all three of these magazines.

13. According to state tax officials, $\frac{1}{8}$ of the mistakes made on tax returns are in the arithmetic calculations, $\frac{1}{9}$ of the mistakes are because people use the wrong forms, and $\frac{1}{15}$ of the mistakes involve both arithmetic calculations and wrong tax forms. If a rejected tax form is selected at random, find the probability that it is because the wrong tax form was used or because it contains an arithmetic mistake.

14. At Jackmore College, all business majors must complete the following three math-related courses:

Math 23   Introduction to College Algebra
Math 41   Introduction to Computer Science
Math 57   Introduction to Finite Math

The probabilities that a student registers for one or more of these three courses in the year are given in the accompanying table. Find the probability that a student registers for all three courses in his or her freshman year.

Course	Probability
Math 23	$\frac{2}{3}$
Math 41	$\frac{37}{60}$
Math 57	$\frac{13}{30}$
Math 23 and 41	$\frac{19}{60}$
Math 23 and 57	$\frac{3}{10}$
Math 41 and 57	$\frac{1}{3}$
Any of Math 23, 41, or 57	1

5.3 CONDITIONAL PROBABILITY

Although the addition rule given in Section 5.2 applies to many different situations, there are still other problems that cannot be solved by that formula. It is for this reason that we introduce conditional probability. Let us first consider the following problem:

● **Example 1** *Anti-Smoking Campaign*
In an effort to reduce the amount of smoking, the administration of Podunk University is considering the establishment of a smoking clinic to help students

"break the habit." However, not all the students at the school favor the proposal. As a result, a survey of the 1000 students at the school was conducted to determine student opinion about the proposal. The following table summarizes the results of the survey:

	Against Smoking Clinic	For Smoking Clinic	No Opinion	Total
Freshmen	23	122	18	163
Sophomores	39	165	27	231
Juniors	58	238	46	342
Seniors	71	127	66	264
Total	191	652	157	1000

Answer the following questions:

a. What is the probability that a student selected at random voted against the establishment of the smoking clinic?
b. If a student is a freshman, what is the probability that the student voted for the smoking clinic?
c. If a senior is selected at random, what is the probability that the senior has no opinion about the clinic?

Solution

a. Since there was a total of 191 students who voted against the establishment of the smoking clinic out of a possible 1000 students, we apply the definition of probability and get

$$p(\text{student voted against the smoking clinic}) = \frac{191}{1000}.$$

Thus, the probability that a student selected at random voted against the establishment of the smoking clinic is $\frac{191}{1000}$.

b. There are 163 freshmen in the school. One hundred twenty-two of them voted for the smoking clinic. Since we are concerned with freshmen only, the number of possible outcomes of interest to us is 163, not 1000. Out of these, 122 are favorable. Thus the probability that a student voted for the smoking clinic given that the student is a freshman is $\frac{122}{163}$.

c. In this case the information given narrows the sample space to the 264 seniors, 66 of which had no opinion. Thus the probability that a student has no opinion given that the student is a senior is $\frac{66}{264}$, or $\frac{1}{4}$.

The situation of part (b) or that of part (c) in Example 1 is called a **conditional probability** because we are interested in the probability of a student voting in favor of the establishment of the smoking clinic given that, or conditional upon the fact that, the student is a freshman. We express this condition mathematically by using a vertical line "|" to stand for the words "given that" or "if we know that." We then write

$$p(\text{student voted in favor of smoking clinic} \mid \text{student is a freshman}) = \frac{122}{163}.$$

Similarly for part (c) we write

$$p(\text{student had no opinion} \mid \text{student is a senior}) = \frac{66}{264} = \frac{1}{4}.$$

● Example 2

Sherman is repairing his car. He has removed the six spark plugs. Four are good and two are defective. He now selects one plug and then, without replacing it, selects a second plug. What is the probability that both spark plugs selected are good?

Solution

We will list the possible outcomes and then count all the favorable ones. To do this we label the good spark plugs as g_1, g_2, g_3, and g_4, and the defective ones as d_1 and d_2. The possible outcomes are

g_1, g_2	g_2, g_1	g_3, g_1	g_4, g_1	d_1, g_1	d_2, g_1
g_1, g_3	g_2, g_3	g_3, g_2	g_4, g_2	d_1, g_2	d_2, g_2
g_1, g_4	g_2, g_4	g_3, g_4	g_4, g_3	d_1, g_3	d_2, g_3
g_1, d_1	g_2, d_1	g_3, d_1	g_4, d_1	d_1, g_4	d_2, g_4
g_1, d_2	g_2, d_2	g_3, d_2	g_4, d_2	d_1, d_2	d_2, d_1

There are 30 possible outcomes. Twelve of these are favorable. These are the circled ones. They represent the outcome that both spark plugs are good. Thus

$$p(\text{both spark plugs selected are good}) = \frac{12}{30} = \frac{2}{5}.$$

● Example 3

In Example 2, what is the probability that both spark plugs selected are good if we know that the first plug selected is good?

Solution

Again we list all the possible outcomes and count the number of favorable ones.

(g_1, g_2)	(g_2, g_1)	(g_3, g_1)	(g_4, g_1)
(g_1, g_3)	(g_2, g_3)	(g_3, g_2)	(g_4, g_2)
(g_1, g_4)	(g_2, g_4)	(g_3, g_4)	(g_4, g_3)
g_1, d_1	g_2, d_1	g_3, d_1	g_4, d_1
g_1, d_2	g_2, d_2	g_3, d_2	g_4, d_2

Since we know that the first plug selected is good, there are only 20 possible outcomes. Of these, 12 are favorable. These are the circled ones. Thus the probability that both spark plugs are good if we know that the first plug is good is

$$\frac{12}{20} = \frac{3}{5}$$

Using the conditional probability notation, this result can be written as

$$p(\text{both spark plugs are good} \mid \text{first spark plug is good}) = \frac{3}{5}$$

COMMENT   Example 3 differs from Example 2 since in Example 3 we are interested in determining the probability of getting two good spark plugs once we know that the first one selected is good. On the other hand, in Example 2 we were interested in determining the probability of getting two good plugs without knowing whether the first plug is defective or not.

Let us analyze the problem discussed at the beginning of this section in detail. There are a total of 163 freshmen out of a possible 1000 students in the school. Thus

$$p(\text{freshman}) = \frac{163}{1000}$$

Also, there were 122 freshmen who voted in favor of the clinic. Thus,

$$p(\text{freshman and voted in favor of clinic}) = \frac{122}{1000}$$

Summarizing these results we have

$$p(\text{freshman}) = \frac{163}{1000}$$

and

$$p(\text{freshman and voted in favor of clinic}) = \frac{122}{1000}$$

Let us now divide p(freshman *and* voted in favor of clinic) by p(freshman). We get

$$\frac{p(\text{freshman } and \text{ voted in favor of clinic})}{p(\text{freshman})} = \frac{122/1000}{163/1000}$$

$$= \frac{122}{1000} \div \frac{163}{1000}$$

$$= \frac{122}{1000} \cdot \frac{1000}{163}$$

$$= \frac{122}{163}$$

This is the same result as p(student voted in favor of smoking clinic | student is a freshman). In both cases the answer is $\frac{122}{163}$.

If we let A stand for "student voted in favor of smoking clinic" and B stand for "student is a freshman," then the previous result suggests that

$$p(A \mid B) = \frac{p(A \text{ and } B)}{p(B)}$$

We can apply the same analysis for part (c) of the problem. We have

$$p(\text{senior}) = \frac{264}{1000} \quad \text{and} \quad p(\text{senior and no opinion}) = \frac{66}{1000}$$

If we divide p(senior and no opinion) by p(senior) we get

$$\frac{p(\text{senior and no opinion})}{p(\text{senior})} = \frac{66/1000}{264/1000}$$

$$= \frac{66}{1000} \div \frac{264}{1000}$$

$$= \frac{66}{1000} \cdot \frac{1000}{264} = \frac{66}{264} = \frac{1}{4}$$

Thus,

$$p(\text{student had no opinion} \mid \text{student is a senior}) = \frac{p(\text{senior and no opinion})}{p(\text{senior})}.$$

We can generalize our discussion by using a formula that is called the **conditional probability formula.**

FORMULA 5.3   *Conditional Probability Formula*

If A and B are any events, then

$$p(A \mid B) = \frac{p(A \text{ and } B)}{p(B)}, \quad \text{provided } p(B) \neq 0$$

We illustrate the use of Formula 5.3 with several examples.

● **Example 4**

In Ashville the probability that a married man drives is 0.90. If the probability that a married man *and* his wife both drive is 0.85, what is the probability that his wife drives given that he drives?

Solution
We will use Formula 5.3. We are told that

p(husband drives) $= 0.90$

and

p(husband and wife drive) $= 0.85$

Thus,

$$p(\text{wife drives} \mid \text{husband drives}) = \frac{p(\text{husband and wife drive})}{p(\text{husband drives})}$$

$$= \frac{0.85}{0.90}$$

$$= \frac{85}{90} \quad \text{(We multiply numerator and denominator by 100)}$$

$$= \frac{17}{18}$$

Thus, p(wife drives | husband drives) $= \dfrac{17}{18}$

● **Example 5**

Joe often speeds while driving to school in order to arrive on time. The probability that he will speed to school is 0.75. If the probability that he speeds and gets stopped by a police officer is 0.25, find the probability that he is stopped, given that he is speeding.

Solution

We use Formula 5.3. We are told that $p($Joe speeds$)$ is 0.75 and $p($speeds and is stopped$)$ is 0.25. Thus

$$p(\text{he is stopped} \mid \text{he speeds}) = \frac{p(\text{speeds and is stopped})}{p(\text{speeds})}$$

$$= \frac{0.25}{0.75}$$

$$= \frac{25}{75}$$

$$= \frac{1}{3}$$

Thus $p(\text{he is stopped} \mid \text{he speeds}) = \dfrac{1}{3}$.

● **Example 6**

Janet likes to study. The probability that she studies *and* passes her math test is 0.80. If the probability that she studies is 0.83, what is the probability that she passes the math test, given that she has studied?

Solution

We use Formula 5.3. We have

$$p(\text{passes math test} \mid \text{she studied}) = \frac{p(\text{studies and passes math test})}{p(\text{she studies})}$$

$$= \frac{0.80}{0.83}$$

$$= \frac{80}{83}$$

Thus, $p(\text{she passes math test} \mid \text{she has studied}) = \dfrac{80}{83}$.

EXERCISES

1. Sixty-eight percent of all purchases made at Fred's Department Store are paid for by credit card. Additionally, 11% of all purchases are made by credit card and are returned for one reason or another. If a customer is seen paying for an item by credit card, what is the probability that the customer will return it?

$$p(A \mid B) = \frac{p(A' \text{ and } B)}{p(B)} \qquad \frac{11/100}{68/100} = \frac{11}{68}$$

2. According to a survey conducted by a local affiliate of the American Automobile Association, the attendant at 47% of the stations surveyed will check your oil when purchasing gas. Additionally, 23% of the attendants *will also* wash your windows. If an attendant is observed checking the oil in Maryanne's car when she is buying gas, find the probability that the attendant will also wash the windows.

3. *Nuclear Disarmament*   An environmentalist interviewed 1200 individuals in the Houston area to determine their opinion on nuclear disarmament. The results of the survey are shown in the following table:

Age Group	In Favor of Nuclear Disarmament	Opposed to Nuclear Disarmament
Under 25 years	58	306
26–35 years	94	184
36–50 years	161	102
Over 50 years	203	92

684

Find the probability that a randomly selected individual in the group is
a. under 25 years of age and is opposed to nuclear disarmament.   306/1200
b. opposed to nuclear disarmament given that he or she is under 25 years.   306/364
c. under 25 years of age given that he or she is opposed to nuclear disarmament.   306/684

4. A marketing director interviewed 1600 randomly selected people to get their reaction to a new type of cologne that was being marketed. The results of the survey are shown in the following table:

Age Group	Opinion		
	Didn't Like It	Liked It	No Opinion
Under 30 years	302	183	161
31–50 years	202	212	143
Over 51 years	168	36	193

Find the probability that a randomly selected individual in the group
a. is under 30 years of age.   646/1200
b. didn't like the cologne given that the individual is under 30 years of age.   302/646
c. is under 30 years of age given that the individual didn't like the cologne.   302/672

5. In a certain community the probability that a family has a VCR (video cassette recorder) *and* a personal computer is 0.26. The probability that the family has a VCR is 0.58. Find the probability that a randomly selected family in this community has a personal computer given that the family has a VCR.

$$P(\text{VCR and PC}) = \frac{.26}{.58}$$

$$P(\text{PC/VCR})$$

$$p(B/A) = \frac{p(BA)}{p(A)} = \frac{.27}{.53}$$

A   B

6. The probability that a student at State University is a coed *and* wears contact lenses is 0.27. Fifty-three percent of all students at State University are coeds. What is the probability that a randomly selected student wears contact lenses given that the student is a coed? $p(A) = .53$ $p(B) = ?$ $p(AB) = .27$

7. In an effort to conserve energy, Elizabeth often goes to work by means of public transportation. Since such transportation is unpredictable, she is sometimes late for her job. The probability that she will go to work by public transportation is 0.73, and the probability that she will come to work by public transportation *and* be late for her job is 0.11. If Elizabeth is observed going to work by public transportation, what is the probability that she will be late for her job? $\frac{11}{73}$

8. Fifty-three percent of the members of a particular health club exercises regularly (at least three times a week) *and* watch their diet. Seventy-four percent of the members of the health club watch their diet. If a member who is known to watch his or her diet is randomly selected, find the probability that the member exercises regularly.

9. Sixty-nine percent of all the cars insured by a major insurance company for theft have both an ignition-type cutoff switch *and* an alarm to discourage auto theft. Furthermore, 84% of the insured cars have an alarm. If a randomly selected car that is insured by this company is known to have an alarm, find the probability that it also has an ignition-type cutoff switch.

10. Thirty-seven percent of all the members of the Staten Country Club play tennis *and* also play racquet ball. Fifty-eight percent of the members of the club play tennis. If a club member who plays tennis is randomly selected, find the probability that the member does *not* play racquet ball.

11. Red Cross officials estimate that about 40% of the people of Crescent have type A blood. Moreover, only 8% of the population with type A blood participate in the annual blood program drive. If a randomly selected individual who is participating in the blood drive is selected, find the probability that the individual has type A blood. $\frac{8}{40} = \frac{A}{5}$

12. *Death Penalty*   A sociologist interviewed 1000 individuals from minority groups in Los Angeles to determine their opinion on the death penalty. The results of the survey are shown in the accompanying table.

Group	In Favor of Death Penalty	Opposed to Death Penalty
Black	176	221
Puerto Rican	118	147
Oriental	94	62
Other minorities	151	31

Find the probability that an individual in the group is
a. black and opposes the death penalty.

$\frac{221}{1000}$

b. opposed to the death penalty given that he or she is black. $\frac{24}{397}$

c. black given that he or she is opposed to the death penalty. $\frac{221}{461}$

13. *Religion and Politics* The sociologist described in Exercise 12 then interviewed 1000 randomly selected individuals to determine whether there is a relationship between a person's religion and the person's political beliefs. The results of the survey are shown in the accompanying table.

Political Preference

Religion	Democrat	Republican	Conservative
Protestant	310	360	60
Catholic	120	69	11
Jewish	29	13	8
Other	10	6	4

Find the probability that a randomly selected individual in the group is a

a. Republican. $\frac{448}{1000}$

b. Republican, given that he or she is a Catholic. $\frac{69}{200}$

c. Catholic, given that he or she is a Republican. $\frac{69}{448}$

5.4 MULTIPLICATION RULES

In Section 5.3 we discussed the conditional probability formula and how it is used. In this section we will discuss a variation of the conditional probability formula known as the multiplication rule.

Consider a large electric company in the northeastern United States. In recent years it has been unable to meet the demand for electricity. To prevent any cable damage and blackouts as a result of overload, that is, too much electrical demand, it has installed two special switching devices to automatically shut off the flow of electricity and thus prevent cable damage when an overload occurs. The probability that the first switch will not work properly is 0.4, and the probability that the second switch will not work properly, given that the first switch fails is 0.3. What is the probability that both switches will fail?

Let us look at Formula 5.3 in Section 5.3. It says that for any events *A* and *B*

$$p(A \mid B) = \frac{p(A \text{ and } B)}{p(B)}$$

If we multiply both sides of this equation by $p(B)$ we get

$$p(A \mid B) \cdot p(B) = p(A \text{ and } B)$$

This equation is called the **multiplication rule.** We state this formally as follows:

FORMULA 5.4   *Multiplication Rule*

If A and B are any events, then

$$p(A \text{ and } B) = p(A \mid B) \cdot p(B)$$

If we now apply Formula 5.4 to our example we get

p(both switches fail)
  = p(switch 2 fails | switch 1 has failed) · p(switch 1 fails)
  = (0.3)(0.4)
  = 0.12

Thus, the probability that both switches fail is 0.12.

● **Example 1**

In a certain community the probability that a man over 40 years old is over-weight is 0.42. The probability that his blood pressure is high given that he is overweight is 0.67. If a man over 40 years of age is selected at random, what is the probability that he is overweight and that he has high blood pressure?

Solution

We use Formula 5.4. We have

p(overweight and high blood pressure)
  = p(high blood pressure | overweight) · p(overweight)
  = (0.67)(0.42)
  = 0.2814

Thus, the probability that a man over 40 is overweight and has high blood pressure is approximately 0.28.

● **Example 2**   *TV Commercials*

A new cleansing product has recently been introduced and is being advertised on television as having remarkable cleansing qualities. The manufacturer be-lieves that if a homemaker is selected at random, the probability that the homemaker watches television and sees the commercial between the hours of 12 noon and 4 P.M. is $\frac{4}{11}$. Furthermore, if the homemaker sees the commercial, then the probability that the homemaker buys the cleanser is $\frac{22}{36}$. What is the probability that a homemaker selected at random will watch television *and* buy the product?

Solution
We use Formula 5.4. We have

p(watches TV and buys product)
    $= p$(buys product | watches TV) $\cdot$ p(watches TV)

$$= \frac{22}{36} \cdot \frac{4}{11}$$

$$= \frac{88}{396} = \frac{2}{9}$$

Thus, the probability that a homemaker selected at random watches television and buys the cleanser is $\frac{2}{9}$.

Independent Events

In many cases it turns out that whether or not one event happens does not affect whether another will happen. For example, if two cards are drawn from a deck and the first card is replaced before the second card is drawn, the outcome on the first draw has nothing to do with the outcome on the second draw. Also, if two dice are rolled, the outcome for one die has nothing to do with the outcome for the second die. Such events are called **independent events.**

DEFINITION 5.2

*Two events A and B are said to be **independent** if the likelihood of the occurrence of event B is in no way affected by the occurrence or non-occurrence of event A.*

When dealing with the independent events, we can simplify Formula 5.4. The following example will show this.

● **Example 3**
Two cards are drawn from a deck of 52 cards. Find the probability that both cards drawn are aces if the first card

a. is *not* replaced before the second card is drawn. NOT INDEPENDENT
b. is replaced before the second card is drawn. INDEPENDENT

Solution
a. Since the first card is not replaced, we use the multiplication rule. We have

p(both cards are aces)
    $= p$(2nd card is ace | 1st card is ace) $\cdot$ p(1st card is ace).

Notice that since the first card is not replaced, there are only 3 aces remaining out of a possible 51 cards. This is because the first card removed was an ace. Thus

$$p(\text{both cards are aces}) = \frac{3}{51} \cdot \frac{4}{52}$$

$$= \frac{12}{2652} = \frac{1}{221}$$

Thus, the probability that both cards are aces is $\frac{1}{221}$.

b. Since the first card is replaced before the second card is drawn, then whether or not an ace appeared on the first card in no way affects what happens on the second draw. The events "ace on second draw" and "ace on first draw" are independent. Thus $p(\text{2nd card is ace} \mid \text{1st card is ace})$ is exactly the same as $p(\text{2nd card is ace})$. Therefore

$p(\text{both cards are aces})$
    $= p(\text{2nd card is ace} \mid \text{1st card is ace}) \cdot p(\text{1st card is ace})$
    $= p(\text{2nd card is ace}) \cdot p(\text{1st card is ace})$

$$= \frac{4}{52} \cdot \frac{4}{52}$$

$$= \frac{16}{2704} = \frac{1}{169}.$$

Hence the probability that both cards are aces in this case is $\frac{1}{169}$.

Example 3 suggests that if two events A and B are independent, we can substitute $p(B)$ for $p(B \mid A)$ since B is in no way affected by what happens with A. We then get a special multiplication rule for independent events.

FORMULA 5.5

If A and B are independent events, then

$p(A \text{ and } B) = p(A) \cdot p(B)$

● **Example 4**
Two randomly selected travelers, Carlos and Pedro, who do not know each other, are at the information desk at Kennedy International Airport. The probability that Carlos speaks Spanish is 0.86 and the probability that Pedro speaks Spanish is 0.73. What is the probability that they both speak Spanish?

Solution

Since both travelers do not know each other, the events "Carlos speaks Spanish" and "Pedro speaks Spanish" are independent. We therefore use Formula 5.5. We have

$$p(\text{both speak Spanish}) = p(\text{Carlos speaks Spanish}) \cdot p(\text{Pedro speaks Spanish})$$
$$= (0.86)(0.73)$$
$$= 0.6278$$

Thus the probability that they both speak Spanish is approximately 0.63.

● **Example 5**

If the probability of a skin diver in a certain community having untreated diabetes is 0.15, what is the probability that two totally unrelated skin divers from the community do *not* have untreated diabetes? (Assume independence.)

Solution

These are independent events, so we use Formula 5.5. The probability of a skin diver having untreated diabetes is 0.15. Thus the probability that the diver does not have untreated diabetes is $1 - 0.15$, or 0.85. Therefore,

$$p(\text{both divers do not have diabetes}) = p(\text{diver 1 does not have diabetes})$$
$$\cdot p(\text{diver 2 does not have diabetes})$$
$$= (0.85)(0.85)$$
$$= 0.7225$$

Hence, the probability that neither of two totally unrelated skin divers have untreated diabetes is approximately 0.72.

COMMENT The multiplication rule for independent events can be generalized for more than two independent events. We simply multiply all the respective probabilities. Thus if event A has probability 0.7 of occurring, event B has probability 0.6 of occurring, and event C has probability 0.5 of occurring, and if these events are independent, then the probability that all three occur is

$$p(A \text{ and } B \text{ and } C) = p(A) \cdot p(B) \cdot p(C)$$
$$= (0.7)(0.6)(0.5)$$
$$= 0.21$$

Therefore, the probability that all three occur is 0.21.

EXERCISES

1. *Uninsured Motorist* Due to the rising costs of auto insurance in a certain city the probability that a randomly selected driver is driving an uninsured motor vehicle is 0.12. Moreover, the probability that a driver is under 30 years of age given that the car is uninsured is 0.38. If a driver is randomly selected, find the probability that the driver is under 30 years of age *and* that the car driven is uninsured.

Statistical Probability of Divorce for Persons Born 1945–1949 Correlated with Years of Education

Years of School Completed	Percent Whose First Marriage Had Ended in Divorce by 1975		Percent Whose First Marriage May Eventually End in Divorce	
	Men	Women	Men	Women
0 to 11 *H.S. drop*	15	24	34	44
12 *HS*	15	17	36	37
13 to 15 *College drop*	15	19	42	49
16 *Coll grad*	8	8	29	29
17 or more *masters*	8	9	30	33
All ever-married persons born 1945–1949	13	17	34	38

Source: U.S. Bureau of the Census, 1975.

2. Consider the table above. Rosalita Juarez was born in 1947 and holds a master's degree from Princeton University. Find the probability that her marriage will *not end* in divorce.

3. According to government statistics, the probability that a randomly selected individual in Dover has his or her annual federal income tax form prepared by a professional tax consultant is 0.72. Find the probability that three totally unrelated people have their taxes prepared by a professional tax consultant. $p(A) = (.72)(.72)(.72) = .36288$

4. The probability that a male employee of the Calvington Corp. smokes is 0.82. The probability that a female employee of the Calvington Corp. smokes is 0.63. Find the probability that two randomly selected and totally unrelated employees (one male and one female) both *do not* smoke.

5. The Casco Corp. uses many different delivery services. The probability that any given parcel will be sent with the ABC Speedy Delivery Service is 0.71. The probability that the parcel will arrive on time given that the ABC Speedy Delivery Service was used is 0.93. If a parcel is randomly selected, find the probability that it will be sent with the ABC Speedy Delivery Service *and* that it will arrive on time. $p(A \text{ and } B) = .71 \cdot .93$ $.6603$

6. Medical records indicate that the probability that a man over 40 years of age in a certain mining community has black lung disease is 0.37. The probability that the individual has high blood pressure given that the individual has black lung disease is 0.23. If an individual in this mining community over 40 years of age is randomly selected, find the probability that the individual has high blood pressure *and* black lung disease.

7. *Tuition Increase* The board of trustees of a college is considering increasing tuition for the coming school year by 9%. The probability that they vote to raise tuition is 0.82. Past experience has shown that whenever tuition is increased by 9%, student enrollment drops by approximately 5% with a probability of 0.95. Find the probability that the board votes to increase tuition by 9% *and* that the enrollment drops by 5%.

8. *Joyriding* Jake is considering taking his old jalopy out for a drive into the countryside this weekend. The probability that he takes the car out is 0.72. The probability that the car breaks down given that he takes it out for a drive is 0.12. Find the probability that Jake takes the car out *and* that it breaks down.

9. A shipment of 100 CB radio sets is received by an electronics store. It is known that 5 of the sets are defective. Two sets are randomly selected and inspected. Find the probability that both sets are defective under the following conditions.
 a. The first set is not replaced before the second is selected.
 b. The first set is replaced before the second set is selected.

10. *Strike* The workers of the Babet Dairy Corporation are on strike, demanding a large pay raise and a cost of living adjustment clause in their contract (see the accompanying newspaper clipping). There is a 0.98 probability that the cost of milk will go up at least 2¢ a quart if the workers' demands are met. Furthermore, there is a 0.19 probability that their demands will be met. Find the probability that their demands are met *and* that the price of milk goes up at least 2¢ a quart.

> ## MILK TRUCK DRIVERS STILL ON STRIKE
>
> *New York (April 11)* – The strike by the milk drivers enters its third week today with no end in sight. State mediators have been unable to get both sides together to negotiate on a new contract. The union is demanding a 30% raise spread out over two years and a cost of living adjustment clause. Management considers such demands irresponsible and refuses to negotiate with the union.

Reprinted by permission.

11. A recent survey in Alexville showed that 86% of the population have some form of health insurance. Furthermore, the survey showed that 21% of the people who have health insurance also have some form of life insurance. If an adult resident of Alexville is randomly selected, find the probability that the individual has both health and life insurance.

12. A leading department store has had bad experiences with customers who pay by check. The probability that a customer will pay for some purchases with a check that "bounces" is 0.35. Furthermore, the probability that a customer will pay by check is 0.71. Bob is about to pay for some purchases. What is the probability that Bob pays for them with a check that will *not* bounce?

13. The Jackson Company is planning to hire 12 new engineers this month. Company personnel have interviewed many candidates. The probability that Doris is hired is 0.59, the probability that Stephanie is hired is 0.47, and the probability that Morgan is hired is 0.78. What is the probability that all three are hired? (Assume independence.)

14. Refer back to Exercise 13. What is the probability that Stephanie and Doris will be hired and that Morgan is not hired?

15. *Nuclear Safeguards*   A particular nuclear reactor has two safety valves that are designed to prevent a nuclear accident similar to the one that occurred at the Three Mile Island facility in Pennsylvania. The probability that the first safety valve will function satisfactorily in an emergency is 0.98, and the probability that the second backup valve will function properly <u>if the first fails</u> is 0.99. Find the probability that both valves fail and that a nuclear mishap occurs.

Not independent

*5.5 BAYES' FORMULA

Let us look in on Dr. Carey, who has two bottles of sample pills on his desk for the treatment of arthritic pain. One day he gives Madeline a few pills from one of the bottles. (All other treatments have failed.) However, he does not remember from which bottle he took the pills. The pills in bottle B_1 are effective 70% of the time, with no known side effects. The pills in bottle B_2 are effective 90% of the time, with some possible side effects. Bottle B_1 is closer to Dr. Carey on his desk and the probability is $\frac{2}{3}$ that he selected the pills from this bottle. On the other hand, bottle B_2 is farther away from Dr. Carey and the probability is therefore $\frac{1}{3}$ that he selected the pills from this bottle. The problem is to determine the bottle from which the pills were taken.

In many problems we are given situations such as this one, where we know the outcome of the experiment and are interested in concluding that the outcome happened because of a particular event. Figure 5.1 is an example of this. For these situations, we need a formula.

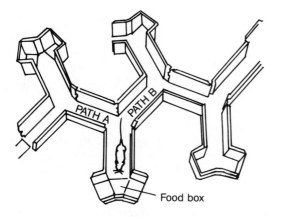

Food box

FIGURE 5.1

Consider the situation pictured at the left. Often we read about psychologists experimenting with rats to determine how quickly they learn maze patterns. Mazes used to study human learning are similar, in principle, to those used with animals. We notice that the rat in the picture has already arrived at the food box. What is the probability that it came there from Path A, and not from Path B? By using Bayes' formula we will be able to answer this and similar such questions.

*An asterisk indicates that the section requires more time and thought than other sections.

- **Example 1** *Medicine*
In the situation we are discussing, find the probability that the pills are effective in relieving Madeline's pain.

Solution
Madeline can be relieved of her pain by taking the pills from either bottle B_1 or bottle B_2. Let A represent the event *Madeline's pain is relieved*, let B_1 represent the event that *the pills were taken from bottle* B_1, and let B_2 represent the event that *the pills were taken from bottle* B_2. The accompanying tree diagram illustrates how Madeline's pain can be relieved. From the given information

$$\text{Prob}(A \mid B_1) = 0.70 \quad \text{and} \quad \text{Prob}(B_1) = \frac{2}{3}$$

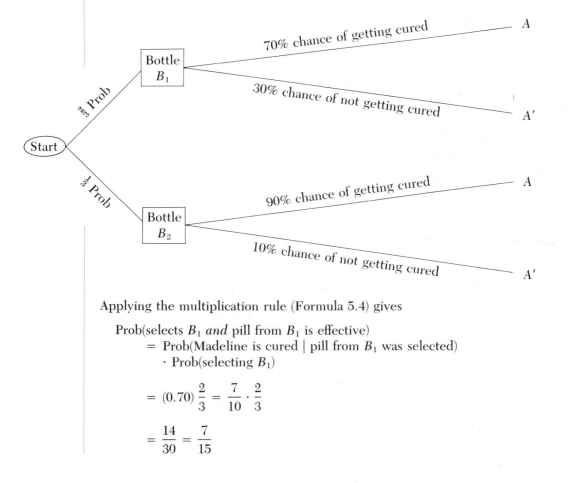

Applying the multiplication rule (Formula 5.4) gives

Prob(selects B_1 *and* pill from B_1 is effective)
$\quad = \text{Prob}(\text{Madeline is cured} \mid \text{pill from } B_1 \text{ was selected})$
$\quad\quad \cdot \text{Prob}(\text{selecting } B_1)$

$$= (0.70)\frac{2}{3} = \frac{7}{10} \cdot \frac{2}{3}$$

$$= \frac{14}{30} = \frac{7}{15}$$

Symbolically,

$$\text{Prob}(A \text{ and } B_1) = \text{Prob}(A \mid B_1) \cdot \text{Prob}(B_1) = \frac{7}{15}$$

Also, from the given information,

$$\text{Prob}(A \mid B_2) = \frac{9}{10} \quad \text{and} \quad \text{Prob}(B_2) = \frac{1}{3},$$

so that

$$\text{Prob}(A \text{ and } B_2) = \text{Prob}(A \mid B_2) \cdot \text{Prob}(B_2)$$

$$= \frac{9}{10} \cdot \frac{1}{3} = \frac{3}{10}$$

Madeline can be cured in one of two mutually exclusive ways:

1. Bottle B_1 is selected and a pill from it is effective.
2. Bottle B_2 is selected and a pill from it is effective.

Thus, using the addition rule for mutually exclusive events (Formula 5.1), we have

Prob(pills are effective) = Prob(selects B_1 and pills are effective)
$$+ \text{Prob(selects } B_2 \text{ and pills are effective)}$$

Symbolically,

$$\text{Prob(pills are effective)} = \text{Prob}(A \text{ and } B_1) + \text{Prob}(A \text{ and } B_2)$$

$$\text{Prob}(A) = \text{Prob}(A \mid B_1) \cdot \text{Prob}(B_1) + \text{Prob}(A \mid B_2) \cdot \text{Prob}(B_2)$$

$$= \frac{7}{15} + \frac{3}{10}$$

$$= \frac{14}{30} + \frac{9}{30} = \frac{23}{30}$$

Therefore, the probability that the pills are effective in relieving Madeline of her pain is $\frac{23}{30}$.

- **Example 2**
  Two weeks later Madeline returns to Dr. Carey and reports that the pills were extremely effective. Dr. Carey would now like to recommend the same medicine for his other patients who suffer from the same pain. What is the probability that the pills came from

  a. Bottle B_1?
  b. Bottle B_2?

Solution

a. Using the conditional probability formula (Formula 5.3), we have

Prob(pills came from B_1 | pills are effective)

$$= \frac{\text{Prob(selects } B_1 \text{ and pills are effective)}}{\text{Prob(pills are effective)}}$$

Symbolically,

$$\text{Prob}(B_1 \mid A) = \frac{\text{Prob}(A \text{ and } B_1)}{\text{Prob}(A)}$$

Substituting the values obtained in Example 1 gives

$$\text{Prob}(B_1 \mid A) = \frac{7/15}{23/30} = \frac{7}{15} \div \frac{23}{30}$$

$$= \frac{7}{15} \cdot \frac{30}{23} = \frac{14}{23}$$

Thus, the probability that the pills came from bottle B_1 is $\dfrac{14}{23}$.

b. To find the probability that the pills came from bottle B_2, we use a procedure similar to the one used in part (a). We have

Prob(pills came from B_2 | pills are effective)

$$= \frac{\text{Prob(selects } B_2 \text{ and pills are effective)}}{\text{Prob(pills are effective)}}$$

In symbols,

$$\text{Prob}(B_2 \mid A) = \frac{\text{Prob}(A \text{ and } B_2)}{\text{Prob}(A)}$$

$$= \frac{3/10}{23/30} = \frac{3}{10} \div \frac{23}{30}$$

$$= \frac{3}{10} \cdot \frac{30}{23} = \frac{9}{23}$$

Thus, the probability that the pills came from bottle B_2 is $\dfrac{9}{23}$.

We can combine the results of the preceding two examples as follows:

$$\text{Prob}(B_1 \mid A) = \frac{\text{Prob}(A \text{ and } B_1)}{\text{Prob}(A)}$$

$$= \frac{\text{Prob}(A \text{ and } B_1)}{\text{Prob}(A \text{ and } B_1) + \text{Prob}(A \text{ and } B_2)}$$

$$= \frac{\text{Prob}(A \mid B_1) \cdot \text{Prob}(B_1)}{\text{Prob}(A \mid B_1) \cdot \text{Prob}(B_1) + \text{Prob}(A \mid B_2) \cdot \text{Prob}(B_2)}$$

and

$$\text{Prob}(B_2 \mid A) = \frac{\text{Prob}(A \text{ and } B_2)}{\text{Prob}(A)}$$

$$= \frac{\text{Prob}(A \text{ and } B_2)}{\text{Prob}(A \text{ and } B_1) + \text{Prob}(A \text{ and } B_2)}$$

$$= \frac{\text{Prob}(A \mid B_2) \cdot \text{Prob}(B_2)}{\text{Prob}(A \mid B_1) \cdot \text{Prob}(B_1) + \text{Prob}(A \mid B_2) \cdot \text{Prob}(B_2)}.$$

When these results are generalized we have Bayes' rule.

FORMULA 5.5   *Bayes' Rule*

Consider a sample space that is composed of the mutually exclusive events A_1, A_2, A_3, . . . , A_n. Suppose each event has a nonzero probability of occurring and that one must definitely occur. If B is any event in the sample space, then

$$Prob(A_1 \mid B) = \frac{Prob(B \mid A_1) \cdot Prob(A_1)}{Prob(B \mid A_1) \cdot Prob(A_1) + Prob(B \mid A_2) \cdot Prob(A_2) + \cdots + Prob(B \mid A_n) \cdot Prob(A_n)}$$

$$Prob(A_2 \mid B) = \frac{Prob(B \mid A_2) \cdot Prob(A_2)}{Prob(B \mid A_1) \cdot Prob(A_1) + Prob(B \mid A_2) \cdot Prob(A_2) + \cdots + Prob(B \mid A_n) \cdot Prob(A_n)}$$

$$\vdots \qquad\qquad\qquad \vdots$$

$$Prob(A_n \mid B) = \frac{Prob(B \mid A_n) \cdot Prob(A_n)}{Prob(B \mid A_1) \cdot Prob(A_1) + Prob(B \mid A_2) \cdot Prob(A_2) + \cdots + Prob(B \mid A_n) \cdot Prob(A_n)}$$

Bayes' rule may seem rather complicated but it is easy to use, as the following examples illustrate.

● **Example 3**

If $p(A \mid B) = \frac{1}{5}$, $p(A \mid C) = \frac{2}{7}$, $p(B) = \frac{1}{2}$, and $p(C) = \frac{1}{2}$, find

a. $p(B \mid A)$.
b. $p(C \mid A)$.

Solution
a. Using Bayes' rule, we have

$$p(B \mid A) = \frac{p(A \mid B) \cdot p(B)}{p(A \mid B) \cdot p(B) + p(A \mid C) \cdot p(C)}$$

$$= \frac{\left(\frac{1}{5}\right)\left(\frac{1}{2}\right)}{\left(\frac{1}{5}\right)\left(\frac{1}{2}\right) + \left(\frac{2}{7}\right)\left(\frac{1}{2}\right)}$$

$$= \frac{\dfrac{1}{10}}{\left(\dfrac{1}{10}\right) + \left(\dfrac{1}{7}\right)}$$

$$= \frac{1/10}{17/70} = \frac{1}{10} \div \frac{17}{70}$$

$$= \frac{1}{10} \cdot \frac{70}{17} = \frac{7}{17}$$

Thus, $p(B \mid A) = \dfrac{7}{17}$.

b. Again we use Bayes' formula. We have

$$p(C \mid A) = \frac{p(A \mid C) \cdot p(C)}{p(A \mid B) \cdot p(B) + p(A \mid C) \cdot p(C)}$$

$$= \frac{\left(\frac{2}{7}\right)\left(\frac{1}{2}\right)}{\left(\frac{1}{5}\right)\left(\frac{1}{2}\right) + \left(\frac{2}{7}\right)\left(\frac{1}{2}\right)} = \frac{\dfrac{1}{7}}{\left(\dfrac{1}{10}\right) + \left(\dfrac{1}{7}\right)} = \frac{\dfrac{1}{7}}{\dfrac{17}{70}}$$

$$= \frac{10}{17}$$

Thus $p(C \mid A) = \dfrac{10}{17}$.

● **Example 4**

A prisoner has just escaped from jail. There are three roads leading away from the jail. If the prisoner selects road A to make good her escape, the probability that she succeeds is $\dfrac{1}{4}$. If she selects road B, the probability that she succeeds is $\dfrac{1}{5}$. If she selects road C, the probability that she succeeds is $\dfrac{1}{6}$. Furthermore, the probability that she selects each of these roads is the same. It is $\dfrac{1}{3}$. If the prisoner succeeds in her escape, what is the probability that she made good her escape by using road B?

Solution
We use Bayes' rule. We have

$$(p \text{ used road B } | \text{ succeeded}) = \frac{p(\text{succeeds } | \text{ uses road B}) \cdot p(\text{uses road B})}{p(\text{succeeds } | \text{ uses road A}) \cdot p(\text{uses road A}) + p(\text{succeeds } | \text{ uses road B}) \cdot p \text{ uses road B} \\ + p(\text{succeeds } | \text{ uses road C}) \cdot p(\text{uses road C})}$$

$$= \frac{\left(\dfrac{1}{5}\right)\left(\dfrac{1}{3}\right)}{\left(\dfrac{1}{4}\right)\left(\dfrac{1}{3}\right) + \left(\dfrac{1}{5}\right)\left(\dfrac{1}{3}\right) + \left(\dfrac{1}{6}\right)\left(\dfrac{1}{3}\right)}$$

$$= \frac{\dfrac{1}{15}}{\left(\dfrac{1}{12}\right) + \left(\dfrac{1}{15}\right) + \left(\dfrac{1}{18}\right)}$$

$$= \frac{12}{37}$$

Thus, the probability that she made good her escape by using road B is $\dfrac{12}{37}$.

● **Example 5**   *Smoke Detectors*

A large real estate manager purchased 50,000 smoke detectors to comply with new city ordinances. She purchased 25,000 from company A, 15,000 from company B, and 10,000 from company C. It is known that some of the smoke detectors malfunction and go off spontaneously. It is also known that 4% of the detectors produced by company A are defective, 5% of the detectors produced by company B are defective, and 6% of the detectors produced by company C are defective. A call is received by the management office that one of the detectors is malfunctioning. Find the probability that it was produced by company A.

Solution
From the given information,

$$\text{Prob(produced by company A)} = \frac{25,000}{50,000} = 0.5$$

$$\text{Prob(produced by company B)} = \frac{15,000}{50,000} = 0.3$$

$$\text{Prob(produced by company C)} = \frac{10,000}{50,000} = 0.2$$

$$\text{Prob(defective } | \text{ produced by company A)} = 0.04$$

Prob(defective | produced by company B) = 0.05

Prob(defective | produced by company C) = 0.06.

Here we must determine Prob(produced by company A | defective). We apply Bayes' rule:

Prob(produced by A | defective)

$$
= \frac{\text{Prob(defective | prod. by A)} \cdot \text{Prob(prod. by A)}}{\text{Prob(def. | prod. by A)} \cdot \text{Prob(prod. by A)} + \text{Prob(def. | prod. by B)} \cdot \text{Prob(prod. by B)} + \text{Prob(def. | prod. by C)} \cdot \text{Prob(prod. by C)}}
$$

$$
= \frac{(0.04)(0.5)}{(0.04)(0.5) + (0.05)(0.3) + (0.06)(0.2)}
$$

$$
= \frac{0.02}{0.02 + 0.015 + 0.012} = \frac{0.02}{0.047}
$$

$$
= \frac{20}{47}
$$

Thus, the probability that the defective smoke detector was produced by company A is $\frac{20}{47}$.

- ● **Example 6**

  There are four photocopying machines I, II, III, and IV on the third floor of a large office building. The probabilities that the copies produced from each of these machines will be blurred are 0.2, 0.5, 0.3, and 0.1, respectively. Furthermore, because of the location of these machines, management estimates that the probabilities that a worker will use any one of these machines are 0.6, 0.2, 0.1, and 0.1 respectively. The president of the company receives a blurred memo that was photocopied on one of these machines. What is the probability that it was photocopied on machine I?

  Solution

  We use Bayes' formula. Let A represent the event "copy is blurred," B_1, represent the event "Machine I is used", B_2 represent the event "Machine II is used", B_3 represent the event "Machine III is used" and B_4 represent the event "Machine IV is used". Then we are interested in $p(B_1 \mid A)$. Using Bayes' rule we have

$$
p(B_1 \mid A) = \frac{p(A \mid B_1) \cdot p(B_1)}{p(A \mid B_1) \cdot p(B_1) + p(A \mid B_2) \cdot p(B_2) + p(A \mid B_3) \cdot p(B_3) + p(A \mid B_4) \cdot p(B_4)}
$$

$$
= \frac{(0.6)(0.2)}{(0.6)(0.2) + (0.2)(0.5) + (0.1)(0.3) + (0.1)(0.1)}
$$

$$
= \frac{0.12}{0.26} = \frac{12}{26} = \frac{6}{13}
$$

Thus the probability that the blurred memo was photocopied on Machine I is $\dfrac{6}{13}$.

EXERCISES

1. *Tracking an Escaped Prisoner* A prisoner has just escaped from jail. There are three roads leading away from the jail. If the prisoner selects road A to make good his escape, the probability that he succeeds is $\frac{1}{8}$. If he selects road B, the probability is $\frac{1}{9}$. If he selects road C, the probability that he succeeds is $\frac{1}{10}$. Furthermore, the probability that he selects each of these roads is the same; it is $\frac{1}{3}$. If the prisoner succeeds in his escape, what is the probability that he made good his escape by using road C?

2. *Medicine* The American Cancer Society as well as the medical profession recommends that people have themselves checked annually for any cancerous growths. If a person has cancer, then the probability is 0.99 that it will be detected by a test. Furthermore, the probability that the test results will be positive (meaning that cancer is possible) when no cancer actually exists is 0.10. Government records indicate that 8% of the population in the vicinity of a chemical corporation that produces asbestos have some form of cancer. Donald Williams takes the test and the results are positive. What is the probability that he does *not* have cancer?

3. A large supermarket received two milk deliveries yesterday. In the morning shipment, 30 of the 1000 containers delivered were spoiled. In the afternoon shipment, 40 of the 600 containers delivered were spoiled. All the milk that is delivered is immediately placed on the shelves in a random manner. Before any other customer purchases any milk from either delivery, Ginette purchases a container of milk from the supermarket and discovers that the milk is sour. What is the probability that Ginette's sour milk came from the morning shipment?

4. A doctor receives 10,000 hypodermic needles a year: 3,000 from company A; 4,000 from company B; 2,000 from company C; and 1,000 from company D. Based on past experience, the doctor knows that 7% of company A's needles are defective, 6% of company B's needles are defective, 8% of company C's needles are defective, and 9% of company D's needles are defective. If a defective needle is found, what is the probability that it was produced by company A?

5. A speech clinic is experimenting with three different approaches, A, B, and C, to cure a particular speech defect. Thus far, method A has been successful with 10 out of 15 patients, method B has been effective with 12 out of 18 patients, and method C has been successful with 8 out of 15 patients. If a little boy who had a speech defect was cured, what is the probability that treatment A was used? (Assume that each method is equally likely to be selected to treat a patient.)

6. A gambler has two dice in his hand. One of them is an ordinary die and the other die has three dots painted on each of its six faces. One of the dice is selected at random and

rolled twice. Both times the face with three dots on it comes up. What is the probability that the altered die was used?

7. An author who lives on the East Coast must send a manuscript to a publisher on the West Coast. The manuscript can be sent by any one of four delivery services, A, B, C, or D. The probabilities that the manuscript will arrive at its destination safely by 9:00 A.M. on the next business day when shipped by these four delivery services are 0.8, 0.9, 0.7, and 0.85, respectively. The author randomly selects a delivery service. If the manuscript *does not* arrive safely at its destination by 9:00 A.M. on the next business day, what is the probability that delivery service B was used?

8. In Exercise 7, if the manuscript does arrive safely, find the probability that delivery service D was used.

9. A psychologist is conducting experiments with rats. A rat can enter maze A, maze B, or maze C with probabilities 0.5, 0.4, and 0.1, respectively. Furthermore, the probability that the rat finds its way through each maze is 0.7, 0.2, and 0.1, respectively. If a rat actually finds its way through a maze, what is the probability that it went through maze B?

10. *Ecology* Environmentalists have developed a test for determining when the mercury level in fish is above the permissible levels. If the fish actually contain an excessive amount of mercury, then the test is 98% effective in determining this, and only 2% will escape undetected. On the other hand, if the mercury content is within permissible limits, then the test will correctly indicate this 95% of the time. Only 5% of the time will the test incorrectly indicate that the mercury content is not within permissible limits. The test is to be used on fish from a river into which a chemical company has been dumping its wastes. It is estimated that 20% of the fish in the river contain excessive amounts of mercury. A fish is caught and tested by this procedure. The test indicates that the mercury level is within permissible limits. Find the probability that the mercury content is actually greater than the permissible level.

5.6 SUMMARY

In this chapter we discussed many different rules concerning the calculation of probabilities. Each formula given applies to different situations. Thus the addition rule allows us to determine the probability of event *A* or event *B* or both events happening. We distinguished between mutually exclusive and non-mutually exclusive events and their effect on the addition rule.

We then discussed conditional probability and how the probability of one event is affected by the occurrence or non-occurrence of a second event. This led us to the multiplication rule. When one event is in no way affected by the occurrence or non-occurrence of a second event, we have independent events and a simplified multiplication rule.

Finally we discussed Bayes' formula, which is used when we know the outcome of some experiment and are interested in determining the probability that it was caused by or is the result of some other event. In each case many applications of all the formulas introduced were given.

STUDY GUIDE

You should now be able to demonstrate your knowledge of the following ideas presented in this chapter by giving definitions, descriptions, or specific examples. Page references are given for each term so that you can check your answer.

Mutually exclusive events (page 212)
Complement of an event (page 215)
Addition rule (page 216)
Conditional probability (page 223)
Conditional probability formula (page 227)
Multiplication rule (page 232)
Independent events (page 233)
Bayes' rule (page 242)

FORMULAS TO REMEMBER

You should be able to identify each symbol in the following formulas, understand the relationships among the symbols expressed in each formula, understand the significance of each formula, and use the formulas in solving problems.

1. Addition rule, for mutually exclusive events:

$$p(A \text{ or } B) = p(A) + p(B)$$

2. Addition rule, general case: $p(A \text{ or } B) = p(A) + p(B) - p(A \text{ and } B)$

3. Addition rule, for three events: $p(A \text{ or } B \text{ or } C) = p(A) + p(B) + p(C) - p(A \text{ and } B) - p(A \text{ and } C) - p(B \text{ and } C) + p(A \text{ and } B \text{ and } C)$

4. Complement of event A: $p(A') = 1 - p(A)$

5. Conditional probability formula: $p(A \mid B) = \dfrac{p(A \text{ and } B)}{p(B)}$

6. Multiplication rule, general: $p(A \text{ and } B) = p(A \mid B) \cdot p(B)$

7. Multiplication rule, for independent events: $p(A \text{ and } B) = p(A) \cdot p(B)$

8. Bayes' rule:

$$p(A_i \mid B) = \frac{p(B \mid A_i) \cdot p(A_i)}{p(B \mid A_1) \cdot p(A_1) + p(B \mid A_2) \cdot p(A_2) + \cdots + p(B \mid A_n) \cdot p(A_n)}$$

The following tests will be more useful if you take them after you have studied the examples and solved the exercises given in this chapter.

1. Assume that we have three mutually exclusive events which partition a sample space. If $p(A \mid B) = \frac{1}{7}$, $p(A \mid C) = \frac{1}{8}$, $p(A \mid D) = \frac{1}{9}$, $p(B) = \frac{1}{5}$, $p(C) = \frac{2}{5}$, and $p(D) = \frac{2}{5}$, find

   a. $p(B \mid A)$ **b.** $p(C \mid A)$ **c.** $p(D \mid A)$.

2. A card is selected from a deck of 52 cards and observed to be a picture card. Find the probability that it is the king of clubs.

3. One hundred pints of blood were collected by the American Red Cross at the Crossview Shopping Center during a recent blood donor drive. The blood groups were classified as follows.

	Type A	Type B	Type AB	Type O
Rh positive	35	8	4	39
Rh negative	5	2	1	6

   If a pint of collected type O blood is randomly selected, what is the probability that it is Rh negative?

4. Refer to question 3. If a pint of Rh negative blood is randomly selected, what is the probability that it is type O blood?

5. In Calvington, the probability that a man drinks is 0.71. If the probability that a man drinks *and* his wife smokes is 0.56, what is the probability that his wife smokes, given that the man drinks?

6. Insurance company records indicate the following about the driving records of the people of Midville. The probability that both the husband *and* wife drive a car is 0.43, and the probability that only the husband drives is 0.79. If it is known that Mr. Federow drives, find the probability that Mrs. Federow drives also.

7. *Fertilizer* The Salmo Fertilizer Corporation manufactures eight different kinds of fertilizer, each containing different concentrations of nitrogen, phosphorus, and potassium as follows:

4– 4– 2	5–10– 5
4– 4– 1	4– 8– 3
4– 6– 2	4– 2– 2
4– 6– 3	15–30–15

Pete McDonald, who uses equal amounts of these fertilizers during the year, discovers a bag of fertilizer produced by the Salmo Corp. remaining from last summer in his workshed. The label indicating the nitrogen content concentration is missing. If Pete remembers that the first digit was a 4, find the probability that the bag contains a 4–4–2 mixture.

8. If the probability of an adult having untreated hypertension (high blood pressure) is 0.15, find the probability that three totally unrelated people *do not* have untreated hypertension.

9. *Weather Emergency* A hospital administration is arranging emergency plans. The weather bureau has forecast a heavy snowfall with a 0.90 probability. The administration knows that if the snowfall is heavy, then there is an 80% probability that all hospital personnel will not be able to get to the hospital. Find the probability that the snowfall is heavy *and* that all hospital personnel cannot get to the hospital.

10. *Smuggling* According to one U.S. customs agent, there is a 0.13 chance that a smuggler can transport drugs across the border between the United States and Mexico without being apprehended, and a 0.06 chance that drugs coming from South American countries can be smuggled into Florida by sea without the smuggler being apprehended. A drug dealer is planning to smuggle two shipments of drugs into the United States, one by way of Mexico and the other by sea into Florida. Find the probability that both shipments will be confiscated by authorities.

MASTERY TESTS
Form B

1. Leon has volunteered to roast a turkey for the family dinner. The probability that he forgets to use spice and tenderizer is $\frac{1}{3}$. The probability that he forgets to use only spice is $\frac{4}{9}$. If we know that Leon forgot to use spice, what is the probability that he also forgot to use tenderizer?

2. *Drugs on Campus* Twenty percent of all students at Whipple University are dormitory students who have tried drugs at one time or another. Thirty-seven percent of the students are dormitory students. If a dormitory student is selected at random, what is the probability that the student has tried drugs?

3. Mary Smith is reading the classified section of a newspaper in search of a job that is near her home *and* that has a good health plan. The probability that she finds such a job is 0.23. The probability that she finds a job near her home is 0.71. After making several phone calls Mary gets a job with a brokerage firm near her home. What is the probability that the firm has a good health plan?

4. Three people who do not know each other have volunteered to search independently for a lost cat. The probability that the first volunteer finds the cat is 0.38, the probability that the second volunteer finds the cat is 0.52, and the probability that the third volunteer finds the cat is 0.46. Find the probability that none of the volunteers find the cat.

5. Three mountain climbers are independently attempting to scale a high mountain peak. Their probabilities of succeeding are 0.32, 0.21, and 0.19. Find the probability that exactly one of them succeeds.

6. Refer to Question 5. Find the probability that *at least* one of them succeeds.

*7. Three people, who do not know each other, are standing on line at a movie theater waiting for the box office to open. What is the probability that their birthdays (month and day) are different?

Would you believe that we need only 23 people in a crowd to have a 50% probability that at least two of these people would have the same birthday? The probability increases to about 1, almost a certainty, when we have a crowd of 63 people.

*8. Refer back to Question 7. If four people who do not know each other are standing on line, what is the probability that none of them have the same birthday (month and day)? Assume that no one was born on February 29.

9. At Widmark College, the probability that a student will graduate with a master's degree in computer technology is 0.45. The probability that a graduate who has such a degree will get a well-paying job is 0.91. What is the probability that a randomly selected student will graduate with a master's degree in computer technology and will *not* get a well-paying job?

10. A survey of 400 students in a college dormitory revealed the following information on the reading habits of those students.

Magazines Read	Probability
Time	0.47
Newsweek	0.44
Sports Illustrated	0.46
Time and *Newsweek*	0.19
Time and *Sports Illustrated*	0.09
Newsweek and *Sports Illustrated*	0.11

If it is known that a student definitely reads at least one of these magazines, what is the probability that the student reads all three magazines?

CASE STUDY Law

In June of 1964 an elderly woman was mugged in San Pedro, California. In the vicinity of the crime a bearded black man sat waiting in a yellow car. Shortly after the crime was committed, a young white woman, wearing her blonde hair in a ponytail, was seen running from the scene of the crime and getting into the car, which sped off. The police broadcast a description of the suspected muggers. Soon afterwards, a couple fitting the description was arrested and convicted of the crime. Although the evidence in the case was largely circumstantial, the prosecutor based his case on probability and the unlikeliness of another couple having such characteristics. He assumed the following probabilities.

Characteristic	Assumed Probability
Drives yellow car	$\dfrac{1}{10}$
Black-white couple	$\dfrac{1}{1000}$
Black man	$\dfrac{1}{3}$
Man with beard	$\dfrac{1}{10}$
Blonde woman	$\dfrac{1}{4}$
Woman wears her hair in ponytail	$\dfrac{1}{10}$

He then multiplied the individual probabilities:

$$\frac{1}{10} \cdot \frac{1}{1000} \cdot \frac{1}{3} \cdot \frac{1}{10} \cdot \frac{1}{4} \cdot \frac{1}{10} = \frac{1}{12,000,000}$$

He claimed that the probability is $\dfrac{1}{12,000,000}$ that another couple has such characteristics. The jury agreed and convicted the couple. The conviction was overturned by the California Supreme Court in 1968. The defense attorneys got some professional advice on probability. Serious errors were found in the prosecutor's probability calculations. Some of these involved assumptions about independent events. As a matter of fact, it was demonstrated that the probability is 0.41 that another couple with the same characteristics existed in the area once it was known that there was at least one such couple.

For a complete discussion of this probability case, read "Trial by Mathematics," which appeared in *Time*, January 8, 1965, p. 42, and April 26, 1968, p. 41.

SUGGESTED FURTHER READING

1. Freund, J. *Statistics, A First Course*. Englewood Cliffs, NJ: Prentice-Hall, 1970.
2. Glass, G. V., and J. C. Stanley. *Statistical Methods in Education and Psychology*. Englewood Cliffs, NJ: Prentice-Hall, 1970.
3. Guilford, J. P., and B. Fruchter. *Fundamental Statistics in Psychology and Education*. New York: McGraw-Hill, 1973.
4. Hoel, P. G. *Elementary Statistics*. New York: Wiley, 1966.
5. "Mathematics in the Modern World." Readings from *Scientific American*, article 22, San Francisco, CA: Freeman, 1968.
6. Newmark, Joseph. *Using Finite Mathematics*. New York: Harper and Row, 1982.

CHAPTER 6

Some Discrete Probability Distributions and Their Properties

OBJECTIVES

- *To introduce* the concept of a random variable, where the outcome of some experiment is of interest to us.

- *To see* how probability functions assign probabilities to the different values of the random variable.

- *To understand* how the mean and variance of a probability function give us the expected value of a probability function as well as the spread of the distribution.

- *To study* Bernoulli or binomial experiments, which are experiments having only two possible outcomes, success or failure.

- *To discuss* the binomial probability distribution, which gives us the probability of obtaining a specified number of successes when an experiment is performed *n* times.

- *To learn* about several important binomial distribution properties.

- *To analyze* the Poisson probability distribution, which can be applied in very specific cases.

- *To use* the hypergeometric probability distribution when the absence of certain assumptions makes the binomial distribution inappropriate.

Life Expectation by Sex for Some Countries
Average Lifetime (in years)

Country	Male	Female
North America		
Canada	68.7	75.1
United States	64.7	75.2
Mexico	59.4	63.4
Puerto Rico	68.9	75.2
Europe		
Belgium	67.7	73.5
Denmark	70.7	75.9
England and Wales	68.9	75.1
France	68.5	76.1
Italy	67.9	73.4
Norway	71.1	76.8
Poland	66.8	73.8
Sweden	72.0	77.4
Switzerland	69.2	75.0
U.S.S.R	65.0	74.0
Yugoslavia	65.3	70.1
Asia		
India	41.9	40.6
Israel	70.1	72.8
Japan	70.5	75.9
Jordan	52.6	52.0
Korea	59.7	64.1
Africa		
Egypt	51.6	53.8
Nigeria	37.2	36.7

Source: Information Please Yearbook.

Consider the information contained in the above table. How do we calculate the average life expectancy for both males and females? Does life expectancy depend upon the sex of the person and upon the country in which the person lives? Such information has important implications for demographers.

6.1 INTRODUCTION

In Chapter 2 we discussed frequency distributions of sets of data. Using these distributions we were able to analyze data more intelligently to determine which outcomes occurred most often, least often, and so on. In Chapters 4 and 5 we discussed the various rules of probability and how they can be applied to many different situations. These rules enable us to predict how often something will happen in the long run. In this chapter we will combine these ideas.

In any given experiment there may be many different things of interest. For example, if a scientist decides to mate a white rat with a black rat, she may be interested in the number of offspring that are white, black, gray, and so on. We will therefore have to define what is meant by a random variable and we will then discuss its probability function. Special emphasis will be given in this chapter to the binomial random variable and its probability distribution.

6.2 DISCRETE PROBABILITY FUNCTIONS

To understand what is meant by a discrete probability function, let us analyze the following examples.

Valerie is a dentist and keeps accurate records on the number of cavities of each of her patients. Her records indicate that each patient has anywhere from zero to five cavities. Based upon past experience, she has compiled the data given in Table 6.1.

TABLE 6.1
The Number of Cavities and Their Probabilities

Number of Cavities	Probability
0	$\frac{1}{16}$
1	$\frac{4}{16}$
2	$\frac{5}{16}$
3	$\frac{3}{16}$
4	$\frac{2}{16}$
5	$\frac{1}{16}$
	$Total = \frac{16}{16} = 1$

Notice that each patient has anywhere from zero to five cavities. Thus, the number of cavities that each patient has is somehow dependent upon chance, as the probabilities in the table indicate.

Now consider Eric who is an elevator operator. The number of people who enter the elevator at exactly 9:00 A.M. varies from zero to ten people. The capacity of the elevator is ten people. From past experience, Eric has been able to set up the chart shown in Table 6.2.

In each of the two previous examples the values assumed by the item of interest, that is, the number of cavities or the number of people entering the elevator, were whole numbers and were somehow dependent upon chance. We refer to such a quantity as a **discrete random variable.**

The term discrete random variable applies to many different situations. Thus, it may represent the number of people buying tickets to a movie, the number of mistakes made by a secretary in typing a letter, the number of telephone calls received by the school switchboard during the month of September, the number of games that the Green Bay Packers will win next season, or the number of students that will enroll in a particular course, Music 161, to be offered for the first time in the spring.

Basically, if an experiment is performed and some quantitative variable, denoted by x, is measured or observed, then the quantitative variable x is called a **random variable** since the values that x may assume in the given experiment depend on chance. It is a random outcome. Whenever all the possible values that a random variable may assume can be listed (or counted), then the random variable is said to be **discrete.** As opposed to this, the time required to complete a transaction at a bank is a continuous random variable because it could theoretically assume any one of an infinite number of values—namely, any value 0 seconds or more. Random variables that can assume values corresponding to any of the points contained in one or more intervals on a line are called **continuous random variables.**

Other examples of continuous random variables are

1. the time it takes for a drug to take effect.
2. the height (in cm) of a player on a basketball team.
3. the serum blood cholesterol level of a person.
4. the weight of a bag of sugar or of a large jar of coffee.
5. the length of time between births in the maternity ward of a hospital.

In this chapter we will discuss discrete random variables and their probability distributions. In the next chapter, we will discuss continuous random variables and their distributions. *To summarize:* A **random variable** is a *numerical* quantity, the *value* of which is determined by an experiment. In other words, its value is determined by chance.

TABLE 6.2
The Number of People Entering the
Elevator and Their Probabilities

Number of People Entering Elevator	Probability
0	$\dfrac{1}{50}$
1	$\dfrac{3}{50}$
2	$\dfrac{4}{50}$
3	$\dfrac{5}{50}$
4	$\dfrac{7}{50}$
5	$\dfrac{8}{50}$
6	$\dfrac{10}{50}$
7	$\dfrac{6}{50}$
8	$\dfrac{3}{50}$
9	$\dfrac{2}{50}$
10	$\dfrac{1}{50}$
	$Total = \dfrac{50}{50} = 1$

The following examples further illustrate the idea of a random variable.

● **Example 1**

Three cards are selected, without replacement, from a deck of 52 cards. The random variable may be the number of aces obtained. It would then have values of 0, 1, 2, or 3, depending upon the number of aces actually obtained. This is a discrete random variable.

● **Example 2**

Calvin drives his car over some nails. The random variable is the number of flat tires that Calvin gets. The values of the random variable are 0, 1, 2, 3, 4, corresponding to zero flats, one flat, two flats, three flats, and four flats, respectively. This is a discrete random variable.

● **Example 3**

Let the number of people who will attend the next concert at the Hollywood Bowl be the random variable of interest. Then this random variable can assume values ranging from 0 to the seating capacity of the Hollywood Bowl. This is a discrete random variable.

● **Example 4**

A nurse is taking Chuck's blood pressure. Let the random variable be Chuck's systolic blood pressure. What values can the random variable assume? This represents a continuous random variable.

● **Example 5**

Get on a scale and weigh yourself. Let the random variable be your weight. What values can the random variable assume? This represents a continuous random variable.

Let us now return to the two examples discussed at the beginning of this section. You will notice that the probabilities associated with the different values of the random variable are indicated. Thus, Table 6.1 tells us that the probability of having one cavity is $\frac{4}{16}$ and that the probability of having three cavities is $\frac{3}{16}$. Similarly, Table 6.2 tells us that the probability of three people entering the elevator is $\frac{5}{50}$.

When discussing a random variable, we are almost always interested in assigning probabilities to the various values of the random variable. For this reason we now discuss probability functions.

> DEFINITION 6.1
>
> A **probability function** or **probability distribution** is a correspondence that assigns probabilities to the values of a random variable.

● **Example 6**

If a pair of dice is rolled, the random variable that may be of interest to us is the number of dots appearing on both of the dice together. When a pair of dice is rolled, there are 36 possible outcomes. These are shown in Figure 6.1 on the next page.

We can then set up the following chart:

Sum on Both Dice	Number of Different Ways in Which Sum Can Be Obtained		Probability
2	1,1	1	$\dfrac{1}{36}$
3	1,2 ; 2,1	2	$\dfrac{2}{36}$
4		3	$\dfrac{3}{36}$
5		4	$\dfrac{4}{36}$
6		5	$\dfrac{5}{36}$
7		6	$\dfrac{6}{36}$
8		5	$\dfrac{5}{36}$
9		4	$\dfrac{4}{36}$
10		3	$\dfrac{3}{36}$
11		2	$\dfrac{2}{36}$
12		1	$\dfrac{1}{36}$

$$Total = \frac{36}{36} = 1$$

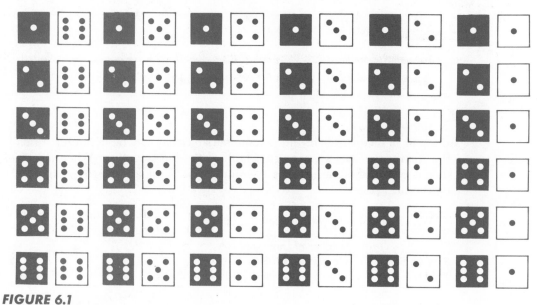

FIGURE 6.1
Thirty-six possible outcomes when a pair of dice is rolled.

- **Example 7**

An airplane has four engines, each of which operates independently of the other. If the random variable is the number of engines that are functioning properly, the random variable has values 0, 1, 2, 3, 4. For a particular airplane we have the following probability function:

Number of Engines Functioning Properly	Probability
0	$\dfrac{1}{19}$
1	$\dfrac{2}{19}$
2	$\dfrac{4}{19}$
3	$\dfrac{5}{19}$
4	$\dfrac{7}{19}$
Total =	$\dfrac{19}{19} = 1$

Notice that in Example 7, as well as in Examples 1 to 6, the sum of all the probabilities is 1. This will be true for any probability function. Also, the probability that the random variable assumes any one particular value is between 0 and 1. Again, this is true in every case. We state this formally as

> **RULE**
>
> a. The sum of all the probabilities of a probability function is always 1.
> b. The probability that a random variable assumes any one value in particular is between 0 and 1 inclusive. Zero means that it can never happen and 1 means that it must happen.

COMMENT   Some authors distinguish between **probability functions** and **probability distributions.** In this book we will use these terms interchangeably.

In all examples mentioned thus far, some of the random variables assumed many values and some took on only two values. The following random variables take on many values:

1. The number of flat tires that Calvin gets when he drives over nails.
2. The sum obtained when rolling a pair of dice. There are 11 possible values, as indicated on page 261.
3. The number of people in an elevator.

Consider the following events:

1. The possible outcome when one coin is flipped (heads or tails).
2. The outcome of either having a cavity or not having a cavity (yes or no).
3. The sex of a newborn child (male or female).
4. The results of an exam (pass or fail only).

The outcomes "head" and "tail" are not the values of random variables because they are not in numerical form. However, if we let "head" equal 0 and "tail" equal 1, the result is a random variable. Remember, a random variable *must* take on numerical values only. We now have a random variable that can take on only two values.

If a random variable has only two possible values, and if the probability of these values remains the same for each trial regardless of what happened on any previous trial, (that is, the trials are independent) the variable is called a **binomial variable** and its probability distribution is called a **binomial distribution.** Actually, this is rather a special case of the binomial distribution. Since the binomial distribution is so important in statistics, we will discuss it in great detail in Section 6.5.

EXERCISES

1. For each of the following situations, find the possible values of the indicated random variable.

Situation	Random Variable of Interest
a. A birthday party that occurs in April	The day of the month that it occurs.
b. An airport runway	The number of airplane takeoffs on a particular day.
c. A telephone switchboard	The number of incoming calls received per day.
d. Crimes in a city	The number of violent crimes committed in the city during a week.
e. Unemployed people	The number of unemployed people in a particular city.
f. A polluted lake	The number of coliform bacteria per cubic centimeter of water in the lake.
g. A polluted lake	The number of inches of acid rain that fell on the lake.

2. Identify the following random variables as discrete or continuous.
   a. The length of time for recovery from a heart bypass operation.
   b. The number of people in Chicago who use a kidney dialysis machine.
   c. The number of programs that will "run" on a particular personal computer.
   d. The number of college credits required by a university for a student to receive an M.B.A. degree.

3. Can the following be a probability distribution for the random variable, number of arrests for robbery? If your answer is no, explain why.

Number of Arrests for Robbery x	Probability
0	0.32
1	0.04
2	0.38
3	0.17
4	0.13
5	0.01

no — because the sum of the probabilities is greater than one

4. Given the following probability distribution for the random variable x, what is the probability that the random variable has a value of 3?

Random Variable x	Probability
1	0.23
2	0.09
3	?
4	0.27
5	0.02
6	0.19

5. A family is known to have three children. Let x be the number of boys in the family. Find the probability distribution of x. (Assume that the probabilities of each sex are equal.)

6. A pair of dice is thrown once. Let x be the number of 4's that shows on either or both of the dice. Find the probability distribution of x.

7. Mr. and Mrs. Jones have four television sets in their house. Let x be the number that are color television sets. Find the probability distribution of x. (Assume that a television set can either be a color set or a black-and-white set with equal likelihood.)

8. A coin box contains a penny, a nickel, a dime, a quarter, a half-dollar, and a dollar piece. The box is shaken and a coin falls out. The coin is returned to the box and the box is shaken again. A coin falls out for a second time. Let x denote the sum of money obtained using both coins. Find the probability distribution of x.

* 9. A swimming team consists of five females and three males. The coach writes the name of each swimmer on a separate piece of paper and places the papers in a hat. The coach then selects two names at random, with the first one replaced before the second is selected. Let x be the number of females selected. Find the probability distribution of x.

*10. The locker room of the swimming team mentioned in the previous exercise contains 7 white towels, 4 pink towels, 5 blue towels and 3 brown towels. Three towels are selected at random, with each towel replaced before the next towel is taken. Let x denote the number of pink towels obtained. Find the probability distribution of x.

6.3 THE MEAN OF A RANDOM VARIABLE

Imagine that the traffic department of a city is considering installing traffic signals at the intersection of Main Street and Broadway. The department's statisticians have kept accurate records over the past year on the number of accidents reported per day at this particularly dangerous intersection. They have submitted the report on the number of accidents per day and their respective probabilities shown in Table 6.3.

Let us multiply each of the possible values for the random variable x, which represents the number of accidents given in Table 6.3, by the respective probabilities. The results of these multiplications are shown in the third column of Table 6.3. If we now add the products we get

$$0 + \left(\frac{1}{32}\right) + \left(\frac{18}{32}\right) + \left(\frac{30}{32}\right) + \left(\frac{32}{32}\right) + \left(\frac{15}{32}\right) = \frac{96}{32} = 3$$

This result is known as the mean of the random variable. It tells us that on the average there are three accidents per day at this dangerous intersection.

Recall that in Chapter 3, when we discussed measures of central tendency and measures of variation, we distinguished between sample statistics and population parameters. Thus, we used the symbols $\bar{x}$, s^2, and s as symbols for

TABLE 6.3
Report on Accidents at Intersection of Main Street
and Broadway

Number of Accidents x	Probability $p(x)$	Product $x \cdot p(x)$
0	$\dfrac{1}{32}$	$0\left(\dfrac{1}{32}\right) = 0$
1	$\dfrac{1}{32}$	$1\left(\dfrac{1}{32}\right) = \dfrac{1}{32}$
2	$\dfrac{9}{32}$	$2\left(\dfrac{9}{32}\right) = \dfrac{18}{32}$
3	$\dfrac{10}{32}$	$3\left(\dfrac{10}{32}\right) = \dfrac{30}{32}$
4	$\dfrac{8}{32}$	$4\left(\dfrac{8}{32}\right) = \dfrac{32}{32}$
5	$\dfrac{3}{32}$	$5\left(\dfrac{3}{32}\right) = \dfrac{15}{32}$

the sample mean, sample variance, and sample standard deviation, respectively. Also we used the symbols μ, σ^2, and σ as symbols for the population mean, population variance, and population standard deviation, respectively. In our case, the mean number of accidents represents the mean of the entire population of observed data. Therefore, we will use the symbol μ to represent the average number of accidents. Usually, when dealing with probability distributions, we work with the entire population. Hence we use μ as the symbol for mean.

Now consider the following. Suppose we intend to flip a coin four times. What is the average number of heads that we can expect to get? To answer this question we will first find the probability distribution. Let x represent the random variable "the number of heads obtained in four flips of the coin." Since the coin is flipped four times, we may get 0, 1, 2, 3, or 4 heads. Thus, the random variable x can have the values 0, 1, 2, 3, or 4. The probabilities associated with each of these values is indicated in Table 6.4. You should verify that the probabilities given in this table are correct.

We now multiply each possible outcome by its probability, the results of which are shown in the third column of Table 6.4. If we now add these products, the result is again called the mean of the probability distribution. In our

Table 6.4
Number of Heads That Can Be Obtained When a
Coin Is Flipped Four Times

Random Variable x	Probability $p(x)$	Product $x \cdot p(x)$
0	$\dfrac{1}{16}$	$0\left(\dfrac{1}{16}\right) = 0$
1	$\dfrac{4}{16}$	$1\left(\dfrac{4}{16}\right) = \dfrac{4}{16}$
2	$\dfrac{6}{16}$	$2\left(\dfrac{6}{16}\right) = \dfrac{12}{16}$
3	$\dfrac{4}{16}$	$3\left(\dfrac{4}{16}\right) = \dfrac{12}{16}$
4	$\dfrac{1}{16}$	$4\left(\dfrac{1}{16}\right) = \dfrac{4}{16}$

case the mean is

$$0 + \left(\frac{4}{16}\right) + \left(\frac{12}{16}\right) + \left(\frac{12}{16}\right) + \left(\frac{4}{16}\right) = \frac{32}{16} = 2$$

This tells us that on the average we can expect to get two heads.

We generalize the results of the previous two examples as follows:

DEFINITION 6.2

*The **mean** of a random variable for a given probability distribution is the number obtained by multiplying all the possible values of the random variable having this particular distribution by their respective probabilities and adding these products together.*

FORMULA 6.1

The mean of a random variable for a given probability distribution is denoted by the Greek letter μ, read as mu. Thus,

$$\mu = \Sigma\, x \cdot p(x)$$

where this summation is taken over all the values that the random variable x can assume and the quantities $p(x)$ are the corresponding probabilities.

COMMENT   Many books refer to the mean of a random variable for a given probability distribution as its **mathematical expectation** or its **expected value.** We will use the word **mean.** The mean is the mathematical expectation that was discussed in Section 4.6 of Chapter 4.

We illustrate the use of Formula 6.1 with several examples.

● **Example 1**
Matthew is a waiter. The following table gives the probabilities that customers will give tips of varying amounts of money:

Amounts of Money (in cents), x	30	35	40	45	50	55	60
Probability, p(x)	0.45	0.25	0.12	0.08	0.05	0.03	0.02

Find the mean for this distribution.

Solution
Applying Formula 6.1 we have

$$\mu = \Sigma \, x \cdot p(x) = 30(0.45) + 35(0.25) + 40(0.12) + 45(0.08)$$
$$+ \, 50(0.05) + 55(0.03) + 60(0.02)$$
$$= 13.5 + 8.75 + 4.8 + 3.6 + 2.5 + 1.65 + 1.2$$
$$= 36$$

The mean is 36. Thus, Matthew can expect to receive an average tip of 36 cents.

● **Example 2**
Rosemary works for the Census Bureau in Washington. For a particular midwestern town, the number of children per family and their respective probabilities is as follows:

Number of Children, x	0	1	2	3	4	5	6
Probability, p(x)	0.07	0.17	0.31	0.27	0.11	0.06	0.01

Find the mean for this distribution.

Solution
Applying Formula 6.1 we have

$$\mu = \Sigma x \cdot p(x) = 0(0.07) + 1(0.17) + 2(0.31) + 3(0.27)$$

$$+ 4(0.11) + 5(0.06) + 6(0.01)$$

$$= 0 + 0.17 + 0.62 + 0.81 + 0.44 + 0.30 + 0.06$$

$$= 2.4$$

The mean is 2.4. How can the average number of children per family be 2.4? Should it not be a whole number such as 2 or 3, not 2.4?

COMMENT   It may seem to you that Formula 6.1 is a new formula for calculating the mean. Actually this is not the case. Recall that in Chapter 3 we defined the mean of a frequency distribution as mean value $= \Sigma \frac{xf}{n}$ or $\Sigma x \frac{f}{n}$. In Chapter 4 we defined the probability of an event as the relative frequency of the event, that is: $p(x) = \frac{f}{n}$. If we substitute this value for $\frac{f}{n}$ in the formula above, we get mean value $= \Sigma x \, p(x)$.

6.4 MEASURING CHANCE VARIATION

Suppose a manufacturer guarantees that a tire will last 20,000 miles under normal driving conditions. If a tire is selected at random and lasts only 12,000 miles, can the difference between what was expected and what actually happened be reasonably attributed to chance, or is there something wrong with the claim?

Similarly, if a coin is flipped 100 times, we would expect the average number of heads to be 50. If a coin was actually flipped 100 times and resulted in only 25 heads, can we conclude that the difference between what was expected and what actually happened is to be attributed to chance, or is it possible that the coin is loaded?

To answer these questions we need some method of measuring the variations of a random variable that are due to chance. Thus, we will discuss the variance and standard deviation of a probability distribution.

You will recall that in Chapter 3 (page 112) we discussed variation of a set of numbers. We now extend this idea to variation of a probability distribution. We let μ represent the mean, $x - \mu$ represent the difference of any number from the mean, and $(x - \mu)^2$ represent the square of the difference. The difference of a number from the mean is called the **deviation from the mean.** We multiply each of the squared deviations from the mean by their respective probabilities. The sum of these products is called the **variance of a random variable with the given probability distribution.** Formally we have

DEFINITION 6.3

*The **variance** of a random variable with a given probability distribution is the number obtained by multiplying each of the squared deviations from the mean by their respective probabilities and adding these products.*

FORMULA 6.2

The variance of a random variable with a given probability distribution is denoted by the Greek letter σ^2 (read as sigma squared). Thus

$$\sigma^2 = \Sigma(x - \mu)^2 \cdot p(x),$$

where this summation is taken over all the values that random variable x can take on. The quantities p(x) are the corresponding probabilities and $(x - \mu)^2$ is the square of the deviations from the mean.

DEFINITION 6.4

*The **standard deviation** of a random variable with a given probability distribution is the square root of the variance of the probability distribution. We denote the standard deviation by the symbol σ (sigma). Thus*

$$\sigma = \sqrt{variance}$$

● **Example 1**

A random variable has the following probability distribution:

x	0	1	2	3	4	5
p(x)	$\frac{7}{24}$	$\frac{5}{24}$	$\frac{1}{8}$	$\frac{1}{8}$	$\frac{1}{12}$	$\frac{1}{6}$

Find the mean, variance, and standard deviation for this distribution.

Solution

We first find μ by using Formula 6.1 of Section 6.3, and then proceed to use Formula 6.2. We arrange the computations in the form of a chart, as follows:

x	p(x)	x · p(x)	x − μ	(x − μ)²	(x − μ)² · p(x)
0	$\dfrac{7}{24}$	$0\left(\dfrac{7}{24}\right) = 0$	$0 - 2 = -2$	$(-2)^2 = 4$	$4\left(\dfrac{7}{24}\right) = \dfrac{28}{24}$
1	$\dfrac{5}{24}$	$1\left(\dfrac{5}{24}\right) = \dfrac{5}{24}$	$1 - 2 = -1$	$(-1)^2 = 1$	$1\left(\dfrac{5}{24}\right) = \dfrac{5}{24}$
2	$\dfrac{1}{8}$	$2\left(\dfrac{1}{8}\right) = \dfrac{2}{8}$	$2 - 2 = 0$	$0^2 = 0$	$0\left(\dfrac{1}{8}\right) = 0$
3	$\dfrac{1}{8}$	$3\left(\dfrac{1}{8}\right) = \dfrac{3}{8}$	$3 - 2 = 1$	$1^2 = 1$	$1\left(\dfrac{1}{8}\right) = \dfrac{1}{8}$
4	$\dfrac{1}{12}$	$4\left(\dfrac{1}{12}\right) = \dfrac{4}{12}$	$4 - 2 = 2$	$2^2 = 4$	$4\left(\dfrac{1}{12}\right) = \dfrac{4}{12}$
5	$\dfrac{1}{6}$	$5\left(\dfrac{1}{6}\right) = \dfrac{5}{6}$	$5 - 2 = 3$	$3^2 = 9$	$9\left(\dfrac{1}{6}\right) = \dfrac{9}{6}$

We then have

$$\mu = \Sigma x \cdot p(x)$$

$$= 0\left(\frac{7}{24}\right) + 1\left(\frac{5}{24}\right) + 2\left(\frac{1}{8}\right) + 3\left(\frac{1}{8}\right) + 4\left(\frac{1}{12}\right) + 5\left(\frac{1}{6}\right)$$

$$= 0 + \left(\frac{5}{24}\right) + \left(\frac{2}{8}\right) + \left(\frac{3}{8}\right) + \left(\frac{4}{12}\right) + \left(\frac{5}{6}\right)$$

$$= 0 + \left(\frac{5}{24}\right) + \left(\frac{6}{24}\right) + \left(\frac{9}{24}\right) + \left(\frac{8}{24}\right) + \left(\frac{20}{24}\right)$$

$$\mu = \frac{48}{24} = 2$$

Also

$$\sigma^2 = \Sigma(x - \mu)^2 \cdot p(x)$$

$$= \left(\frac{28}{24}\right) + \left(\frac{5}{24}\right) + 0 + \left(\frac{1}{8}\right) + \left(\frac{4}{12}\right) + \left(\frac{9}{6}\right)$$

$$= \frac{80}{24} \approx 3.33$$

Thus, the mean is 2, the variance is 3.33, and the standard deviation is $\sqrt{3.33}$, or approximately 1.82.

- **Example 2**

A dress manufacturer claims that the probability that a customer will buy a particular size dress is as follows:

Size, x	8	10	12	14	16	18
Probability, p(x)	0.11	0.21	0.28	0.17	0.13	0.10

Find the mean, variance, and standard deviation for this distribution.

Solution
We first find μ by using Formula 6.2 and then arrange the data in tabular form as follows:

x	p(x)	x · p(x)	x − μ	(x − μ)²	(x − μ)² · p(x)
8	0.11	0.88	8 − 12.6 = −4.6	21.16	(21.16)(0.11) = 2.33
10	0.21	2.10	10 − 12.6 = −2.6	6.76	(6.76)(0.21) = 1.42
12	0.28	3.36	12 − 12.6 = −0.6	0.36	(0.36)(0.28) = 0.10
14	0.17	2.38	14 − 12.6 = ₁1.4	1.96	(1.96)(0.17) = 0.33
16	0.13	2.08	16 − 12.6 = 3.4	11.56	(11.56)(0.13) = 1.50
18	0.10	1.80	18 − 12.6 = 5.4	29.16	(29.16)(0.10) = 2.92

$$\mu = \Sigma x \cdot p(x) = 0.88 + 2.10 + 3.36 + 2.38 + 2.08 + 1.80 = 12.6$$

$$\sigma^2 = 2.33 + 1.42 + 0.10 + 0.33 + 1.50 + 2.92 = 8.6$$

Thus the mean is 12.6, the variance is 8.6, and the standard deviation is $\sqrt{8.6}$, or approximately 2.93.

Formula 6.2, like the formula for the variance of a set of numbers (see page 112), requires us to first compute the mean and then to find the square of the deviations from the mean. In many cases we do not wish to do this. For such situations we can use an alternate formula to calculate the variance of a probability distribution.

FORMULA 6.3

The variance of a discrete random variable with a given probability distribution is

$$\sigma^2 = \Sigma x^2 \cdot p(x) - [\Sigma x \cdot p(x)]^2$$

Formula 6.3 may seem strange but it is similar to Formula 3.5 on page 115. Let us see how Formula 6.3 is used.

● **Example 3**
Calculate the variance for the probability distribution given in Example 2 by using Formula 6.3.

Solution

We arrange the data as follows:

x	p(x)	x · p(x)	x^2	x^2 · p(x)
8	0.11	0.88	64	64(0.11) = 7.04
10	0.21	2.10	100	100(0.21) = 21.00
12	0.28	3.36	144	144(0.28) = 40.32
14	0.17	2.38	196	196(0.17) = 33.32
16	0.13	2.08	256	256(0.13) = 33.28
18	0.10	1.80	324	324(0.10) = 32.4
		$Total = 12.60$		$Total = 167.36$

Using Formula 6.3, we find the variance is

$$\sigma^2 = \Sigma x^2 \cdot p(x) - [\Sigma x \cdot p(x)]^2$$
$$= 167.36 - (12.6)^2$$
$$= 167.36 - 158.76$$
$$= 8.6$$

The variance is 8.6. This is the same result that we got using Formula 6.2.

EXERCISES

1. Refer to the newspaper clipping shown below. The state officials will be testing for the presence of up to seven possible toxic chemical pollutants. The probability that they will

STATE OFFICIALS TO INVESTIGATE THE DUMPING OF CHEMICAL WASTES

Atlantic City (Aug. 1): Officials of the New Jersey Environment Protection Bureau announced yesterday that they would begin testing several dumping sites not too far from Atlantic City for the presence of different toxic chemical wastes. These chemical wastes are now beginning to work their way into the water supply of several neighboring communities.

Saturday, August 1, 1981

find different types of chemical pollutants is as follows:

Number of Toxic Pollutants Found x	Probability p(x)	$x \cdot p(x)$	x^2	$x - \mu$
0	0.08	0	0	0 - 3.04 -3.04
1	0.14	.14	1	1 -3.04 -2.04
2	0.16	.32	4	2 -1.04
3	0.18	.54	9	3 -.04
4	0.21	.84	16	4 .96
5	0.19	.95	25	5 1.96
6	0.03	.14	36	6 2.96
7	0.01	.07	49	7 3. 3.96

$\mu = \Sigma x \cdot p(x)$

Find the mean, variance, and standard deviation for this distribution.

2. A survey of the number of television sets that families in Dover own produced the following results:

Number of Television Sets x	Probability p(x)
0	0.07
1	0.36
2	0.29
3	0.13
4	0.09
5	0.06

Find the mean, variance, and standard deviation for this distribution.

3. A counselor at a drug rehabilitation center for young juveniles questioned each participant in the program to determine how many years, x, each had been on drugs before enrolling in the program. The relative frequencies corresponding to x are given in the following probability distribution.

Number of Years x	Probability p(x)
0	0.06
1	0.17
2	0.19
3	0.20
4	0.23
5	0.12
6	0.03

Find the mean, variance, and standard deviation for this distribution.

4. False alarms reported to the fire department are not only a waste of money but can also be life-threatening, as the unit nearest to a real fire may be out on a false alarm. For a particular district, the following distribution is available on the number of false alarms reported during a 24-hour period.

Number of False Alarms Reported x	Probability p(x)
0	0.07
1	0.19
2	0.22
3	0.16
4	0.12
5	0.10
6	0.08
7	0.06

Find the mean, variance, and standard deviation for this distribution.

5. Francisco Benoit is in charge of maintenance for a large Florida car rental agency. The following is the probability distribution for the number of customers who will call the car rental agency daily because of malfunctioning cars.

Number of Customers x	Probability p(x)
5	0.11
6	0.19
7	0.18
8	0.07
9	0.12
10	0.06
11	0.10
12	0.14
13	0.03

Find the mean of this distribution.

6. The following is the probability distribution for the number of child custody dispute cases that the courts in Brownfield handle daily:

Number of Child Custody Cases x	Probability $p(x)$
0	0.03
1	0.11
2	0.14
3	0.16
4	0.17
5	0.15
6	0.12
7	0.06
8	0.04
9	0.02

Find the mean, variance, and standard deviation for this distribution.

7. The following table gives the probabilities that a photocopying machine will malfunction 0, 1, 2, 3, 4, 5 or 6 times on any given day:

Number of Malfunctions x	Probability $p(x)$
0	0.07
1	0.14
2	0.17
3	0.21
4	0.18
5	0.19
6	0.04

Find the mean, variance, and standard deviation for this distribution.

* 8. Using the rules for summation given in Chapter 4, show how Formula 6.3 can be derived from Formula 6.2.

9. Given the probability function

$$p(x) = \frac{7 - x}{21} \quad \text{for } x = 1, 2, 3, 4, 5, 6,$$

find the mean, and standard deviation of this distribution.

10. A psychologist is conducting experiments with mice. In one cage, it is known that four of the ten mice are gray. The remainder are black. A sample of three is selected with replacement. Let x be the number of gray mice in the sample.

    a. Find the probability distribution of x.

    b. Find the mean and standard deviation for this distribution.

11. The following is the probability distribution for the number of daily requests for blood analysis that the laboratory of a district hospital receives:

Number of Requests x	Probability $p(x)$
1	0.05
2	0.43
3	0.17
4	0.25
5	0.06
6	0.03
7	0.01

    a. Find μ and σ.

  *b. Find the probability that at least x number of requests will be in the interval $\mu \pm 2\sigma$. (*Hint:* Use Chebyshev's Theorem, discussed in Chapter 3.)

12. A die is altered by painting an additional dot on the face that originally had one dot (see diagram below). A pair of such altered dice is rolled. Let x be the number of dots showing on both dice together.

    a. Find the probability distribution of x.

    b. Find μ, σ^2, and σ for this distribution.

6.5 THE BINOMIAL DISTRIBUTION

Consider the following probability problem:

● **Example 1**

Paula is about to take a five-question true–false quiz. She is not prepared for the exam and decides to guess the answers without reading the questions.

> *ANSWER SHEET*
>
> *Directions:* For each question darken the appropriate space.
>
> 1. | True | | False |
> 2. | True | | False |
> 3. | True | | False |
> 4. | True | | False |
> 5. | True | | False |

Find the probability that she gets

a. all the answers correct.
b. all the answers wrong.
c. three out of the five answers correct.

Solution

Let us denote a correct answer by the letter "c" and a wrong answer by the letter "w." There are 2 equally likely possible outcomes for question 1, c or w. Similarly, there are 2 equally likely possible outcomes for question 2 regardless of whether the first question was correct or incorrect. There are 2 equally likely possible outcomes for each of the remaining questions 3, 4, and 5. Thus, there are 32 equally likely possible outcomes since

$$2 \times 2 \times 2 \times 2 \times 2 = 32$$

These outcomes we list as follows:

ccccc	cwccc	wcccc	wwccc
cccwc	cwcwc	wccwc	wwcwc
ccccw	cwccw	wcccw	wwccw
cccww	cwcww	wccww	wwcww
ccwcc	cwwcc	wcwcc	wwwcc
ccwcw	cwwcw	wcwcw	wwwcw
ccwwc	cwwwc	wcwwc	wwwwc
ccwww	cwwww	wcwww	wwwww

Once we have listed all the possible outcomes, we can construct a chart similar to Table 6.5.

TABLE 6.5
Number of Correct Answers

0	1	2	3	4	5
wwwww	cwwww	ccwww	cccww	wcccc	ccccc
	wcwww	cwcww	ccwcw	cwccc	
	wwcww	cwwcw	ccwwc	ccwcc	
	wwwcw	cwwwc	cwcwc	cccwc	
	wwwwc	wccww	cwccw	ccccw	
		wcwcw	cwwcc		
		wcwwc	wccwc		
		wwcwc	wcccw		
		wwccw	wcwcc		
		wwwcc	wwccc		

Now we can calculate the probability associated with each outcome. We have

$$p(0 \text{ correct}) = \frac{1}{32}$$

$$p(1 \text{ correct}) = \frac{5}{32}$$

$$p(2 \text{ correct}) = \frac{10}{32}$$

$$p(3 \text{ correct}) = \frac{10}{32}$$

$$p(4 \text{ correct}) = \frac{5}{32}$$

$$p(5 \text{ correct}) = \frac{1}{32}$$

We can picture these results in the form of a histogram, as shown in Figure 6.2. The relative frequency histogram for Figure 6.2 is shown in Figure 6.3.

We can now answer the question raised at the beginning of the problem. We have

a. $p(\text{all correct answers}) = \dfrac{1}{32}$

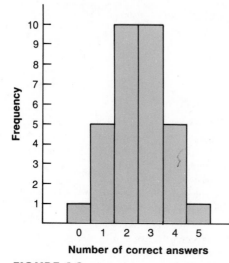

FIGURE 6.2

Histogram of the number of correct answers obtained by guessing at five true-false questions.

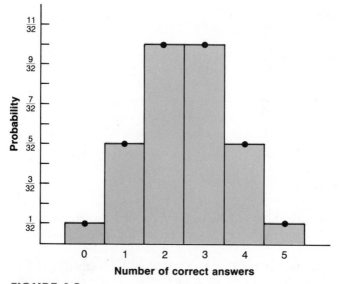

FIGURE 6.3

Relative frequency histogram for the frequency distribution of Figure 6.2.

b. p(all wrong answers) $= \dfrac{1}{32}$

c. p(3 out of 5 correct answers) $= \dfrac{10}{32}$

COMMENT These answers assume that since Paula is not prepared, then p(correct answer) $= p$(wrong answer) $= \dfrac{1}{2}$. On the other hand, these answers would be wrong if Paula were better prepared and had a better probability of, say 0.8, of getting a correct answer.

Many experiments or probability problems result in outcomes that can be grouped into two categories, success or failure. For example, when a coin is flipped, there are two possible outcomes, heads or tails; when a hunter shoots at a target there are two possible outcomes, hit or miss; and when a baseball player is at bat the result is get on base or not get on base.

Statisticians apply this idea of success and failure to wide-ranging problems. For instance, if a quality-control engineer is interested in determining the life of a typical light bulb in a large shipment, each time a bulb burns out he has a "success." Similarly, if we were interested in determining the probability of a family having ten boys, assuming they planned to have ten children, then each time a boy is born we have a "success."

As we mentioned in Section 6.2 (page 263), experiments that have only two possible outcomes are referred to as **binomial probability experiments** or **Bernoulli experiments** in honor of the mathematician Jacob Bernoulli (1654–1705), who studied them in detail. His contributions to probability theory are contained in his book *Ars Conjectandi* published after his death in 1713. This book also contains a reprint of an earlier treatise of Huygens. In 1657 the great Dutch mathematician Christian Huygens had written the first formal book on probability based upon the Pascal–Fermat correspondences discussed in Chapter 1 (page 11). Huygens also introduced the important ideas of mathematical expectation discussed in Chapter 4. All these ideas are contained in Bernoulli's book. Binomial probability experiments are characterized by the following:

DEFINITION 6.5

A *binomial probability experiment* is an experiment that satisfies the following properties:

1. *There is a fixed number, n, of repeated trials whose outcomes are independent.*
2. *Each trial results in one of two possible outcomes. We call one outcome a success and denote it by the letter S and the other outcome a failure, denoted by F.*

3. *The probability of success on a single trial equals p and remains the same from trial to trial. The probability of a failure equals* $1 - p = q$. *Symbolically,*

$$p(success) = p \quad and \quad p(failure) = q = 1 - p$$

4. *We are interested in the number of successes obtained in n trials of the experiment.*

COMMENT Although very few real-life situations satisfy all the above requirements, many of them can be thought of as satisfying, at least approximately, these requirements. Thus, we can apply the binomial distribution to many different problems.

Since all binomial probability experiments are similar in nature and result in either success or failure for each trial of the experiment, we seek a formula for determining the probability of obtaining x successes out of n trials of the experiment, where the probability of success on any one trial is p and the probability of failure is q. To achieve this goal let us consider the following.

A coin is tossed 4 times. What is the probability of getting exactly one head? We could list all the possible outcomes and count the number of favorable ones. This is shown below with the favorable outcomes circled.

HHHH HHTT THHH (THTT)
HHHT HHTH THHT THTH
HTHH (HTTT) TTHH TTTT
HTTH HTHT (TTTH) (TTHT)

Thus, the probability of getting exactly one head is $\dfrac{4}{16}$.

However, in many cases it is not possible or advisable to list all the possible outcomes. It is for this reason that we consider an alternate approach.

When a coin is tossed there are two possible outcomes, heads or tails. Thus,

$$p(\text{heads}) = \frac{1}{2} \quad and \quad p(\text{tails}) = \frac{1}{2}$$

Since we are interested in getting only a head, we classify this outcome as a success and write

$$p(\text{head}) = p(\text{success}) = \frac{1}{2} \quad and \quad p(\text{tail}) = p(\text{failure}) = \frac{1}{2}$$

Therefore,

$$p = \frac{1}{2}, q = \frac{1}{2} \quad and \quad p + q = \frac{1}{2} + \frac{1}{2} = 1$$

Each toss is independent of what happened on the preceding toss. We are interested in obtaining one head in four tosses. One possible way in which this can happen, along with the corresponding probabilities, is as follows:

Outcome	head	tail	tail	tail
Success or Failure	Success	Failure	Failure	Failure
Probability	$\frac{1}{2}$	$\frac{1}{2}$	$\frac{1}{2}$	$\frac{1}{2}$
Symbolically	p	q	q	q

Since the probability of success is p and the probability of failure is q, we can summarize this as $p \cdot q^3$. Remember q^3 means $q \cdot q \cdot q$. We would then say that the probability of getting one head is

$$\left(\frac{1}{2}\right) \cdot \left(\frac{1}{2}\right)^3 = \left(\frac{1}{2}\right) \cdot \left(\frac{1}{2}\right) \cdot \left(\frac{1}{2}\right) \cdot \left(\frac{1}{2}\right) = \frac{1}{16}$$

However, we have forgotten one thing. The head may occur on the second, third, or fourth toss:

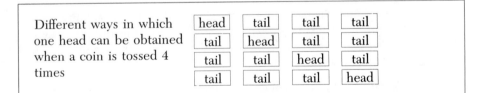

There are then four ways in which we can get one head. Thus the $\frac{1}{16}$ that we calculated before can occur in four different ways. Therefore,

$$p(\text{exactly 1 head}) = 4\left(\frac{1}{16}\right) = \frac{4}{16}$$

Notice that this is exactly the same answer we obtained by listing all the possible outcomes.

Similarly, if we were interested in the probability of getting exactly two heads, we could consider one particular way in which this can happen:

Outcome	head	head	tail	tail
Success or Failure	Success	Success	Failure	Failure
Probability	$\dfrac{1}{2}$	$\dfrac{1}{2}$	$\dfrac{1}{2}$	$\dfrac{1}{2}$
Symbolically	p	p	q	q

The probability is thus

$$p^2 \cdot q^2 = \left(\frac{1}{2}\right)^2 \left(\frac{1}{2}\right)^2 = \left(\frac{1}{2}\right)\left(\frac{1}{2}\right)\left(\frac{1}{2}\right)\left(\frac{1}{2}\right) = \frac{1}{16}$$

Again, we must multiply this result by the number of ways that these two heads can occur in the four trials. This is the number of combinations of four things taken two at a time. We can use Formula 4.4 of Chapter 4 and get

$$_4C_2 = \frac{4!}{2!(4-2)!} = \frac{4!}{2!2!} = 6$$

Thus, the probability of getting two heads in four flips of a coin is

$$6\left(\frac{1}{16}\right) = \frac{6}{16} = \frac{3}{8}$$

More generally, if we are interested in the probability of getting x successes out of n trials of an experiment, then we consider one way in which this can happen. Here we have assumed that all the x successes occur first and all the failures occur on the remaining $n - x$ trials.

Success or Failure   Success · Success · Success · · · Failure · Failure · Failure
Probability $\underbrace{p \cdot p \cdot p}_{x \text{ of them}}$ · · · $\underbrace{q \cdot q \cdot q}_{n - x \text{ of them}}$

This gives $p^x q^{n-x}$. We then multiply this result by the number of ways that exactly x successes can occur in n trials.

The number of ways that exactly x successes can occur in a set of n trials is given by

$$_nC_x = \frac{n!}{x!(n-x)!}$$

Here we have replaced r by x in the number of possible combinations formula (Formula 4.4). This leads us to the following **binomial distribution formula.**

FORMULA 6.4   *Binomial Distribution Formula*

Consider a binomial experiment that has two possible outcomes, success or failure. Let p(success) = *p and p(failure)* = *q. If this experiment is performed n times, then the probability of getting x successes out of the n trials is*

$$p(x \text{ successes}) = {}_nC_x \, p^x q^{n-x} = \frac{n!}{x!(n-x)!} \, p^x q^{n-x}$$

We illustrate the use of Formula 6.4 with numerous examples.

● **Example 1**   *Admission to Medical School*
Ninety percent of the graduates of State University who apply to a particular medical school are admitted. This year six graduates from State University have applied for admission to the medical school. Find the probability that only four of them will be accepted.

Solution
Since six students have applied for admission, $n = 6$. We are interested in the probability that four are accepted, so $x = 4$. Also, 90% of the graduates who apply are admitted, so $p = 0.90$, and 10% of the graduates who apply are not admitted, so $q = 0.10$. Now we apply Formula 6.4. We have

$$\text{Prob(4 are accepted)} = \text{Prob}(x = 4)$$

$$= \frac{6!}{4!(6-4)!} (0.90)^4 (0.10)^{6-4}$$

$$= \frac{6!}{4!2!} (0.9)^4 (0.1)^2$$

$$= \frac{6 \cdot 5 \cdot 4 \cdot 3 \cdot 2 \cdot 1}{4 \cdot 3 \cdot 2 \cdot 1 \cdot 2 \cdot 1} (0.9)(0.9)(0.9)(0.9)(0.1)(0.1)$$

$$= 0.0984$$

Thus, the probability that only four of them will be accepted is 0.0984.

● **Example 2**
Consider the accompanying newspaper article. Bill and his friends own five cars, which they park in front of his house. Find the probability that none of them will be stolen this year.

CAR THEFTS ON THE RISE AGAIN

Washington (April 7) — Look out your window. Is your car still in your driveway or in front of your house? If it is there, then you're lucky. According to the latest FBI study released yesterday, there are an average of 2300 vehicles stolen per day in the United States. This puts the chances of your car being stolen at about 1 in 120. According to the survey, about 60% of the vehicles are stolen from private residences, apartments, or streets in residential areas between the hours of 6:00 P.M. and 6:00 A.M.

Most of the cars are stolen to be stripped for their parts or to be used for joyriding.

Wednesday — April 7, 1982

Reprinted by permission.

Solution

Since Bill and his friends own five cars, $n = 5$. We are interested in the probability that none of the cars will be stolen, so that $x = 0$. According to the newspaper article, 1 out of every 120 cars is stolen, so that $p = \dfrac{1}{120} = 0.008$ and $q = \dfrac{119}{120} = 0.992$. Applying Formula 6.4 gives

$$\text{Prob}(0 \text{ stolen cars}) = \text{Prob}(x = 0) = \frac{5!}{0!5!}(0.008)^0(0.992)^5$$

Remember that $(0.008)^0 = 1$, so that

$$\text{Prob}(0 \text{ stolen cars}) = 1(0.992)^5 = 0.9606$$

Thus, the probability that none of these cars will be stolen this year is 0.9606.

● Example 3

Mario is taking a multiple-choice examination that consists of five questions. Each question has four possible answers. Mario guesses at every answer. What is the probability that he passes the exam if he needs *at least* four correct answers to pass?

Solution

In order to pass, Mario needs to get at least four correct answers. Thus, he passes if he gets four answers correct or five answers correct. Each question has four possible answers so that $p(\text{correct answer}) = \dfrac{1}{4}$ and $p(\text{wrong}$

answer) $= \dfrac{3}{4}$. Also, there are five questions, so $n = 5$. Therefore,

$$p(4 \text{ answers correct}) = \frac{5!}{4!1!} \left(\frac{1}{4}\right)^4 \left(\frac{3}{4}\right)$$

$$= \frac{5 \cdot \cancel{4} \cdot \cancel{3} \cdot \cancel{2} \cdot \cancel{1}}{\cancel{4} \cdot \cancel{3} \cdot \cancel{2} \cdot \cancel{1} \cdot \cancel{1}} \left(\frac{1}{4}\right)\left(\frac{1}{4}\right)\left(\frac{1}{4}\right)\left(\frac{1}{4}\right)\left(\frac{3}{4}\right)$$

$$= \frac{15}{1024}$$

and

$$p(5 \text{ answers correct}) = \frac{5!}{5!0!} \left(\frac{1}{4}\right)^5 \left(\frac{3}{4}\right)^0 \quad \text{(Any number to the 0 power is 1.)}$$

$$= \frac{\cancel{5} \cdot \cancel{4} \cdot \cancel{3} \cdot \cancel{2} \cdot 1}{\cancel{5} \cdot \cancel{4} \cdot \cancel{3} \cdot \cancel{2} \cdot \cancel{1} \cdot \cancel{1}} \left(\frac{1}{4}\right)\left(\frac{1}{4}\right)\left(\frac{1}{4}\right)\left(\frac{1}{4}\right)\left(\frac{1}{4}\right) \cdot 1$$

$$= \frac{1}{1024}$$

Adding the two probabilities, we get

$$p(\text{at least 4 correct answers}) = p(4 \text{ answers correct}) + p(5 \text{ answers correct})$$

$$= \frac{15}{1024} + \frac{1}{1024}$$

$$= \frac{16}{1024} = \frac{1}{64}$$

Hence, the probability that Mario passes is $\dfrac{1}{64}$.

● **Example 4**
A shipment of 100 tires from the Apex Tire Corporation is known to contain 20 defective tires. Five tires are selected at random and each tire is replaced before the next tire is selected. What is the probability of getting *at most* 2 defective tires?

Solution
We are interested in the probability of getting *at most* two defective tires. This means zero defective tires, one defective tire, or two defective tires. Thus the probability of at most two defective tires equals

$$p(0 \text{ defective}) + p(1 \text{ defective}) + p(2 \text{ defective})$$

The probability of a defective tire is $\dfrac{20}{100}$, or $\dfrac{1}{5}$. Therefore, the probability of getting a non-defective tire is $\dfrac{4}{5}$. Now

$$p(0 \text{ defective}) = \frac{5!}{0!5!} \left(\frac{1}{5}\right)^0 \left(\frac{4}{5}\right)^5 = \frac{1024}{3125}$$

$$p(1 \text{ defective}) = \frac{5!}{1!4!} \left(\frac{1}{5}\right)^1 \left(\frac{4}{5}\right)^4 = \frac{1280}{3125}$$

$$p(2 \text{ defective}) = \frac{5!}{2!3!} \left(\frac{1}{5}\right)^2 \left(\frac{4}{5}\right)^3 = \frac{640}{3125}$$

Adding, we get

$$p(\text{at most 2 defectives}) = \frac{1024}{3125} + \frac{1280}{3125} + \frac{640}{3125} = \frac{2944}{3125}$$

Hence the probability of getting at most 2 defective tires is $\dfrac{2944}{3125}$, or approximately 0.94.

- **Example 5**

If the conditions are the same as in the previous problem except that now 15 tires are selected, what is the probability of getting *at least* 1 defective tire?

Solution

We could proceed as we did in Example 4. Thus

$$p(\text{at least 1 defective}) = p(1 \text{ defective}) + p(2 \text{ defective})$$

$$+ \cdots + p(15 \text{ defective})$$

However this involves a tremendous amount of computation. Recall (see the Rule on page 263) that the sum of all the values of a probability function must be 1. Thus

$$p(0 \text{ defective}) + p(1 \text{ defective}) + p(2 \text{ defective})$$

$$+ \cdots + p(15 \text{ defective}) = 1$$

Therefore, if we subtract $p(0 \text{ defective})$ from both sides we have

$$p(1 \text{ defective}) + p(2 \text{ defective}) + \cdots + p(15 \text{ defective})$$

$$= 1 - p(0 \text{ defective})$$

Now

$$p(0 \text{ defective}) = \frac{15!}{0!15!} \left(\frac{1}{5}\right)^0 \left(\frac{4}{5}\right)^{15} = 0.035$$

Consequently, the probability of obtaining at least 1 defective tire is $1 - 0.035$, or 0.965.

COMMENT Calculating binomial probabilities can sometimes be quite a time-consuming task. To make the job a little easier, we can use Table IV of the Appendix which gives us the binomial probabilities for different values of n, x, and p. No computations are needed. We only need to know the values of n, x, and p.

● **Example 6**

Given a binomial distribution with $n = 11$ and $p = 0.4$, use Table IV of the Appendix to find the probability of getting

a. exactly four successes.
b. at most three successes.
c. five or more successes.

Solution

We use Table IV of the Appendix with $n = 11$ and $p = 0.4$. We first locate $n = 11$ and then move across the top of the table until we reach the $p = 0.4$ column.

a. To find the probability of exactly four successes, we look for the value given in the table for $x = 4$. It is 0.236. Thus when $n = 11$ and $p = 0.4$, the probability of exactly four successes is 0.236.
b. To find the probability of getting at most three successes, we look in the table for the values given for $x = 0$, $x = 1$, $x = 2$, and $x = 3$. These probabilities are 0.004, 0.027, 0.089, and 0.177, respectively. We add these (Why?) and get

$$0.004 + 0.027 + 0.089 + 0.177 = 0.297$$

Thus, the probability of at most three successes is 0.297.
c. To find the probability of five or more successes, we look in the chart for the values given for $x = 5$, $x = 6$, $x = 7$, $x = 8$, $x = 9$, $x = 10$, and $x = 11$. These probabilities are 0.221, 0.147, 0.070, 0.023, 0.005, 0.001, and 0. We add these and get

$$0.221 + 0.147 + 0.070 + 0.023 + 0.005 + 0.001 + 0 = 0.467$$

Thus, the probability of five or more successes is 0.467.

COMMENT In the previous example you will notice that there is no value given in Table IV when $x = 11$. It is left blank. Whenever there is a blank in the chart, this means that the probability is approximately 0. This is the reason that we used 0 as the probability in our calculations.

When $p = 0.5$, then the binomial distribution is symmetrical and begins to resemble the normal distribution as the value of n gets larger. This is shown in Figure 6.4.

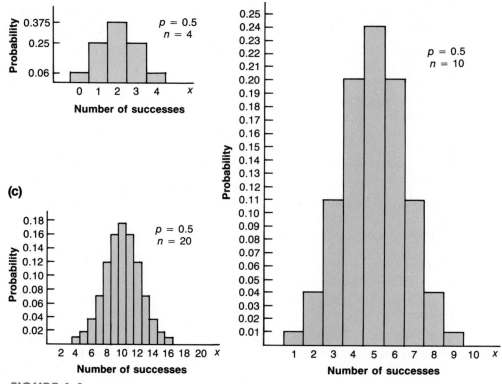

FIGURE 6.4
Binomial distributions with $p = 0.50$ and different values of n.

As a matter of fact, we can use the values given in Table IV and draw the graphs of binomial probabilities. In Figure 6.5, we have drawn several such graphs using different values of p. In each case, however, $n = 5$. What do the graphs look like? We will have more to say about this in the next chapter.

EXERCISES

1. Forty-three percent of all parking tickets issued in Morgantown are for cars illegally parked next to fire hydrants. If six parking tickets issued are randomly selected, find the probability that exactly three of them are for cars illegally parked next to fire hydrants.

2. The president of the Collex Savings Bank claims that 60% of all applicants for credit cards are women. If the bank receives eight applications for credit cards on January 4, find the probability that three of the applicants are women.

3. A researcher has found that 80% of all the families in Brownsville have at least one pet dog or cat. If seven families in Brownsville are randomly selected, find the probability that *all* seven have at least one pet dog or cat.

(Exercises continue on p. 292)

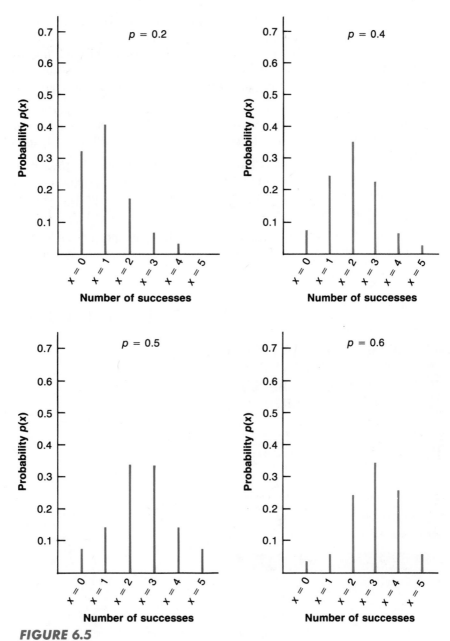

FIGURE 6.5

Graphs of binomial probabilities for $n = 5$.

4. A reservations agent for a large airline claims that 15% of the flying passengers request seats in the smoking section of an airplane when making reservations. If ten people who have made reservations for a particular flight are randomly selected, find the probability that five of these people requested seats in the smoking section. (Assume that the number of people with reservations is large, so p remains constant as seats are selected.)

5. Animal researchers have found that 90% of the animals of a particular species will reproduce in captivity. If eight female animals of this species have been obtained by a zoo, find the probability that *at least* six of these animals will reproduce in captivity.

6. *Medicine*   Medical researchers have found that approximately 15% of the population who take a particular antihistamine drug will develop some form of reaction to the drug. If nine people have taken the drug, find the probability that *at most* two of these people will develop some form of reaction to the drug.

7. Consider the accompanying newspaper article. If ten people are randomly selected, find the probability that *at least* eight of them believe nuclear energy should *not* be abandoned.

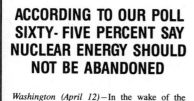

ACCORDING TO OUR POLL SIXTY-FIVE PERCENT SAY NUCLEAR ENERGY SHOULD NOT BE ABANDONED

Washington (April 12) — In the wake of the recent nuclear mishap at Three Mile Island in Pennsylvania, the *Daily Press* conducted a nationwide survey to determine the people's grass-roots sentiment about nuclear energy. The survey reported that 65% of those responding said that nuclear energy should not be abandoned but rather that more safeguards should be found to protect the public.

Tuesday — April 12, 1983

Reprinted by permission.

8. *Baseball*   Reggie Jones is a baseball player whose batting average is 0.325. In tomorrow's game, he will be at bat four times. Find the probability that he gets two hits.

9. *Decaying Bridges*   According to a federal investigatory commission, 80% of the bridges in a state are in desperate need of repair. If ten bridges in the state are selected at random, find the probability that fewer than four of them will be in desperate need of repair.

10. *On-time Arrivals* An executive of a particular airline company claims that 98% of the company's planes arrive on time. If the company has 12 planes scheduled to land tomorrow, find the probability that *none* of them will arrive late.

11. Based upon past experience, the president of Joe's Transmission Service knows that about 5% of the rebuilt transmissions installed in cars by this company will not function properly. If 15 rebuilt transmissions will be installed today, find the probability that *at least* 4 of them will not function properly.

12. Twenty-eight percent of the cars on the state's highways cannot pass the state's tough antipollution exhaust emission tests. If 12 cars on the state's highways are randomly selected and tested, find the probability that *at most* 10 of them cannot pass the state's exhaust emission test.

13. According to the Mayoca Bridge Authority, 24% of all drivers use the $1.25 exact change lanes when paying the bridge toll. If eight cars pull up to the toll gates, find the probability that
   a. exactly two of them will use the exact change lane.
   b. at most two of them will use the exact change lane.
   c. at least two of them will use the exact change lane.

14. Market research indicates that 75% of all U.S. homes have at least one live houseplant. Nine homes are randomly selected. Find the probability that five or more of these homes will have *at least* one live houseplant.

15. Felicia is a quality-control engineer for a large electronics company. Her job is to thoroughly test each stereo set manufactured by the company and to classify it as acceptable or unacceptable. Due to the company's rigorous quality standards, each set has an equally likely chance of being acceptable or unacceptable. On a given morning, Felicia inspects six sets. Find the probability that she
   a. rejects all the sets.
   b. accepts all the sets.
   c. accepts at least three of the sets.
   d. rejects at most four of the sets.

6.6 THE MEAN AND STANDARD DEVIATION OF THE BINOMIAL DISTRIBUTION

Consider the binomial distribution given on page 279, which is also repeated on the next page. Recall that x represents the number of correct answers that Paula obtained on an exam of five questions, where the probability of a correct answer is $\frac{1}{2}$.

x	p(x)	x · p(x)	x²	x² · p(x)
0	$\dfrac{1}{32}$	0	0	0
1	$\dfrac{5}{32}$	$\dfrac{5}{32}$	1	$\dfrac{5}{32}$
2	$\dfrac{10}{32}$	$\dfrac{20}{32}$	4	$\dfrac{40}{32}$
3	$\dfrac{10}{32}$	$\dfrac{30}{32}$	9	$\dfrac{90}{32}$
4	$\dfrac{5}{32}$	$\dfrac{20}{32}$	16	$\dfrac{80}{32}$
5	$\dfrac{1}{32}$	$\dfrac{5}{32}$	25	$\dfrac{25}{32}$
		$Total = \dfrac{80}{32} = \dfrac{5}{2}$		$Total = \dfrac{240}{32} = \dfrac{15}{2}$

Let us calculate the mean and variance for this distribution. Using Formula 6.1 of Section 6.4 we get

$$\mu = \Sigma x \cdot p(x) = 0 + \left(\frac{5}{32}\right) + \left(\frac{20}{32}\right) + \left(\frac{30}{32}\right) + \left(\frac{20}{32}\right) + \left(\frac{5}{32}\right)$$

$$= \frac{80}{32} = \frac{5}{2}$$

To calculate the variance we use Formula 6.3 of Section 6.3. From the preceding table we have

$$\sigma^2 = \Sigma x^2 \cdot p(x) - [\Sigma x \cdot p(x)]^2$$

$$= \frac{15}{2} - \left(\frac{5}{2}\right)^2$$

$$= \frac{30}{4} - \frac{25}{4} = \frac{5}{4}$$

Thus $\mu = \dfrac{5}{2}$, or 2.5, and $\sigma^2 = \dfrac{5}{4}$, or 1.25.

Notice that if we multiply the total number of exam questions, which is 5, by the probability of a correct answer, which is $\dfrac{1}{2}$, we get $5\left(\dfrac{1}{2}\right) = 2.5$. This is exactly the same answer we get for the mean by applying Formula 6.1. We might be tempted to conclude that $\mu = np$. This is indeed the case.

Similarly, if we multiply the total number of questions with the probability of a correct answer and with the probability of a wrong answer we get

$$5\left(\frac{1}{2}\right)\left(\frac{1}{2}\right) = 1.25.$$

Here we might conclude that $\sigma^2 = npq$. Again this is indeed the case. We can generalize theses ideas in the following:

FORMULA 6.5

*The **mean of a binomial distribution,** μ, is found by multiplying the total number of trials with the probability of success on each trial. If there are n trials of the experiment and if the probability of success on each trial is p, then*

$$\mu = np$$

FORMULA 6.6

*The **variance of a binomial distribution** is given by*

$$\sigma^2 = npq$$

*The **standard deviation** σ is the square root of the variance. Thus*

$$\sigma = \sqrt{variance} = \sqrt{npq}$$

The proofs of Formulas 6.5 and 6.6 will be given as an exercise at the end of this section.

● **Example 1**

A die is rolled 600 times. Find the mean and standard deviation of the number of 1's that show.

Solution

This is a binomial distribution. Since the die is rolled 600 times, $n = 600$. Also there are six possible outcomes so that $p = \dfrac{1}{6}$ and $q = \dfrac{5}{6}$. Thus, using Formulas 6.5 and 6.6 we have

$$\mu = np$$

$$= 600 \left(\frac{1}{6}\right) = 100$$

and

$$\sigma^2 = npq$$

$$= 600 \left(\frac{1}{6}\right)\left(\frac{5}{6}\right) = 83.33$$

Therefore, the mean is 100 and the standard deviation, which is the square root of the variance, is $\sqrt{83.33}$, or approximately 9.13. This tells us that if this experiment were to be repeated many times, we could expect an average of 100 1's per trial with a standard deviation of $\sqrt{83.33}$, or 9.13.

● **Example 2**

A large mail-order department store finds that approximately 17% of all purchases are returned for credit. If the store sells 100,000 different items this year, about how many items will be returned? Find the standard deviation.

Solution

This is a binomial distribution. Since 100,000 items were sold, $n = 100,000$. Also the probability that a customer will return the item is 0.17 so that $p = 0.17$. Therefore the probability that the customer will not return the item is 0.83. Thus, using Formulas 6.5 and 6.6 we have

$$\mu = np$$
$$= (100,000)(0.17)$$
$$= 17,000$$

and

$$\sigma = \sqrt{npq}$$
$$= \sqrt{(100,000)(0.17)(0.83)}$$
$$= \sqrt{14110}$$
$$\approx 118.79$$

Thus, the department store can expect about 17,000 items to be returned with a standard deviation of 118.79.

Formulas 6.5 and 6.6 will be applied in greater detail in Chapter 7.

EXERCISES

1. A hotel owner finds that approximately 43% of all people pay for their room with a credit card. This week, 300 people will pay for their room. How many people can be expected to pay by credit card? Find the standard deviation.

2. A weight-reduction club finds that the dropout rate is 24% per semester (a six-month program). As a result of an intensive advertising campaign, 340 new members joined the club. How many people can be expected to withdraw from the program? Find the standard deviation.

3. Consider the newspaper clipping below. If 400 randomly selected restaurants are visited, about how many of them can be expected to have unsanitary health conditions? Find the standard deviation.

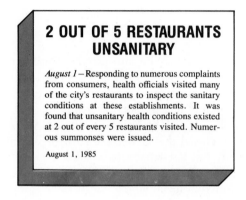

2 OUT OF 5 RESTAURANTS UNSANITARY

August 1 – Responding to numerous complaints from consumers, health officials visited many of the city's restaurants to inspect the sanitary conditions at these establishments. It was found that unsanitary health conditions existed at 2 out of every 5 restaurants visited. Numerous summonses were issued.

August 1, 1985

4. An insecticide is advertised as being 95% effective in killing moths and bugs. If 1000 moths and bugs are sprayed, find the mean number of bugs killed. What is the standard deviation?

5. A restaurant owner finds that approximately 14% of the people who make reservations do not show. Three hundred twenty people have made reservations for Mother's Day parties. About how many people can actually be expected to show up? Find the standard deviation.

6. Refer back to Exercise 5. The restaurant owner plans to overbook, estimating that 310 people will actually show up. How many reservations should the owner accept?

7. A large airline company finds that about 9% of passenger reservations are involved in travel plan changes of one type or another during a busy holiday travel period. If 20,000 passengers have confirmed reservations for the holiday travel period, about how many travel plan changes will be made? Find the standard deviation.

8. Because of rising imports and sluggish domestic sales, the president of the ABCO Textile Corporations announces that 157 workers will have to be laid off during the next fiscal year. The company employs 8378 workers at its various divisions. Past experience indicates that about 2% of the workers die, resign, or retire during a fiscal year. Neglecting any other considerations, can these cutbacks be accomplished through attrition (that is, by deaths, resignations, or retirements)?

9. A physician finds that about 15% of infants injected with the whooping cough vaccine develop some form of minor reaction. If the doctor injects 300 infants during a month, about how many can be expected *not* to develop any form of reaction? Find the standard deviation.

10. A large mail-order department store finds that approximately 18% of all purchases are returned for credit. If it sells 70,000 different items this year, about how many items will *not* be returned?

*11. Using the formulas for summation, verify that Formulas 6.5 and 6.6 are indeed valid for a binomial distribution.

*6.7 THE POISSON DISTRIBUTION

There are many practical problems where we may be interested in finding the probability that x "successes" will occur over a given interval of time or a region of space. This is especially true when we do not expect many successes to occur over the time interval (which may be of any length, such as a minute, a day, a week, a month, or a year). For example, we may be interested in determining the number of days that school will be closed due to snowstorms, or we may be interested in determining the number of days that a baseball game will be postponed in a given season because of rain. For these and similar problems we can use the Poisson Probability Function formula.

Before giving this formula, we wish to emphasize that certain underlying assumptions must be satisfied for this formula to be applicable. Among these assumptions are the following:

1. Each "success" occurs independently of the others.
2. The probability of "success" in any interval or region or that most of the probability is concentrated at the low end of the domain of x is very small.
3. The probability of a success in any one small interval is the same as that for any other small interval of the same size.
4. The number of successes in any interval is independent of the number of successes in any other nonoverlapping interval.

When the above assumptions are satisfied we can use the following:

FORMULA 6.7   *Poisson Distribution*

The Poisson Probability distribution representing the number of successes occurring in a given time interval or specified region is given by

$$p(x \text{ successes}) = \frac{e^{-\mu}\mu^x}{x!} \qquad x = 0, 1, 2, 3, \dots$$

where μ is the average number of successes occurring in the given time interval or specified region. Thus for the Poisson distribution, the mean is μ, the variance is μ, and the standard deviation is $\sqrt{\mu}$.

COMMENT The symbol e that appears in the formula is used often in mathematics to represent an irrational number whose value is approximately equal to 2.71828. Table XIII in the Appendix gives the values of $e^{-\mu}$ for different values of μ.

Let us illustrate the use of this formula with several examples.

● **Example 1**

An animal trainer finds that the number, x, of animal bites per month that her crew experiences follows an approximate Poisson distribution with a mean of 7.5. Find the variance and standard deviation of x, the number of animal bites per month.

Solution

For a Poisson distribution, the mean and variance are both equal to μ. Thus, in our case, we have

$$\mu = 7.5 \quad \text{and} \quad \sigma^2 = 7.5$$

The standard deviation is $\sqrt{7.5}$, or approximately 2.7.

● **Example 2**   *School Closings*

Official records in a particular city show that the average number of school closings in a school year due to snowstorms is four. What is the probability that there will be six school closings this year because of snowstorms?

Solution

Based upon the given information, $\mu = 4$. We are interested in the probability that $x = 6$. Using the Poisson Distribution formula, we get

$$p(x = 6) = \frac{e^{-4}(4)^6}{6!}$$

From Table XIII $e^{-4} = 0.0183156$ so that

$$p(x = 6) = \frac{e^{-4}(4)^6}{6!}$$

$$= \frac{(0.0183156)(4096)}{720} = 0.1042$$

Thus, the probability that there will be six school closings this year because of snowstorms is 0.1042.

● **Example 3**   *Rainouts*

From past experience, a baseball club owner knows that about six games, on average, will have to be postponed during the season because of rain. Find the probability that this season

a. three games will have to be postponed because of rain.
b. no games will have to be postponed because of rain.

Solution
Based upon the given information, $\mu = 6$

a. Here we are interested in the probability that $x = 3$. We have

$$p(x = 3) = \frac{e^{-6}(6)^3}{3!}$$

From Table XIII $e^{-6} = 0.00247875$, so that

$$p(x = 3) = \frac{(0.00247875)(216)}{6} = 0.0892$$

Thus, the probability that three games will have to be postponed during the season because of rain is 0.0892.

b. Here we are interested in the probability that $x = 0$. We have

$$p(x = 0) = \frac{e^{-6}(6)^0}{0!} \qquad \text{(Remember } 6^0 = 1 \text{ and } 0! = 1\text{)}$$

$$= \frac{(0.00247875)1}{1}$$

$$= 0.00247875$$

Thus, the probability that no games will have to be postponed is 0.00247875.

COMMENT   The Poisson Probability Distribution can also be used as an approximation to the Binomial probability function. However, such considerations are beyond the scope of this book.

EXERCISES

1. Assume that the number of vacant beds in a nursing home on any given day follows a Poisson distribution with $\mu = 7$. What is the probability that the number of vacant beds on a given day will be less than or equal to five?

2. Bill Rogers is a medical technician. He receives an average of about seven emergency calls per day. What is the probability that on a particular day, Bill will receive
   a. at least two emergency calls?
   b. at most two emergency calls?
   c. exactly two emergency calls?

3. The probability that a skier will have an accident on a particularly dangerous slope is 0.002. What is the probability that among 2000 skiers, there will be 4 accidents? (*Hint:* $\mu = 2000 \times 0.002 = 4$.)

4. On the average, five "slugs" (counterfeit coins) are deposited daily into a particular automatic toll-collecting machine on a state highway. What is the probability that on a given day exactly six counterfeit coins will be deposited into this toll-collecting machine?

5. An expert statistical typist averages about one error for every four pages typed, or about 0.25 errors per page. What is the probability that the typist will make no errors at all while typing the next page?

6. The police department of a large city receives an average of three suicide-threat calls per day. What is the probability that on a particular day the police department will receive
   a. four or more suicide-threat calls?
   b. no suicide-threat calls?

7. A helicopter pilot on traffic patrol claims that the number of accidents occurring per hour during the morning drive-to-work period follows a Poisson distribution with an average of 0.6 accidents per hour. What is the probability of no traffic accidents occurring in a given hour during the morning drive-to-work period?

* 8. A computer dealer finds that the demand for a particular disk drive during a week follows a Poisson distribution with $\mu = 5$. The dealer wishes to have a sufficient supply of the disk so as to satisfy customer demand. How large a stock of the disk should the dealer have on hand to be able to supply the customer demand with a probability of at least 0.90?

9. The number of bacteria found on a microscope slide follows a Poisson distribution with an average of three bacteria per square centimeter. What is the probability that at most two bacteria per square centimeter will be found?

10. The number of requests for financial assistance received daily by a particular welfare agency follows a Poisson distribution with $\mu = 4$. What is the probability that the welfare agency will receive at most three requests for financial assistance on any given day?

*6.8 HYPERGEOMETRIC DISTRIBUTION

In our discussion of the binomial distribution, we noted that when we select objects without replacement we cannot use the binomial distribution formula since the probability of success from trial to trial is not constant and the trials are not independent.

Before proceeding with our discussion, let us pause for a moment to review some notation. Recall that when we discussed combinations we used the notation $_nC_r$ to represent the number of combinations of n things taken r at a time. As mentioned, another notation that is often used to represent the same idea is $\binom{n}{r}$. Thus

$$_5C_3 = \binom{5}{3} = \frac{5!}{3!2!},$$

$$_{10}C_8 = \binom{10}{8} = \frac{10!}{8!2!},$$

and

$$_nC_r = \binom{n}{r} = \frac{n!}{r!(n-r)!}$$

Suppose 5 cards are randomly drawn without replacement from a 52-card deck. Can we find the probability of obtaining 3 red and 2 black cards? The answer is yes if we proceed as follows. The 3 red cards can be selected from the 26 available red cards in $\binom{26}{3}$ possible ways. Similarly the 2 black cards can be selected from the 26 available black cards in $\binom{26}{2}$ possible ways. Thus, the total number of ways of selecting 3 red cards and 2 black cards in five draws is $\binom{26}{3} \cdot \binom{26}{2}$. Now we determine how many different ways there are for selecting 5 cards from a 52–card deck. This is simply $\binom{52}{5}$. Using the definition of probability, we have

$$\text{Prob} \begin{pmatrix} 3 \text{ red cards and 2 black} \\ \text{cards in 5 draws from a} \\ 52\text{–card deck} \end{pmatrix} = \frac{\binom{26}{3} \cdot \binom{26}{2}}{\binom{52}{5}}$$

$$= \frac{\left(\dfrac{26!}{3!23!}\right) \cdot \left(\dfrac{26!}{2!24!}\right)}{\left(\dfrac{52!}{5!47!}\right)}$$

$$= \frac{(2600)(325)}{2,598,960} = 0.3251$$

More generally, suppose we are interested in the probability of selecting x successes from k items labelled success and $n - x$ failures from $N - k$ items labelled failure when a random sample of size n is selected from N items. We can then apply the hypergeometric probability function to determine this probability. We have:

FORMULA 6.8   *Hypergeometric Probability Function*

The probability of obtaining x successes when a sample of size n is selected without replacement from N items of which k are labelled success and N − k are labelled failure is given by

$$\frac{\dbinom{k}{x}\dbinom{N-k}{n-x}}{\dbinom{N}{n}} \qquad x = 0, 1, 2, \ldots, n$$

Let us illustrate the use of this formula with several examples.

● **Example 1**
A production run of 100 radios is received by the shipping department. It is known that 10 of the radios in the production are defective. The quality control engineer randomly selects 8 of the radios from the production run. Find the probability that 6 of the radios selected are defective.

Solution
We apply the hypergeometric probability function. Here $N = 100$, $n = 8$, $k = 10$, and $N - k = 100 - 10 = 90$. We are interested in the probability that $x = 6$. Thus

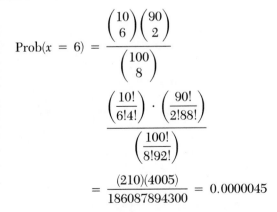

$$\text{Prob}(x = 6) = \frac{\dbinom{10}{6}\dbinom{90}{2}}{\dbinom{100}{8}}$$

$$= \frac{\left(\dfrac{10!}{6!4!}\right) \cdot \left(\dfrac{90!}{2!88!}\right)}{\left(\dfrac{100!}{8!92!}\right)}$$

$$= \frac{(210)(4005)}{186087894300} = 0.0000045$$

Thus, the probability that the quality-control engineer selects 6 defective radios is 0.0000045.

● **Example 2**

A faculty–student committee is to be selected at random from three students and six faculty members. The committee is to consist of five people. Find the probability that the committee will contain

a. 0 students.
b. 1 student.
c. 2 students.
d. 3 students.

Solution

Let x be the number of students selected to be on the committee.

a. Here $x = 0$, $N = 9$, $n = 5$, and $k = 3$. Thus

$$\text{Prob(0 students)} = \text{Prob}(x = 0) = \frac{\binom{3}{0}\binom{6}{5}}{\binom{9}{5}}$$

$$= \frac{(1)(6)}{126} = \frac{6}{126}$$

b. Here $x = 1$, $N = 9$, $n = 5$, and $k = 3$. Thus

$$\text{Prob(1 student)} = \text{Prob}(x = 1) = \frac{\binom{3}{1}\binom{6}{4}}{\binom{9}{5}}$$

$$= \frac{(3)(15)}{126} = \frac{45}{126}$$

c. Here $x = 2$, $N = 9$, $n = 5$, and $k = 3$. Thus

$$\text{Prob(2 students)} = \text{Prob}(x = 2) = \frac{\binom{3}{2}\binom{6}{3}}{\binom{9}{5}}$$

$$= \frac{(3)(20)}{126} = \frac{60}{126}$$

d. Here $x = 3$, $N = 9$, $n = 5$, and $k = 3$. Thus

$$\text{Prob(3 students)} = \text{Prob}(x = 3) = \frac{\binom{3}{3}\binom{6}{2}}{\binom{9}{5}}$$

$$= \frac{(1)(15)}{126} = \frac{15}{126}$$

EXERCISES

1. Seven cards are randomly selected from a deck of cards without replacement. Find the probability that three of the cards are spades.

2. It is known that of 50 people in a gymnasium 14 are left-handed. If a sample of 8 people in the gymnasium is selected without replacement, find the probability that the sample contains at most 3 people that are left-handed.

3. A committee of five people is to be selected from eight men and eight women. Find the probability that the committee will contain
   a. 0 men.
   b. 1 man.
   c. 2 men.
   d. 3 men.
   e. 4 men.

4. There are 60 turkeys in the freezer of a large supermarket. It is known that 6 of these turkeys are spoiled. If a sample of 9 turkeys is selected without replacement, find the probability that the sample will contain at most 2 spoiled turkeys.

5. An environmental protection club has 55 members. It is known that 25 of the members are opposed to any further construction of nuclear generating stations. A sample of 8 members from the club is randomly selected. Find the probability that half of the people in the sample are opposed to any further construction of nuclear generating stations.

6. Fourteen of the 100 students in a school have been exposed to air contaminated with excessive levels of asbestos particles from falling pieces of asbestos in their classrooms. What is the probability that in a random sample of 10 of these 100 students, 6 of them will be found *not* to have been exposed to air contaminated with excessive levels of asbestos particles?

7. A random sample of 4 homes is taken from a region in which it is known that 15 of the 25 homes have high levels of radon gases accumulated within them. What is the probability that none of the homes has a high level of radon gases?

8. A bank officer is analyzing the 75 loan applications received today. Thirty-three are for home improvement and the remainder are for a variety of personal needs. If a sample

of 10 of these loan applications is randomly selected without replacement, what is the probability that it will contain all home improvement loan applications?

9. Find the probability of being dealt a bridge hand of 13 cards that contains exactly 5 picture cards.

10. Refer back to the previous exercise. Find the probability of being dealt a bridge hand of 13 cards that contains 9 red cards (hearts or diamonds) and 4 black cards (spades or clubs).

6.9 SUMMARY

In this chapter we discussed how the ideas of probability can be combined with frequency distributions. Specifically, we introduced the idea of a random variable and its probability distribution. These enable an experimenter to analyze outcomes of experiments and to speak about the probability of different outcomes.

We then discussed the mean and variance of a probability distribution. These allow us to determine the expected number of favorable outcomes of an experiment.

Although we did not emphasize the point, all the events discussed were mutually exclusive. Thus, we were able to add probabilities by the addition rule for probabilities. Also, the events were independent. This allowed us to multiply probabilities by the multiplication rule for probabilities of independent events.

We discussed one particular distribution in detail, the binomial distribution, since it is one of the most widely used distributions in statistics. In addition to the binomial distribution formula itself, which allows us to calculate the probability of getting a specified number of successes in repeated trials of an experiment, formulas for calculating its mean, variance, and standard deviation were given. These formulas were applied to numerous examples. Because of its importance, we will discuss the binomial distribution further in Chapter 7.

We also discussed the Poisson distribution, which can be used only when certain assumptions are satisfied. These were given on page 298.

Finally we presented the hypergeometric probability function which is used when the sampling is without replacement. Thus, we cannot use the binomial distribution since the probability of success is not the same from trial to trial.

STUDY GUIDE

You should now be able to demonstrate your knowledge of the following ideas presented in this chapter by giving definitions, descriptions, or specific examples. Page references are given for each term so that you can check your answer.

FORMULAS TO REMEMBER

You should be able to identify each symbol in the following formulas, understand the relationships among the symbols expressed in each formula, understand the significance of each formula, and use the formulas in solving problems.

1. Mean of a Probability distribution: $\mu = \Sigma x \cdot p(x)$

2. Variance of a Probability distribution: $\sigma^2 = \Sigma(x - \mu)^2 \cdot p(x)$

3. Variance of a Probability distribution: $\sigma^2 = \Sigma x^2 \cdot p(x) - [\Sigma x \cdot p(x)]^2$

4. Standard Deviation of a Probability distribution: $\sigma = \sqrt{\text{variance}}$

5. Binomial distribution

$$p(x \text{ successes out of } n \text{ trials}) = \frac{n!}{x!(n - x)!} p^x q^{n-x}$$

6. Mean of the Binomial distribution: $\mu = np$

7. Variance of the Binomial distribution: $\sigma^2 = npq$

8. Poisson distribution: $p(x \text{ successes}) = \dfrac{e^{-\mu}\mu^x}{x!}$ $\qquad x = 0, 1, 2, 3, \ldots$

9. Mean of the Poisson distribution is μ.

10. Variance of a Poisson distribution is μ.

11. $_nC_x = \begin{pmatrix} n \\ x \end{pmatrix} = \dfrac{n!}{x!(n-x)!}$

12. Hypergeometric distribution

$$p(x \text{ successes}) = \dfrac{\begin{pmatrix} k \\ x \end{pmatrix} \begin{pmatrix} N-k \\ n-x \end{pmatrix}}{\begin{pmatrix} N \\ n \end{pmatrix}} \qquad x = 0, 1, 2, \ldots, n$$

The tests of the following section will be more useful if you take them after you have studied the examples and solved the exercises given in this chapter.

MASTERY TESTS

Form A

In working the following exercises, assume that all the necessary conditions for using the binomial distribution are satisfied (unless told otherwise).

1. It has been estimated that about 50% of the state's bars are obeying the state's new law against serving alcoholic beverages to anyone under 21 years of age. If there are 100 bars in a certain district of the state, about how many of these can be expected to obey the state's new law against serving alcoholic beverages to anyone under 21 years of age?
   a. 50   **b.** 25   **c.** 5   **d.** 100   **e.** none of these

2. Refer back to the previous question. Find the standard deviation.
   a. 50   **b.** 25   **c.** 5   **d.** 100   **e.** none of these

3. In the following probability distribution, find the probability that $x = 9$.

x	7	8	9	10	11	12
p(x)	0.12	0.37	?	0.09	0.25	0.11

   a. 0.94   **b.** 0.81   **c.** 0.69   **d.** 0.06   **e.** none of these

4. A hospital administrator finds that 8% of all claims filed with Medicaid are rejected for one reason or another. During November, her staff filed 300 claims with Medicaid. About how many claims can be expected *not* to be rejected?
   a. 24   **b.** 276   **c.** 274   **d.** 237   **e.** none of these

5. Environmentalists estimate that two out of every seven deer on a particular reservation die from starvation because of overpopulation. If ten deer are randomly selected, what is the probability that three of them will die from starvation?

a. $\dfrac{10!}{3!7!}\left(\dfrac{2}{7}\right)^3\left(\dfrac{5}{7}\right)^7$ **b.** $\dfrac{10!}{3!7!}\left(\dfrac{2}{7}\right)^7\left(\dfrac{5}{7}\right)^3$

c. $\dfrac{10!}{3!}\left(\dfrac{2}{7}\right)^3\left(\dfrac{5}{7}\right)^7$ **d.** $\dfrac{10!}{7!}\left(\dfrac{2}{7}\right)^3\left(\dfrac{5}{7}\right)^7$

e. none of these

6. Consider the following probability distribution, which gives the probability of an accident occurring daily on the state freeway. Find the variance.

Number of Accidents Occurring on Freeway x	Probability p(x)
0	0.13
1	0.18
2	0.23
3	0.17
4	0.19
5	0.10

a. 2.41   **b.** 5.8081   **c.** 8.17   **d.** 2.3619   **e.** none of these

7. What is the probability that a family that has four children will have *at least* three boys?

a. $\dfrac{1}{2}$ **b.** $\dfrac{3}{8}$ **c.** $\dfrac{1}{4}$ **d.** $\dfrac{5}{16}$ **e.** none of these

8. Consider the following probability distribution. Find its mean.

x	0	1	2	3	4	5
p(x)	0.16	0.18	0.28	0.14	0.19	0.05

a. 2.50   **b.** 0.36   **c.** 2.17   **d.** 2.33   **e.** none of these

9. It has been estimated that about 50% of the adult population in Tompkins go to a dentist only when they have a toothache. If there are 400 adults in the town of Tompkins, about how many of the adult population can be expected to go to a dentist only when they have a toothache?
a. 200   **b.** 100   **c.** 50   **d.** 10   **e.** none of these

10. Refer back to the previous question. Find the variance.
a. 200   **b.** 100   **c.** 50   **d.** 10   **e.** none of these

1. A carton of eggs contains eight hard-boiled eggs and four one-minute-boiled eggs. Three eggs are randomly selected without replacement. Find the probability that all three eggs are hard boiled.

2. A cookie company makes chocolate chip cookies with the following distribution of chocolate chips:

x	0	1	2	3	4	5
p(x)	0.05	0.22	0.38	0.25	0.07	0.03

   Find the expected number of chocolate chips in a cookie.

3. Refer back to the previous question. Find the standard deviation.

4. An airline company claims that 96% of its scheduled plane flights arrive on time. This Sunday the airline has 12 flights scheduled. Assuming independence, find the probability that *at least* eight of these flights arrive on time.

5. The child-abuse hotline in a large city receives calls for help at an average rate of about 3.2 calls per day. Assuming these calls are received independently, what is the probability that the hotline will receive three or more calls on a given day? (*Hint:* Poisson.)

6. If the probability that the average freshman will complete college is $\frac{2}{7}$, what is the probability that *at most* three of ten randomly selected freshmen will complete college?

7. According to the dean, 4 out of every 13 students change their academic major within their first two years of college. If 12 students are randomly selected, what is the probability that exactly 3 of them will change their academic major within their first two years of college? (*Hint:* Binomial.)

8. A typesetter of a highly technical book averages about two typesetting errors per page set. What is the probability that the typesetter will make *at least* one error on the next page that is typeset? (*Hint:* Poisson.)

* 9. An auto supply store owner finds that the demand for a particular type of carburetor follows a Poisson distribution with $\mu = 5$. The store owner is anxious to have sufficient stock to satisfy customer demand or the customers will go elsewhere for all of their needs. How large a stock of the item should the store owner have on hand so as to be able to supply the customer demand with a probability of *at least* 0.90?

10. *Cancer*   According to medical researchers, about 30% of individuals diagnosed as having one particular form of cancer die within a year. If 12 individuals are diagnosed as having this form of cancer, find the probability that *at most* 3 of them will die within a year.

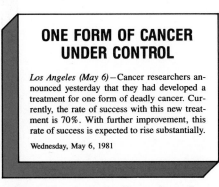

**ONE FORM OF CANCER
UNDER CONTROL**

Los Angeles (May 6) — Cancer researchers announced yesterday that they had developed a treatment for one form of deadly cancer. Currently, the rate of success with this new treatment is 70%. With further improvement, this rate of success is expected to rise substantially.

Wednesday, May 6, 1981

If 20 people who are suffering from the particular form of cancer mentioned are randomly selected and are subjected to this new treatment, what is the probability that 8 of them will recover? Can we determine such probabilities?

11. A used-car dealer has 100 cars to sell. It is known that the odometer (which indicates the number of miles that the car has been driven) on 12 of these cars has been tampered with. If a random sample of 10 of the cars is selected without replacement, find the probability that the sample contains *at most* 2 cars whose odometers have been tampered with.

SUGGESTED FURTHER READING

1. "Computers and Computation." Readings from *Scientific American*, Article 16. San Francisco, CA: Freeman, 1971.
2. Hoel, Paul G., Sidney C. Port, and Charles J. Stone. *Introduction to Probability Theory*. Boston: Houghton Mifflin, 1972.
3. Mendenhall, William. *Introduction to Probability and Statistics*, 6th ed. North Scituate, MA: Duxbury Press, 1983.
4. Kruskal, W., et al, eds. *Statistics by Example: Winning Chances*. Reading, MA: Addison-Wesley, 1973.
5. Mosteller, Frederick, R. E. K. Rourke, and G. B. Thomas, Jr. *Probability with Statistical Applications*, 2nd ed. Reading, MA: Addison-Wesley, 1970.
6. National Bureau of Standards. *Tables of the Binomial Probability Distribution*. Applied Mathematics Series 6. Washington, D.C.: U.S. Department of Commerce, 1949.
7. Runyon, R. P. and Audrey Haber. *Fundamentals of Behavioral Statistics*. Reading, MA: Addison-Wesley, 1967.

CHAPTER 7

The Normal Distribution

OBJECTIVES

- *To discuss* in detail a probability distribution which is bell-shaped or mound-shaped, called the normal distribution.

- *To indicate* how the normal distribution can be used to calculate probabilities.

- *To apply* the normal distribution to many different situations. This will be accomplished by converting to z-scores or standard scores and using the standard normal distribution.

- *To point out* how statistical quality-control charts are used in industry. This represents an additional application of the normal distribution.

- *To learn* how the normal distribution can be used to simplify lengthy computations involving the binomial distribution.

- *To mention* briefly some historical facts about some of the mathematicians who used the normal distribution.

The 1984 Age Distribution of the U.S. Population:

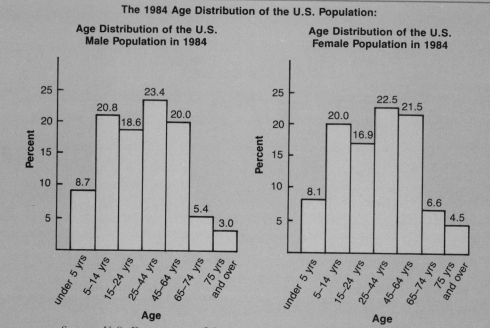

Source: U.S. Department of Commerce, Bureau of the Census.

The above graph gives the age distribution of the U.S. population for both males and females. Contrary to expectation, the ages for both sexes are not normally distributed. Many people believe that in order to achieve zero population growth, the age distribution should look like the following graph, where we have drawn the bars side by side horizontally for ease in comparison.

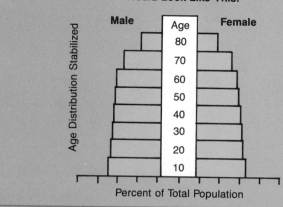

The Age Distribution in a Stable Population Would Look Like This:

Source: ZPG HAS NOT YET BEEN REACHED, Zero Population Growth, Inc., Washington, D.C.

7.1 INTRODUCTION

Until now the random variables that we discussed in detail assumed only the limited values of 0, 1, 2, ... Thus, when a coin is flipped many times, the number of heads that comes up is 0, 1, 2, 3, ... Also the number of defective bulbs in a shipment of 100 bulbs is 0, 1, 2, 3, . . . , 100. Since these random variables can assume only the values 0, 1, 2, . . . , they are called **discrete random variables.**

As opposed to the preceding examples, consider the following random variables: the length of a page of this book, the height of your pet dog, the temperature at noon on New Year's Day, or the weight of a bag of sugar on your grocer's shelf. Since each of these variables can assume an infinite number of values on a measuring scale, they are called **continuous random variables.** Thus, the weight of a bag of sugar can be 5 pounds, 5.1 pounds, 5.161 pounds, 5.16158 pounds, 5.161581 pounds, and so forth, depending upon the accuracy of the scale. Similarly, the temperature at noon on New Year's Day may be 38°, 38.2°, 38.216°, and so forth.

Among the many different kinds of distributions of random variables that are used by statisticians, the normal distribution is by far the most important. This type of distribution was first discovered by the English mathematician Abraham De Moivre (1667–1754). De Moivre spent many hours with London gamblers. In his *Annuities Upon Lives*, which played an important role in the history of actuarial mathematics, and his *Doctrine of Chances*, which is a manual for gamblers, he essentially developed the first treatment of the normal probability curve, which is important in the study of statistics. De Moivre also developed a formula, known as Stirling's formula, that is used for approximating factorials of large numbers.

A rather interesting story is told of De Moivre's death. According to the story, De Moivre was ill and each day he noticed that he slept a quarter of an hour longer than on the preceding day. Using progressions, he computed that he would die in his sleep on the day after he slept 23 hours and 45 minutes. On the day following a sleep of 24 hours De Moivre died.

Many years later the French mathematician Pierre-Simon Laplace (1749–1827) applied the normal distribution to astronomy and other practical problems. The normal distribution was also used extensively by the German mathematician Carl Friedrich Gauss (1777–1855) in his studies of physics and astronomy. Gauss is considered by many as the greatest mathematician of the nineteenth century. At the age of three he is alleged to have detected an error in his father's bookkeeping records.

The normal distribution is sometimes known as the **bell-shaped** or **Gaussian distribution** in honor of Gauss, who studied it extensively.

Although there are other distributions of continuous random variables that are important in statistics, the normal distribution is by far the most important. In this chapter we will discuss in detail the nature of the normal distribution, its properties, and its uses.

7.2 THE GENERAL NORMAL CURVE

Refer back to the frequency polygon given on page 52 of Chapter 2. It is reproduced below. Experience has taught us that for many frequency distributions drawn from large populations, the frequency polygons approximate what is known as a **normal** or **bell-shaped curve** as shown in Figure 7.1.

Heights and weights of people, IQ scores, waist sizes, or even life expectancy of cars, to name but a few, are all examples of distributions whose frequency polygons approach a normal curve when the samples taken are from large populations. When the graph of a frequency distribution resembles the bell-shaped curve shown in Figure 7.2, the graph is called a **normal curve** and its frequency distribution is known as a **normal distribution.** The word normal is simply a name for this particular distribution. It does not indicate that this distribution is more typical than any other.

Since the normal distribution has wide-ranging applications, we need a careful description of a normal curve and some of its properties.

As mentioned before, the graph of a normal distribution is a bell-shaped curve. It extends in both directions. Although the curve gets closer and closer to the horizontal axis, it never really crosses it, no matter how far it is extended. The normal distribution is a probability distribution satisfying the following properties:

1. The mean is at the center of the distribution and the curve is symmetrical about the mean. This tells us that we can fold the curve along the dotted

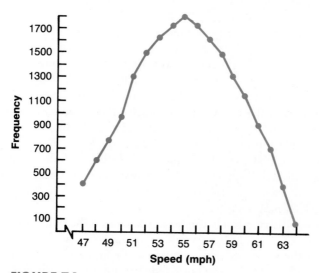

FIGURE 7.1

Distribution of the speeds of cars as they passed through a radar trap.

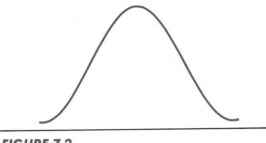

FIGURE 7.2
Normal curve.

line shown in Figure 7.3 and either portion of the curve will correspond with the other portion.

2. Its graph is the bell-shaped curve, most often referred to as the normal curve.
3. The mean equals the median.
4. The scores that make up the normal distribution tend to cluster around the middle, with very few values more than three standard deviations away from the mean on either side.

Normal distributions can come in different sizes and shapes. Some are tall and skinny or flat and spread out as shown in Figure 7.4. However, for a given mean and a given standard deviation, there is one and only one normal distribution. The normal distribution is completely specified once we know its mean and standard deviation.

We mentioned in Chapter 2 (see page 27) that the area under a particular rectangle of a histogram gives us the relative frequency and hence the probability of obtaining values within that rectangle. We can generalize this idea to any distribution. We say that *the area under the curve between any two points a and b gives us the probability that the random variable having this particular continuous distribution will assume values between a and b.* This idea is very important since calculating probabilities for the normal distribution

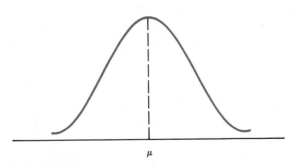

FIGURE 7.3

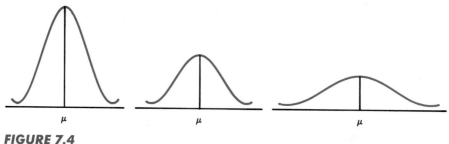

FIGURE 7.4
Different normal distributions.

will depend upon the areas under the curve. Also since the sum of the probabilities of a random variable assuming all possible values must be 1 (see page 263), the total area under its probability curve must also be 1.

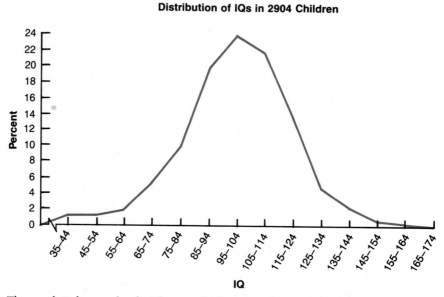

The graph indicates the distribution of IQ scores of 2904 children. Notice the type of frequency distribution that is illustrated. This is approximately a normal distribution.

7.3 THE STANDARD NORMAL CURVE

A normal distribution is completely specified by its mean and standard deviation. Thus, although all normal distributions are basically mound-shaped, different means and different standard deviations will describe different bell-shaped curves. However, it is possible to convert each of these different normal distributions into one standardized form. You may be wondering, why bother? The answer is rather simple.

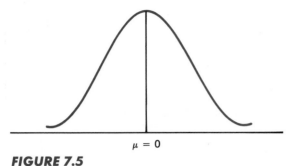

$\mu = 0$

FIGURE 7.5
Curve of a typical standard distribution.

Since areas under a normal curve are related to probability, we can use special normal distribution tables for calculating probabilities. Such a table is given in the Appendix at the end of this book. However, since the mean and standard deviation can be any values, it would seem that we need an endless number of tables. Fortunately, this is not the case. We only need one standardized table. Thus, the area under the curve between 40 and 60 of a normal distribution with a mean of 50 and a standard deviation of 10 will be the same as the area between 70 and 80 of another normally distributed random variable with mean 75 and standard deviation 5. They are both within one standard deviation unit from the mean. It is for this reason that statisticians use a standard normal distribution.

DEFINITION 7.1

A **standardized normal distribution** *is a normal distribution with a mean of 0 and a standard deviation of 1. The curve of a typical standard distribution is shown in Figure 7.5.*

Table V in the Appendix gives us the areas of a standard normal distribution between $z = 0$ and $z = 3.09$. We read this table as follows: The first two digits of the z-score are under the column headed by z, the third digit heads the other columns. Thus, to find the area from $z = 0$ to $z = 2.43$ we first look under z to 2.4 and then read across from $z = 2.4$ to the column headed by 0.03. The area is 0.4925, or 49.25%.

Similarly, to find the area from $z = 0$ to $z = 1.69$ we first look under $z = 1.6$ and then read across from $z = 1.6$ to the column headed by 0.09. The area is 0.4545.

● **Example 1**
Find the area between $z = 0$ and $z = 1$ in a standard normal curve.

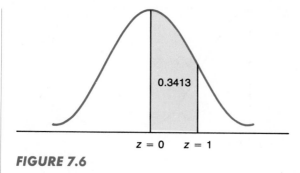

$z = 0$ $z = 1$

FIGURE 7.6

Solution

We first draw a sketch as shown in Figure 7.6. Then using Table V for $z = 1.00$ we find that the area between $z = 0$ and $z = 1$ is 0.3413. This means that the probability of a score with this normal distribution falling between $z = 0$ and $z = 1$ is 0.3413.

● **Example 2**

Find the area between $z = -1.15$ and $z = 0$ in a standard normal curve.

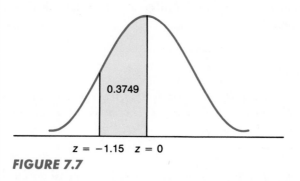

$z = -1.15$ $z = 0$

FIGURE 7.7

Solution

We first draw a sketch as shown in Figure 7.7. Then using Table V we look up the area between $z = 0$ and $z = 1.15$. The area is 0.3749, not -0.3749. A negative value of z just tells us that the value is to the left of the mean. The area under the curve (and the resulting probability) is *always* a positive number. Thus, the probability of getting a z-score between 0 and -1.15 is 0.3749.

● **Example 3**

Find the area between $z = -1.63$ and $z = 2.22$ in a standard normal curve.

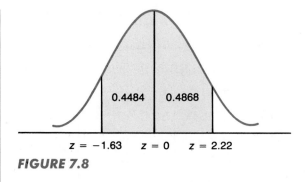

FIGURE 7.8

Solution

We draw the sketch shown in Figure 7.8. Since Table V gives the area only from $z = 0$ on, we first look under the normal curve from $z = 0$ to $z = 1.63$. We get 0.4484. Then we look up the area between $z = 0$ and $z = 2.22$. We get 0.4868. Finally we add these two together and get

$$0.4484 + 0.4868 = 0.9352$$

Thus, the probability that a z-score is between $z = -1.63$ and $z = 2.22$ is 0.9352.

By following a procedure similar to that used in Example 3, you should verify the following:

PROPERTIES OF STANDARD NORMAL DISTRIBUTION

1. *The probability that a z-score falls within one standard deviation of the mean on either side, that is, between $z = -1$ and $z = 1$, is approximately 68%.*
2. *The probability that a z-score falls within two standard deviations of the mean, that is, between $z = -2$ and $z = 2$, is approximately 95%.*
3. *The probability that a z-score falls within three standard deviations of the mean is approximately 99.7%.*

Thus, approximately 99.7% of the z-scores fall within $z = -3$ and $z = 3$ (see Figure 7.9).

It should be noted that the above three statements are true because we are dealing with a normal distribution. A more general result, which is true for any set of measurements—population or sample—and regardless of the shape of the frequency distribution is known as Chebyshev's Theorem, which we discussed in Chapter 4. We restate it here:

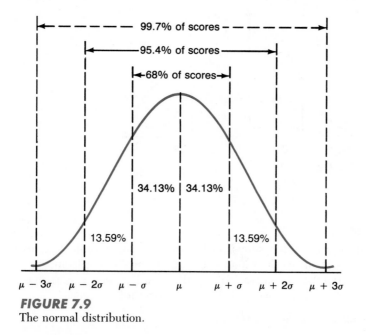

FIGURE 7.9
The normal distribution.

CHEBYSHEV'S THEOREM

Let K be any number equal to or greater than 1. Then the proportion of any distribution that lies within K standard deviations of the mean is at least $1 - \dfrac{1}{K^2}$.

Whereas Chebyshev's theorem is more general in scope, if we know that we have a normal distribution, we use the results given above. Since in this chapter we are interested in the normal distribution, we will not use Chebyshev's theorem.

In many cases we have to find areas between two given values of z or areas to the right or left of some value of z. Finding these areas is an easy task provided we remember that the area under the entire normal distribution is 1. Thus, since the normal distribution is symmetrical about $z = 0$, we conclude that the area to the right of $z = 0$ and the area to the left of $z = 0$ are both equal to 0.5000.

● **Example 4**
Find the area between $z = 0.87$ and $z = 2.57$ in a standard normal distribution (see Figure 7.10).

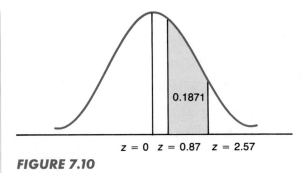

$z = 0$ $z = 0.87$ $z = 2.57$

FIGURE 7.10

Solution

We cannot look this up directly since the chart starts at 0, not at 0.87. However, we can look up the area between $z = 0$ and $z = 2.57$ and get 0.4949, then look up the area between $z = 0$ and $z = 0.87$ and get 0.3078. We then take the difference between the two and get

$0.4949 - 0.3078 = 0.1871$

● **Example 5**

Find the probability of getting a z-value less than 0.43 in a standard normal distribution.

Solution

The probability of getting a z-value less than 0.43 really means the area under the curve to the left of $z = 0.43$. This represents the shaded portion of Figure 7.11. We look up the area from $z = 0$ to $z = 0.43$ and get 0.1664 and add this to 0.5000 to get

$0.5000 + 0.1664 = 0.6664$

Thus the probability of getting a z-value less than 0.43 is 0.6664.

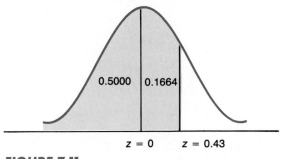

$z = 0$ $z = 0.43$

FIGURE 7.11

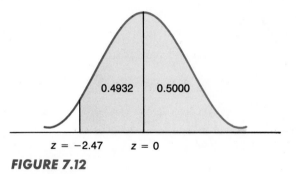

$z = -2.47$ $\quad$ $z = 0$

FIGURE 7.12

● **Example 6**

Find the probability of getting a z-value in a standard normal distribution which is

a. greater than -2.47.
b. greater than 1.82.
c. less than -1.53.

Solution

a. Using Table V we first find the area between $z = 0$ and $z = 2.47$ (see Figure 7.12). We get 0.4932. Then we add this to 0.5000, which is the area to the right of $z = 0$, and get

$$0.4932 + 0.5000 = 0.9932$$

b. Here we are interested in finding the area to the right of $z = 1.82$. See Figure 7.13. We find the area from $z = 0$ to $z = 1.82$. It is 0.4656. Since we are interested in the area to the right of $z = 1.82$, we must *subtract* 0.4656 from 0.5000, which represents the *entire* area to the right of $z = 0$. We get

$$0.5000 - 0.4656 = 0.0344$$

Thus, the probability that a z-score is greater than $z = 1.82$ is 0.0344.

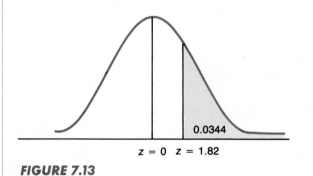

$z = 0$ $\quad$ $z = 1.82$

FIGURE 7.13

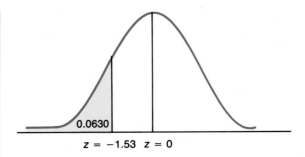

0.0630

$z = -1.53$ $z = 0$

FIGURE 7.14

c. Here we are interested in the area to the left of $z = -1.53$. See Figure 7.14. We calculate the area between $z = 0$ and $z = -1.53$. It is 0.4370. We subtract this from 0.5000. Our result is

$$0.5000 - 0.4370 = 0.0630$$

Thus, the probability that a z-score is less than $z = -1.53$ is 0.0630.

In the preceding examples we interpreted the area under the normal curve as a probability. If we know the probability of an event, we can look at the probability chart and find the z-value that corresponds to this probability.

● **Example 7**
If the probability of getting less than a certain z-value is 0.1190, what is the z-value?

Solution
We first draw the sketch shown in Figure 7.15. Since the probability is 0.1190, which is less than 0.5000, we know that the z-value must be to the left of the mean. We subtract 0.1190 from 0.5000 and get

$$0.5000 - 0.1190 = 0.3810$$

See Figure 7.15. This means that the area between $z = 0$ and some z-value is 0.3810. Table V tells us that the z-value is 1.18. However, this is to the left of the mean. Therefore, our answer is $z = -1.18$.

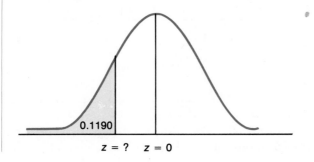

0.1190

$z = ?$ $z = 0$

FIGURE 7.15

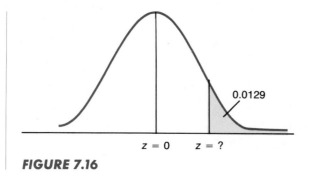

FIGURE 7.16

● **Example 8**

If the probability of getting larger than a certain z-value is 0.0129, what is the z-value?

Solution

In this case we are told that the area to the right of some z-value is 0.0129. This z-value must be on the right side. If it were on the left side, the area would have to be at least 0.5000. Why? See Figure 7.16. Thus, we subtract 0.0129 from 0.5000 and get 0.4871. Then we look up this probability in Table V and find the z-value that gives this probability. It is $z = 2.23$. If $z = 2.23$, the probability of getting a z-value greater than 2.23 is 0.0129.

The General Normal Distribution

If we are given a normal distribution with a mean different from 0 and a standard deviation different from 1, we can convert this normal distribution into a standardized normal distribution by converting each of its scores into standard scores. To accomplish this we use Formula 3.7 of Chapter 3, which we will now call Formula 7.1:

FORMULA 7.1

$$z = \frac{x - \mu}{\sigma}$$

Expressing the scores of a normal distribution as standard scores allows us to calculate different probabilities, as the following examples will show.

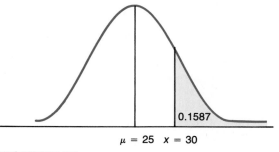

$\mu = 25 \quad x = 30$

FIGURE 7.17

● **Example 9**

In a normal distribution, $\mu = 25$ and $\sigma = 5$. What is the probability of obtaining a value

a. greater than 30?
b. less than 15?

Solution

a. We use Formula 7.1. We have $\mu = 25$, $x = 30$, and $\sigma = 5$, so that

$$z = \frac{x - \mu}{\sigma} = \frac{30 - 25}{5} = \frac{5}{5} = 1$$

See Figure 7.17. Thus we are really interested in the area to the right of $z = 1$ of a standardized normal curve. The area from $z = 0$ to $z = 1$ is 0.3413. The area to the right of $z = 1$ is then

$$0.5000 - 0.3413 = 0.1587$$

Therefore, the probability of obtaining a value greater than 30 is 0.1587.

b. We use Formula 7.1. We have $\mu = 25$, $x = 15$, and $\sigma = 5$, so that

$$z = \frac{x - \mu}{\sigma} = \frac{15 - 25}{5} = \frac{-10}{5} = -2$$

See Figure 7.18. Thus, we are interested in the area to the left of $z = -2$.

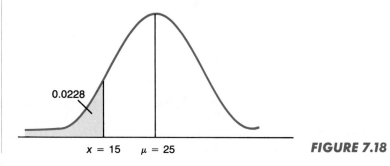

$x = 15 \qquad \mu = 25$

FIGURE 7.18

The area from $z = 0$ to $z = -2$ is 0.4772. Thus the area to the left of $z = -2$ is

$$0.5000 - 0.4772 = 0.0228$$

The probability of obtaining a value less than 15 is therefore 0.0228.

● **Example 10**
Find the percentile rank of 20 in a normal distribution with $\mu = 15$ and $\sigma = 2.3$.

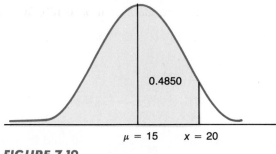

FIGURE 7.19

Solution
The problem is to find the area to the left of 20 in a normal distribution with $\mu = 15$ and $\sigma = 2.3$. We use Formula 7.1 with $\mu = 15$, $x = 20$, and $\sigma = 2.3$, so that

$$z = \frac{x - \mu}{\sigma} = \frac{20 - 15}{2.3} = \frac{5}{2.3} = 2.17$$

The area between $z = 0$ and $z = 2.17$ is 0.4850. See Figure 7.19. Thus, the area to the left of 20 is

$$0.5000 + 0.4850 = 0.9850$$

The percentile rank of 20 is therefore 98.5.

● **Example 11**
In a certain club, heights of members are normally distributed with $\mu = 63$ inches and $\sigma = 2$ inches. If Sam is in the 90th percentile, find his height.

Solution
Since Sam is in the 90th percentile, this means that 90% of the club members are shorter than he is. So, the problem here is to find a z-value that has 90 percent of the area to the left of z. Therefore, we look in the area portion of Table V to find a z-value that has 0.4000 of the area to its left. See Figure 7.20. We use 0.4000, not 0.9000, since 0.5000 of this is to the left of $z = 0$. The

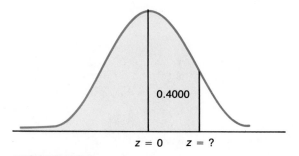

FIGURE 7.20

closest entry is 0.3997, which corresponds to $z = 1.28$. Now we convert this score into a raw score by using Formula 3.8, on page 136. We have

$$x = \mu + z\sigma$$
$$= 63 + 1.28(2)$$
$$= 63 + 2.56$$
$$= 65.56$$

Thus, Sam's height is approximately 65.56 inches.

● **Example 12**

Use the same information as Example 11 of this section except Bill's percentile rank is 40. Find his height.

Solution

Since Bill's percentile rank is 40, this means that 40% of the club members are shorter than he is. The problem here is to find a z-value that has 40 percent of the area to the left of z. See Figure 7.21. Since we are given the area to the left of z, we must subtract 0.4000 from 0.5000:

$$0.5000 - 0.4000 = 0.1000$$

Then we look in the area portion of Table V to find a z-value that has 0.1000 as the area between $z = 0$ and that z-value. The closest entry is 0.0987, which corresponds to $z = 0.25$. Since this z-value is to the left of the mean, we have $z = -0.25$. When we convert this score into a raw score, we have

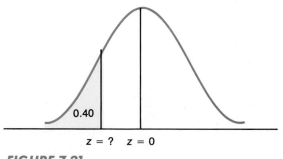

FIGURE 7.21

$$x = \mu + z\sigma$$
$$= 63 + (-0.25)2$$
$$= 63 - 0.50 = 62.50$$

Therefore, Bill's height is approximately 62.50 inches.

EXERCISES

1. In a standard normal distribution, find the area that lies
  a. between $z = 0$ and $z = 1.83$.
  b. between $z = -0.63$ and $z = 0$.
  c. to the right of $z = 1.98$.
  d. to the left of $z = -0.86$.
  e. to the right of $z = 2.47$.
  f. between $z = -1.68$ and $z = 1.54$.
  g. between $z = -1.62$ and $z = 2.28$.
  h. between $z = -2.87$ and $z = -1.47$.

2. In a standard normal distribution, find the percentage of scores that are
  a. above $z = -1.87$.
  b. below $z = 0.97$.
  c. between $z = 1.58$ and $z = 2.07$.
  d. above $z = 2.83$.
  e. above $z = 3.42$.
  f. between $z = -2.23$ and $z = -1.38$.
  g. between $z = -1.42$ and $z = 1.08$.
  h. between $z = -2.92$ and $z = -1.83$.

3. In a normal distribution, find the z-score that cuts off the bottom
  a. 28 percent.
  b. 17 percent.
  c. 35 percent.
  d. 5 percent.

4. In a normal distribution, find the z-score(s) that cut off the middle
  a. 18 percent.
  b. 28 percent.
  c. 35 percent.

5. Find z, if the area under a standard normal curve
  a. between $z = 0$ and z is 0.4808.
  b. to the left of z is 0.9854.
  c. to the left of z is 0.2033.
  d. to the right of z is 0.0037.
  e. between $z = 1.14$ and z is 0.1155.
  f. between $z = 2.06$ and z is 0.0180.

6. In a normal distribution with $\mu = 60$ and $\sigma = 7$, find the percentage of scores that are
  a. between 55 and 70.
  b. between 51 and 58.
  c. between 48 and 62.
  d. greater than 58.
  e. less than 65.

7. In a normal distribution with $\mu = 52$ and $\sigma = 9$, find the percentile rank of a score of
  a. 42.
  b. 65.
  c. 45.
  d. 70.

8. All candidates for a public relations job at a large company must take a special personality exam. It is known that the test scores are approximately normally distributed with a mean of 150 and a standard deviation of 17. Find the score of
  a. Melissa, if she is in the 85th percentile.
  b. Louis, if he is in the 47th percentile.
  c. Ken, if he is in the 93rd percentile.

9. The following is a standard normal curve with various z-values marked on it. If this curve also represents a normal distribution with $\mu = 48$ and $\sigma = 7$, replace the z-values with raw scores.

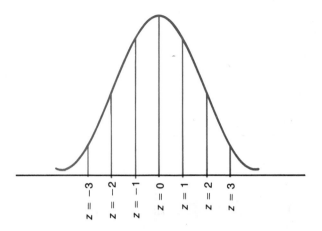

$$z = -3 \quad z = -2 \quad z = -1 \quad z = 0 \quad z = 1 \quad z = 2 \quad z = 3$$

10. A normal distribution has mean $\mu = 50$ and unknown standard deviation σ. However, it is known that 30% of the area lies to the right of 60. Find σ.

11. A normal distribution has an unknown mean, μ, with a standard deviation of 12.37. However, it is known that the probability that a score is less than 85 is 0.7123. Find μ.

*12. A certain normal distribution has unknown mean μ and unknown standard deviation σ. However, it is known that 15.87% of the scores are less than 29 and 2.28% of the scores are more than 67. Find μ and σ.

7.4 SOME APPLICATIONS OF THE NORMAL DISTRIBUTION

We mentioned earlier that the importance of the normal distribution lies in its wide-ranging applications. In this section we will apply the normal distribution to some concrete examples.

● **Example 1**
From past experience it has been found that the weight of a newborn infant at a maternity hospital is normally distributed with mean $7\frac{1}{2}$ pounds (which equals 120 ounces) and standard deviation 21 ounces. If a newborn baby is selected at random, what is the probability that the infant weighs less than 4 pounds 15 ounces (which equals 79 ounces)?

Solution
We use Formula 7.1 of Section 7.3. Here $\mu = 120$, $\sigma = 21$, and $x = 79$, so that

$$z = \frac{x - \mu}{\sigma} = \frac{79 - 120}{21} = \frac{-41}{21}, \text{ or } -1.95$$

Thus, we are interested in the area to the left of $z = -1.95$. The area from $z = 0$ to $z = -1.95$ is 0.4744, so that the area to the left of $z = -1.95$ is

from table

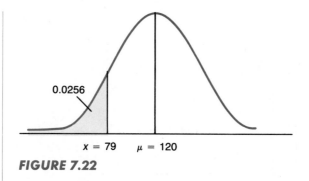

0.0256

x = 79   μ = 120

FIGURE 7.22

$0.5000 - 0.4744$, or 0.0256. See Figure 7.22. Therefore, the probability that a randomly selected baby weighs less than 4 pounds 15 ounces is 0.0256.

● **Example 2**

The Flatt Tire Corporation claims that the useful life of its tires is normally distributed with a mean life of 28,000 miles and with a standard deviation of 4000 miles. What percentage of the tires are expected to last more than 35,000 miles?

Solution

Here $\mu = 28,000$, $\sigma = 4000$, and $x = 35,000$. Using Formula 7.1 we get

$$z = \frac{x - \mu}{\sigma} = \frac{35,000 - 28,000}{4000} = 1.75$$

See Figure 7.23. We are interested in the area to the right of $z = 1.75$. The area between $z = 0$ and $z = 1.75$ is 0.4599, so the area to the right of $z = 1.75$ is 0.0401. Thus approximately 4% of the tires can be expected to last more than 35,000 miles.

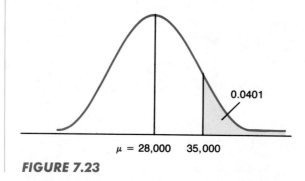

0.0401

μ = 28,000   35,000

FIGURE 7.23

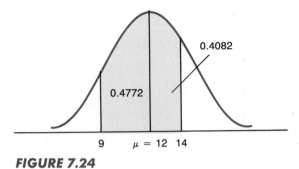

FIGURE 7.24

● **Example 3**
In a recent study it was found that in one town the number of hours that a typical ten-year-old child watches television per week is normally distributed with a mean of 12 hours and with a standard deviation of 1.5 hours. If Gary is a typical ten-year-old child in this town, what is the probability that he watches between 9 and 14 hours of television per week?

Solution
We first find the probability that Gary will watch television between 12 and 14 hours per week and add to this the probability that he will watch television between 9 and 12 hours per week. Using Formula 7.1 we have

$$z = \frac{x - \mu}{\sigma} = \frac{14 - 12}{1.5} = 1.33$$

The area between $z = 0$ and $z = 1.33$ is 0.4082. See Figure 7.24. Similarly,

$$z = \frac{x - \mu}{\sigma} = \frac{9 - 12}{1.5} = \frac{-3}{1.5} = -2$$

The area between $z = 0$ and $z = -2$ is 0.4772. See Figure 7.24. Adding these two probabilities we get $0.4082 + 0.4772 = 0.8854$. Thus, the probability that Gary watches between 9 and 14 hours of television per week is 0.8854.

● **Example 4**
Daisy discovers that the amount of time it takes her to drive to work is normally distributed with a mean of 35 minutes and a standard deviation of 7 minutes. At what time should Daisy leave her home so that she has a 95% chance of arriving at work by 9 A.M.?

Solution
We first find a z-value that has 95% of the area to the left of z. See Figure 7.25. Thus we look in the area portion of Table V to find a z-value that has 0.4500 of the area to its left. (Remember 0.5000 of the area is to the left of $z = 0$.) From Table V we find that z is midway between 1.64 and 1.65. We will

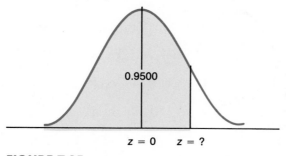

z = 0   z = ?

FIGURE 7.25

use 1.65. We then find the raw score corresponding to $\mu = 35$, $\sigma = 7$, and $z = 1.65$. We have

$$x = \mu + z\sigma$$
$$= 35 + (1.65)7$$
$$= 35 + 11.55$$
$$= 46.55$$

So, if Daisy leaves her house 46.55 minutes before 9 A.M., she will arrive on time about 95% of the time. She should leave her home at 8:13 A.M.

*● **Example 5**

In one study, a major television manufacturing corporation found that the life of a typical color television tube is normally distributed with a standard deviation of 1.53 years. If 7% of these tubes last more than 6.9944 years, find the mean life of a television tube.

Solution

We are told that 7% of the tubes last more than 6.9944 years. See Figure 7.26. Thus, approximately 43% of the tubes last between μ and 6.9944 years. This means that on a standardized normal distribution the area between $z = 0$ and z is 0.43. Using Table V we find that 0.43 of the area is between $z = 0$ and $z = 1.48$. Then

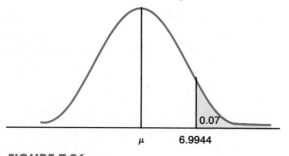

0.07

μ 6.9944

FIGURE 7.26

$$x = \mu + z\sigma$$
$$6.9944 = \mu + (1.48)(1.53)$$
$$6.9944 = \mu + 2.2644$$
$$6.9944 - 2.2644 = \mu$$
$$4.73 = \mu$$

Thus, the mean life of a television tube is 4.73 years.

EXERCISES

1. The life of a washing machine produced by one major company is known to be normally distributed with a mean life of 8.3 years and a standard deviation of 1.86 years. Find the probability that a randomly selected machine produced by this company will last more than 10 years.

2. The XYZ Corporation sells and services photocopying machines. The number of copies that a machine will give before requiring new toner is normally distributed with a mean of 14,000 copies and a standard deviation of 1200 copies. What is the probability that a typical machine will give between 13,000 and 15,000 copies before new toner is required?

3. The distribution of the lengths of the playing career of a major-league baseball player is approximately normal with a mean of 8.2 years and a standard deviation of 3.26 years. What is the probability that a randomly selected player will have a playing career of more than 11 years?

* 4. Assume that the life of a special type of movie projector bulb is normally distributed with a mean of 300 hours. If a theater manager requires at least 90% of the bulbs to have lives exceeding 250 hours, what is the largest value that σ can have and still keep the manager satisfied?

5. The life of a new energy-efficient burner for an oil furnace is normally distributed with a mean of five years and a standard deviation of one year. The manufacturer will replace any defective burner free of charge (including labor) while the furnace is under warranty. For how many years should the company guarantee the burners, if the manufacturer does not wish to replace more than 4% of them?

6. Professor Turner has found that the grades on the calculus final exam are normally distributed with a mean of 67 and a standard deviation of 8.
   a. If the passing grade is 52, what percent of the class will pass?
   b. If Professor Turner wants only 85% of the class to pass, what should be the passing grade?
   c. If Professor Turner wants only 4% of the class to get grades of A, what grade must a student have in order to get an A?

7. On a recent physical endurance test, the mean score was 143 with a standard deviation of 7. It has been decided that the top 10% of the contestants will receive a trophy. What is the minimum score that a contestant must obtain in order to receive a trophy?

8. In an effort to become more cost-efficient, several oil distributors have been experimenting with self-service automatic gas pumping machines that dispense a particular number of gallons of gas after a specified amount of money has been deposited. One such machine has been set so that it will dispense 4 gallons of unleaded high-test gasoline when a $5 bill is inserted. However, the pump is not functioning properly and can be adjusted according to the vendor's specifications. It is known that the pumping and filling process are approximately normally distributed with a standard deviation of 0.35 gallons. At what level should the mean be set so that

a. only 5% of the time will the pump deliver less than 4 gallons of gas when $5 is deposited?

b. at most 5% of the time will the pump deliver more than 4.5 gallons of gas when $5 is deposited?

9. Refer back to Exercise 8. The gas pump has been adjusted by mechanics so that the number of gallons of gas dispensed when $5 is deposited is approximately normally distributed with a standard deviation of 0.23 gallons. If 10% of the time more than 4.44 gallons of gas are dispensed, what is the new mean amount of gasoline dispensed by this pump when $5 is deposited?

10. Mary brings her car into an auto repair shop to have the air-conditioning system serviced. Based upon past experience, it is known the time it takes a mechanic to service the air-conditioning system of the type of car that Mary owns is approximately normally distributed with a mean of 35 minutes and a standard deviation of 6 minutes. At what time should the mechanic start working on Mary's car so that he has a 95% probability of completing the job properly by 5:00 P.M. when Mary will return?

11. A state is administering a qualifying law exam that will enable successful candidates who pass the bar to practice law within the state. It is known that the time required to complete the exam is approximately normally distributed with a mean of 240 minutes and a standard deviation of 35 minutes. If the judges wish to assure enough time so that only 85% of the candidates complete the exam, how much time should the judges allow?

*12. It is known that lengths of the sleeves of a shirt produced by a certain machine are normally distributed. Furthermore, 8.08% of the shirts have sleeve lengths that are more than 15.948 inches, and 18.67% of the shirts have sleeve lengths that are less than 15.2152 inches. What is the mean sleeve length and standard deviation of the sleeve length of the shirts produced by this machine?

13. A medical research company administers a special exam to its employees. The scores of the employees are normally distributed with a mean of 175 and a standard deviation of 17. Furthermore, it is known that a particular chemical research job bores people who score over 190 and demands a minimum score of 165 because of the danger of working with the chemicals. What percentage of the company's employees who took the exam can be used for this particular job (on the basis of the exam results)?

14. Mortgage statistics collected by a large bank indicate that the number of years that the average homeowner will occupy a house before moving or selling is normally distributed with a mean of 6.8 years and a standard deviation of 2.78 years. If a homeowner is selected at random, what is the probability that the owner will keep the house between 4 and 6 years before selling?

7.5 THE NORMAL CURVE APPROXIMATION TO THE BINOMIAL DISTRIBUTION

An important application of the normal distribution is the approximation of the binomial distribution. (The binomial distribution, which we discussed in Chapter 6, is often used in sampling without replacement situations if the sample size is small in comparison to the population size.) To see why such an approximation is needed, suppose we were interested in determining the probability of getting 9 heads when a coin is flipped 20 times. This is a binomial distribution problem where $n = 20$, $x = 9$, $p = \frac{1}{2}$, and $q = \frac{1}{2}$. We use Formula 6.4 of Chapter 6 (page 285) to determine this probability. We get

$$p(9 \text{ heads}) = \frac{20!}{9!11!} \left(\frac{1}{2}\right)^9 \left(\frac{1}{2}\right)^{11}$$

$$= \frac{167,960}{1,048,576} = 0.1602$$

Although evaluating this expression is not difficult, the calculations involved are time consuming. If we actually compute the answer, we get a value of 0.1602. Thus, the probability of getting 9 heads when tossing a coin 20 times is 0.1602.

It turns out that we can obtain a fairly good approximation to the binomial distribution by using the normal curve. To accomplish this, we need the mean and standard deviation of the binomial distribution. Recall that for a binomial distribution the mean is $\mu = np$ and that the standard deviation is $\sigma = \sqrt{npq}$. These formulas were given in Section 6.6 of Chapter 6. Applying them to our example we get

$$\mu = np = 20 \left(\frac{1}{2}\right) = 10$$

and

$$\sigma = \sqrt{npq} = \sqrt{20 \left(\frac{1}{2}\right) \left(\frac{1}{2}\right)} = \sqrt{5} \approx 2.236$$

We can now approximate this binomial distribution by a normal curve with $\mu = 10$ and $\sigma = 2.24$. We get this approximation by calculating the area under the normal curve between $x = 8.5$ and $x = 9.5$. See Figure 7.27. Any time we use the normal curve as an approximation for the binomial, we must calculate

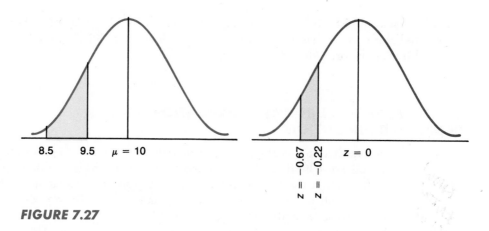

FIGURE 7.27

probabilities using an extra 0.5 either added to or subtracted from the number as a correction factor.

COMMENT A complete justification as to why we add or subtract 0.5 is beyond the scope of this text. (Suffice it to say that we are representing an integer value in the binomial distribution by those values of a normal distribution that round off to the integer.)

Returning to our example we have

$$z = \frac{x - \mu}{\sigma} = \frac{8.5 - 10}{2.24} = -0.67$$

and

$$z = \frac{x - \mu}{\sigma} = \frac{9.5 - 10}{2.24} = -0.22$$

The area between $z = 0$ and $z = -0.67$ is 0.2486 and the area between $z = 0$ and $z = -0.22$ is 0.0871, so that the area between $z = -0.67$ and $z = -0.22$ is $0.2486 - 0.0871 = 0.1615$.

Using the binomial distribution, we find the probability of getting 9 heads is 0.1602. Using the normal curve approximation to the binomial distribution, we find the probability of getting 9 heads in 20 tosses of a coin is 0.1615. Although the answers differ slightly, the answer we get using the normal curve approximation is accurate enough for most applied problems. Furthermore, it is considerably easier to calculate.

More generally, if we were interested in the probability of getting 13 heads in 20 tosses of a coin, we can approximate this by calculating the area between $x = 12.5$ and $x = 13.5$.

COMMENT Any time you approximate a binomial probability with the normal distribution, depending upon the situation make sure to add or subtract 0.5 from the number.

The normal curve approximation to the binomial distribution is especially helpful when we must calculate the probability of many different values. The following examples will illustrate its usefulness.

• Example 1

Melissa is a nurse at Maternity Hospital. From past experience she determines that the probability that a newborn child is a boy is $\frac{1}{2}$. (In the United States today, the probability that a newborn child is a boy is approximately 0.53, not 0.50.) What is the probability that among 100 newborn babies, there are at least 60 boys?

Solution

We can determine the probability *exactly* by using the binomial distribution or we can get an *approximation* by using the normal curve approximation. To determine the answer exactly, we say that a newborn child is either a boy or a girl with equal probability. Thus, $p = \frac{1}{2}$ and $q = \frac{1}{2}$. The probability that there are at least 60 boys means that we must calculate the probability of having 60 boys, 61 boys, . . . , and finally the probability of having 100 boys. Using the binomial distribution formula, we have

$$p(\text{at least 60 boys}) = p(\text{60 boys}) + p(\text{61 boys}) + \cdots + p(\text{100 boys})$$

$$= \frac{100!}{60!(40)!} \left(\frac{1}{2}\right)^{60} \left(\frac{1}{2}\right)^{40} + \frac{100!}{61!39!} \left(\frac{1}{2}\right)^{61} \left(\frac{1}{2}\right)^{39}$$

$$+ \cdots + \frac{100!}{100!0!} \left(\frac{1}{2}\right)^{100} \left(\frac{1}{2}\right)^{0}$$

Calculating these probabilities requires some lengthy computations. However, the same answer can be closely approximated, and more quickly so, by using the normal curve approximation. We first determine the mean and standard deviation:

$$\mu = np = 100 \left(\frac{1}{2}\right) = 50$$

$$\sigma = \sqrt{npq} = \sqrt{100 \left(\frac{1}{2}\right) \left(\frac{1}{2}\right)} = \sqrt{25} = 5$$

Then we find the area to the right of 59.5 as shown in Figure 7.28. We use 59.5 rather than 60.5 in our approximation since we want to include exactly

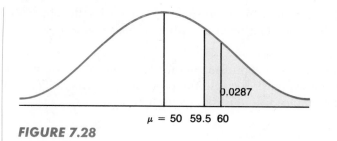

FIGURE 7.28

60 boys in our calculations. If the problem had specified more than 60 boys, we would have used 60.5 rather than 59.5 since more than 60 means do not include 60.

In our case we have

$$z = \frac{x - \mu}{\sigma} = \frac{59.5 - 50}{5} = \frac{9.5}{5} = 1.9$$

From Table V we find that the area to the right of $z = 1.9$ is $0.5000 - 0.4713$, or 0.0287.

Thus, the probability that among 100 newborn children there are at least 60 boys is 0.0287.

● **Example 2**

Refer back to Example 1 of this section. Find the probability that there will be between 45 and 60 boys (not including these numbers) among 100 newborn babies at Maternity Hospital.

Solution

The probability is approximated by the area under the normal curve between 45.5 and 59.5. We do not use 44.5 or 60.5 since 45 and 60 are not to be included. To find the area between 45.5 and 50 we have

$$z = \frac{x - \mu}{\sigma} = \frac{45.5 - 50}{5} = -0.90$$

The area is thus 0.3159.

Also to find the area between 50 and 59.5, we have

$$z = \frac{x - \mu}{\sigma} = \frac{59.5 - 50}{5} = 1.9$$

This area is 0.4713.

Adding these two areas we get

$$0.3159 + 0.4713 = 0.7872$$

Therefore, the probability that there are between 45 and 60 boys among 100 newborn babies at Maternity Hospital is 0.7872.

● Example 3

A large television network is considering canceling its weekly 7:30 P.M. comedy show because of a decrease in the show's viewing audience. The network decides to randomly phone 5000 viewers and to cancel the show if fewer than 1900 viewers are actually watching the show. What is the probability that the show will be canceled if

a. only 40% of all television viewers actually watch the comedy show?
b. only 39% of all television viewers actually watch the comedy show?

Solution

a. Since a randomly selected television viewer that is phoned either watches the show or does not watch the show, we can consider this as a binomial distribution with $n = 5000$ and $p = 0.40$. We first calculate μ and σ:

$$\mu = 5000(0.40) = 2000$$

$$\sigma = \sqrt{5000(0.40)(0.60)} = \sqrt{1200} \approx 34.64$$

Since the show will be canceled only if fewer than 1900 people watch it, we are interested in the probability of having 0, 1, 2, . . . , 1899 viewers. Using a normal curve approximation, we calculate the area to the left of 1899.5. We have

$$z = \frac{x - \mu}{\sigma} = \frac{1899.5 - 2000}{34.64} = -2.90$$

The area to the left of $z = -2.90$ is

$$0.5000 - 0.4981 = 0.0019$$

See Figure 7.29. Thus, the probability that the show is canceled is 0.0019.

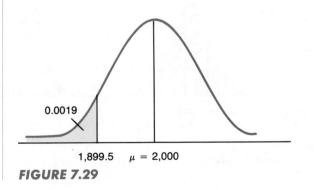

0.0019

1,899.5 $\mu = 2,000$

FIGURE 7.29

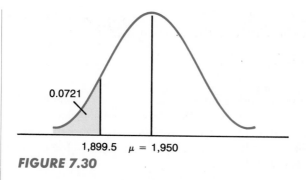

FIGURE 7.30

b. In this case the values of μ and σ are different, since the value of p is 0.39. We have

$$\mu = 5000(0.39) = 1950$$

$$\sigma = \sqrt{5000(0.39)(0.61)} = \sqrt{1189.5} \approx 34.49$$

Since the show will be canceled if fewer than 1900 people watch it, we calculate the area to the left of 1899.5. We have

$$z = \frac{1899.5 - 1950}{34.49} = -1.46$$

The area to the left of $z = -1.46$ is

$$0.5000 - 0.4279 = 0.0721$$

See Figure 7.30. Thus, the probability that the show is canceled is 0.0721.

COMMENT   The degree of accuracy of the normal curve approximation to the binomial distribution depends upon the values of n and p. Figure 7.31 indicates how fast the histogram for a binomial distribution approaches that of a normal distribution as n gets larger. As a rule, the normal curve approximation can be used with fairly accurate results when *both np* and *nq* are greater than 5.

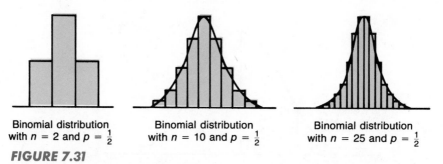

| Binomial distribution with $n = 2$ and $p = \frac{1}{2}$ | Binomial distribution with $n = 10$ and $p = \frac{1}{2}$ | Binomial distribution with $n = 25$ and $p = \frac{1}{2}$ |

FIGURE 7.31
Histogram for a binomial distribution approaches that of a normal distribution as n gets larger.

COMMENT   The normal curve approximation to the binomial distribution actually depends upon a very important theorem in statistics known as the Central Limit Theorem. This theorem will be discussed in a later chapter.

EXERCISES

In each of the following exercises, use the normal curve approximation to the binomial distribution.

1. A veterinarian claims that 15% of the families in Dearburn have one or more animal pets. If a sample of 250 families residing in Dearburn is randomly selected, find the probability that
   a. fewer than 40 of them have one or more animal pets.
   b. more than 50 of them have one or more animal pets.

2. If 60% of all the adult residents in Denver are in favor of tax reform, what is the probability that a sample of size 400 will contain fewer than 220 people who favor tax reform?

3. School officials claim that about 12% of the students who take out government-backed loans to pay for college tuition default on their payments. What is the probability that a sample of 125 randomly selected students who have taken out government-backed loans will contain at most 10 students who will default on their payments?

4. A charity organization finds that the probability that a person will respond to a telephone call solicitation for charity is only 0.18. If 800 people are called, what is the probability that at least 120 of them will actually respond?

5. A gas station attendant has found that, on the average, the probability that a customer will ask to have the oil checked is 0.23. If 50 customers are observed buying gas, what is the probability that
   a. at least 10 of them will ask to have their oil checked?
   b. exactly 10 of them will ask to have their oil checked?

6. *Overbooking*   Airline company officials find that 14% of all people who make reservations either do not show up or change their flights. As a result many airlines overbook flights. If an airline has accepted 240 reservations for a particular flight and if there are 213 available seats, find the probability that the airline will have a seat for each person who has reserved one and who shows up.

7. An automated production process produces computer disks that are defective about 3% of the time. If a sample of 200 randomly selected computer disks is taken from a day's production run, what is the probability that the sample will contain
   a. at most 6 defective disks?
   b. exactly 4 defective disks?

8. An insurance agent claims that 65% of all cars insured by her company are equipped with a stereo AM–FM radio and a cassette tape deck. If 220 cars insured by the company are randomly selected, what is the probability that

    a. more than 160 of the cars are equipped with a stereo AM–FM radio and a cassette tape deck?

    b. exactly 155 of the cars are equipped with a stereo AM–FM radio and a cassette tape deck?

9. A welfare agency interviewer claims that 18 percent of all applications for welfare received by the welfare agency are rejected. What is the probability that among a random sample of 175 applications for welfare,

    a. exactly 30 of them will be rejected?

    b. not more than 50 of them will be rejected?

10. A stockbroker claims that she averages 1 sale out of every 12 telephone calls that she makes. What is the probability that in 36 randomly selected telephone calls, she will make between 4 and 7 sales, inclusive?

11. A public library claims that approximately 33% of all its books that are borrowed are fiction. Yesterday the library lent out 2341 books. What is the probability that the number of the fiction books borrowed yesterday is

    a. between 710 and 800?   **b.** exactly 770?   **c.** no more than 750?

12. According to government records, 15% of all female inmates in U.S. jails are between the ages of 55 and 64 years.* If 375 female inmates are randomly selected, what is the probability that fewer than 56 of them will be between the ages of 55 and 64 years?

7.6 APPLICATION TO STATISTICAL QUALITY CONTROL

In recent years numerous articles and books have been written on how statistical quality controls operate. This is an important branch of applied statistics. What are quality-control charts? To answer this question we must consider the mass production process.

Industrial experience shows that most production processes can be thought of as normally distributed. So, when a manufacturer adjusts the machines to fill a jar with 10 ounces of coffee, although not all the jars will actually weigh 10 ounces, the weight of a typical jar will be very close to 10 ounces. When too many jars weigh more than 10 ounces, the manufacturer will lose money. When too many jars weigh less than 10 ounces, he or she will lose customers. The manufacturer is therefore interested in maintaining the weight of the jars as close to 10 ounces as possible.

If the production process behaves in the manner described above, the weight of a typical jar of coffee is either acceptable or not acceptable. Thus, it can be thought of as a binomial variable. We can then use the normal approximation.

Rather than weigh each individual jar of coffee, the manufacturer can use **quality-control charts.** This is a simple graphical method that has been found to be highly useful in the solution to problems of this type.

*Source: U.S. Department of Justice. *Profile of Jail Inmates: Sociodemographic Findings from the 1978 Survey of Inmates of Local Jails.* Washington, D.C.: U.S. Government Printing Office, 1981.

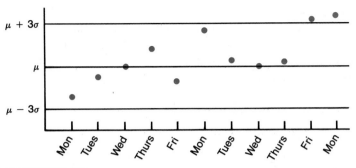

FIGURE 7.32

Typical quality control chart.

Figure 7.32 is a typical quality-control chart. The horizontal line represents the time scale. The vertical line has three markings: μ, $\mu + 3\sigma$, and $\mu - 3\sigma$.

The middle line is thought of as the mean of the production process although in reality it is usually the mean weight of past daily samples. The two other lines serve as control limits for the daily production process. These lines have been spaced three standard deviation units above and below the mean. From the normal distribution table, we find that approximately 99.7% of the area should be between $\mu - 3\sigma$ and $\mu + 3\sigma$. Thus, the probability that the average weight of many jars of coffee falls outside the control bands is only 0.003. This is a relatively small probability. Therefore, if a sample of sufficient size is taken and the average weight is outside the control bands, the manufacturer can then assume that the production process is not operating properly and that immediate adjustment is necessary to avoid losing money or customers. Each of the dots plotted in Figure 7.32 represents the mean of the weights of n jars of coffee determined on different days.

Figure 7.32 indicates that the production process went out of control on a Friday.

7.7 SUMMARY

In this chapter we discussed the difference between a discrete random variable and a continuous random variable. We studied the normal distribution in detail since it is the most important probability distribution of a continuous random variable. Because of their usefulness, normal curve area charts have been constructed. These charts allow us to calculate the area under the standard normal curve. Thus, we can determine the probability that a random variable will fall within a specified range.

Not only can these charts be used to calculate probabilities for variables that are normally distributed, but they can also be used to obtain a fairly good approximation to binomial probabilities. This is especially helpful when we must calculate the probability that a binomial random variable assumes many different values. Numerous applications of these ideas were given.

The normal distribution was then applied to the construction of quality-control charts, which are so important in many industrial processes. Today statistical quality control is an important branch of applied statistics. Without discussing them in detail, we indicated the usefulness of these quality-control charts.

STUDY GUIDE

You should now be able to demonstrate your knowledge of the following ideas presented in this chapter by giving definitions, descriptions, or specific examples. Page references are given for each term so that you can check your answer.

Discrete random variable (page 315).
Continuous random variable (page 315).
Bell-shaped distribution (page 315).
Gaussian distribution (page 315).
Normal curve (page 316).
Normal distribution (page 316).
Standardized normal distribution (page 319).
Chebyshev's Theorem (page 322).
Quality-control charts (page 344).

FORMULAS TO REMEMBER

You should be able to identify each symbol in the following formulas, understand the relationships among the symbols expressed in each formula, understand the significance of each formula, and use the formulas in solving problems.

1. $z = \dfrac{x - \mu}{\sigma}$

2. $x = \mu + z\sigma$

3. At least $1 - \dfrac{1}{k^2}$ of a set of measurements will lie within k standard deviations of the mean: $k = 1, 2, \ldots$

The tests of the following section will be more useful if you take them after you have studied the examples and solved the exercises given in this chapter.

MASTERY TESTS

Form A

1. If 50% of all full-time employees of the Lite Corporation have a second part-time job, what is the probability that a survey of 100 full-time employees will contain 70 or more people who have a second part-time job?
   a. 0.5000   **b.** approximately 0   **c.** 0.4832
   d. 0.9832   **e.** none of these

2. The number of video tapes reproduced by one particular machine during a week of operation is normally distributed with μ (mean) of 75 and a standard deviation of 8. What is the probability that the number of video tapes reproduced next week by this machine will be between 71 and 81?
   a. 0.3944   **b.** 0.0351   **c.** 0.4649   **d.** 0.2734   **e.** none of these

3. Find the area under the normal curve from $z = -0.8$ to $z = 0.9$.
   a. 0.8159   **b.** 0.7881   **c.** 0.0278   **d.** 0.6040   **e.** none of these

4. Find the area under the normal curve to the right of $z = 1.83$.
   a. 0.4664   **b.** 0.9664   **c.** 0.0336   **d.** 0.3485   **e.** none of these

5. In a standard normal distribution, find the closest z-score corresponding to the 85th percentile.
   a. 0.85   **b.** 0.39   **c.** 0.30   **d.** 1.04   **e.** none of these

6. A farmer's cooperative packages potatoes in bags with weights known to be normally distributed with a mean of 10 pounds and a standard deviation of 0.84 pounds. What percentage of such bags of potatoes weigh between 10.84 and 11.68 pounds?
   a. 15.87%   **b.** 34.13%   **c.** 13.59%
   d. 47.63%   **e.** none of these

For questions 7–9 assume that the length of time required by an auto mechanic to completely service a car is normally distributed with a mean of 40 minutes and a standard deviation of 9 minutes.

7. The percent of cars that require more than 49 minutes to be completely serviced is approximately
   a. 16   **b.** 84   **c.** 34   **d.** 33   **e.** none of these

8. The percent of cars that require fewer than 49 minutes to be completely serviced is approximately
   a. 16   **b.** 84   **c.** 34   **d.** 33   **e.** none of these

9. If a car is randomly selected, what is the probability that it will require between 31 and 49 minutes to be completely serviced?
   a. 0.8413   **b.** 0.6667   **c.** 0.6826   **d.** 0.1587   **e.** none of these

10. The number of child abuse cases reported to one particular welfare agency

during a week of operation is normally distributed with a mean of 75 and a standard deviation of 8. What is the probability that the number of child abuse cases reported next week will be between 71 and 81?

a. 0.3944   **b.** 0.2734   **c.** 0.0351   **d.** 0.4969   **e.** none of these

MASTERY TESTS
Form B

1. The breaking strength of a certain material is normally distributed with a mean of 90 pounds and a standard deviation of 20 pounds. What is the probability that a randomly selected piece of this material will have a breaking strength between 95 and 110 pounds?

   a. 0.4013   **b.** 0.2426   **c.** 0.4417   **d.** 0.1587   **e.** none of these

2. If 60% of all members of a swimming group have Type O blood, what is the probability that a random sample of 700 members will contain 450 or more members with Type O blood?

3. Studies show that the length of time required for a patient to wait in the emergency room at St. Vincent's Hospital before seeing a doctor is approximately normally distributed with a mean of 50 minutes and a standard deviation of 15 minutes. What percent of the patients must wait less than 38 minutes before seeing a doctor?

4. Refer back to the previous question. If a patient is randomly selected, what is the probability that the patient will have to wait between 40 minutes and 1 hour before seeing a doctor?

5. Consider the newspaper article below. One researcher claims that despite medical claims to the contrary, two out of seven potential donors are reluctant to donate blood because of their fear of contracting AIDS. Assuming this researcher's claim is true, find the probability that in a random survey of 200 potential donors, at most 50 of them will be reluctant to donate blood because of their fear of contracting AIDS.

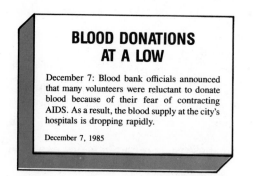

BLOOD DONATIONS AT A LOW

December 7: Blood bank officials announced that many volunteers were reluctant to donate blood because of their fear of contracting AIDS. As a result, the blood supply at the city's hospitals is dropping rapidly.

December 7, 1985

6. An Internal Revenue Service (IRS) auditor claims that 35% of the taxpayers do not accurately report their true income. A supervisor decides to randomly select 200 tax returns and to thoroughly investigate them for accurate reporting of income. What is the probability that the supervisor will find that the number of taxpayers in this group not accurately reporting their income is between 50 and 80?

7. *Uninsured Motorists*   It is claimed that 9% of all the cars registered in a certain state are driven by uninsured motorists. Assuming this claim to be true, a random sample is taken of 150 cars registered in the state. What is the probability that the number of cars driven by uninsured motorists is
a. less than 10?   **b.** exactly 8?   **c.** at least 7?

8. The amount of soda dispensed by a soft drink vending machine in a college dormitory is normally distributed with a mean of 9 ounces and a standard deviation of 0.82 ounces.
a. What percentage of the cups will contain more than 9.2 ounces?
b. What is the probability that a cup will overflow if it can hold exactly 10.5 ounces?

9. Housing industry officials claim that 70% of all the two-family homes in Springfield have at least one smoke detector. If 300 two-family homes in Springfield are randomly selected, find the probability that at least 200 of them have a smoke detector.

10. The city maintenance department, which oversees the safe operation of the city's fleet of 1200 cars, has been instructed by the mayor to replace the brakes on all the cars before they wear out. It is known that the life of the brakes on the city cars is normally distributed with a mean life of 14,000 miles and a standard deviation of 850 miles. After how many miles driven should the brakes on these cars be replaced so that no more than 8% of them wear out while in use?

11. It is known that the time required to complete a particular state nursing exam is normally distributed with a mean of 140 minutes and a standard deviation of 15 minutes. How much time should be allowed, if it is desired that at most 90% of the people taking the exam have sufficient time to complete it?

SUGGESTED FURTHER READING

1. Hoel, Paul G. *Elementary Statistics*, 4th ed. Chap. 4. New York: Wiley, 1976.
2. Huntsberger, D. V. *Elements of Statistical Inference*, 3rd ed. Boston, MA: Allyn and Bacon, 1973.
3. Mendenhall, W. *Introduction to Probability and Statistics*, 6th ed. Belmont, CA: Duxbury Press, 1983.
4. Walpole, R. E. *Introduction to Statistics*. New York: Macmillan, 1968.

CHAPTER 8

Sampling

OBJECTIVES

- *To discuss* what a random sample is and how it is obtained.

- *To learn* about a table of random digits where each number that appears is obtained by a process that gives every digit an equally likely chance of being selected.

- *To introduce* stratified sampling, which is a sampling procedure that is used when we want to obtain a sample with a specified number of people from different categories.

- *To see* that when repeated samples are taken from a population, the frequency distribution of the values of the sample means is called the distribution of the sample means.

- *To analyze* the standard error of the mean, which represents the standard deviation of the distribution of sample means.

- *To study* the Central Limit Theorem. This tells us that the distribution of the sample means is basically a normal distribution. We discuss how we can use this theorem to make predictions about and calculate probabilities for the sample means.

Gary Settle/NYT Pictures.

On April 6, 1976, both ABC and NBC television networks projected that Morris Udall would win the Democratic primary in Wisconsin. Their predictions were based upon samples from selected precincts, and did not take certain districts into consideration. When all the rural votes were counted, Jimmy Carter came out on top. Many newspapers were so confident of their predictions that they printed, erroneously, the morning editions of their newspapers with the headline "CARTER UPSET BY UDALL."

8.1 INTRODUCTION

Suppose we are interested in determining how many people in the United States believe in the fiscal soundness of the Social Security system. Must we ask each person in the country before making any statement concerning the fiscal soundness of the Social Security system? As there are so many people in the United States, it would require an enormous amount of work to interview each adult and gather all the data.

Do we actually need the complete population data? Can a properly selected sample give us enough information to make predictions about the entire population? In many cases, obtaining the complete population data may be quite costly or even impossible.

Similarly, suppose we are interested in purchasing an electric bulb. Must we use (or test) all electric bulbs produced by a particular company in order to determine the bulb's average life? This is very impractical. Maybe we can estimate the average life of a bulb by testing a sample of only 100 bulbs.

As we noticed earlier in Chapter 1, Definition 1.2, this is exactly what inferential statistics involves. Samples are studied to obtain valuable information about a larger group called the **population.** Any part of the population is called a **sample.** The purpose of sampling is to select that part which truly represents the entire population.

Any sample provides only partial information about the population from which it is selected. Thus, it follows that any statement we make (based on a sample) concerning the population may be subject to error. One way of minimizing this error is to make sure that the sample is randomly selected. How is this to be done?

In this chapter we will discuss how to select a random sample and how to interpret different sample results.

CITY TO CALL $5 MILLION OF THE 1995 SERIES 14 BOND

New Dorp (June 30) — City officials announced yesterday that their financial condition had improved sufficiently to enable it to call $5 million of the 1995 series 14 bonds. These bonds, which were originally issued in 1975, have a 12% annual interest rate and are costing the city $7 million in annual interest charges. The bonds to be recalled will be selected in a random manner and will be made public tomorrow. The holders of a called bonds will no longer earn any interest as of tomorrow.

Tuesday, June 30, 1981

The article on the left specifies that the bonds to be recalled will be selected in a random process. How are such random selections made?

8.2 SELECTING A RANDOM SAMPLE

The purpose of most statistical studies is to make generalizations from samples about the entire population. Yet, not all samples lend themselves to such generalizations. Thus we cannot generalize about the average income of a working person in the United States by sampling only lawyers and doctors. Similarly, we cannot make any generalizations about the Social Security system by sampling only people who are receiving Social Security benefits.

Over the years many incorrect predictions have been made on the basis of nonrandom samples. For example, in 1936 the *Literary Digest* was interested in determining who would win the coming presidential election. It decided to poll the voters by mailing 10 million ballots. On the basis of the approximately 2 million ballots returned, it predicted that Alfred E. Landon would be elected. An October 31 headline read

Landon	1,293,669
Roosevelt	972,897

Final returns in the Digest's poll of ten million voters

Source: The Literary Digest, 1936.

Actually Franklin Roosevelt carried 46 of the 48 states and many of them by a landslide. The ten million people to whom the *Digest* sent ballots were selected from telephone listings and from the list of its own subscribers. The year 1936 was a depression year and many people could not afford telephones or magazine subscriptions. Thus, the *Digest* did not select a random sample of the voters of the United States. The *Literary Digest* soon went out of business. In 1976, Maurice Bryson argued persuasively that the major problem was the *Digest's* reliance on voluntary response.*

Again, in 1948, the polls predicted that Dewey would win the presidential election. One newspaper even printed the morning edition of its newspaper with the headline "DEWEY WINS BY A LANDSLIDE." Of course, Truman won the election and laughed when presented with a copy of the newspaper predicting his defeat.

In both examples the reason for the incorrect prediction is that it was based on information obtained from poor samples. It is for this reason that statisticians insist that samples be randomly selected.

*M. Bryson, "The *Literary Digest* Poll: Making of a Statistical Myth," *American Statistician*, November 1976.

DEFINITION 8.1

A *random sample* of n items is a sample selected from a population in such a way that every different sample of size n from the population has an equal chance of being selected.

DEFINITION 8.2

Random sampling is the procedure by which a random sample is obtained.

It may seem that the selection of a random sample is an easy task. Unfortunately this is not the case. You may think that we can get a random sample of voters by opening a telephone book and selecting every tenth name. This will not give a random sample since many voters either do not have phones or else have unlisted numbers. Furthermore, many young voters and most women are not listed. These people, who are members of the voting population, do not have an equal chance of being selected.

To further illustrate the nature of random sampling, suppose that the administration of a large southern college with an enrollment of 30,000 students is considering revising its grading system. The administration is interested in replacing its present grading system with a pass–fail system. Since not all students agree with this proposed change, the administration has decided to poll 1000 students. How is this to be done? Polling a thousand students in the school cafeteria or in the student lounge will not result in a random sample since there may be many students who neither eat in the cafeteria nor go to the student lounge.

One way of obtaining a random sample is to write each student's name on a separate piece of paper and then put all the pieces in a large bowl where they can be thoroughly mixed. A paper is then selected from the bowl. This procedure is repeated until 1000 names are obtained. In this manner a random sample of 1000 names can be obtained. Great care must be exercised to make sure that the bowl is thoroughly mixed after a piece of paper is selected. Otherwise, the papers on the bottom of the bowl do not have an equal chance of being selected and the sample will not be random.

Although using slips of paper in a bowl will result in a random sample if properly done, this fish bowl method becomes unmanageable as the number of people in the population increases. The job of numbering slips of paper can be completely avoided by using a **table of random numbers** or a **table of random digits.**

What are random numbers? They are the digits 0, 1, 2, . . . , 9 arranged in a random fashion, that is, in such a way that all the digits appear with approximately the same frequency. Table VII in the Appendix is an example of a table of random numbers. How are such tables constructed? Today most tables of random numbers are constructed with the help of electronic computers. Yet a simple spinner such as the one shown in Figure 8.1 will also generate such a table of random numbers. After each spin we use the digit selected by the arrow. The resulting sequence of random digits could then be used to construct random number tables of five digits each, for example, as shown in Table VII. Although this method will generate random numbers, it is not practical. After a while some numbers will be favored over others as the spinner begins to wear out.

Thus, the best way to select a random sample is to use a table of random digits obtained with the help of an electronic computer. Table VII in the Appendix is such a table. In this table the various digits are scattered at random and arranged in groups of five for greater legibility. Nowadays, even some calculators can generate random digits.

Let us now return to our example. As a first step in using this table, each student is assigned a number from 00001 to 30000. To obtain a sample of 1000 students we merely read down column 1 and select the first 1000 students whose numbers are listed. Thus the following students would be selected:

   10,480   22,368   24,130   28,918   09,429 . . .

Notice that we skip the number 42,167 since no student has this number. The same is true for the numbers 37,570; 77,921; 99,562. . . . We disregard any numbers larger than 30,000 that are obtained from this table since there are no students associated with these numbers.

To further illustrate the proper use of this table, suppose that a hotel has 150 guests registered for a weekend. The management wishes to select a sample of 15 people at random to rate the quality of its service. It should proceed as follows: Assign each guest a different number from 1 to 150 as 001, 002, . . . , 150. Then select any column in the table of random numbers. Suppose the fourth column is selected. Three-digit numbers are then read off the table by reading down the column. If necessary they continue on to another column or another page. Starting on top of column 4, they get

   020   853   972   616   166   427   699, . . .

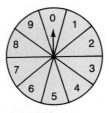

FIGURE 8.1
A spinner.

Since the guests have numbers between 001 and 150 only, they ignore any number larger than 150. From column 4 they get

020   079   102   034   081   099   143   073   129

From column 5 they get

078   061   091   133   040   023

Thus, the management should interview the guests whose numbers are 20, 79, 102, 34, 81, 99, 143, 73, 129, 78, 61, 91, 133, 40, and 23.

COMMENT   Whenever we speak of a random sample in this text we will assume that it has been selected in the manner just described.

EXERCISES

1. Suppose that the Yellow Pages of a local telephone directory lists 100 roofers alphabetically. Suppose also that your roof is leaking and that you wish to call five roofers for an estimate on the cost of repairing the roof. By using column 12 of Table VII of the Appendix, decide which roofers you should call.

2. Each member of the security force of a local department store carries a badge. The badges are numbered, in order, from 1 to 211. A random sample of 17 officers is to be selected for training in new antishoplifting techniques. By using columns 10 and 11 of Table VII, how would you decide which officers should be selected?

3. A computer manufacturer attaches a special serial number from 1 to 2000 to each disk drive that is manufactured during a week. A random sample of 20 disk drives is to be selected and inspected for quality, workmanship, etc. By using columns 5 and 6 of Table VII, decide which disk drives should be inspected.

4. A certain prison has 937 prisoners, each of whom has been assigned a number from 1 to 937. A committee of 15 prisoners is to be selected from among them to discuss their grievances with the warden. By using column 4 of Table VII, how would you decide which prisoners should be selected?

5. *Bond Recalls*   Many bonds issued by various localities have a provision which states that they are subject to call before the stated maturity date. In 1975, Greensburg issued 10,000 bonds, each numbered in order, 1, 2, 3, . . . , 10,000. In 1990, town officials decide to redeem 12 of these bonds ahead of their maturity date by a random selection process. If they use columns 5, 6, and 7 of Table VII in the Appendix, which bonds will they select to be redeemed?

6. *Insurance Claims*   An auto insurance company processed 759 claims during the first week of December. Company officials would like to determine whether the claims were handled courteously and efficiently. The processed claims are numbered 1, 2, 3, 4, . . . , 759. Twenty of these claimants will be randomly selected and contacted. If the officials use column 8 of Table VII in the Appendix, which claim numbers will be reviewed?

7. A large department store has 91,878 credit card customers. The sales department has decided to invite 30 of these customers, randomly selected, to an advance Columbus Day Sale. If column 10 of Table VII in the Appendix is used, which credit card customers will be invited?

8. A state tax department has decided to randomly select 30 individual income tax returns for 1987 and to thoroughly audit them. Each return has been filed according to the taxpayer's Social Security number. If the last four digits of the Social Security number and column 2 of Table VII in the Appendix are used, find which returns will be audited.

9. There are 473 restaurants in a large northeastern city, each of which has a permit for operation from the health department. The restaurants are numbered 1, 2, 3, . . . , 473. Officials of the health department have decided to check on the sanitary conditions at 18 of these restaurants. If column 4 of Table VII in the Appendix is used, which restaurants will be selected to be checked?

10. Each car that is manufactured in the United States has a serial number. After a series of accidents involving a sticking accelerator rod, one of the auto manufacturers has decided to randomly check 40 cars sold on the west coast. According to company records, cars sold on the west coast during 1986 had serial numbers whose last five digits were between 20,000 and 61,000. If columns 11 and 12 of Table VII in the Appendix are used, which cars will be checked?

8.3 STRATIFIED SAMPLING

Although random sampling, as discussed in Section 8.2, is the most popular way of selecting a sample, there are times when **stratified samples** are preferred. To obtain a stratified sample, we divide the entire population into a number of groups or strata. The purpose of such stratification is to obtain groups of people that are more or less equal in some respect. We select a random sample, as discussed in Section 8.2 from each of these groups or stratum. This stratified sampling procedure ensures that no group is missed and improves the precision of our estimates. If we use stratified sampling, then in order to estimate the population mean, we must use *weighted averages* of the strata means, weighted by the population size for that stratum.

Thus, in the example discussed in the beginning of Section 8.2, the administration of the college may first divide the entire student body into four groups: freshmen, sophomores, juniors, and seniors. Then they can select a random sample from each of these groups. The groups are often sampled in proportion to their actual percentages. In this manner, the administration can obtain a more accurate poll of student opinion by stratified sampling. However, the cost of obtaining a stratified sample is often higher than that of obtaining a random sample since the administration must spend money to research dividing the student body into four groups.

The method of stratifying samples is especially useful in pre-election polls.

Past experience indicates that different subpopulations often demonstrate particularly different voting preferences.

COMMENT   Statistical analyses and tests based upon data obtained from stratified samples are somewhat different from what we have discussed in this book. We will not analyze such procedures here.

8.4 CHANCE VARIATION AMONG SAMPLES

Imagine that a cigarette manufacturer is interested in knowing the average tar content of a new brand of cigarettes that is about to be sold. The Food and Drug Administration requires such information to be indicated alongside all advertisements that appear in magazines, newspapers, and so on.

The manufacturer decides to send random samples of 100 cigarettes each to 20 different testing laboratories. With the information obtained from these samples, the manufacturer hopes to be able to estimate the mean or average milligram tar content of the cigarette.

Since we will be discussing both samples and populations, let us pause for a moment to indicate the notation that we will use to distinguish between samples and populations. See Table 8.1.

TABLE 8.1
Notation for Sample and Population

Term	Sample	Population
Mean	$\bar{x}$	μ
Standard deviation	s	σ
Number	n	N

From Chapter 3 we have the following formulas:

FORMULA 8.1   *Mean*

Sample	Population
$\bar{x} = \dfrac{\Sigma x}{n}$	$\mu = \dfrac{\Sigma x}{N}$

FORMULA 8.2   *Standard Deviation*

Sample	Population
$s = \sqrt{\dfrac{\Sigma(x - \bar{x})^2}{n - 1}}$	$\sigma = \sqrt{\dfrac{\Sigma(x - \mu)^2}{N}}$

Let us now return to our example. Since each sample sent to a laboratory is randomly selected, it is reasonably safe to assume that there will be differences among the means of each sample. The 20 laboratories report the following average milligram content per cigarette:

14.8   16.2   14.8   15.8   15.3   13.9   16.9   15.9   14.3   15.2
14.9   16.2   15.6   15.5   13.4   15.1   15.7   14.8   14.4   15.3

These figures indicate that the sample means vary considerably from sample to sample.

The manufacturer decides to take the average of these 20 sample means and gets

$$\text{Average of 20 sample means} = \frac{\Sigma\bar{x}}{n} = \frac{14.8 + 16.2 + \cdots + 15.3}{20} = \frac{304}{20} = 15.2$$

The manufacturer now uses this overall average of the sample means, 15.2, as an estimate of the true population mean.

How reliable is this estimate? Although we cannot claim for certain that the population mean is 15.2, we can feel reasonably confident that 15.2 is not a bad estimate of the population mean since it is based on 20 × 100, or 2000, observations.

Thus we can obtain a fairly good estimate of the population mean by calculating the mean of samples. If we let $\mu_{\bar{x}}$, read as mu sub x bar, represent the mean of the samples, then we say that $\mu_{\bar{x}}$ is a good estimate of μ. Generally speaking, if a random sample of size n is taken from a population with mean μ, then the mean of $\bar{x}$ will always equal the mean of the population (regardless of sample size), that is $\mu_{\bar{x}} = \mu$.

What about the standard deviation? Let us calculate the standard deviation of the sample means. Recall that the formula for the population standard deviation is

$$\sqrt{\frac{\Sigma(x - \mu)^2}{N}}$$

Since μ is unknown, we have to replace it with an estimate. The most obvious replacement is $\mu_{\bar{x}}$. To account for this replacement, we divide by $N - 1$ instead of by N. Thus the formula for the standard deviation for the sample means is given by Formula 8.3.

FORMULA 8.3

*The **standard deviation of the sample means** is given by*

$$\sqrt{\frac{\Sigma(\bar{x} - \mu_{\bar{x}})^2}{n - 1}}$$

where n is the number of sample means.

● **Example 1**

Calculate the standard deviation of the sample means for the data of the average tar content of the 20 laboratories.

Solution

We arrange the data as follows:

$\bar{x}$	$\bar{x} - \mu_{\bar{x}}$	$(\bar{x} - \mu_{\bar{x}})^2$
14.8	$14.8 - 15.2 = -0.4$	0.16
16.2	$16.2 - 15.2 = 1$	1.00
14.8	$14.8 - 15.2 = -0.4$	0.16
15.8	$15.8 - 15.2 = 0.6$	0.36
15.3	$15.3 - 15.2 = 0.1$	0.01
13.9	$13.9 - 15.2 = -1.3$	1.69
16.9	$16.9 - 15.2 = 1.7$	2.89
15.9	$15.9 - 15.2 = 0.7$	0.49
14.3	$14.3 - 15.2 = -0.9$	0.81
15.2	$15.2 - 15.2 = 0$	0
14.9	$14.9 - 15.2 = -0.3$	0.09
16.2	$16.2 - 15.2 = 1$	1.00
15.6	$15.6 - 15.2 = 0.4$	0.16
15.5	$15.5 - 15.2 = 0.3$	0.09
13.4	$13.4 - 15.2 = -1.8$	3.24
15.1	$15.1 - 15.2 = -0.1$	0.01
15.7	$15.7 - 15.2 = 0.5$	0.25
14.8	$14.8 - 15.2 = -0.4$	0.16
14.4	$14.4 - 15.2 = -0.8$	0.64
<u>15.3</u>	$15.3 - 15.2 = 0.1$	<u>0.01</u>
304		13.22
$\Sigma\bar{x} = 304$		$\Sigma(\bar{x} - \mu_{\bar{x}})^2 = 13.22$

Using Formula 8.3 we have

$$\text{Standard deviation of sample means} = \sqrt{\frac{\Sigma(\bar{x} - \mu_{\bar{x}})^2}{n - 1}}$$

$$= \sqrt{\frac{13.22}{20 - 1}} = \sqrt{\frac{13.22}{19}}$$

$$= \sqrt{0.696}$$

$$\approx 0.83$$

Thus, the standard deviation of the sample means is approximately 0.83.

In practice the standard deviation is not calculated by using Formula 8.3 since the computations required are time consuming. Instead we can use a shortcut formula given as Formula 8.4. The advantage in using Formula 8.4 is that we do not have to calculate $\mu_{\bar{x}}$ and $\bar{x} - \mu_{\bar{x}}$ and square $\bar{x} - \mu_{\bar{x}}$. We only have to calculate $\Sigma\bar{x}$ and $\Sigma\bar{x}^2$. These represent the sum of the $\bar{x}$'s and the sum of the squares of the $\bar{x}$'s, respectively. Then we use Formula 8.4:

FORMULA 8.4

*The **standard deviation of the sample means** is given by*

$$\sqrt{\frac{n(\Sigma\bar{x}^2) - (\Sigma\bar{x})^2}{n(n - 1)}}$$

where n is the number of sample means.

● **Example 2**

Using Formula 8.4, find the standard deviation of the sample means for the data of Example 1 in this section.

Solution

We arrange the data as follows:

$\bar{x}$	$\bar{x}^2$
14.8	219.04
16.2	262.44
14.8	219.04
15.8	249.64
15.3	234.09
13.9	193.21
16.9	285.61
15.9	252.81
14.3	204.49

$\bar{x}$	$\bar{x}^2$
15.2	231.04
14.9	222.01
16.2	262.44
15.6	243.36
15.5	240.25
13.4	179.56
15.1	228.01
15.7	246.49
14.8	219.04
14.4	207.36
15.3	234.09
304	4634.02
$\Sigma \bar{x} = 304$	$\Sigma \bar{x}^2 = 4634.02$

Using Formula 8.4, we have

$$\begin{aligned}\text{Standard deviations}\atop\text{of sample means} &= \sqrt{\frac{n(\Sigma \bar{x}^2) - (\Sigma \bar{x})^2}{n(n-1)}} \\[2mm] &= \sqrt{\frac{20(4634.02) - (304)^2}{20(19)}} \\[2mm] &= \sqrt{\frac{92680.4 - 92416}{380}} \\[2mm] &= \sqrt{\frac{264.4}{380}} = \sqrt{0.696} \approx 0.83\end{aligned}$$

Thus, the standard deviation of the sample means is approximately 0.83. This is the same result we obtained using Formula 8.3.

● **Example 3**
A large office building has six elevators, each with a capacity for 10 people. The operator of each elevator has determined the average weight of the people in the elevators when operating at full capacity. The results follow:

Elevator	1	2	3	4	5	6
Average Weight (lbs.)	125	138	145	137	155	140

Find the overall average of these sample means. Also find the standard deviation of these sample means by first using Formula 8.3 and then Formula 8.4.

Solution
We arrange the data as follows:

$\bar{x}$	$\bar{x} - \mu_{\bar{x}}$	$(\bar{x} - \mu_{\bar{x}})^2$	$\bar{x}^2$
125	$125 - 140 = -15$	225	15,625
138	$138 - 140 = -2$	4	19,044
145	$145 - 140 = 5$	25	21,025
137	$137 - 140 = -3$	9	18,769
155	$155 - 140 = 15$	225	24,025
140	$140 - 140 = 0$	0	19,600
840		488	118,088
$\Sigma\bar{x} = 840$		$\Sigma(\bar{x} - \mu_{\bar{x}})^2 = 488$	$\Sigma\bar{x}^2 = 118,088$

Then

$$\mu_{\bar{x}} = \frac{\Sigma\bar{x}}{n} = \frac{840}{6} = 140$$

Using Formula 8.3 we get

$$\text{Standard deviation of sample means} = \sqrt{\frac{\Sigma(\bar{x} - \mu_{\bar{x}})^2}{n-1}} = \sqrt{\frac{488}{5}} = \sqrt{97.6} \approx 9.88$$

Using Formula 8.4 we get

$$\text{Standard deviation of sample means} = \sqrt{\frac{n(\Sigma\bar{x}^2) - (\Sigma\bar{x})^2}{n(n-1)}} = \sqrt{\frac{6(118,088) - (840)^2}{6(5)}}$$

$$= \sqrt{\frac{708528 - 705600}{30}} = \sqrt{97.6} \approx 9.88$$

Thus the mean of the samples is 140 and the standard deviation of the sample means (by Formula 8.3 or Formula 8.4) is approximately 9.88.

8.5 DISTRIBUTION OF SAMPLE MEANS

Let us refer back to the example discussed at the beginning of Section 8.4. The manufacturer decides to draw the histogram for the average cigarette tar content that was obtained from the 20 laboratories. This is shown in Figure 8.2. Notice that the value of $\bar{x}$ is actually a random variable since its value is different from sample to sample. In repeated samples different values of $\bar{x}$ were obtained. Yet they are all close to the 15.2 we obtained as the average of the sample means. Moreover, exactly 70% of the sample means are between 14.37 and 16.03, which represents 1 standard deviation away from the mean in either direction. Also, 90% of the sample means are between 13.54 and 16.86, which represents 2 standard deviations away from the mean in either direction. Thus Figure 8.2 actually represents the distribution of $\bar{x}$ since it tells us how the

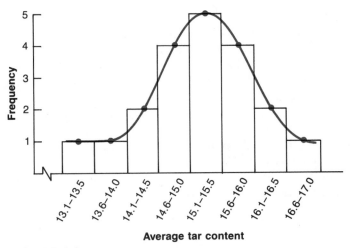

FIGURE 8.2

means of the samples vary from sample to sample. We refer to the distribution of $\bar{x}$ as the **distribution of sample means** or as **the sampling distribution of the mean.** Although the first terminology is much clearer, the second is more commonly used.

Strictly speaking, Figure 8.2 is not a complete distribution of $\bar{x}$ since it is based on only 20 sample means. To obtain the complete distribution of sample means, we would have to take thousands of samples of 100 cigarettes each. Of course, in practice we do not take thousands of samples from the same population.

COMMENT Notice that the sample means form an approximate normal distribution. We will have more to say about this in Section 8.6.

What can we say about this distribution? What is its mean? Its standard deviation? How does this distribution compare with the distribution of *all* the cigarettes? To answer this question, the manufacturer decides to draw the frequency polygon for the tar content of all the 2000 cigarettes. This is shown in Figure 8.3.

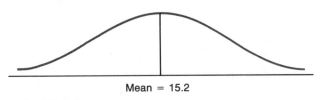

Mean = 15.2

FIGURE 8.3

Let us now compare these two distributions. Notice that both distributions are centered around the same number, 15.2. Thus it is reasonable to assume that $\mu_{\bar{x}} = \mu$. Also notice that the distribution of the sample means is not spread out as much as (that is, has a smaller standard deviation than) the distribution of the tar content of all the cigarettes. The reason for this should be obvious. When *all* the cigarettes are considered, several have a very high tar content and several have a very low tar content. These appear on the tail ends of the distribution of Figure 8.3. However, it is unlikely that an entire sample of 100 cigarettes will have a tar content of 18.5. Thus, the distribution of $\bar{x}$ has very little frequency at large distances from the mean.

We use the symbol $\sigma_{\bar{x}}$ to represent the standard deviation of the sampling distribution of the mean. We have the following formula for $\sigma_{\bar{x}}$.

FORMULA 8.5

*The standard deviation of the sampling distribution of the mean is referred to as the **standard error of the mean**. If random samples of size n are selected from a population whose mean is μ and whose standard deviation is σ, then the theoretical sampling distribution of $\bar{x}$ has mean $\mu_{\bar{x}} = \mu$ and standard deviation of*

$$\sigma_{\bar{x}} = \frac{\sigma}{\sqrt{n}} \cdot \sqrt{\frac{N-n}{N-1}} \qquad \begin{array}{l} \text{Standard error of the mean for} \\ \text{finite populations of size N} \end{array}$$

and

$$\sigma_{\bar{x}} = \frac{\sigma}{\sqrt{n}} \qquad \begin{array}{l} \text{Standard error of the mean for} \\ \text{infinite populations} \end{array}$$

COMMENT   The factor $\sqrt{\dfrac{N-n}{N-1}}$ in the first formula for $\sigma_{\bar{x}}$ is referred to as the **finite population correction factor.** It is usually ignored, that is, it has very little effect in the calculation of $\sigma_{\bar{x}}$, unless the sample constitutes at least 5% of the population.

COMMENT   It should be obvious from Formula 8.5 that the larger the sample size, the smaller will be the variation of the means. Thus, as we take larger and larger samples, we can expect the mean of the samples, $\mu_{\bar{x}}$, to be close to the mean of the population, μ. This is illustrated in Figure 8.4.

The standard error of the mean, $\sigma_{\bar{x}}$, plays a very important rule in statistics as will be illustrated in the remainder of this chapter.

To illustrate the concept of distribution of sample means and to see how

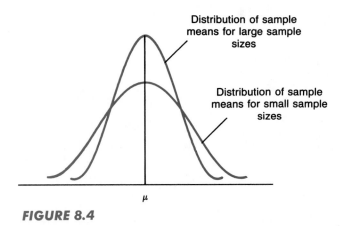

Distribution of sample
means for large sample
sizes

Distribution of sample
means for small sample
sizes

μ

FIGURE 8.4

Formula 8.5 is used, let us consider Mr. and Mrs. Avery, who have five children. The ages of the children are 2, 5, 8, 11, and 14 years. Suppose we first calculate the mean and standard deviation of these ages. We have

Age x	$x - \mu$	$(x - \mu)^2$
2	$2 - 8 = -6$	36
5	$5 - 8 = -3$	9
8	$8 - 8 = 0$	0
11	$11 - 8 = 3$	9
14	$14 - 8 = 6$	36
40		90
$\Sigma x = 40$		$\Sigma(x - \mu)^2 = 90$

Then

$$\mu = \frac{\Sigma x}{N} = \frac{40}{5} = 8 \quad \text{and} \quad \sigma = \sqrt{\frac{\Sigma(x - \mu)^2}{N}}$$

$$= \sqrt{\frac{90}{5}}$$

$$= \sqrt{18} \approx 4.2426$$

Thus the mean age is 8 and the standard deviation is approximately 4.2426 years.

Let us now calculate the mean of each sample of size two that can be formed from these ages. We have

If Ages Selected Are	Sample Mean, $\bar{x}$ Is
2 and 5 years	3.5
2 and 8 years	5
2 and 11 years	6.5
2 and 14 years	8
5 and 8 years	6.5
5 and 11 years	8
5 and 14 years	9.5
8 and 11 years	9.5
8 and 14 years	11
11 and 14 years	12.5

There are ten possible samples of size two that can be formed from these five ages. What is the average and standard deviation of these sample means? To determine these, we set up the following chart using the sample means as data:

$\bar{x}$	$\bar{x} - \mu_{\bar{x}}$	$(\bar{x} - \mu_{\bar{x}})^2$
3.5	$3.5 - 8 = -4.5$	20.25
5.0	$5.0 - 8 = -3$	9.00
6.5	$6.5 - 8 = -1.5$	2.25
8.0	$8.0 - 8 = 0$	0.00
6.5	$6.5 - 8 = -1.5$	2.25
8.0	$8.0 - 8 = 0$	0.00
9.5	$9.5 - 8 = 1.5$	2.25
9.5	$9.5 - 8 = 1.5$	2.25
11.0	$11.0 - 8 = 3$	9.00
12.5	$12.5 - 8 = 4.5$	20.25
80		67.5
$\Sigma\bar{x} = 80$		$\Sigma(\bar{x} - \mu_{\bar{x}})^2 = 67.5$

Thus, the average of the sample means is $\mu_{\bar{x}} = \dfrac{\Sigma\bar{x}}{\text{No. of samples}} = \dfrac{80}{10} = 8$ and the standard deviation of the sample means is

$$\sigma_{\bar{x}} = \sqrt{\frac{\Sigma(\bar{x} - \mu_{\bar{x}})^2}{\text{No. of samples}}} = \sqrt{\frac{67.5}{10}} = \sqrt{6.75} \approx 2.598.$$

In our case the population size, N, is 5 and the sample size n, is 2 so that the sample size is $\dfrac{2}{5}$ or 40% of the population size. Using Formula 8.5,

$$\sigma_{\bar{x}} = \frac{\sigma}{\sqrt{n}} \sqrt{\frac{N - n}{N - 1}}$$

$$= \frac{4.2426}{\sqrt{2}} \sqrt{\frac{5-2}{5-1}} \qquad \text{(Remember } \sigma = 4.2426, \text{ as previously calculated on page 367.)}$$

$$= \frac{4.2426}{1.414} \sqrt{\frac{3}{4}} \approx \frac{4.2426}{1.414} (0.866)$$

$$\approx 2.598$$

This is exactly the value that we obtained previously. Thus the average of the sample means is exactly the same as the population mean, that is, $\mu_{\bar{x}} = \mu$, and the standard deviation of the sample means is considerably less than the population standard deviation.

COMMENT In the balance of this chapter, we will always assume that our sample size is less than 5% of the population size. Consequently we will use $\sigma_{\bar{x}} = \dfrac{\sigma}{\sqrt{n}}$ as the standard error of the mean.

Let us summarize our discussion up to this point. Using the distribution of the sample means of the laboratories and the distribution of the tar content of all the 2000 cigarettes, or the distribution of the sample means of the ages, we conclude the following:

1. The mean of the distribution of sample means and the mean of the original population are the same.
2. The standard deviation of the distribution of the sample means is less than the standard deviation of the original population. The exact relationship is referred to as the standard error of the mean and is found by using Formula 8.5.
3. The distribution of the sample means is approximately normally distributed.

COMMENT The last statement is so important that it is referred to as the **Central Limit Theorem.** Since much of the work of statistical inference is based on this theorem, we will discuss its importance, as well as its applications in detail, in the following sections.

EXERCISES

1. The Lincoln school district operates six school buses. The buses were new when purchased and have been operating for 6, 10, 4, 7, 9, and 8 years.
   a. Make a list of all the possible samples of size two that can be drawn from this list of numbers.
   b. Determine the mean of each of these samples and form a sampling distribution of these sampling means.
   c. Find the mean, $\mu_{\bar{x}}$, of this sampling distribution.
   d. Find the standard deviation of this sampling distribution.

2. The Globe Electronics Corporation has five photocopying machines scattered throughout its building. The number of photocopies made by these machines on February 7 was 53, 117, 61, 78, and 91.
   a. Make a list of all the possible samples of size two that can be drawn from this set of numbers.
   b. Determine the mean of these samples and form a sampling distribution of these sampling means.
   c. Find the mean, $\mu_{\bar{x}}$, of this sampling distribution.
   d. Find the standard deviation $\sigma_{\bar{x}}$ of this sampling distribution.

3. The Marvel Travel Agency arranges and operates tours of the Orient. During one week, the following number of requests were received for tour information for a particular packaged tour.

Mon	Tues	Wed	Thurs	Fri	Sat
5	8	9	12	7	6

   a. Make a list of all the possible samples of size three that can be drawn from these numbers.
   b. Determine the mean of each of these samples and form a sampling distribution of these sampling means.
   c. Find the mean, $\mu_{\bar{x}}$, of this sampling distribution.
   d. Find the standard deviation, $\sigma_{\bar{x}}$, of this sampling distribution.

4. There are 20 brokerage houses in a large office building on Wall Street in New York City. The average age of the workers in each of the brokerage houses is: 33, 42, 37, 53, 47, 41, 55, 38, 29, 38, 38, 45, 53, 58, 27, 45, 52, 31, 46, and 32.
   a. Calculate the mean of the 20 sample means.
   b. Draw the histogram for these sample means (similar to what was done in Figure 8.2).

5. There are five agents that work for the state's Department of Environmental Protection. The number of cases involving illegal dumping of chemicals and other industrial wastes that these agents investigated last year was 8, 17, 12, 11, and 22.
   a. Determine the mean, μ, and standard deviation, σ, of the population of the number of cases that these agents investigated last year.
   b. List all the possible samples of size two *and* of size three that can be selected from these numbers, and determine the mean for each of these samples.
   c. Find the mean, $\mu_{\bar{x}}$, and standard deviation, $\sigma_{\bar{x}}$, for the sample means in each case.
   d. Show that for both the samples of size two and the samples of size three

$$\sigma_{\bar{x}} = \frac{\sigma}{\sqrt{n}} \sqrt{\left(\frac{N-n}{N-1}\right)}$$

6. A huge shipment (entire population) of special batteries is known to have a mean life of 30 hours and a standard deviation of 9 hours. Many samples of size 64 are taken. Find the mean of these samples and the standard error of the mean.

7. Refer back to Exercise 6. What would the standard error of the mean be if the samples are of
a. size 49 each?   **b.** size 100 each?

8.6 *THE CENTRAL LIMIT THEOREM*

One of the most important theorems in probability is the **Central Limit Theorem.** This theorem, first established by De Moivre in 1733 (see the discussion on page 315), was named "The Central Limit Theorem of Probability" by G. Polya in 1920. The theorem may be summarized as follows:

THE CENTRAL LIMIT THEOREM

If large random samples of size n (usually samples of size $n > 30$) are taken from a population with mean μ and standard deviation σ, and if a sample mean $\bar{x}$ is computed for each sample, then the following three things will be true about the distribution of sample means.

1. The distribution of the sample means will be approximately normally distributed.
2. The mean of the sampling distribution will be equal to the mean of the population. Symbolically,

$$\mu_{\bar{x}} = \mu$$

3. The standard deviation of the sampling distribution will be equal to the standard deviation of the population divided by the square root of the number of items in each sample. Symbolically,

$$\sigma_{\bar{x}} = \frac{\sigma}{\sqrt{n}}$$

COMMENT   If the sample size is large enough, the sampling distribution will be normal, even if the original distribution is not. "Large enough" usually means larger than 30 items in the sample.

COMMENT   The n referred to in the theorem refers to the size of each sample, not to the number of samples.

Since the Central Limit Theorem is so important, we will discuss its applications in the next section.

8.7 *APPLICATIONS OF THE CENTRAL LIMIT THEOREM*

In this section we use the Central Limit Theorem to predict the behavior of sample means. To apply the standardized normal distribution discussed in Chapter 7, we have to change Formula 7.1 somewhat. Recall that

$$z = \frac{x - \mu}{\sigma}$$

It can be shown that when dealing with sample means this formula becomes that given as Fomula 8.6.

FORMULA 8.6

$$z = \frac{\bar{x} - \mu}{\sigma/\sqrt{n}}$$

The following examples will illustrate how the Central Limit Theorem is applied.

● **Example 1**

The average height of all the workers in a hospital is known to be 65 inches with a standard deviation of 2.3 inches. If a sample of 36 people is selected at random, what is the probability that the average height of these 36 people will be between 64 and 65.5 inches?

Solution

We use Formula 8.6. Here $\mu = 65$, $\sigma = 2.3$, and $n = 36$. Thus $\bar{x} = 64$ corresponds to

$$z = \frac{64 - 65}{2.3/\sqrt{36}} = \frac{-1}{0.3833} = -2.61$$

and $\bar{x} = 65.5$ corresponds to

$$z = \frac{65.5 - 65}{2.3/\sqrt{36}} = \frac{0.5}{0.3833} = 1.30$$

Thus, we are interested in the area of a standard normal distribution between $z = -2.61$ and $z = 1.31$. See Figure 8.5.

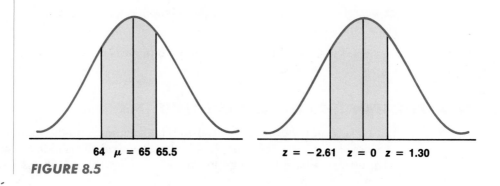

64 $\mu = 65$ 65.5 $z = -2.61$ $z = 0$ $z = 1.30$

FIGURE 8.5

From Table V in the Appendix we find that the area between $z = 0$ and $z = -2.61$ is 0.4955 and that the area between $z = 0$ and $z = 1.30$ is 0.4032. Adding we get

$$0.4955 + 0.4032 = 0.8987$$

Thus, the probability that the average height of the sample of 36 people is between 64 and 65.5 inches is 0.8987.

● Example 2

The average amount of money that a depositor of the Second National City Bank has in an account is $5000 with a standard deviation of $650. A random sample of 36 accounts is taken. What is the probability that the average amount of money that these 36 depositors have in their accounts is between $4800 and $5300?

Solution
We use Formula 8.6. Here $\mu = 5000$, $\sigma = 650$, and $n = 36$. Thus, $\bar{x} = 4800$ corresponds to

$$z = \frac{4800 - 5000}{650/\sqrt{36}} = \frac{-200}{108.33} = -1.85$$

and $\bar{x} = 5300$ corresponds to

$$z = \frac{5300 - 5000}{650/\sqrt{36}} = \frac{300}{108.33} = 2.77$$

Thus we are interested in the area between $z = -1.85$ and $z = 2.77$. See Figure 8.6.

From Table V in the Appendix we find that the area between $z = 0$ and $z = -1.85$ is 0.4678 and that the area between $z = 0$ and $z = 2.77$ is 0.4972. Adding these two we get

$$0.4678 + 0.4972 = 0.9650$$

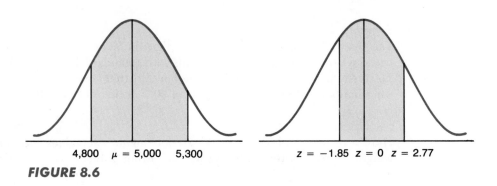

4,800 $\mu = 5,000$ 5,300 $z = -1.85$ $z = 0$ $z = 2.77$

FIGURE 8.6

Thus, the probability is 0.9650 that the average amount of money these depositors have in their accounts is between $4800 and $5300.

● Example 3

The average purchase by a customer in a large novelty store is $4.00 with a standard deviation of $0.85. If 49 customers are selected at random, what is the probability that their average purchases will be less than $3.70?

Solution

We use Formula 8.6. Here $\mu = 4.00$, $\sigma = 0.85$, and $n = 49$. Thus, $\bar{x} = 3.70$ corresponds to

$$z = \frac{3.70 - 4.00}{0.85/\sqrt{49}} = \frac{-0.30}{0.1214} = -2.47$$

Therefore we are interested in the area to the left of $z = -2.47$ (Fig. 8.7).

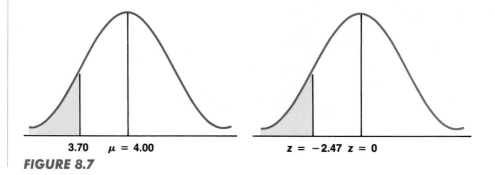

3.70 $\mu = 4.00$ $z = -2.47$ $z = 0$

FIGURE 8.7

From Table V in the Appendix we find that the area from $z = 0$ to $z = -2.47$ is 0.4932. Thus, the area to the left of $z = -2.47$ is $0.5000 - 0.4932$, or 0.0068. Therefore, the probability that the average purchase will be less than $3.70 is 0.0068.

● Example 4

The Smith Trucking Company claims that the average weight of its delivery trucks when fully loaded is 6000 pounds with a standard deviation of 120 pounds. Thirty-six trucks are selected at random and their weights recorded. Within what limits will the average weights of 90% of the 36 trucks lie?

Solution

Here $\mu = 6000$ and $\sigma = 120$. We are looking for two values within which the weights of 90% of the 36 trucks will lie. From Table V we find that the area

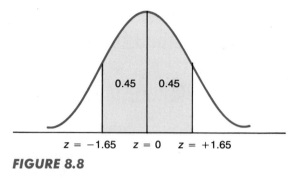

FIGURE 8.8

between $z = 0$ and $z = 1.65$ is approximately 0.45. Similarly, the area between $z = 0$ and $z = -1.65$ is approximately 0.45. See Figure 8.8. Using **Formula 8.6** we have

$$z = \frac{\bar{x} - \mu}{\sigma/\sqrt{n}}$$

If $z = 1.65$, then

$$1.65 = \frac{\bar{x} - 6000}{120/\sqrt{36}}$$

$$= \frac{\bar{x} - 6000}{20}$$

$$1.65(20) = \bar{x} - 6000$$

$$33 + 6000 = \bar{x}$$

$$6033 = \bar{x}$$

If $z = -1.65$, then

$$-1.65 = \frac{\bar{x} - 6000}{120/\sqrt{36}}$$

$$= \frac{\bar{x} - 6000}{20}$$

$$-1.65(20) = \bar{x} - 6000$$

$$6000 - 33 = \bar{x}$$

$$5967 = \bar{x}$$

Thus, 90% of the trucks will weigh between 5967 and 6033 pounds.

EXERCISES

1. The average length of a suitcase carried by a passenger boarding an airplane is 30 inches with a standard deviation of 3.4 inches. If a sample of 64 passengers is selected at random, what is the probability that the average length will be less than 29 inches?

2. A recent survey found that the average cost of a two-head wireless remote-control video cassette recorder (VCR) in a large city was $248 with a standard deviation of $32. If a survey of 36 stores selling these VCRs is taken, what is the probability that the average cost of this type of VCR will be less than $240?

3. The average diastolic blood pressure of young women in Hampton under 25 years of age was found to be 110 with a standard deviation of 8. A sample of 40 young women in Hampton is taken. What is the probability that the average diastolic blood pressure of these women is between 110 and 113?

4. The average length of a newborn baby is 20.1 inches with a standard deviation of 1.12 inches.* A random sample of 45 newborn babies is taken. What is the probability that the average length of these newborn babies will be more than 20.5 inches?

5. The average number of pages in a statistics book sold by a large bookstore is 453 pages with a standard deviation of 23 pages. A sample of 36 statistics books sold by this bookstore is taken. What is the probability that the average number of pages that these books contain is between 450 and 460?

6. Police records indicate that the average age at which children in a certain community first try drugs is 12.24 years with a standard deviation of 2.71 years. A sample of 40 young drug addicts is taken. What is the probability that the average age at which these young addicts first tried drugs is between 11 and 13 years of age?

7. The Gold Baking Company claims that the average weight of all of its crumb coffeecakes is 12 ounces with a standard deviation of 1.01 ounces. To check on the accuracy of the stated weight, an agent from the Consumer's Fraud Bureau selects a random sample of 40 of these cakes. Within what limits should the weights of 95% of these cakes lie?

8. The average life of a special rechargeable battery for a calculator that performs 46 functions is 800 hours (of continuous use) with a standard deviation of 38 hours. A sample of 60 such batteries is randomly selected. What percentage of the batteries will last more than 815 hours?

9. The average cost of a taxi cab ride from the airport to the center of the city is $14.28 with a standard deviation of $1.76. A dispatcher selects a random sample of 45 trips from the airport to the center of town. What is the probability that the average fare will be between $14 and $15?

10. A welfare claim investigator states that the average one-time emergency grant is $134 with a standard deviation of $17. A random sample of 36 such emergency grants is

Source: Bureau of Vital Statistics.

randomly selected. Within what limits will the amount of money for 90% of such emergency grants lie?

8.8 USING COMPUTER PACKAGES

The MINITAB software system can be used to generate, for example, a sampling distribution of $\bar{x}$ based on 25 samples of size n = 4 measurements, each drawn from a normal population with μ = 10 and σ = 3. We first have to use NRANDOM and AVERAGE to generate each sample of size n = 4 measurements and to compute the sample mean. To accomplish this, we generate one sample of size n = 4 as follows:

Sample 1

```
MTB > NRANDOM 4 OBSN, MU = 10, SIGMA = 3, PUT IN C1
   4 NORMAL OBS. WITH MU = 10.0000 AND SIGMA = 3.0000
      8.9321     11.6106     10.1234     9.7862

MTB > AVERAGE THE OBSERVATIONS IN C1
MEAN = 10.113
```

To generate a second sample, we must repeat the NRANDOM and AVERAGE statements until we obtain the 25 samples of size n = 4 measurements that we desire. This can be very time consuming. Nevertheless, after the 25 sample means are obtained, we enter these into MINITAB to obtain some numerical descriptive measures as was done in Chapter 3. For our example, we enter the 25 sample means into MINITAB as follows:

```
MTB > SET THE FOLLOWING DATA INTO C1
DATA > 10.113    10.101    9.996    10.002    10.123
DATA >  9.468     9.823    9.982    10.176    10.082
    :
    :
```

Then, we have MINITAB obtain numerical descriptive measures for the data. We get

```
MTB > DESCRIBE C1
```

The computer output for our example will be

```
          C1
N         25
MEAN      10.016
MEDIAN    10.183
T MEAN    10.213
STDEV     1.386
SEMEAN    0.812
MAX       10.816
MIN       7.938
```

```
Q3          11.547
Q1           8.096
MTB > STOP
```

COMMENT   The mean of the sample means is 10.016. This is fairly close to the theoretical value of $\mu = 10$. Similarly, the standard deviation of the sample means is 1.386 which is close to the theoretical value of $\dfrac{\sigma}{\sqrt{n}} = \dfrac{3}{\sqrt{4}} = 1.5$.

These experimental values would be closer to their respective theoretical values had we taken many more samples of size n = 4 rather than only the 25 samples that we generated.

8.9 SUMMARY

In this chapter we discussed the nature of random sampling and how to go about selecting a random sample. The most convenient way of selecting a random sample is to use a table of random digits.

In some situations, as we pointed out, stratified samples are preferred.

When repeated random samples are taken from the same population, different sample means are obtained. The average of these sample means can be used as an estimate of the population mean. Of course, these sample means form a distribution. If the samples are large enough, the Central Limit Theorem tells us that they will be normally distributed. Furthermore, the mean of the sampling distribution is the same as the population mean. The standard deviation of the sampling distribution is less than the population standard deviation. The Central Limit Theorem led to many useful applications.

STUDY GUIDE

You should now be able to demonstrate your knowledge of the following ideas presented in this chapter by giving definitions, descriptions, or specific examples. Page references are given for each term so that you can check your answer.

Population (page 353)
Sample (page 353)
Random sample (page 355)
Random sampling (page 355)
Table of random numbers (page 355
Table of random digits (page 355)
Stratified sample (page 358)
Distribution of sample means (page 365)
Sampling distribution of the mean (page 365)
Standard error of the mean (page 366)
Finite population correction factor (page 366)
The Central Limit Theorem (page 371)

FORMULAS TO REMEMBER

You should be able to identify each symbol in the following formulas, understand the relationships among the symbols expressed in each formula, understand the significance of each formula, and use the formulas in solving problems.

1. Population mean: $\mu = \dfrac{\Sigma x}{N}$

2. Sample mean: $\bar{x} = \dfrac{\Sigma x}{n}$

3. Population standard deviation: $\alpha = \sqrt{\dfrac{\Sigma(x - \mu)^2}{N}}$

4. Standard deviation of sample means: $\sqrt{\dfrac{\Sigma(\bar{x} - \mu_{\bar{x}})^2}{n - 1}}$

5. Computational formula for standard deviation of sample means:

$$\sqrt{\dfrac{n(\Sigma \bar{x}^2) - (\Sigma \bar{x})^2}{n(n - 1)}}$$

6. Standard error of the mean when sample size is less than 5% of the population size: $\sigma_{\bar{x}} = \dfrac{\sigma}{\sqrt{n}}$

7. When dealing with sample means: $z = \dfrac{\bar{x} - \mu}{\sigma/\sqrt{n}}$

The tests of the following section will be more useful if you take them after you have studied the examples and solved the exercises given in this chapter.

MASTERY TESTS

Form A

1. The Yellow Pages of a local telephone directory list 78 gardeners in alphabetical order. A homeowner wishes to randomly select six gardeners from the list to get an estimate for the cost of seeding her lawn. Use column 12 of Table VII to help the homeowner decide which gardeners to call.

For questions 2–5, use the following information: There are six members in the applied statistics department of a small college. Their ages are 42, 36, 53, 64, 37, and 38.

2. Make a list of all the possible samples of size two that can be drawn from this set of numbers.

3. Determine the mean of each of these samples and form a sampling distribution of these sampling means.

4. Find the mean, $\mu_{\bar{x}}$, of this sampling distribution.

5. Find the standard deviation, $\sigma_{\bar{x}}$, of this sampling distribution.

6. The average life of a particular brand of washing machine is 11.4 years with a standard deviation of 2.1 years. A random sample of 81 such washing machines is selected. Find the probability that the average life will be less than 11 years.

7. Assume that we are selecting a random sample from an infinite population. What happens to the standard error of the mean if the size of the sample is increased from 36 to 3600?

8. A two-family house in a particular city uses an average of 1200 gallons of heating oil per season for heating purposes. The standard deviation is 95 gallons. The heating bills for 50 two-family houses in this city are randomly selected. What is the probability that the average number of gallons of heating oil used is less than 1160?

9. There are 281 licensed real-estate agents authorized to sell farms in a certain county. Their licenses are numbered consecutively from 1 to 281. Mack Jones wishes to select eight of these agents and to interview each one personally to determine who should "list" his farm. Using column 4 of Table VIII, which real-estate agents should Mack select?

10. The length of time between car arrivals at a toll booth on a rural highway and the length of time required by the toll collector to make change are two random variables that are important to management in helping decide how many toll booths to build and how many toll collectors are needed for efficient operations. Studies of one toll road show that the interarrival time (the time between the arrivals of two consecutive cars) has a mean of 4.79 minutes with a standard deviation of 2.32 minutes. A random sample of the interarrival times of 49 cars is selected. What is the probability that the average interarrival time will exceed 5 minutes?

MASTERY TESTS

Form B

1. A local college newspaper claims that a college student staying in a college dormitory spends an average of $7.61 at the local fast-food store per week with a standard deviation of $1.26. In order to verify this claim, a reporter randomly selects 80 college dormitory students. What is the probability that these students spend between an average of $7.25 and $8 per week at the local fast-food store?

2. Studies show that the average amount of time that a patient spends in a doctor's office is 55 minutes (including examination time) with a standard deviation of 18 minutes. What is the probability that a random sample of

49 patients will show that these patients spent more than an hour in their doctor's office?

3. The average diameter of a pizza made by Mario's Pizzeria is 16.1 inches with a standard deviation of 0.82 inches. Seventy-five pizzas made by this pizzeria are randomly selected. Within what limits will the diameters of 90% of these pizzas lie?

4. A new technique to teach reading comprehension is being tried. At the end of the school year, the usual standardized test, which in the past has produced scores with a mean of 140 and a standard deviation of 12, will be given. What is the probability that the scores of 36 randomly selected students taught by this new method will be greater than 150?

For Questions 5–7 use the following information: FBI statistics indicate that the number of violent crimes in a particular city averages about 14.7 per day with a standard deviation of 2.3. A sample of 36 days is randomly selected and the average daily number of violent crimes occurring, $\bar{x}$, is calculated.

5. Find the mean and standard deviation of the sampling distribution of $\bar{x}$.

6. What is the probability that $\bar{x}$ will be less than 14?

7. What is the probability that $\bar{x}$ will be between 14 and 15?

8. The average weights of deer killed by hunters in a certain region is normally distributed with a mean of 85 pounds and a standard deviation of 12 pounds.
   a. What is the probability that a deer killed by Christopher will weigh more than 88 pounds?
   b. If a sample of 49 deer killed by hunters is randomly selected and their average weight determined, what is the probability that the mean weight will be greater than 88 pounds?

9. Bill Kenny is in charge of the mailroom. He oversees the operation of the postage meter machine. Based upon past experience, Bill knows that the postage cost of mailing a certain type of package averages $1.30 with a standard deviation of $0.12. Bill realizes that he has to mail 50 additional such packages. The postage meter machine indicates that there is only $60 remaining for postage before the machine has to be reset at a special post office. If the postage for the remaining packages averages less than $1.25 per package, then he will have sufficient postage money before the machine has to be reset. What is the probability that Bill will have sufficient postage for all of the remaining packages before having to reset the postage meter machine?

10. A study conducted by the Newton Medical Group found that the average cost of medical malpractice insurance for a doctor in that city was $9100 with a standard deviation of $875. A subsequent random sample of 36 doctors in this city found that the average cost for malpractice insurance was $10,000. Are these new results consistent with the findings of the earlier study? (*Hint:* Determine the probability of this event happening.)

SUGGESTED FURTHER READING

1. Hicks, C. R. *Fundamental Concepts in the Design of Experiments*, 2nd ed. New York: Holt, Rinehart and Winston, 1973.
2. Johnson, N. L. and F. C. Leone. *Statistics and Experimental Design in Engineering and the Physical Sciences*. New York: Wiley, 1964.
3. Lapin, L. *Statistics: Meaning and Method*. New York: Harcourt, Brace, Jovanovich, 1975. (Section 5.5 and all of Chapter 6)
4. Mendenhall, W. L., L. Ott, and R. L. Scheaffer. *Elementary Survey Sampling*. Belmont, CA. Wadsworth, 1968.
5. Mendenhall, W. L. *Introduction to Probability and Statistics*, 6th ed. North Scituate, MA.: Duxbury Press, 1983.
6. Slonim, M. *Sampling*. New York: Simon & Schuster, 1960.
7. "For Better Polls." *Business Week*, June 25, 1949, p. 24.

CHAPTER 9

Estimation

OBJECTIVES

- *To discuss* how sample data can often be used to estimate certain unknown quantities. This use of samples is called statistical estimation.

- *To point out* that population parameters are statistical descriptions of the population.

- *To learn* that sample data can be used to obtain both point and interval estimates. Point estimates give us a single number whereas interval estimates set up an interval within which the parameter is expected to lie.

- *To see* that degrees of confidence give us the probability that the interval will actually contain the quantity that we are trying to estimate.

- *To apply* the Central Limit Theorem to set up confidence intervals for the mean and standard deviation.

- *To introduce* the Student's *t*-distribution to set up confidence intervals when the sample size is small.

- *To indicate* how the Central Limit Theorem is also used to set up confidence intervals for population proportions.

- *To analyze* how we determine the correct size of a sample for a given allowable error.

OCCUPATIONAL OUTLOOK HIGHLIGHTS, PROFESSIONAL WORKERS, 1986–1988

Professional Occupations	Employment		
	1986	1987*	1988
Total, all occupations	8,529,054	8,653, 380	8,777,829
Professional and technical workers	1,720,916	1,750,897	1,780,794
Engineers	94,110	95,804	97,490
Life and physical scientists	15,740	15,848	15,988
Mathematical specialists	4,589	4,654	4,716
Engineering, science technicians	87,236	89,138	91,031
Medical workers, excluding technicians	241,867	246,283	250,701
Other health technologists and technicians	112,370	115,475	118,568
Technicians, excluding health or science and engineering	37,643	38,003	38,360
Computer specialists	63,181	64,563	65,944
Social scientists	24,441	25,194	25,943
Teachers	385,454	390,516	395,572
Teachers, adult education	8,761	8,887	9,013
College and university teachers	102,996	104,322	105,643
Teachers, elementary	141,281	143,163	145,045
Teachers, secondary	122,058	123,585	125,112
Teachers, nursery/early childhood	10,358	10,559	10,759
Writers, artists, entertainers	109,162	112,509	115,847
Other professional and technical workers	545,159	552,910	560,634

*Figures for 1987 are simple linear interpolations of 1986 and 1988 data, and should not be interpreted as representing cyclical fluctuations in the economy.

- An estimated 952,000 job openings are expected annually in the forecast period.
- Separations from the labor force and occupational mobility will account for approximately 8 in every 10 openings.
- Growth in employment will make available over 124,000 job opportunities annually.
- Clerical jobs will require the largest number of workers, totalling more than 244,000 openings annually.
- Over 184,000 professional and technical workers are needed to fill positions which will become available each year in the projection period.
- Service occupations will provide the third largest source of employment with about 150,000 openings expected annually.
- Managers and officials will account for over 10 percent of the available jobs, totaling nearly 96,000 each year.
- Blue collar jobs, including the crafts, operatives, and nonfarm laborer occupational groups, will provide about 198,000 openings per year or 1 in every 5 occupational opportunities.
- An anticipated 74,000 jobs will be available for filling within the sales occupations yearly.
- Demand for farm workers will exceed 6,000 openings annually through 1988.

Source: "Occupational Needs in the 1980s. New York State, 1986–1988." NYS Department of Labor.

The clippings on p. 384 present some information on the occupational outlook for the years 1986–1988. How are such estimates on the expected number of job openings obtained? Moreover, how accurate are they?

Very often we use currently available information to make predictions about the future. Great care must be exercised in making such predictions as there are many factors that can offset the reliability of such estimates. The same is true for the population forecasts presented in the article below.

U.S. POPULATION – NEW FORECAST

The U.S. population will rise from 238.2 million now to 267.5 million by the year 2000 – the result of a growth rate slowed from 1.1 percent annually in the '70s to 0.8 for the rest of the century.

That's the latest projection of the National Planning Association, which also predicts that 84 percent of the population expansion will occur in the West and South. Arizona, Nevada and Florida will have the fastest growth rates, and California, Texas and Florida will gain the most people.

	1985	2000		1985	2000
California	25.8 mil.	30.4 mil.	Oklahoma	3.2 mil.	3.7 mil.
New York	17.6 mil.	17.5 mil.	Connecticut	3.2 mil.	3.3 mil.
Texas	15.9 mil.	20.0 mil.	Iowa	2.9 mil.	3.0 mil.
Pennsylvania	11.9 mil.	12.1 mil.	Oregon	2.8 mil.	3.4 mil.
Florida	11.6 mil.	15.6 mil.	Mississippi	2.6 mil.	2.9 mil.
Illinois	11.6 mil.	11.9 mil.	Arkansas	2.4 mil.	2.8 mil.
Ohio	10.9 mil.	11.2 mil.	Kansas	2.4 mil.	2.6 mil.
Michigan	9.3 mil.	9.8 mil.	West Virginia	1.9 mil.	1.8 mil.
New Jersey	7.5 mil.	7.8 mil.	Utah	1.6 mil.	2.2 mil.
North Carolina	6.3 mil.	7.6 mil.	Nebraska	1.6 mil.	1.6 mil.
Georgia	5.9 mil.	6.9 mil.	New Mexico	1.4 mil.	1.7 mil.
Massachusetts	5.9 mil.	6.3 mil.	Maine	1.2 mil.	1.3 mil.
Virginia	5.6 mil.	6.4 mil.	New Hampshire	1.0 mil.	1.3 mil.
Indiana	5.6 mil.	5.9 mil.	Idaho	1.0 mil.	1.2 mil.
Missouri	5.0 mil.	5.2 mil.	Hawaii	1.0 mil.	1.1 mil.
Tennessee	4.8 mil.	5.5 mil.	Rhode Island	997,000	1.2 mil.
Wisconsin	4.8 mil.	5.1 mil.	Nevada	944,000	1.3 mil.
Washington	4.5 mil.	5.3 mil.	Montana	801,000	807,000
Louisiana	4.4 mil.	4.7 mil.	South Dakota	692,000	713,000
Maryland	4.3 mil.	4.7 mil.	North Dakota	666,000	672,000
Minnesota	4.2 mil.	4.4 mil.	Delaware	615,000	657,000
Alabama	4.1 mil.	4.6 mil.	Dist. of Columbia	597,000	535,000
Kentucky	3.7 mil.	4.1 mil.	Vermont	541,000	625,000
South Carolina	3.4 mil.	4.0 mil.	Wyoming	510,000	616,000
Arizona	3.2 mil.	4.7 mil.	Alaska	460,000	583,000
Colorado	3.2 mil.	4.1 mil.	**U.S. Total**	**238.2 mil.**	**267.5 mil.**

June 17, 1985

Source: *U.S. News & World Report*, June 17, 1985.

9.1 INTRODUCTION

We have mentioned on several occasions that statistical inference is the process by which statisticians make predictions about a population on the basis of samples. Much information can be gained from a sample. As we mentioned in Chapter 8, the average of the sample means can be used as an estimate of the population mean. Also, we can obtain an estimate of the population standard deviation on the basis of samples. Thus, one use of sample data is to *estimate* certain unknown quantities of the population. This use of samples is referred to as **statistical estimation.**

On the other hand, sample data can also be used to either accept or reject specific claims about populations. To illustrate this use of sample data, suppose a manufacturer claims that the average milligram tar content per cigarette of a particular brand is 15 with a standard deviation of 0.5. Repeated samples are taken to test this claim. If these samples show that the average tar content per cigarette is 22, then the manufacturer's claim is incorrect. If these samples show that the average tar content is within "predictable limits," then the claim is accepted. Thus, sample data can also be used to either accept or reject specific claims about populations. This use of samples is referred to as **hypothesis testing.**

Statistical inference can be divided into two main categories: problems of estimation and tests of hypotheses. In this chapter we will discuss statistical estimation. In Chapter 10 we will analyze the nature of hypothesis testing.

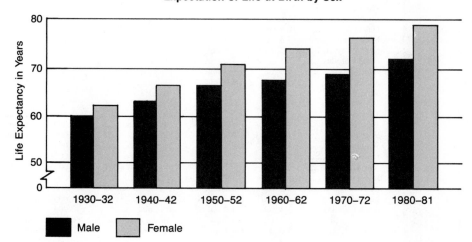

Expectation of Life at Birth by Sex

FIGURE 9.1

The above graph gives us an estimate of the life expectancy of men and women. Such information is used by insurance companies in determining the premium to charge for life insurance. Courtesy of The Metropolitan Life Insurance Company and the Population Reference Bureau (Washington).

9.2 POINT AND INTERVAL ESTIMATES

In most statistical problems we do not know certain population values such as the mean and the standard deviation. Somehow we want to use the information obtained from samples to estimate their values. These values, which really are statistical descriptions of the population, are often referred to as **population parameters.**

Suppose we are interested in determining the average life of an electric refrigerator under normal operating conditions. A sample of 50 refrigerators is taken and their lives recorded as shown below:

6.9	7.6	5.7	3.6	7.7	6.6	7.2	7.3	10.6	5.9
8.2	5.7	7.6	8.7	7.9	8.8	7.0	8.1	7.3	6.8
5.7	11.1	8.5	8.9	7.6	5.6	9.0	9.2	6.8	8.3
6.1	9.7	9.8	7.4	6.8	7.3	8.3	9.9	7.5	7.8
7.7	7.4	9.1	7.3	5.5	8.1	6.7	8.8	7.6	5.3

The average life of these refrigerators is 7.6 years. Since this is the only information available to us, we would logically say that the mean life of *all* similar refrigerators is 7.6 years. This estimate of 7.6 years for the population mean is called a **point estimate** since this estimate is a single number. Of course, this estimate may be a poor estimate, but it is the best we can get under the circumstances.

Our confidence in this estimate would be improved considerably if the sample size were larger. Thus, we would have much greater confidence in an estimate that is based on 5000 refrigerators or 50,000 refrigerators than in one that is based only on 50 refrigerators.

One major disadvantage with a point estimate is that the estimate does not indicate the extent of the possible error. Furthermore, a point estimate does not specify how confident we can be that the estimate is close in value to the parameter that it is estimating. Yet, point estimates are often used to estimate population parameters.

Another type of estimate that is often used by statisticians, which overcomes the disadvantages mentioned in the previous paragraph, is **interval estimation.** In this method we first find a point estimate. Then we use this estimate to construct an interval within which we can be reasonably sure that the true parameter will lie. Thus, in our example a statistician may say that the mean life of the refrigerators will be between 7.2 and 8.0 years with a 95% degree of confidence. An interval such as this is called a **confidence interval.** The lower and upper boundaries, 7.2 and 8.0 respectively, of the interval are called **confidence limits.** The probability that the procedure used will give a correct interval is called the **degree of confidence.**

Generally speaking, as we increase the degree of certainty, namely, the degree of confidence, the confidence interval will become wider. Thus, if the length of an interval is very small (with a specific degree of confidence), then a fairly accurate estimate has been obtained.

When estimating the parameters of a population, statisticians use both point and interval estimates. In the next few sections we will indicate how point and interval estimates are obtained.

9.3 ESTIMATING THE POPULATION MEAN ON THE BASIS OF A LARGE SAMPLE

In Chapter 8 we indicated that the average of the sample means can be used as an estimate of the population mean, μ. Moreover, the larger the sample size the better the estimate. Yet, as we pointed out in Section 9.2 there are some disadvantages with using point estimates.

The Central Limit Theorem (see page 371) says that the sample means will be normally distributed if the sample sizes are large enough. Generally speaking, statisticians say that a sample size is considered large if it is greater than 30. We can use the Central Limit Theorem to help us construct confidence intervals. This is done as follows.

Since the sample means are approximately normally distributed, we can expect 95% of the $\bar{x}$'s to fall between

$$\mu - 1.96\sigma_{\bar{x}} \quad \text{and} \quad \mu + 1.96\sigma_{\bar{x}}$$

(since from a normal distribution chart we note that 0.95 probability implies that $z = 1.96$ or -1.96).

Recall (Formula 8.5, page 366) that

$$\sigma_{\bar{x}} = \frac{\sigma}{\sqrt{n}}$$

Thus, 95% of the $\bar{x}$'s are expected to fall between

$$\mu - 1.96\frac{\sigma}{\sqrt{n}} \quad \text{and} \quad \mu + 1.96\frac{\sigma}{\sqrt{n}}$$

This is shown in Figure 9.2. If all possible samples of size n are selected, and the interval $\bar{x} \pm 1.96\frac{\sigma}{\sqrt{n}}$ is established for each sample, then 95% of all such intervals are expected to contain μ. Thus a 95% confidence interval for μ is $\bar{x} \pm 1.96\frac{\sigma}{\sqrt{n}}$.

In order to determine the interval estimate of μ, we must first know the value of the population standard deviation, σ. Although this value is usually unknown, since the sample size is large we can use the sample standard deviation as an approximation for σ. We then have the following confidence interval for μ:

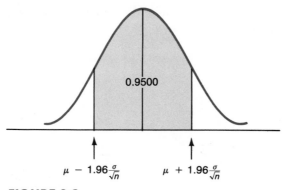

FIGURE 9.2

FORMULA 9.1

*Let $\bar{x}$ be a sample mean and s be the sample standard deviation. Then an interval is called a **95% confidence interval for μ** if the lower boundary of the confidence interval is*

$$\bar{x} - 1.96 \frac{s}{\sqrt{n}}$$

and if the upper boundary of the confidence interval is

$$\bar{x} + 1.96 \frac{s}{\sqrt{n}}.$$

COMMENT Formula 9.1 tells us how to find a 95% confidence interval for μ. This means that, in the long run, 95% of such intervals will contain μ. We can be 95% confident that μ lies within the specified interval. We must still realize that 5% of the time the population mean will fall outside this interval. This is true because the sample means are normally distributed and 5% of the values of a random variable in a normal distribution will fall further away than 2 standard deviations from the mean (see page 321).

COMMENT Depending upon the nature of the problem, some statisticians will often prefer a 99% confidence interval for μ or a 90% confidence interval for μ. The boundaries for these intervals are as follows:

	Lower Boundary	Upper Boundary
99% Confidence interval	$\bar{x} - 2.58 \frac{s}{\sqrt{n}}$	$\bar{x} + 2.58 \frac{s}{\sqrt{n}}$
90% Confidence interval	$\bar{x} - 1.64 \frac{s}{\sqrt{n}}$	$\bar{x} + 1.64 \frac{s}{\sqrt{n}}$

Thus, as we reduce the size of the interval, we reduce our confidence that the true mean will fall within that interval.

The following examples will illustrate how we establish confidence intervals.

● **Example 1**

A coffee vending machine fills 100 cups of coffee before it has to be refilled. On Monday the mean number of ounces in a filled cup of coffee was 7.5. The population standard deviation is known to be 0.25 ounces. Find 95% and 99% confidence intervals for the mean number of ounces of coffee dispensed by this machine.

Solution

We use Formula 9.1. Here $n = 100$, $\sigma = 0.25$, and $\bar{x} = 7.5$. To construct a 95% confidence interval for μ, we have

Lower Boundary	Upper Boundary
$= \bar{x} - 1.96 \dfrac{\sigma}{\sqrt{n}}$	$= \bar{x} + 1.96 \dfrac{\sigma}{\sqrt{n}}$
$= 7.5 - 1.96 \left(\dfrac{0.25}{\sqrt{100}} \right)$	$= 7.5 + 1.96 \left(\dfrac{0.25}{\sqrt{100}} \right)$
$= 7.5 - 0.05$	$= 7.5 + 0.05$
$= 7.45$	$= 7.55$

Thus, we conclude that the population mean will lie between 7.45 and 7.55 ounces with a confidence of 0.95.

To construct a 99% confidence interval for μ, we have

Lower Boundary	Upper Boundary
$= \bar{x} - 2.58 \dfrac{\sigma}{\sqrt{n}}$	$= \bar{x} + 2.58 \dfrac{\sigma}{\sqrt{n}}$
$= 7.5 - 2.58 \left(\dfrac{0.25}{\sqrt{100}} \right)$	$= 7.5 + 2.58 \left(\dfrac{0.25}{\sqrt{100}} \right)$
$= 7.5 - 0.06$	$= 7.5 + 0.06$
$= 7.44$	$= 7.56$

Thus, we conclude with a 99-percent confidence that the population mean will lie between 7.44 and 7.56 ounces. In this example we did not have to use s as an estimate of σ since we were told that the population standard deviation was known to be 0.25.

- **Example 2**

A sample survey of 81 movie theaters showed that the average length of the main feature film was 90 minutes with a standard deviation of 20 minutes. Find a

a. 90% confidence interval for the mean of the population.
b. 95% confidence interval for the mean of the population.

Solution
We use Formula 9.1. Here $n = 81$, $s = 20$, and $\bar{x} = 90$.

a. To construct a 90% confidence interval for μ, we have

Lower Boundary	Upper Boundary
$= \bar{x} - 1.64 \dfrac{s}{\sqrt{n}}$	$= \bar{x} + 1.64 \dfrac{s}{\sqrt{n}}$
$= 90 - 1.64 \left(\dfrac{20}{\sqrt{81}} \right)$	$= 90 + 1.64 \left(\dfrac{20}{\sqrt{81}} \right)$
$= 90 - 3.64$	$= 90 + 3.64$
$= 86.36$	$= 93.64$

Thus, a 90% confidence interval for μ is 86.36 to 93.64 minutes.

b. To construct a 95% confidence interval for μ, we have

Lower Boundary	Upper Boundary
$= \bar{x} - 1.96 \dfrac{s}{\sqrt{n}}$	$= \bar{x} + 1.96 \dfrac{s}{\sqrt{n}}$
$= 90 - 1.96 \left(\dfrac{20}{\sqrt{81}} \right)$	$= 90 + 1.96 \left(\dfrac{20}{\sqrt{81}} \right)$
$= 90 - 4.36$	$= 90 + 4.36$
$= 85.64$	$= 94.36$

Thus, a 95% confidence interval for μ is 85.64 to 94.36 minutes.

Notice that as we increase the size of the confidence interval, our confidence that this interval contains μ also increases.

- **Example 3**

The management of the Night-All Corporation recently conducted a survey of 196 of its employees to determine the average number of hours that each employee sleeps at night. The company statistician submitted the following information to the management:

$$\Sigma x = 1479.8 \quad \text{and} \quad \Sigma(x - \bar{x})^2 = 1755$$

where x is the number of hours slept by each employee. Find a 95% confidence interval estimate for the average number of hours that each employee sleeps at night.

Solution

In order to use Formula 9.1, we must first calculate $\bar{x}$ and s. Using the given information, we have

$$\bar{x} = \frac{\Sigma x}{n}$$

$$= \frac{1479.8}{196}$$

$$= 7.55$$

and

$$s = \sqrt{\frac{\Sigma(x - \bar{x})^2}{n - 1}}$$

$$= \sqrt{\frac{1755}{195}}$$

$$= \sqrt{9} = 3$$

Now we can use Formula 9.1, with $\bar{x} = 7.55$, $s = 3$, and $n = 196$. To find the 95% confidence for μ, we have

Lower Boundary	Upper Boundary
$= \bar{x} - 1.96 \dfrac{s}{\sqrt{n}}$	$= \bar{x} + 1.96 \dfrac{s}{\sqrt{n}}$
$= 7.55 - 1.96 \left(\dfrac{3}{\sqrt{196}}\right)$	$= 7.55 + 196 \left(\dfrac{3}{\sqrt{196}}\right)$
$= 7.55 - 0.42$	$= 7.55 + 0.42$
$= 7.13$	$= 7.97$

Thus, the management can conclude with a 95% confidence that the average number of hours that an employee sleeps at night is between 7.13 and 7.97 hours.

EXERCISES

1. Consider the newspaper article at the top of the next page. If the standard deviation is 8.7 gallons of water, find a 95% confidence interval for the average amount of water wasted.

CITY STILL IN WATER CRISIS

New York (Aug. 10) – Despite the heavy rains that fell over the weekend, the city is still in the grip of a water emergency. The reservoirs stand at only 51.4% of capacity.

A survey of 100 randomly selected city dwellers found that they were wasting an average of 61 gallons of water per day in a variety of ways including letting the water run needlessly, leaky faucets and pipes, unnecessary flushes of the toilet, etc.

The mayor is pleading with the citizens to conserve water.

Monday, August 10, 1981

2. A computer programming teacher has been keeping records on how long it takes a student to execute a particular program. For a group of 100 students, the average time was 18 minutes with a standard deviation of 3.2 minutes. Find a 99% confidence interval for the mean time needed by a student to execute the particular program.

3. A sample of 49 mortgages at the First State Bank showed that the average amount of initial home mortgage loans was $48,500 with a standard deviation of $3600. Find a 90% confidence interval for the average amount of the initial home mortgage loan.

4. A consumer's group sampled 36 different stores in the city and found that the average price of a particular model and brand of scientific calculator was $22.95 with a standard deviation of $0.278. Find a 95% confidence interval for the average price of this particular calculator.

5. Consider the newspaper article below. If the standard deviation was 0.09, find a 90% confidence interval for the average cost of a gallon of home heating oil.

HEATING OIL PRICES TO REMAIN STEADY

Washington (October 22) – Energy Department officials predicted yesterday that the current supply of home heating oil would be adequate for this country's winter needs. Furthermore, the price was expected to remain relatively stable. A nationwide random survey of 50 oil distributors conducted by Energy Department officials found that the average current price of a gallon of number 2 home heating oil in the Northeastern part of the U.S. was $1.259.

The average price was expected to prevail throughout the 1985–86 heating season.

Thursday, October 22, 1985

6. A sample of 49 check-cashing businesses in Middletown showed that the average amount of money involved in a "bounced" check is $227 with a standard deviation of $19. Find a 95% confidence interval for the average amount of money involved when a check is bounced.

7. The average cost of a gallon of unleaded high-test gasoline at 39 service stations was $1.059 with a standard deviation of 0.08. Find a 99% confidence interval for the average cost of a gallon of unleaded high-test gasoline.

8. A statistician is at a fund-raising drive for a local charity. She wishes to determine the average amount of money pledged. She randomly samples 81 guests and finds that the average amount of money pledged by these guests is $18 with a standard deviation of $1.05. Find a 95% confidence interval for the average amount of money pledged.

* **9.** A survey of 65 homes by a fire marshal in the Boro Hall region of a city found that there were an average of 2.1 smoke detectors in these homes with a standard deviation of 0.31. Find an 85% confidence interval for the average number of smoke detectors in this region of the city.

*10. In one study of 81 vacationers, it was found that the average amount of money in the form of traveller's checks purchased by visitors prior to embarking on their trip to Disneyland was $225 with a standard deviation of $8.17. Find an 88% confidence interval for the average amount of money in the form of traveller's checks purchased by visitors prior to embarking on their trip to Disneyland.

11. A random sample of n measurements was taken from a population with unknown mean μ and standard deviation σ. The following data was obtained: $\Sigma x = 800$, $\Sigma x^2 = 17050$, and $n = 50$.
   a. Find a 95% confidence interval for μ.
   b. Find a 99% confidence interval for μ.

9.4 ESTIMATING THE POPULATION MEAN ON THE BASIS OF A SMALL SAMPLE

In Section 9.3 we indicated how to determine confidence intervals for μ when the sample size is larger than 30. Unfortunately, this is not always the case. Suppose a sample of 16 bulbs is randomly selected from a large shipment and has a mean life of 100 hours with a standard deviation of 5 hours. Using only the methods of Section 9.3, we cannot determine confidence intervals for the mean life of a bulb since the sample size is less than 30.

Fortunately, in such situations we can base confidence intervals for μ on a distribution that is in many respects similar to the normal distribution. This is the **Student's t-distribution.** This distribution was first studied by William S. Gosset, who was a statistician for Guinness, an Irish brewing company. Gosset was the first to develop methods for interpreting information obtained from small samples. Yet his company did not allow any of its employees to publish anything. So, Gosset secretly published his findings in 1907 under the name

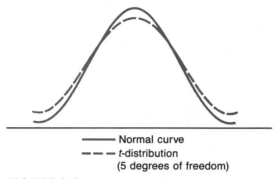

—— Normal curve
– – – *t*-distribution
(5 degrees of freedom)

FIGURE 9.3
Relationship between the normal distribution and
the *t*-distribution.

"Student." To this day, this distribution is referred to as the Student's *t*-distribution.

Figure 9.3 indicates the relationship between the normal distribution and the *t*-distribution. Notice that the *t*-distribution is also symmetrical about zero, which is its mean. However, the shape of the *t*-distribution depends upon a parameter called the **number of degrees of freedom.** In our case the number of degrees of freedom, abbreviated as d.f., is equal to the sample size minus 1. If the population sampled is normally distributed, then $\dfrac{\bar{x} - \mu}{s/\sqrt{n}}$ has a *t*-distribution. This standardized *t*-distribution is symmetrical, bell shaped, and has zero as its mean.

Table VIII in the Appendix indicates the value of *t* for different degrees of freedom. Thus, the 1.96 of Formula 9.1 of Section 9.3 can be replaced by the $t_{0.025}$ value as listed in this table, depending upon the number of degrees of freedom. When using the $t_{0.025}$ value of Table VIII, 95% of the area under the curve of the *t*-distribution will fall between $-t_{0.025}$ and $t_{0.025}$, as shown in Figure 9.4.

We then have the following formula:

FORMULA 9.2

Let $\bar{x}$ be a sample mean and let s be the sample standard deviation. We have the following 95% small-sample confidence interval for μ:

$$\textbf{Lower boundary} = \bar{x} - t_{0.025}\,\frac{s}{\sqrt{n}}$$

$$\textbf{Upper boundary} = \bar{x} + t_{0.025}\,\frac{s}{\sqrt{n}}$$

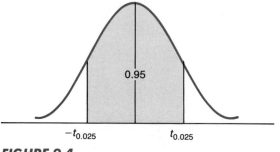

FIGURE 9.4

COMMENT In addition to the $t_{0.025}$ values, Table VIII in the Appendix also lists many other values of t. Thus the $t_{0.005}$ values are used when we want a 99% confidence interval for μ, the $t_{0.050}$ values are used when we want a 90% confidence interval for μ, the $t_{0.10}$ values are used when we want an 80% confidence interval for μ, and so on.

● **Example 1**

A survey of 16 taxi drivers found that the average tip they receive is 40 cents with a standard deviation of 8 cents. Find a 95% confidence interval for estimating the average amount of money that a taxi driver receives as a tip.

Solution

We will use Formula 9.2. Here we have $n = 16$, $\bar{x} = 40$, and $s = 8$. We first find the number of degrees of freedom, which is $n - 1$, or $16 - 1 = 15$. Then we find the appropriate value of t from Table VIII. The $t_{0.025}$ value with 15 degrees of freedom is 2.131. Finally we establish the confidence interval. We have

Lower Boundary	*Upper Boundary*
$= \bar{x} - t_{0.025}\dfrac{s}{\sqrt{n}}$	$= \bar{x} + t_{0.025}\dfrac{s}{\sqrt{n}}$
$= 40 - 2.131\left(\dfrac{8}{\sqrt{16}}\right)$	$= 40 + 2.131\left(\dfrac{8}{\sqrt{16}}\right)$
$= 40 - 4.26$	$= 40 + 4.26$
$= 35.74$	$= 44.26$

Thus, a 95% confidence interval for the average amount of money that a taxi driver will receive as a tip is 35.74 to 44.26 cents.

● **Example 2**

A survey of the hospital records of 25 randomly selected patients suffering from a particular disease indicated that the average length of stay in the hospital is

10 days. The standard deviation is estimated to be 2.1 days. Find a 99% confidence interval for estimating the mean length of stay in the hospital.

Solution
We use Formula 9.2. Here $n = 25$, $\bar{x} = 10$, and $s = 2.1$. We first find the number of degrees of freedom, which is $n - 1$, or $25 - 1 = 24$. Then we find the appropriate value of t from Table VIII in the Appendix. The $t_{0.005}$ value with 24 degrees of freedom is 2.797. Finally, we establish the confidence interval. We have

Lower Boundary	Upper Boundary
$= \bar{x} - t_{0.005} \dfrac{s}{\sqrt{n}}$	$= \bar{x} + t_{0.005} \dfrac{s}{\sqrt{n}}$
$= 10 - 2.797 \left(\dfrac{2.1}{\sqrt{25}} \right)$	$= 10 + 2.797 \left(\dfrac{2.1}{\sqrt{25}} \right)$
$= 10 - 1.17$	$= 10 + 1.17$
$= 8.83$	$= 11.17$

Thus, a 99% confidence interval for the average length of stay in the hospital is 8.83 to 11.17 days.

● Example 3
A music teacher asks six randomly selected students how many hours a week each practices playing the electric guitar. The teacher receives the following answers: 10, 12, 8, 9, 16, 5. Find a 90% confidence interval for the average length of time that a student practices playing the electric guitar.

Solution
In order to use Formula 9.2 we must first calculate the sample mean and the sample standard deviation. We have

$$\bar{x} = \frac{\Sigma x}{n} = \frac{10 + 12 + 8 + 9 + 16 + 5}{6} = \frac{60}{6} = 10$$

To calculate the sample standard deviation, we arrange the data as follows:

x	x − x̄	(x − x̄)²
10	10 − 10 = 0	0
12	12 − 10 = 2	4
8	8 − 10 = −2	4
9	9 − 10 = −1	1
16	16 − 10 = 6	36
5	5 − 10 = −5	25
	Total =	70

$$s = \sqrt{\frac{\Sigma(x - \bar{x})^2}{n - 1}} = \sqrt{\frac{70}{6 - 1}} = \sqrt{14} \approx 3.74$$

Now we find the appropriate value of t from Table VIII. The $t_{0.05}$ value with $n - 1$, or $6 - 1 = 5$, degrees of freedom is 2.015. Finally we establish the confidence interval. We have

Lower Boundary	Upper Boundary
$= \bar{x} - t_{0.05} \dfrac{s}{\sqrt{n}}$	$= \bar{x} - t_{0.05} \dfrac{s}{\sqrt{n}}$
$= 10 - 2.015 \left(\dfrac{3.74}{\sqrt{6}}\right)$	$= 10 + 2.015 \left(\dfrac{3.74}{\sqrt{6}}\right)$
$= 10 - 3.08$	$= 10 + 3.08$
$= 6.92$	$= 13.08$

Thus, a 90% confidence interval for the average length of time that a student practices playing the electric guitar is 6.92 to 13.08 hours.

COMMENT We wish to emphasize a point made earlier. For small sample inferences to be valid, it is assumed that the sample is obtained from some normally distributed population. Often this may not be true. In this case we have to use other statistical procedures.

EXERCISES

1. A labor union official is analyzing the number of hours per month that union delegates devote to union activities. The following information is available.

Union Delegate	Number of Hours Devoted to Union Activities
Jack	38
Fran	61
Mary	47
April	60
Mark	39
Jennifer	55
Jessica	50

Find a 90% confidence interval for the average number of hours per month that a union delegate devotes to union activities.

2. A publisher wishes to determine the list price to charge for a statistics book. A survey of the list prices of seven competing books sold by other companies showed an average price of $33.95 with a standard deviation of $1.95. Find a 95% confidence interval for the average list price of a statistics book.

3. Eight customers paid an average of $11.50 for the same antibiotic at eight different drug stores in the city. The standard deviation was $1.12. Find a 99% confidence interval for the average cost of this antibiotic.

4. A survey of ten companies providing basic cable television service indicated that the charge for this service was (on a monthly basis):

$7.95 $9.95 $10.20 $8.45 $11.95
$7.65 $7.85 $ 9.40 $16.20 $12.20

Find a 95% confidence interval for the average monthly charge of basic cable television service.

5. Ten photocopying machines produced an average of 8,000 copies before requiring new toner. The standard deviation was 890 copies. Find a 99% confidence interval for the average number of copies that these photocopying machines will yield before requiring new toner.

6. A random sample of the telephone bills of eight companies indicates that the monthly cost for all the collect calls received was:

$426.33 $327.53 $504.88 $702.83
$891.08 $476.32 $621.76 $542.63

Find a 90% confidence interval for the average monthly cost for all collect calls received.

7. Six restaurants in Boulder collected an average of $984.38 per week in sales tax from patrons. The standard deviation was $12.57. Find a 90% confidence interval for the average amount of sales tax collected from patrons of restaurants in Boulder.

8. Ten members of a health club were randomly selected and asked to indicate the number of hours each spends per week at the club. Their answers were 5, 8, 15, 12, 3, 13, 3, 4, 9, and 11 hours. Find a 95% confidence interval for the average amount of time spent by a health club member at the health club.

9. A nutritionist is analyzing the nutritional content (on a particular nutritional scale) of seven leading cereals advertised on children's television shows. The results are as follows:

Cereal	Nutritional Content
A	81
B	57
C	62
D	78
E	31
F	65
G	72

Find a 99% confidence interval for the average nutritional content of cereals advertised on children's television shows.

10. Refer to the article below. Health officials tested the air at seven tunnels to determine the pollution content. The officials found an average of 18 ppm (parts per million) of a certain pollutant in the air at these tunnels. The standard deviation was 3.2 ppm. Find a 95% confidence interval for the average ppm content of this pollutant in the air at these tunnels.

TOLL COLLECTORS PROTEST UNHEALTHY WORKING CONDITIONS

New Dorp (May 17) – Toll collectors staged a 3-hour protest yesterday against the unhealthy working conditions occurring at different times of the day at the toll plazas of the city's seven tunnels. A spokesperson for the toll collectors indicated that the level of sulfur oxide pollutants in the air at the plazas was far in excess of the recommended safety level. Between the hours of 10 A.M. and 1 P.M., motorists were able to use the tunnel without paying any toll. In an effort to resolve the dispute, Bill Sigowsky has been appointed as mediator.

May 17, 1986

9.5 *THE ESTIMATION OF THE STANDARD DEVIATION*

Until now we have been discussing how to obtain point and interval estimates of the population mean μ. In this section we discuss point and interval estimates for the population standard deviation.

Suppose we took a sample of 100 cigarettes of a particular brand and determined that the sample mean tar content was 15.2 with a standard deviation of 1.1. If this procedure were to be repeated many times, each time we would obtain different estimates for the population mean and standard deviation. Nevertheless, the mean of these estimates will approach the true population mean as the sample size gets larger. When this happens we say that the average of the sample means is an **unbiased estimator** of the population mean.

Recall that on page 359 two formulas were given for the variance and standard deviation. Suppose we were to calculate the variance of each sample by using the formula.

$$\frac{\Sigma(x - \bar{x})^2}{n}$$

Then we cannot use the average of these values as an unbiased estimate of the population variance. Specifically, the average of such estimates would most likely always be too small, no matter how many samples are included. We can compensate for this by dividing by $n - 1$ instead of n. When this is done, it can be shown that

$$\frac{\Sigma(x - \bar{x})^2}{n - 1}$$

is an unbiased estimate of σ^2. This means that as more and more samples are included, the average of the variances of these samples will approach the true population variance. Thus a point estimate of σ^2 is

$$\frac{\Sigma(x - \bar{x})^2}{n - 1}$$

We therefore say that $\bar{x}$ is an unbiased estimator of the population mean μ, and that s^2 is an unbiased estimator of σ^2. However, this does not mean that s is an unbiased estimator of σ. Nevertheless, when n is large, this bias is small. Thus we can use s as an estimator of σ.

When dealing with large sample sizes, mathematicians have developed the following 95% confidence interval for σ.

FORMULA 9.3

Let s be the sample standard deviation and let n be the number in the sample. We have the following 95% confidence interval for σ:

$$\textbf{\textit{Lower boundary}} = \frac{s}{1 + \dfrac{1.96}{\sqrt{2n}}}$$

$$\textbf{\textit{Upper boundary}} = \frac{s}{1 - \dfrac{1.96}{\sqrt{2n}}}$$

● **Example 1**
A sample of 50 cigarettes was taken. The sample mean tar content was 15.2 with a standard deviation of 1.1. Find a 95% confidence interval for the population standard deviation.

Solution
We use Formula 9.3. Here $n = 50$ and $s = 1.1$. The lower boundary of the confidence interval is

$$\frac{s}{1 + \dfrac{1.96}{\sqrt{2n}}} = \frac{1.1}{1 + \dfrac{1.96}{\sqrt{2(50)}}}$$

$$= \frac{1.1}{1 + 0.196}$$

$$= 0.92$$

The upper boundary of the confidence interval is

$$\frac{s}{1 - \dfrac{1.96}{\sqrt{2n}}} = \frac{1.1}{1 - \dfrac{1.96}{\sqrt{2(50)}}}$$

$$= \frac{1.1}{1 - 0.196}$$

$$= 1.37$$

Thus, a 95% confidence interval for σ is 0.92 to 1.37.

9.6 DETERMINING THE SAMPLE SIZE

Until now we have been discussing how sample data can be used to estimate various parameters of a population. Selecting a sample usually involves an expenditure of money. The larger the sample size, the greater the cost. Therefore, before selecting a sample, we must determine how large the sample should be.

Generally speaking the size of the sample is determined by the desired degree of accuracy. Most problems specify the maximum allowable error. Yet we must realize that no matter what the size of a sample, any estimate may exceed the maximum allowable error. To be more specific, suppose we are interested in estimating the mean life of a calculator on the basis of a sample. Suppose also that we want our estimate to be within 0.75 years of the true value of the mean. In this case the **maximum allowable error,** denoted as e, is 0.75 years. How large a sample must be taken? Of course, the larger the sample size, the smaller the chance that our estimate will not be within 0.75 years of the value. If we want to be 95% confident that our estimate will be within 0.75 years of the true value, then the sample size must satisfy

$$1.96\sigma_{\bar{x}} = 0.75$$

More generally, if the maximum allowable error is e, then we must have

$$e = 1.96\sigma_{\bar{x}}$$

Since

$$\sigma_{\bar{x}} = \frac{\sigma}{\sqrt{n}}$$

we get

$$e = 1.96\, \frac{\sigma}{\sqrt{n}}$$

Solving this equation for n gives Formula 9.4:

FORMULA 9.4

Let σ be the population standard deviation, e the maximum allowable error, and n the size of the sample that is to be taken from a large population. Then a sample of size

$$n = \left(\frac{1.96\sigma}{e}\right)^2$$

will result in an estimate of μ, which is less than the maximum allowable error 95% of the time.

● **Example 1**
Suppose we wish to estimate the average life of a calculator to within 0.75 years of the true value. Past experience indicates that the standard deviation is 2.6 years. How large a sample must be selected if we want our answer to be within 0.75 years 95% of the time?

Solution
We use Formula 9.4. Here $\sigma = 2.6$ and $e = 0.75$, so that

$$n = \left(\frac{1.96\sigma}{e}\right)^2$$

$$= \left(\frac{1.96(2.6)}{0.75}\right)^2 = (6.79)^2 = 46.10$$

Thus, the sample should consist of 47 calculators. (*Note:* In the determination of sample size, any decimal is always rounded to the next highest number.)

● **Example 2**
The management of a large company in California desires to estimate the average working experience, measured in years, of its 2000 hourly workers. How large a sample should be taken in order to be 95% confident that the

sample mean does not differ from the population mean by more than $\frac{1}{2}$ year? (Past experience indicates that the standard deviation is 2.6 years.)

Solution

We use Formula 9.4. Here $\sigma = 2.6$ and $e = 0.50$. Then

$$n = \left(\frac{1.96\sigma}{e}\right)^2$$

$$= \left(\frac{1.96(2.6)}{0.50}\right)^2$$

$$= (10.19)^2$$

$$= 103.84$$

Thus, the company officials should select 104 hourly workers to determine the average working experience of an hourly worker.

EXERCISES

1. Fifty dentists were asked to indicate their charge for a particular dental procedure. The standard deviation of the charge was $11.78. Find a 95% confidence interval for the standard deviation of the charge by all dentists.

2. A personnel director is analyzing data for the average number of absences of an employee per year from the company. A sample of the data for 75 employees discloses a standard deviation of 3.8 days. Find a 95% confidence interval for the standard deviation of the number of absences of an employee.

3. Thirty-six photography dealers were asked to indicate their price for a wide angle lens for a particular 35-mm camera. The standard deviation of these prices was computed and found to be $2.76. Find a 99% confidence interval for the population standard deviation.

4. A random sample of 60 patients given a new pain-relieving drug indicated that the average time elapsed before relief occurred was 15.2 minutes with a standard deviation of 1.9 minutes. Find a 90% confidence interval for the population standard deviation.

For each of Exercises 5–10, assume that it is required that the maximum allowable error not be exceeded with a 95% degree of confidence.

5. A state income tax bureau wishes to estimate the average cost that a taxpayer pays for income tax preparation. How large a sample should be selected if the bureau wishes that its estimate be within $10 of the true value? (Assume that $\sigma = \$7.75$.)

6. A large auto manufacturer wishes to estimate the average time needed by a driver to stop, now that a new braking light has been mounted on the back of the car above the trunk. How large of a sample should be selected if it is desired that the estimate be within 0.01 seconds of the true value? (Assume that $\sigma = 0.5$.)

7. An economist wishes to estimate the average amount of money spent by a family on entertainment per year. How large a sample should be selected if it is desired that the estimate be within $15 of the true value? (Assume that $\sigma = \$19.75$.)

8. A cosmetics manufacturer wishes to determine how many minutes a woman spends applying facial cosmetics. How large a sample should the manufacturer take so that the estimate be within 4.16 minutes of the true value? (Assume that $\sigma = 5.97$ minutes.)

9. A painter is interested in determining the average drying time of a particular paint. How large a sample of painting jobs must be selected if it is desired that the estimate be within 3 minutes of the true value? (Assume that $\sigma = 8.15$ minutes.)

10. An investor is interested in determining the average cost for an investment advisory service that forecasts the movement of the stock market. The investor wishes that the estimate be within $5 of the true value. How large a sample of investment advisory services should be taken? (Assume that $\sigma = \$7.65$.)

9.7 THE ESTIMATION OF PROPORTIONS

So far, we have discussed the estimation of the population mean and the population standard deviation. Very often statistical problems arise for which the data are available in **proportion** or **count form** rather than in measurement form. For example, suppose a doctor has developed a new technique for predicting the sex of an unborn child. He tests his new method on 1000 pregnant women and correctly predicts the sex of 900 of the children. Is his new technique reliable? The doctor has correctly predicted the sex of 900 unborn children. Thus the **sample proportion** is $\dfrac{900}{1000}$, or 0.90. What is the **true proportion** of unborn children whose sex he can correctly predict?

Since we will be discussing both sample proportions and true population proportions, we will use the following notation:

π true population proportion
p sample proportion

COMMENT   As used when working with population proportions, π represents the true population proportion and is not equal to 3.141592654. . . .

In our case the doctor estimates the true population proportion, π, to be 0.90. This estimate is based upon the sample proportion p. How reliable is the estimate of the true population proportion? Suppose the doctor now tests his technique on 1000 different pregnant women. For what proportion of these 1000 unborn children will he be able to correctly predict the sex?

If the true population proportion is π, repeated sample proportions will be normally distributed. The mean of the sample distribution of proportions will be π. The standard deviation of these sample proportions will equal

$$\sqrt{\frac{\pi(1 - \pi)}{n}}$$

Thus, if the doctor's technique is reliable, he should find that the sample proportion will be normally distributed with a mean of 0.9. The standard deviation will be

$$\sqrt{\frac{(0.9)(1 - 0.9)}{1000}} = \sqrt{0.00009}, \text{ or } 0.0095$$

We can summarize these results as follows:

FORMULA 9.5

Suppose we have a large population, a proportion of which has some particular characteristic. We select random samples of size n and determine the proportion in each sample with this characteristic. Then the sample proportions will be approximately normally distributed with mean π and standard deviation

$$\sigma_p = \sqrt{\frac{\pi(1 - \pi)}{n}}$$

Since the sample proportions are approximately normally distributed, we can apply the standardized normal charts as we did for the sample means. The following examples will illustrate how this is done.

● **Example 1**

From past experience it is known that 70% of all airplane tickets sold by Global Airways are round-trip tickets. A random sample of 100 passengers is taken. What is the probability that at least 75% of these passengers have round-trip tickets?

Solution
We use Formula 9.5. Here $n = 100$, $\pi = 0.70$, and the sample proportion p is 0.75. Since the sample proportions are normally distributed, we are interested in the area to the right of $x = 0.75$ in a normal distribution whose mean is 0.70. See Figure 9.5. We first calculate the standard deviation of the sample proportions, denoted as σ_p.

$$\sigma_p = \sqrt{\frac{\pi(1 - \pi)}{n}}$$

$$= \sqrt{\frac{0.70(1 - 0.70)}{100}}$$

$$= \sqrt{0.0021}$$

$$= 0.046$$

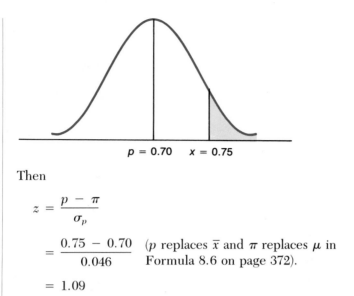

$p = 0.70 \quad x = 0.75$

FIGURE 9.5

Then

$$z = \frac{p - \pi}{\sigma_p}$$

$$= \frac{0.75 - 0.70}{0.046} \quad \begin{array}{l} (p \text{ replaces } \bar{x} \text{ and } \pi \text{ replaces } \mu \text{ in} \\ \text{Formula 8.6 on page 372).} \end{array}$$

$$= 1.09$$

From Table V the area between $z = 0$ and $z = 1.09$ is 0.3621. Therefore the area to the right of $z = 1.09$ is $0.5000 - 0.3621$, or 0.1379. Thus, the probability that at least 75% of the passengers have round trip tickets is 0.1379.

● **Example 2**
Fifty-four percent of all nurses in the day shift at Downtown Hospital have type O blood. Thirty-six nurses from the day shift are selected at random. What is the probability that between 51% and 58% of these members have type O blood?

Solution
We will use Formula 9.5. Here $\pi = 0.54$ and $n = 36$. We first calculate σ_p:

$$\sigma_p = \sqrt{\frac{\pi(1 - \pi)}{n}}$$

$$= \sqrt{\frac{0.54(1 - 0.54)}{36}}$$

$$= \sqrt{0.0069}$$

$$= 0.083$$

Now we can use Formula 9.5 with $\pi = 0.54$, $n = 36$, and $\sigma_p = 0.083$:

$$p = 0.51 \text{ corresponds to} \quad z = \frac{0.51 - 0.54}{0.083} = \frac{-0.03}{0.083} = -0.36$$

$$p = 0.58 \text{ corresponds to} \quad z = \frac{0.58 - 0.54}{0.083} = \frac{0.04}{0.083} = 0.48$$

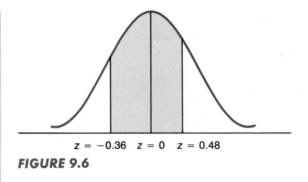

$$z = -0.36 \quad z = 0 \quad z = 0.48$$

FIGURE 9.6

Thus, we are interested in the area between $z = -0.36$ and $z = 0.48$. See Figure 9.6.

From Table V we find that the area between $z = 0$ and $z = -0.36$ is 0.1406, and the area between $z = 0$ and $z = 0.48$ is 0.1844. Adding, we get

$$0.1406 + 0.1844 = 0.3250$$

Thus, the probability that between 51% and 58% of these nurses have type O blood is 0.3250.

In the same way that the sample mean is used to estimate the population mean, we can use the sample proportions, p, as a point estimate of the population proportion, π. However, this estimate does not indicate the probability of its accuracy. Thus we set up interval estimates for the true population proportion, π. We have Formula 9.6.

FORMULA 9.6

Let p be a sample proportion and let π be the true population proportion. Then we have the following 95% confidence interval for π (assuming we have a large sample):

$$\textbf{Lower boundary} = p - 1.96 \sqrt{\frac{p(1-p)}{n}}$$

$$\textbf{Upper boundary} = p + 1.96 \sqrt{\frac{p(1-p)}{n}}$$

● **Example 3**
A union member reported that 80 out of 120 workers interviewed supported some form of work stoppage to further its demands for a shorter work week.

Find a 95% confidence estimate of the true proportion of workers supporting the union's stand on a work stoppage.

Solution

We use Formula 9.6. Here $p = \dfrac{80}{120}$, or 0.67, and $n = 120$. To construct a 95% confidence interval for π, we have

Lower Boundary	Upper Boundary
$= p - 1.96\sqrt{\dfrac{p(1-p)}{n}}$	$= p + 1.96\sqrt{\dfrac{p(1-p)}{n}}$
$= 0.67 - 1.96\sqrt{\dfrac{0.67(1-0.67)}{120}}$	$= 0.67 + 1.96\sqrt{\dfrac{0.67(1-0.67)}{120}}$
$= 0.67 - 1.96\sqrt{0.0018}$	$= 0.67 + 1.96\sqrt{0.0018}$
$= 0.67 - 1.96(0.043)$	$= 0.67 + 1.96(0.043)$
$= 0.67 - 0.084$	$= 0.67 + 0.084$
$= 0.586$	$= 0.754$

Thus, the union official concluded with 95% confidence that the true proportion of workers supporting the union claim is between 0.586 and 0.754. This is the 95% confidence interval.

• Example 4

A new public telephone has been installed in an airport baggage-claim area. A quarter was lost 45 of the first 300 times that it was used. Construct a 95% confidence interval for the true proportion of times that a user will lose a quarter.

Solution

We use Formula 9.6. Here $p = \dfrac{45}{300}$, or 0.15, and $n = 300$. To construct a 95% confidence interval for π, we have

Lower Boundary	Upper Boundary
$= p - 1.96\sqrt{\dfrac{p(1-p)}{n}}$	$= p + 1.96\sqrt{\dfrac{p(1-p)}{n}}$
$= 0.15 - 1.96\sqrt{\dfrac{0.15(1-0.15)}{300}}$	$= 0.15 + 1.96\sqrt{\dfrac{0.15(1-0.15)}{300}}$
$= 0.15 - 1.96\sqrt{0.0004}$	$= 0.15 + 1.96\sqrt{0.0004}$
$= 0.15 - 1.96(0.02)$	$= 0.15 + 1.96(0.02)$
$= 0.15 - 0.039$	$= 0.15 + 0.039$
$= 0.111$	$= 0.189$

Thus, a 95% confidence interval for the true proportion of times that a user will lose a quarter is 0.111 to 0.189.

EXERCISES

1. A senator claims that 40% of the voters are in favor of lowering the gasoline tax. A random survey of 800 voters from his community is taken. What is the probability that the proportion of people favoring this proposal is greater than 45 percent?

2. A cardiologist claims that about 20% of all people who suffer a heart attack often suffer a second heart attack within a year because of damage to heart muscle during the first attack. To check on this claim, a random survey is taken of 275 people who suffered a heart attack. What is the probability that the proportion of people who will suffer a second attack within a year after the first attack is less than 17%?

3. A large power company on the East Coast, which interviewed more than 50,000 of its customers, claims that only 15% of these people are opposed to building the newly proposed nuclear generating station. A local consumer's group is challenging this claim. A random sample of 300 of these customers is taken. What is the probability that the proportion of customers opposed to building the newly proposed nuclear generating station is greater than 18%?

4. A random sample of 500 car owners in Brownsville indicated that 375 of them had full insurance coverage on glass breakage for their car with no deductible amount. Find a 95% confidence interval for the true proportion of car owners in Brownsville who have full insurance coverage on glass breakage for their car.

5. A random sample of 250 voters in Watertown was taken; 195 of the voters polled indicated that they were in favor of reinstating the death penalty for certain crimes. Find a 95% confidence interval for the true proportion of all voters in Watertown who are in favor of reinstating the death penalty for certain crimes.

6. Government officials claim that 45% of the student loans guaranteed by the United States government at a particular college are in default. A random sample of 475 student loans guaranteed by the U.S. government at this college is taken. What is the probability that less than 208 of these loans will be in default?

7. Officials of an insurance company that pays drug reimbursement bills claim that 40% of all druggists in a city will often fill a doctor's prescription with a brand-name drug rather than with a less expensive generic drug even when the doctor does not specify the brand-name drug. A random sample of 425 prescriptions is taken. What is the probability that more than 180 of them will be found to be filled with the more expensive drug rather than with the less expensive generic drug?

8. Many Americans believe that gun control legislation on the federal government level is very difficult because of the power and influence of the gun lobby in Washington. A survey of 450 Americans found that 288 expressed such a view. What is the probability that the proportion of Americans who have such a view is less than 60 percent?

9. A recent Associated Press poll found that 1284 of the 1700 American adults interviewed had no faith in the ability of the Social Security system to be solvent and have funds available to pay for their retirement. Find a 90% confidence interval for the true proportion of American adults who have no faith in the solvency of the Social Security system.

10. Consider the newspaper article below. If a random survey of 425 families across the state is taken, what is the probability that more than 102 of them will be found to be one-person households?

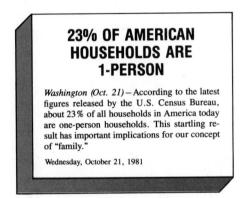

23% OF AMERICAN HOUSEHOLDS ARE 1-PERSON

Washington (Oct. 21) — According to the latest figures released by the U.S. Census Bureau, about 23% of all households in America today are one-person households. This startling result has important implications for our concept of "family."

Wednesday, October 21, 1981

9.8 USING COMPUTER PACKAGES

We can use a MINITAB program to determine confidence intervals quite easily. To illustrate, let us use a MINITAB program to determine a 90% confidence interval for the data on the average length of time that a student practices playing the electric guitar given in Example 3 of Section 9.4 (page 397). We have the following:

```
MTB  > SET THE FOLLOWING DATA INTO C1
DATA > 10   12     8     9     16     5
```

After the data are entered, we instruct the computer to determine a 90% confidence interval. We have the following desired results.

```
MTB > TINTERVAL WITH 90 PERCENT CONFIDENCE FOR DATA IN C1
      N    MEAN    STDEV    SEMEAN          90.0 PERCENT C.I.
C1    6    10      3.74     1.53            (6.92, 13.08)
MTB > STOP
```

COMMENT In a similar manner, we can obtain 95% confidence intervals for μ. Minor changes in the code of the program allows us to obtain any desired confidence intervals for μ.

9.9 SUMMARY

In this chapter we indicated how sample data can be used to estimate the population mean and the population standard deviation. Both point and interval estimates were discussed. In most cases sample data are used to construct confidence intervals within which a given parameter with a specified probability is likely to lie. Sample data can also be used to make estimates about the true population proportion and to construct confidence intervals for the population proportion.

All estimates considered in this chapter were unbiased estimates. This means that the average of these estimates will approach the true population parameter that they are trying to estimate as more and more samples are included. For this reason we divide $\Sigma(x - \bar{x})^2$ by $n - 1$, not by n, in determining the sample variance.

We also indicated how to determine the size of a sample to be used in gathering data. Depending upon the maximum allowable error, Formula 9.4 determines the sample size with a 95% degree of confidence.

STUDY GUIDE

You should now be able to demonstrate your knowledge of the following ideas presented in this chapter by giving definitions, descriptions, or specific examples. Page references are given for each term so that you can check your answer.

Statistical estimation (page 386)
Hypothesis testing (page 386)
Population parameter (page 387)
Point estimate (page 387)
Interval estimate (page 387)
Confidence interval (page 387)
Confidence limits (page 387)
Degree of confidence (page 387)
Student's t-distribution (page 394)
Number of degrees of freedom (page 395)
Unbiased estimator (page 400)
Maximum allowable error (page 402)
Sample proportion (page 405)
True population proportion (page 405)

FORMULAS TO REMEMBER

You should be able to identify each symbol in the following formulas, understand the relationships among the symbols expressed in each formula, understand the significance of each formula, and use the formulas in solving problems.

1.

Size of Sample	Parameter	Degree of Confidence	Lower Boundary	Upper Boundary
Large	Mean	90%	$\bar{x} - 1.64 \dfrac{s}{\sqrt{n}}$	$\bar{x} + 1.64 \dfrac{s}{\sqrt{n}}$
Large	Mean	95%	$\bar{x} - 1.96 \dfrac{s}{\sqrt{n}}$	$\bar{x} + 1.96 \dfrac{s}{\sqrt{n}}$
Large	Mean	99%	$\bar{x} - 2.58 \dfrac{s}{\sqrt{n}}$	$\bar{x} + 2.58 \dfrac{s}{\sqrt{n}}$
Small	Mean	95%	$\bar{x} - t_{0.025} \dfrac{s}{\sqrt{n}}$	$\bar{x} + t_{0.025} \dfrac{s}{\sqrt{n}}$
Large	Standard deviation	95%	$\dfrac{s}{1 + \dfrac{1.96}{\sqrt{2n}}}$	$\dfrac{s}{1 - \dfrac{1.96}{\sqrt{2n}}}$
Large	Proportion	95%	$p - 1.96 \sqrt{\dfrac{p(1-p)}{n}}$	$p + 1.96 \sqrt{\dfrac{p(1-p)}{n}}$

2. Size of sample where maximum allowable error may not be exceeded with a 95% degree of confidence: $\quad n = \left(\dfrac{1.96\sigma}{e}\right)^2$.

3. Mean of sampling proportions: π

4. Standard deviation of sampling proportion: $\quad \sqrt{\dfrac{\pi(1-\pi)}{n}}$

The tests of the following section will be more useful if you take them after you have studied the examples and solved the exercises in this chapter.

MASTERY TESTS

Form A

 1. A manufacturer claims that 7% of the clothing produced by her company is labelled incorrectly with the wrong size. To test this claim, a random sample of 480 coats produced by this company is taken. What is the probability that the proportion of incorrectly labelled coats is between 4% and 8%?

2. A department store clerk randomly selects 180 credit card purchase slips and finds that 54 of the purchases were made with a Visa credit card. Find a 95% confidence interval for the true proportion of credit card purchases made with a Visa credit card.

3. A consumer advocate claims that 30% of all reflex 35-mm automatic cameras sold at discount prices in New York City are gray-market items, that is, they are not guaranteed by an American company. A random sample of 325 cameras sold is taken. What is the probability that the number of such cameras that are gray-market items is less than 85?

4. A sample of 65 college students indicated that a college student spends an average of 9.8 hours per week watching television. The standard deviation was 1.15 hours. Find a 95% confidence interval for the population standard deviation.

5. A computer programming teacher wishes to estimate the average amount of time spent by a computer science student to get a program to run. How large a sample (with a 95% degree of confidence) should the teacher select if she wishes the estimate to be within 0.85 hours of the true value? (Assume that $\sigma = 1.96$ hours.)

6. Fifty lawyers were asked to indicate their charge for an uncontested divorce. The standard deviation of their charges was computed and found to be $88. Find a 99% confidence interval for the population standard deviation.

7. In measuring the reaction time to a sudden skid on wet pavement, a state Motor Vehicle Bureau found that the standard deviation is 0.03 seconds. How large a sample (with 95% degree of confidence) should the bureau take if it wishes to estimate the average reaction time of a driver with a maximum allowable error of 0.005 seconds?

8. A market researcher claims that the percentage of consumers who comparison shop at large supermarkets is 28%. A random sample of 325 shoppers in a large supermarket is taken. What is the probability that more than 95 of these consumers will comparison shop?

9. A random sample of size 200 is selected from a population with unknown mean, μ, and unknown standard deviation, σ. The following values are known: $\Sigma x = 3500$ and $\Sigma x^2 = 120,000$. Find a 95% confidence interval for μ.

10. Refer back to Exercise 9. Find a 99% confidence interval for μ.

1. Consider the newspaper article below. Find the probability that less than 40% of all claims to the 911 number are for nonemergency help.

POLICE APPEAL: USE 911 ONLY IN AN EMERGENCY

New York (Feb. 3) – The Police Department issued an urgent appeal to the citizens of New York to use the 911 emergency number only when there is an actual emergency. For any other police assistance, the local police precinct should be called.

This appeal followed an analysis by the police department which showed that 40% of all calls to the 911 number were for non-emergency help. A survey of 150 randomly selected incoming calls disclosed that only 50 of them were for emergency assistance. A spokesperson for the police department further commented that the 911 emergency number is often busy because of those unnecessary calls.

2. Six airline passengers returning to the United States were randomly selected and asked to indicate how many pictures or slides each had snapped while vacationing in Europe. Their answers were 63, 82, 96, 103, 58, and 72. Find a 95% confidence interval for the average number of pictures taken by a traveller while vacationing in Europe.

3. The average amount of time needed by 100 people to assemble an exercise bicycle (which comes with directions for assembly) is 220 minutes with a standard deviation of 25 minutes. Find a 99% confidence interval for the average length of time needed by a person to assemble an exercise bicycle.

4. A statistician is interested in estimating the population mean weight of a newborn baby. How large a sample must be taken in order to be 95% confident that μ be within 0.3 lb of the true value? (Assume that $\sigma = 1.8$ lb.)

5. Assume that it is known that 55% of all workers at a certain company own their own home. A sample of 375 workers is randomly selected. What is the probability that fewer than 200 of them own their own home?

6. From past experience, it is known that 78% of all chemical waste containers buried in the ground will eventually leak. If a sample of 55 chemical waste containers is randomly selected and the containers are subjected to extensive pressure, what is the probability that between 62% and 68% of these will eventually leak?

7. To determine the feasibility of opening a company cafeteria, the Finex Corporation is conducting a survey of the average amount of money spent by its employees (on a weekly basis) at the local fast-food store. It has received the following results from seven of its employees: $14, $8, $3, $9, $17, $5, and $7. Find a 95% confidence interval for the average amount of money spent by an employee (on a weekly basis) at the local fast-food store.

8. A leading child psychologist claims that fathers are not spending enough time with their children. To support this claim, she selected 14 families in Buffalo and found that the father spent an average of 7.8 hours per week with his children. The standard deviation was 2.13 hours. Find a 90% confidence interval for the average time spent by a Buffalo father with his children.

9. A sample of 36 two-family homes in Scrantoon found that the average heating bill for the month of January was $585 with a standard deviation of $62. Find a 99% confidence interval for the average heating bill for the month of January in Scrantoon.

10. *Do Consumers Use "Cents-Off" Coupons?* A recent Nielsen Poll found that about 75% of those people surveyed used "cents-off" coupons. A random survey of 225 shoppers is taken. What is the probability that the proportion of shoppers who use the cents-off coupons is less than 72%?

SUGGESTED FURTHER READING

1. Adler, H. and E. B. Roessler. *Introduction to Probability and Statistics*, 5th ed. San Francisco, CA: Freeman, 1972.
2. Freund, John. *Statistics, A First Course*, 2nd ed. Englewood Cliffs, NJ: Prentice-Hall, 1976.
3. *Gallup Opinion Index*. Princeton, NJ: American Institute of Public Opinion, January 1971.
4. Hogg, Robert and Allen Craig. *Introduction to Mathematical Statistics*, 3rd ed. New York: Macmillan, 1970.
5. Mood, A. M. *Introduction to the Theory of Statistics*. New York: McGraw-Hill, 1967.
6. Neter, J., W. Wasserman, and G. A. Whitmore. *Applied Statistics*. Boston: Allyn and Bacon, 1978.

CHAPTER 10

Hypothesis Testing

OBJECTIVES

- *To analyze* how sample data can be used to reject or accept a claim about some aspect of a probability distribution. The claim to be tested is called the null hypothesis.

- *To see* that a test statistic is a number that we compute to determine when to reject the null hypothesis.

- *To learn* when critical rejection regions tell us to reject a null hypothesis. This occurs when the test statistic value falls within this region. How these regions are set up depends upon the specifications given within the problem.

- *To discuss* the two errors that can be made when we use sample data to accept or reject a null hypothesis. We may incorrectly reject a true hypothesis or we may incorrectly accept a false hypothesis. In both cases, an error is made.

- *To indicate* what a level of significance is.

- *To distinguish* between tests concerning means, differences between means, and proportions. Thus we discuss the test statistics used when we wish to use sample data to determine whether observed differences between means and proportions are significant.

- *To apply* the hypothesis-testing procedures to wide-ranging problems.

MANY AMERICANS SUPPORT HIGHER TAXES

Washington: The results of a recent nationwide survey by Administration officials indicate that many Americans believe that our nation's public high schools are basically unsound and in drastic need of improvements (such as compulsory basic courses). About 75% of those surveyed said that they would be willing to have their sales tax increased to fund any improvements in education. The President, as well as other congressional members, strongly support the idea of merit pay as an incentive for teachers.

Monday April 6, 1987

NEW PROCESSING MACHINE DEVELOPED

Milford: Researchers claim to have developed a new sorting machine for packing and processing fruit. The new machine can handle an average of 756 pounds of fruit per run as opposed to the old machine, which could process only an average of 740 pounds of fruit per run.

Tuesday Sept. 10, 1985

Often we read about various claims being made. How do we verify the accuracy of such claims? Consider the first article. If we randomly sample 100 Americans, would we expect 75 of them to support this claim? If only 70 of them said that they would be willing to have their sales tax increased to fund any improvements in education, would we say that this is significantly less than the claimed 75 percent?

Now consider the second article. How do we determine whether the observed difference between the average amount of fruit processed by the old machine and the new machine is significant?

10.1 INTRODUCTION

Many television commercials contain unusual performance claims. For example, consider the following TV commercials:

1. Four out of five dentists recommend Brand X sugarless gum for their patients who chew gum.
2. A particular brand of tire will last an average of 40,000 miles before replacement is needed.
3. A certain detergent produces the cleanest wash.
4. Brand X paper towels are stronger and are more absorbent.

How much confidence can one have in such claims? Can they be verified statistically? Fortunately, in many cases the answer is yes. Samples are taken and claims are tested. We can then make a decision on whether to accept or to reject a claim on the basis of sample information. This process is called **hypothesis testing.** Perhaps this is one of the most important uses of samples.

As we indicated in Chapter 9, hypothesis testing is an important branch of statistical inference. Sample data provide us with estimates of population parameters. These estimates are in turn used in arriving at a decision to either accept or reject an hypothesis. By an **hypothesis** we shall mean an assumption about one or more of the population parameters that will either be accepted or rejected on the basis of the information obtained from a sample. In this chapter we will discuss methods for determining whether to accept or reject any hypothesis on the basis of sample data.

10.2 TESTING AGAINST AN ALTERNATIVE HYPOTHESIS

Suppose several players are in a gambling casino rolling a die. A bystander notices that in the first 120 rolls of the die, a 6 showed only 8 times. Is this reasonable or is the die loaded? The management claims that this unusual occurrence is to be attributed purely to chance and that the die is an honest die. The bystander claims otherwise.

If the die is an honest die, then Formula 6.5 (see page 295) for a binomial distribution tells us that the average number of 6s occurring in 120 rolls of the die is 20, as

$$\mu = np$$

$$= 120 \left(\frac{1}{6}\right)$$

$$= 20$$

If the die is loaded, then $\mu \neq 20$. (The symbol $\neq$ means "is not equal to.") Since the die is either an honest die or a loaded die, we must choose between the hypothesis $\mu = 20$ and the hypothesis $\mu \neq 20$. Thus, sample data will be used to either accept or reject the hypothesis $\mu = 20$. Such an hypothesis is called a **null hypothesis** and is denoted by H_0. Any hypothesis that differs from the null hypothesis is called an **alternative hypothesis** and is denoted as H_1, $H_2, \ldots$, and so on. In our example

Null hypothesis, H_0: $\mu = 20$

Alternative hypothesis, H_1: $\mu \neq 20$

Notice that by formulating the alternative hypothesis as $\mu \neq 20$, we are indicating that we wish to perform a **two-sided**, or **two-tailed test**. This means that if the die is not honest, then it may be loaded in favor of obtaining 6s more often than is expected or less often than is expected.

In our example the bystander strongly suspects that the die is loaded against obtaining 6s. Thus, his alternative hypothesis would be $\mu < 20$. (The symbol $<$ stands for "is less than.") Similarly if the bystander suspected that the die was loaded in favor of obtaining 6s more often than is expected, the alternative hypothesis would be $\mu > 20$. (The symbol $>$ stands for "is greater than.") In each of these cases the null hypothesis remains the same, H_0: $\mu = 20$. Such alternative hypotheses indicate that we wish to perform a **one-sided**, or **one-tailed test**.

COMMENT It should be noted that the decision on an alternative hypothesis should be made *before* the results of the sample are known.

A decision as to whether to accept or reject the null hypothesis will be made on the basis of sample data. How is such a decision to be made? We must realize that even if we know for sure that the die is honest, it is very unlikely that we would get exactly twenty 6s in the 120 rolls of the die. Moreover, if we were to roll the die 120 times on many different occasions, we would find that the number of 6s appearing is around 20, sometimes more and sometimes less. It is therefore obvious that we must set up some interval that we will call the **acceptance region.** If the number of 6s appearing in 120 rolls of the die is within this acceptance region, then we will accept the null hypothesis. If the number of 6s obtained is outside this region, then we will reject the null

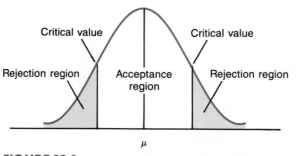

FIGURE 10.1
A two-tailed rejection region.

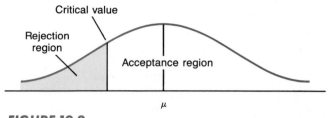

FIGURE 10.2
A one-tailed rejection region.

hypothesis that the die is an honest die. These possibilities are shown in Figures 10.1, 10.2, and 10.3. The value that separates the rejection region from the acceptance region is called the **critical value.**

The type of symbol in the alternative hypothesis tells us what type of rejection region to use as shown in the following table:

If the Symbol in the Alternative Hypothesis Is	<	≠	>
Then the Rejection Region Consists of	one region on the left side.	two regions, one on each side.	one region on the right side.

Suppose we decide to accept the null hypothesis if the number of 6s obtained is between 15 and 25. Since in our case only eight 6s were obtained, we would reject the null hypothesis. In this case the acceptance region is 15 to 25. The two-tailed rejection region corresponds to the two tails, less than 15 and more than 25, as shown in Figure 10.4. When we reject a null hypothesis, we are claiming that the value of the population parameter, that is, the average number of 6s, is some value other than the one specified in the null hypothesis. Also, when the sample data indicate that we should reject a null hypothesis, we say that the observed difference is **significant.**

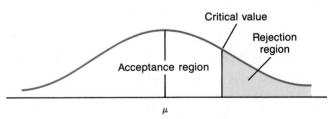

FIGURE 10.3
A one-tailed rejection region.

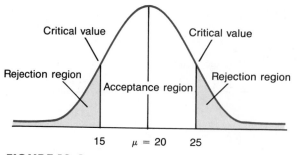

FIGURE 10.4

A two-tailed rejection region.

Our discussions in the last few paragraphs lead us to the following definitions:

DEFINITION 10.1

*The **null hypothesis,** denoted by H_0, is the statistical hypothesis being tested.*

DEFINITION 10.2

*The **alternative hypothesis** denoted by H_1, H_2, . . . , is the hypothesis that will be accepted when the null hypothesis is rejected.*

DEFINITION 10.3

*A **one-sided,** or **one-tailed test** is a statistical test which has the rejection region located in the left tail or the right tail of the distribution.*

DEFINITION 10.4

*A **two-sided,** or **two-tailed test** is a statistical test which has the rejection region located in both tails of the distribution.*

Let us summarize the steps that need to be followed in hypothesis testing:

1. State the null hypothesis, which indicates the value of the population parameter to be tested.
2. State the alternative hypothesis, which indicates the belief that the population parameter has a value other than the one specified in the null hypothesis.

3. Set up rejection and acceptance regions for the null hypothesis. The **region of rejection** is called the **critical region.** The remaining region is called the **acceptance region.**
4. Compute the value of the test statistic. A **test statistic** is a calculated number that is used to decide whether to reject or accept the null hypothesis. The formula for computing the value of the test statistic depends upon the parameter we are testing.
5. Reject the null hypothesis if the test statistic value falls within the rejection region, that is, the critical region. Otherwise, accept the null hypothesis.
6. State the conclusion for the particular problem.

In this chapter we will be discussing various tests that enable us to decide whether to reject to accept a null hypothesis. Such tests are called **statistical tests of hypotheses** or **statistical tests of significance.**

10.3 TWO TYPES OF ERRORS

Since any decision to either accept or reject a null hypothesis is to be made on the basis of information obtained from sample data, there is a chance that we will be making an error. There are two possible errors that we can make. We may reject a null hypothesis when we really should accept it. Thus, returning to the die problem of Section 10.2, we may reject the claim that the die is honest even though it actually is honest. Alternately, we may accept a null hypothesis when we should reject it. Thus, we may say that the die is honest when it really is a loaded die.

These two errors are referred to as a **Type-I** and a **Type-II error,** respectively. In either case we have made a wrong decision. We define these formally as follows:

DEFINITION 10.5

*A **Type-I error** is made when a true null hypothesis is rejected, that is, we reject a null hypothesis when we should accept it.*

DEFINITION 10.6

*A **Type-II error** is made when a false null hypothesis is accepted, that is, we accept a null hypothesis when we should reject it.*

In the following box we indicate how these two errors are made:

	And We Claim That	
	H_0 is True	H_0 is False
If H_0 is True	Correct decision (no error)	Type-I error
H_0 is False	Type-II error	Correct decision (no error)

When deciding whether to accept or reject a null hypothesis, we always wish to minimize the probability of making a Type-I error or a Type-II error. Unfortunately, the relationship between the probabilities of the two types of errors is of such a nature that if we reduce the probability of making one type of error, we usually increase the probability of making the other type. In most applied problems one type of error is more serious than the other. In such situations careful attention is given to the more serious error.

How much of a risk should a statistician take in rejecting a true hypothesis, that is, in making a Type-I error? Generally speaking, statisticians use the limits of 0.05 and 0.01. Each of these limits is called a **level of significance** or **significance level.** We have the following definition:

DEFINITION 10.7

*The **significance level** of a test is the probability that the test statistic falls within the rejection region when the null hypothesis is true.*

The **0.05 level of significance** is used when the statistician wishes that the risk of rejecting a true null hypothesis not exceed 0.05. The **0.01 level of significance** is used when the statistician wishes that the risk of rejecting a true null hypothesis not exceed 0.01.

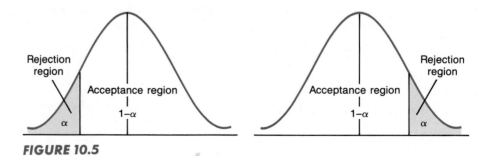

FIGURE 10.5

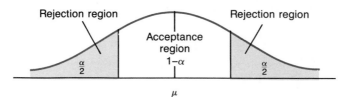

FIGURE 10.6

In this book we will usually assume that we wish to correctly accept the null hypothesis 95% of the time and to incorrectly reject it only 5% of the time. Thus, the maximum probability of a Type-I error that we are willing to accept, that is, the significance level, will be 0.05. *The probability of making a Type-I error when H_0 is true is denoted by the Greek letter α (pronounced alpha).* Therefore, the probability of making a correct decision is $1 - \alpha$.

As we indicated on page 421, when dealing with one-tailed tests, the critical region lies to the left or to the right of the mean. This is shown in Figure 10.5. When dealing with a two-tailed test, one half of the critical region is to the left of the mean and one half is to the right. The probability of making a Type-I error is evenly divided between these two tails, as shown in Figure 10.6.

COMMENT   If the test statistic falls within the acceptance region, we do not reject the null hypothesis. When a null hypothesis is not rejected, this does not mean that what the null hypothesis claims is guaranteed to be true. It simply means that on the basis of the information obtained from the sample data there is not enough evidence to reject the null hypothesis.

10.4 TESTS CONCERNING MEANS FOR LARGE SAMPLES

In this section we will discuss methods for determining whether we should accept or reject a null hypothesis about the mean of a population. We will illustrate the procedure with several examples.

Suppose a manufacturer claims that each family-size bag of pretzels sold weighs 12 ounces, on the average, with a standard deviation of 0.8 ounces. A consumer's group decides to test this claim by accurately weighing 49 randomly selected bags of pretzels. If the mean weight of the sample is considerably different from the population mean, the manufacturer's claim will definitely be rejected. Thus, if the mean weight is 30 ounces or 5 ounces, the manufacturer's claim will be rejected. Only when the sample mean is close to the claimed population mean do we need statistical procedures to determine when to reject or accept a null hypothesis.

Let us assume that the sample mean of the 49 randomly selected bags of pretzels is 11.8 ounces. Since the sample mean, 11.8, is not the same as the population mean, 12, we wish to test the manufacturer's claim at the 5% level of significance.

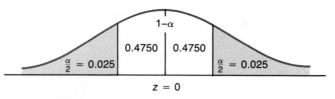

FIGURE 10.7

The population parameter being tested in this case is the mean weight, μ, and the questioned value is 12 ounces. Thus,

Null hypothesis, H_0: $\mu = 12$

Alternative hypothesis, H_1: $\mu \neq 12$

The alternative hypothesis of not equal suggests a two-tailed rejection region. Therefore, the α of 0.05 is split equally between the two tails, as shown in Figure 10.7.

We now look in Table V to find which z-value has 0.4750 of the area between $z = 0$ and this z-value. From Table V we find that the z-value is 1.96. We label this on the diagram in Figure 10.7 and get the acceptance-rejection diagram shown in Figure 10.8.

Since the Central Limit Theorem tells us that the sample means are normally distributed, we use

$$z = \frac{\bar{x} - \mu}{\dfrac{\sigma}{\sqrt{n}}}$$

as the test statistic and reject the null hypothesis if the value of the test statistic falls in the rejection region. In using this test statistic, $\bar{x}$ is the sample mean and μ is the population mean as claimed in the null hypothesis. In our case we have

$$z = \frac{\bar{x} - \mu}{\dfrac{\sigma}{\sqrt{n}}}$$

$$= \frac{11.8 - 12}{\dfrac{0.8}{\sqrt{49}}}$$

$$= \frac{-0.2}{\dfrac{0.8}{7}}$$

$$= -1.75$$

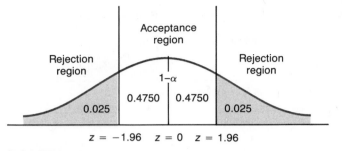

FIGURE 10.8

Since this calculated value of z falls within the acceptance region, our decision is that we cannot reject H_0. The difference between the sample mean and the assumed value of the population mean may be due purely to chance. We say that the difference is *not statistically significant.*

If the level of significance had been 0.01, we would split the 0.01 into two equal tails as shown in Figure 10.9. From Table V we find that the z-value that has 0.4950 of the area between $z = 0$ and this z-value is 2.58. Thus, we reject the null hypothesis if the test statistic falls in the critical region shown in Figure 10.10.

Since we obtained a z-value of -1.75, which is in the acceptance region, we do not reject the null hypothesis.

Let us summarize the testing procedure outlined in the previous paragraphs for a two-tailed rejection region.

1. First convert the sample mean into standard units using

$$z = \frac{\bar{x} - \mu}{\dfrac{\sigma}{\sqrt{n}}}$$

2. Then reject the null hypothesis about the population mean if z is less than -1.96 or greater than 1.96 when using a 5% level of significance.
3. If we are using a 1% level of significance, reject the null hypothesis if z is less than -2.58 or greater than 2.58.

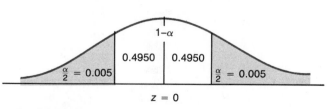

FIGURE 10.9

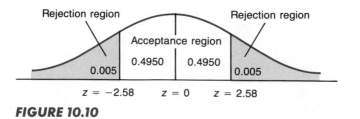

FIGURE 10.10

We further illustrate the procedure with several examples.

● Example 1

A light bulb company claims that the 60-watt light bulb it sells has an average life of 1000 hours with a standard deviation of 75 hours. Sixty-four new bulbs were allowed to burn out to test this claim. The average lifetime of these bulbs was found to be 975 hours. Does this indicate that the average life of a bulb is not 1000 hours? (Use a 5% level of significance.)

Solution

In this case the population parameter being tested is μ, the average life of a bulb, and the value questioned is 1000. Since we are testing whether the average life of a bulb is or is not 1000 hours, we have

H_0: $\mu = 1000$

H_1: $\mu \neq 1000$

We are given that $\bar{x} = 975$, $\sigma = 75$, $\mu = 1000$, and $n = 64$. We first calculate the value of the test statistic, z. We have

$$z = \frac{\bar{x} - \mu}{\dfrac{\sigma}{\sqrt{n}}}$$

$$= \frac{975 - 1000}{\dfrac{75}{\sqrt{64}}}$$

$$= -2.67$$

We use the two-tailed rejection region shown in Figure 10.11. The value of $z = -2.67$ falls in the rejection region. Thus, we reject the null hypothesis that the average life of a bulb is 1000 hours. In this case, the test statistic is sufficiently extreme so that we can reject H_0. The difference is statistically significant.

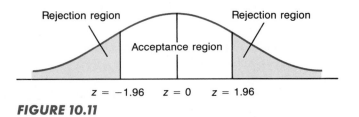

FIGURE 10.11

Perhaps you are wondering why we used $\mu \neq 1000$ as the alternative hypothesis and not $\mu < 1000$. After all, who cares if the average life of a bulb is more than 1000 hours. The answer is that the manufacturer cares. When a manufacturer claims that the average life of a bulb is 1000 hours, the manufacturer is concerned when bulbs last more or less than 1000 hours. If the mean life is less than 1000 hours, the manufacturer will lose business and consumer confidence. If the mean life is more than 1000 hours, the company will lose money.

● Example 2
A bank teller at the Eastern Savings Bank claims that the average amount of money on deposit in a savings account at this bank is $4800 with a standard deviation of $460. A random sample of 36 accounts is taken to test this claim. The average of these accounts is found to be $5000. Does this sample indicate that the average amount of money on deposit is not $4800? (Use a 5% level of significance.)

Solution
In this case the population parameter being tested is μ, the average amount of money on deposit in a savings account. The value questioned is $4800. Since we are testing whether the average amount of money on deposit is $4800 or not, we have

H_0: $\mu = 4800$

H_1: $\mu \neq 4800$

We are given that $\bar{x} = 5000$, $\mu = 4800$, $\sigma = 460$, and $n = 36$ so that

$$z = \frac{\bar{x} - \mu}{\dfrac{\sigma}{\sqrt{n}}}$$

$$= \frac{5000 - 4800}{\dfrac{460}{\sqrt{36}}}$$

$$= 2.61$$

We use the same two-tailed rejection region as shown in Figure 10.11. The value of $z = 2.61$ falls in the rejection region. Thus, we reject the null hypothesis that the average amount of money on deposit is $4800.

● **Example 3**

The average score of all sixth-graders in a certain school district on the 1–2–3 math aptitude exam is 75 with a standard deviation of 8.1. A random sample of 100 students in one school was taken. The mean score of these 100 students was 71. Does this indicate that the students of this school are significantly less skilled in their mathematical abilities than the average student in the district? (Use a 5% level of significance.)

Solution

In this case the population parameter being tested is μ, the mean score on the math aptitude exam. The value questioned is $\mu = 75$. We want to determine if the students of this particular school are significantly less skilled in their mathematical abilities. Thus, it is reasonable to set up a one-sided, or a one-tailed, test with the alternative hypothesis being that the population mean for this school is less than 75. We have

$$H_0: \quad \mu = 75$$

$$H_1: \quad \mu < 75$$

When dealing with one-tail (left-side) alternative hypotheses, we have the rejection regions illustrated in Figure 10.12. These values, like those in the two-tailed tests, are obtained from Table V. You should verify these results.

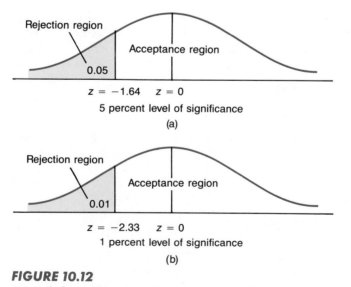

Rejection region

Acceptance region

0.05

$z = -1.64$ $z = 0$

5 percent level of significance

(a)

Rejection region

Acceptance region

0.01

$z = -2.33$ $z = 0$

1 percent level of significance

(b)

FIGURE 10.12

One-tailed tests for μ less than some given value.

Now we can calculate the test statistic, z. Here $\bar{x} = 71$, $\mu = 75$, $\sigma = 8.1$, and $n = 100$. We have

$$z = \frac{\bar{x} - \mu}{\dfrac{\sigma}{\sqrt{n}}}$$

$$= \frac{71 - 75}{\dfrac{8.1}{\sqrt{100}}}$$

$$= -4.94$$

Since the z-value of -4.94 is in the rejection region, we conclude that the students of this school are significantly less skilled in their mathematical abilities than the average student in the district.

● Example 4
The We-Haul Trucking Corp. claims that the average hourly salary of its mechanics is \$9.25 with a standard deviation of \$1.55. A random sample of 81 mechanics showed that the average hourly salary of these mechanics was only \$8.95. Does this indicate that the average hourly salary of a mechanic is less than \$9.25? (Use a 1% level of significance.)

Solution
In this case the population parameter being tested is μ, the mean hourly salary. The questioned value is \$9.25. We wish to know if the hourly salary is less than \$9.25. Thus, we will use a one-tailed test. We have

H_0: $\mu = 9.25$

H_1: $\mu < 9.25$

Here we are given that $\bar{x} = 8.95$, $\mu = 9.25$, $\sigma = 1.55$, and $n = 81$, so that

$$z = \frac{\bar{x} - \mu}{\dfrac{\sigma}{\sqrt{n}}}$$

$$= \frac{8.95 - 9.25}{\dfrac{1.55}{\sqrt{81}}}$$

$$= -1.74$$

We use the same one-tailed rejection region as shown in Figure 10.12. The value of $z = -1.74$ falls in the acceptance region. Thus, the sample data do not provide us with sufficient justification to reject the null hypothesis. We cannot conclude that the average hourly salary of a mechanic is less than \$9.25.

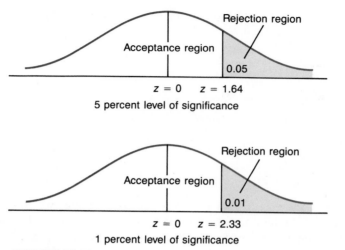

FIGURE 10.13
One-tailed tests for μ greater than some given value.

● **Example 5**

A trash company claims that the average weight of any of its fully loaded garbage trucks is 11,000 pounds with a standard deviation of 800 pounds. A highway department inspector decides to check on this claim. She randomly checks 36 trucks and finds that the average weight of these trucks is 12,500 pounds. Does this indicate that the average weight of a garbage truck is more than 11,000 pounds? (Use a 5% level of significance.)

Solution

In this case the population parameter being tested is μ, the average weight of a garbage truck. The value questioned is 11,000 pounds. The sample data suggest that the mean weight is really more than 11,000 pounds. Thus, the alternative hypothesis will be that the population mean is more than 11,000 pounds. We have

H_0: $\mu = 11,000$

H_1: $\mu > 11,000$

When dealing with a one-tail (right-side) alternative hypothesis, we have the rejection regions shown in Figure 10.13.

Here we are given that $\bar{x} = 12,500$, $\mu = 11,000$, $\sigma = 800$, and $n = 36$, so that

$$z = \frac{\bar{x} - \mu}{\dfrac{\sigma}{\sqrt{n}}}$$

$$= \frac{12{,}500 - 11{,}000}{\dfrac{800}{\sqrt{36}}}$$

$$= 11.25$$

Since the z-value of 11.25 falls within the rejection region, we conclude that the average weight of a garbage truck is not 11,000 pounds. We reject the null hypothesis.

● **Example 6**

An insurance company advertises that it takes 21 days on the average to process an auto accident claim. The standard deviation is 8 days. To check on the truth of this advertisement, a group of investigators randomly selects 35 people who recently filed claims. They find that it took the company an average of 24 days to process these claims. Does this indicate that it takes the insurance company more than 21 days on the average to process a claim? (Use a 1% level of significance.)

Solution

In this case the population parameter being tested is μ, the average number of days needed to process a claim. The questioned value is $\mu = 21$. The sample data suggest that $\mu > 21$. Thus, we will use a one-tailed test. We have

H_0: $\mu = 21$

H_1: $\mu > 21$

Here we are given that $\bar{x} = 24$, $\mu = 21$, $\sigma = 8$, and $n = 35$, so that

$$z = \frac{\bar{x} - \mu}{\dfrac{\sigma}{\sqrt{n}}}$$

$$= \frac{24 - 21}{\dfrac{8}{\sqrt{35}}} = 2.22$$

Since the z-value of 2.22 falls within the acceptance region, we cannot reject the null hypothesis. Thus, we cannot conclude that it takes the insurance company more than 21 days to process a claim.

COMMENT   We wish to emphasize again that the large-sample case tests that we have been discussing work only if the sample size is sufficiently large. This means that n should be at least 30. This is necessary so that the distribution of $\bar{x}$ be approximately normal.

COMMENT Since σ is unknown in many practical applications, we often have no choice but to use the sample standard deviation, s, as an approximation for σ. Again, s provides a good approximation to σ if the sample size is sufficiently large $(n \geq 30)$.

We summarize our discussion as follows:

LARGE-SAMPLE TEST OF HYPOTHESIS ABOUT A POPULATION MEAN

	One-tailed Test	Two-tailed Test	One-tailed Test
Null Hypothesis	H_0: $\mu = \mu_0$	H_0: $\mu = \mu_0$	H_0: $\mu = \mu_0$
Alternative Hypothesis	H_1: $\mu < \mu_0$	H_1: $\mu \neq \mu_0$	H_1: $\mu > \mu_0$
Test Statistic	$z = \dfrac{\bar{x} - \mu_0}{\dfrac{\sigma}{\sqrt{n}}}$	$z = \dfrac{\bar{x} - \mu_0}{\dfrac{\sigma}{\sqrt{n}}}$	$z = \dfrac{\bar{x} - \mu_0}{\dfrac{\sigma}{\sqrt{n}}}$
	$= \dfrac{\bar{x} - \mu_0}{\dfrac{s}{\sqrt{n}}}$	$= \dfrac{\bar{x} - \mu_0}{\dfrac{s}{\sqrt{n}}}$	$= \dfrac{\bar{x} - \mu_0}{\dfrac{s}{\sqrt{n}}}$
Rejection Region	$z < -z_\alpha$	$z < -z_{\alpha/2}$ or $z > z_{\alpha/2}$	$z > z_\alpha$

In this chart, z_α is the z-value such that prob $(z > z_\alpha) = \alpha$.

$z_{\alpha/2}$ is the z-value such that prob $(z > z_{\alpha/2}) = \dfrac{\alpha}{2}$.

μ_0 represents a particular value for μ as specified in the null hypothesis that we are testing.

CURE FOUND FOR DEADLY VIRUS

A treatment has been found for severe cases of a respiratory virus that infects more than 800,000 U.S. infants a year.

The Food and Drug Administration has approved the drug ribavirin for acute cases of respiratory syncytial virus, or RSV. Administered with an inhalation hood in a hospital for up to seven days, ribavirin stops RSV from reproducing.

Taken orally for 10 days, ribavirin also is the first effective therapy for Lassa fever, a deadly virus found primarily in Africa, say doctors from the Centers for Disease Control.

January 27, 1986

When a new drug is developed, how do we determine whether the average number of people cured by the drug is significant? Is one drug better than another in terms of the average number of people cured?

EXERCISES

1. Industry representatives claim that the average price of an introductory college-level math textbook is $34.95. Numerous student groups claim that the average price is considerably higher. A survey of 49 such books disclosed an average price of $36.71 with a standard deviation of $3.61. Using a 5% level of significance, should we reject the industry representative's claim?

2. Welfare department officials of a certain city claim that the average number of cases of child abuse handled daily by its various agencies is 7.62. A newspaper reporter decides to test this claim. The reporter randomly selects 40 days and determines that the average number of child-abuse claims handled on these days was 5.02 with a standard deviation of 1.96. Using a 5% level of significance, should we reject the welfare department officials' claim?

3. The average number of daily accidents at a particular mine during the 1970s was 7.6. In an effort to reduce this number, management instituted a massive safety program. To determine the effectiveness of the new safety program, a random sample of 45 days is taken after the new program has been put into effect. It is found that the average number of accidents per day is now 5.91 with a standard deviation of 1.62. Using a 1% level of significance, can we conclude that the average number of daily accidents has decreased since the institution of the new safety program?

4. Management of a large express-delivery service claims that company employees misroute an average of 19.4 packages per day. After a new sorting system is introduced, a random sample of 41 days' routings is taken. It is found that there were an average of 17.83 misroutings on these days. The standard deviation was 4.62. Using a 5% level of significance, should we accept management's claim that the number of misroutings has decreased?

5. Banking industry officials in a large city claim that the average charge for a "bounced" check is $4.00. To test this claim, a random sample is taken of the charge imposed by 45 banks for a bounced check. It is found that the average charge is $5.25 with a standard deviation of $0.96. Using a 1% level of significance, should we reject the banking industry officials' claim?

6. In an effort to attract new industry to the region, a mayor claims that the average age of a worker in the region is 27 years with a standard deviation of 5.72 years. A prospective company is interested in determining whether the mayor's claim is accurate. A random sample of 60 workers in the region reveals an average age of 29 years. Is there sufficient evidence to conclude that the average age is not 27 years? (Use a 5% level of significance.)

7. A group of urologists claims that the average length of a hospital stay for a patient with one type of kidney disease is 9.2 days. A random sample of 60 patients treated for this particular kidney ailment disclosed an average hospital stay of 9.4 days with a standard deviation of 0.98 days. Using a 5% level of significance, should we reject the urologists' claim?

8. A coffee manufacturer sells jars of coffee supposedly filled with 10 ounces of coffee. The Consumer Fraud Bureau of a state has received numerous complaints that some of the jars contain less than the specified 10 ounces. The Bureau decides to investigate these complaints by sampling 100 jars and accurately measuring the weight of the contents. It is found that the average weight of the coffee in the jars is 9.862 ounces with a standard deviation of 1.44 ounces. Can the coffee manufacturer be accused of "short-changing" the customer? Use a 5% level of significance.

9. The president of a nationwide bank claims that the average auto loan is for $5256 with a standard deviation of $962. A local banker believes that in her city the average auto loan is well above the national average. A random sample of 75 auto loans indicates an auto loan average of $5500. Using a 5% level of significance, can we reject the claim made by the president of the nationwide bank?

10. A manufacturer claims that a particular computer chip component should last an average of 15,000 hours before replacement is needed. To verify the accuracy of this claim, 85 computer chip components are tested. It is found that they lasted an average of 14,800 hours with a standard deviation of 1223 hours. Using a 1% level of significance, can we reject the manufacturer's claim?

*11. An insurance company claims that the average charge by a dentist in a city for bonding a tooth is $95. A consumer samples 80 dentists and calculates the average charge at $101 with a standard deviation of $6. Perform an appropriate hypothesis test at the 10% level of significance to determine whether the insurance company's claim is not acceptable.

10.5 TESTS CONCERNING MEANS FOR SMALL SAMPLES

In the last section, we indicated how sample data can be used to reject or accept a null hypothesis about the mean. The sizes of all the samples discussed were large enough to justify the use of the normal distribution. When the sample size is small, we must use the *t*-distribution discussed in the last chapter instead of the normal distribution. The following examples will illustrate how the *t*-distribution is used in hypothesis testing.

● **Example 1**

A manufacturer claims that each can of mixed nuts sold contains an average of 10 cashew nuts. A sample of 15 cans of these mixed nuts has an average of 8 cashew nuts with a standard deviation of 3. Does this indicate that we should reject the manufacturer's claim? (Use a 5% level of significance.)

Solution

Since the sample size is only 15, the test statistic becomes

FIGURE 10.14
Two-tailed small-sample rejection region (5% level of significance).

$$t = \frac{\bar{x} - \mu}{\dfrac{s}{\sqrt{n}}}$$

instead of z. Depending upon the number of degrees of freedom, we have the acceptance-rejection regions shown in Figures 10.14 and 10.15. In each case the value of t is obtained from Table VIII in the Appendix. It depends upon the number of degrees of freedom, which is $n - 1$.

Let us now return to our problem. The population parameter being tested is μ, the average number of cashew nuts in a can of mixed nuts. The questioned value is 10. We have

H_0: $\mu = 10$

H_1: $\mu \neq 10$

We will use a two-tailed rejection region as shown in Figure 10.14. Here we are given that $\bar{x} = 8$, $\mu = 10$, $s = 3$, and $n = 15$, so that

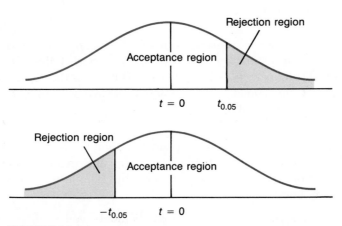

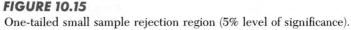

FIGURE 10.15
One-tailed small sample rejection region (5% level of significance).

$$t = \frac{\bar{x} - \mu}{\frac{s}{\sqrt{n}}} = \frac{8 - 10}{\frac{3}{\sqrt{15}}} = -2.58$$

From Table VIII we find that the $t_{0.025}$ value for $15 - 1$, or 14, degrees of freedom is 2.145, which means that we reject the null hypothesis if the test statistic is less than -2.145 or greater than 2.145. Since $t = -2.58$ falls within the critical region, we reject the manufacturer's claim that each can of mixed nuts contains an average of 10 cashew nuts.

● **Example 2**

A new weight-reducing pill is being sold in a midwestern town. The manufacturer claims that any overweight person who takes this pill as directed will lose 15 pounds within a month. To test this claim, a doctor gives this pill to 6 overweight people and finds that they lose an average of only 12 pounds with a standard deviation of 4 pounds. Can we reject the manufacturer's claim? (Use a 5% level of significance.)

Solution

In this case the population parameter being tested is μ, the average number of pounds lost when using this pill. The questioned value is $\mu = 15$. We have

$$H_0: \quad \mu = 15$$

$$H_1: \quad \mu < 15$$

Here we are given that $\bar{x} = 12$, $\mu = 15$, $s = 4$, and $n = 6$, so that

$$t = \frac{\bar{x} - \mu}{\frac{s}{\sqrt{n}}} = \frac{12 - 15}{\frac{4}{\sqrt{6}}} = -1.84$$

From Table VIII we find the $t_{0.05}$ value for $6 - 1$, or 5, degrees of freedom is 2.015, which means that we reject the null hypothesis if the test statistic is less than -2.015. Since $t = -1.84$ does not fall within the rejection region, we cannot reject the manufacturer's claim that the average number of pounds lost when using this pill is 15.

COMMENT Again we wish to emphasize a point made earlier: When testing hypotheses involving small sample sizes, we assume that the relative frequency distribution of the population from which the sample is to be selected is approximately normal.

We summarize the procedure that should be followed when testing hypotheses about a population mean and when we have a small sample size.

SMALL-SAMPLE TEST OF HYPOTHESIS ABOUT A POPULATION MEAN

	One-tailed Test	Two-tailed Test	One-tailed Test
Null Hypothesis	H_0: $\mu = \mu_0$	H_0: $\mu = \mu_0$	H_0: $\mu = \mu_0$
Alternative Hypothesis	H_1: $\mu < \mu_0$	H_1: $\mu \neq \mu_0$	H_1: $\mu > \mu_0$
Test Statistic	$t = \dfrac{\bar{x} - \mu_0}{\dfrac{s}{\sqrt{n}}}$	$t = \dfrac{\bar{x} - \mu_0}{\dfrac{s}{\sqrt{n}}}$	$t = \dfrac{\bar{x} - \mu_0}{\dfrac{s}{\sqrt{n}}}$
Rejection Region	$t < -t_\alpha$	$t < -t_{\alpha/2}$ or $t > t_{\alpha/2}$	$t > t_\alpha$

In this chart, t_α is the t-value such that prob $(t > t_\alpha) = \alpha$.

$t_{\alpha/2}$ is the t-value such that prob $(t > t_{\alpha/2}) = \dfrac{\alpha}{2}$.

The distribution of t has $n - 1$ degrees of freedom.

EXERCISES

1. A manufacturer claims that a special type of movie projector bulb should last an average of 75 hours. However, a random sample of 16 such bulbs lasted an average of only 69 hours with a standard deviation of 3.6 hours. Using a 5% level of significance, can we conclude that the bulbs last less than 75 hours?

2. Medical research indicates that the average reaction time to a certain stimulus is 1.69 seconds. Six adults have been given a new antibiotic drug. Their average reaction time to the stimulus is found to be 1.78 seconds with a standard deviation of 0.82 seconds. Using a 1% level of significance, does the new antibiotic drug affect the average reaction time to the stimulus?

3. It is claimed that the average annual salary of a police officer in Bolton is $29,761. A sample of 16 police officers reveals an average salary of $29,023 with a standard deviation of $946. Using a 5% level of significance, can we conclude that the average annual salary of a police officer in Bolton is less than $29,761?

4. A psychologist claims that rats require an average of 10 minutes to find their way through a maze. Seven rats require 10, 8, 14, 12, 5, 15, and 13 minutes, respectively, to find their way through the maze. Using a 5% level of significance, can we conclude that rats do not require an average of 10 minutes to find their way through a maze?

5. Government officials in a large city claim that a call to the 911 police emergency number will bring an ambulance in an average of 5.1 minutes. Hospital officials claim that ambulance response time has improved since the installation of a new computer. In a sample of ten calls for an ambulance, the average response time was 4.6 minutes with a standard deviation of 1.2 minutes. Does this indicate an improvement of service? (Use a 1% level of significance.)

6. A manufacturer of batteries claims that, on average, a battery will last 1000 hours. The research department claims to have developed a new battery that lasts longer than the old battery. A sample of 20 of the new batteries lasts an average of 1025 hours with a standard deviation of 8.56 hours. At the 5% level of significance, is the new battery longer-lasting than the older battery?

7. A potato chip manufacturer packs 32-ounce bags of potato chips. The manufacturer wants the bags to contain, on the average, 32 ounces of potato chips. A quality-control engineer randomly selects 15 bags of potato chips and determines that the average weight of the bags is 31.6 ounces with a standard deviation of 0.3 ounces. At the 5% level of significance, can we accept the alternative claim that the bags contain less than 32 ounces of potato chips?

8. A forest ranger claims that the average weight of an adult deer on one preserve is 320 pounds with a standard deviation of 16 pounds. A hunter captured 12 deer whose average weight was 325 pounds. At the 1% level of significance, can we conclude that the average weight of an adult deer on the preserve is more than 320 pounds?

9. It is claimed that the average fare for a taxi ride from the airport to the center of the city is $16.20. A random sample of nine trips from the airport to the center of the city had an average fare of $16.80 with a standard deviation of $1.08. At the 5% level of significance, can we conclude that the average fare for a taxi ride from the airport to the center of the city is more than $16.20?

10. A midwestern college claims that its computer science graduates can expect an average starting salary of $21,700 annually. Fourteen recent computer science graduates of that college had an average annual starting salary of $19,900 with a standard deviation of $1100. At the 5% level of significance, can we conclude that the average starting salary of a computer science graduate is less than $21,700?

10.6 TESTS CONCERNING DIFFERENCES BETWEEN MEANS FOR LARGE SAMPLES

There are many instances in which we must decide whether the observed difference between two sample means is due purely to chance or whether the population means from which these samples were selected are really different. For example, suppose a teacher gave an IQ test to 50 girls and 50 boys and obtained the following test scores:

	Boys	Girls
Mean	78	81
Standard Deviation	7	9

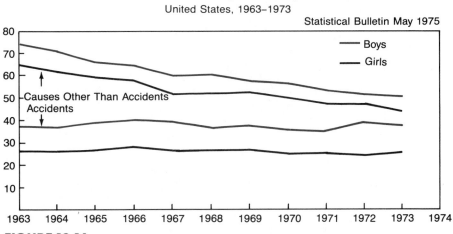

FIGURE 10.16

This clipping indicates the trend in death rates, by sex, among children aged 1–4 years. Would you say that the average number of deaths as a result of accidents is significantly higher for boys than for girls? What about child deaths not resulting from accidents? Is the observed difference significant? (Table courtesy of The Metropolitan Life Insurance Company.) *Source of basic data:* Reports of the Division of Vital Statistics, National Center for Health Statistics.

Is the observed difference between the scores significant? Are the girls smarter?

In problems of this sort the null hypothesis is that there is no difference between the means. Since we will be discussing more than one sample, we use the following notation. We let $\bar{x}_1$, s_1, and n_1 be the mean, standard deviation, and sample size, respectively, of one of the samples, and $\bar{x}_2$, s_2, and n_2 be the mean, standard deviation, and sample size, respectively, of the second sample. Decisions as to whether to reject or accept the null hypotheses are then based upon the test statistic z, where

$$z = \frac{\bar{x}_1 - \bar{x}_2}{\sqrt{\dfrac{s_1^2}{n_1} + \dfrac{s_2^2}{n_2}}}$$

(assuming the samples are independent and both samples are large).

Depending upon whether the alternative hypothesis is $\mu_1 \neq \mu_2$, $\mu_1 < \mu_2$, or $\mu_1 > \mu_2$, we have a two-sided test or a one-sided test as indicated in Section 10.4. The following examples will illustrate how this test statistic is used.

● **Example 1**

Consider the example discussed at the beginning of this section. Is the observed difference between the two IQ scores significant? (Use a 5% level of significance.)

Solution

Let $\bar{x}_1$, s_1, and n_1 represent the boys' mean score, standard deviation, and sample size, and let $\bar{x}_2$, s_2, and n_2 be the corresponding girls' scores. Then the problem is whether the observed difference between the sample means is significant or not. Thus

H_0: $\mu_1 = \mu_2$

H_1: $\mu_1 \neq \mu_2$

Here we are given that $\bar{x}_1 = 78$, $s_1 = 7$, $n_1 = 50$, $\bar{x}_2 = 81$, $s_2 = 9$, and $n_2 = 50$, so that

$$z = \frac{\bar{x}_1 - \bar{x}_2}{\sqrt{\dfrac{s_1^2}{n_1} + \dfrac{s_2^2}{n_2}}} = \frac{78 - 81}{\sqrt{\dfrac{7^2}{50} + \dfrac{9^2}{50}}} = \frac{-3}{\sqrt{\dfrac{49}{50} + \dfrac{81}{50}}}$$

$$= \frac{-3}{\sqrt{2.60}} = -1.86$$

We use the two-tail rejection region of Figure 10.11 (see page 429). The value of $z = -1.86$ falls in the acceptance region. Thus, we cannot conclude that the sample data support the claim that there is a significant difference between the boys' IQ scores and the girls' IQ scores.

● Example 2

An executive who has two secretaries, Jean and Mark, is interested in knowing whether there is any significant difference in their typing abilities. Mark typed a 40-page report and made an average of 4.6 errors per page. The standard deviation was 0.6. Jean typed a 30-page report and made an average of 2.3 errors per page. The standard deviation was 0.8. Is there any significant difference between their performances? (Use a 5% level of significance.)

Solution

Let $\bar{x}_1$, s_1, and n_1 represent Mark's scores and let $\bar{x}_2$, s_2, and n_2 represent Jean's scores. Then the problem is whether the observed difference between the sample means is significant or not. Thus

H_0: $\mu_1 = \mu_2$

H_1: $\mu_1 \neq \mu_2$

Here we are given that $\bar{x}_1 = 4.6$, $s_1 = 0.6$, $n_1 = 40$, $\bar{x}_2 = 2.3$, $s_2 = 0.8$, and $n_2 = 30$, so that

$$z = \frac{\bar{x}_1 - \bar{x}_2}{\sqrt{\dfrac{s_1^2}{n_1} + \dfrac{s_2^2}{n_2}}}$$

$$= \frac{4.6 - 2.3}{\sqrt{\dfrac{(0.6)^2}{40} + \dfrac{(0.8)^2}{30}}}$$

$$= \frac{2.3}{\sqrt{0.009 + 0.0213}}$$

$$= \frac{2.3}{\sqrt{0.03}} = \frac{2.3}{0.174}$$

$$= 13.29$$

We use the two-tail rejection region of Figure 10.11 (see page 429). The value of $z = 13.22$ falls in the rejection region. Thus, there is a significant difference between the typing abilities of the two secretaries.

● Example 3

There are many advertisements on television about toothpastes. One such advertisement claims that children who use Smile toothpaste have fewer cavities than children who use any other brand. To test this claim, a consumer's group selected 100 children and divided them into two groups of 50 each. The children of group I were told to brush daily with only Smile toothpaste. The children of group II were told to brush daily with Vanish toothpaste. The experiment lasted one year. The following number of cavities were identified:

Smile: $\bar{x}_1 = 2.31$ $s_1 = 0.6$

Vanish: $\bar{x}_2 = 2.68$ $s_2 = 0.4$

Is Smile significantly more effective than Vanish in preventing cavities? (Use a 5% level of significance.)

Solution

In this case the question is whether Smile is better than Vanish. This means that people who use Smile toothpaste will have fewer cavities than those who use Vanish. Thus,

H_0: $\mu_1 = \mu_2$

H_1: $\mu_1 < \mu_2$

Here we are given that $\bar{x}_1 = 2.31$, $s_1 = 0.6$, $n_1 = 50$, $\bar{x}_2 = 2.68$, $s_2 = 0.4$, and $n_2 = 50$, so that

$$z = \frac{\bar{x}_1 - \bar{x}_2}{\sqrt{\dfrac{s_1^2}{n_1} + \dfrac{s_2^2}{n_2}}}$$

$$= \frac{2.31 - 2.68}{\sqrt{\dfrac{(0.6)^2}{50} + \dfrac{(0.4)^2}{50}}} = \frac{-0.37}{\sqrt{0.0104}} = \frac{-0.37}{0.102}$$

$$= -3.63$$

We use the one-tail rejection region of Figure 10.13 (see page 432). The value of $z = -3.63$ falls in the rejection region. Thus, we reject the null hypothesis. The sample data would seem to support the manufacturer's claim. Actually, further studies are needed before making any definite decision about the effectiveness of Smile in preventing cavities.

● **Example 4**

The local chapter of a Women's Liberation group claims that a female college graduate earns less than a male college graduate. A survey of 40 men and 30 women indicated the following results:

	Average Starting Salary	Standard Deviation
Women	$29,000	$600
Men	$29,700	$900

Do these figures support the claim that women earn less? (Use a 1% level of significance.)

Solution

Let $\bar{x}_1$, s_1, and n_1 represent the women's scores, and let $\bar{x}_2$, s_2, and n_2 represent the men's scores. The problem is whether the observed difference between the sample means is significant or not. Thus

$$H_0: \quad \mu_1 = \mu_2$$

$$H_1: \quad \mu_1 < \mu_2$$

Here we are given that $\bar{x}_1 = 29,000$, $s_1 = 600$, $n_1 = 30$, $\bar{x}_2 = 29,700$, $s_2 = 900$, and $n_2 = 40$, so that

$$z = \frac{\bar{x}_1 - \bar{x}_2}{\sqrt{\dfrac{s_1^2}{n_1} + \dfrac{s_2^2}{n_2}}}$$

$$= \frac{29{,}000 - 29{,}700}{\sqrt{\dfrac{(600)^2}{30} + \dfrac{(900)^2}{40}}} = \frac{-700}{\sqrt{12{,}000 + 20{,}250}}$$

$$= \frac{-700}{\sqrt{32{,}250}} = \frac{-700}{179.58}$$

$$= -3.9$$

We use the one-tail rejection region of Figure 10.12 (see page 430). The value of $z = -3.9$ falls in the rejection region. Thus, we reject the null hypothesis. There is a significant difference between the starting salary of men and women.

We summarize the procedure to be used in the following chart:

LARGE-SAMPLE TEST OF HYPOTHESIS ABOUT THE DIFFERENCE BETWEEN TWO POPULATION MEANS

	One-tailed Test	Two-tailed Test	One-tailed Test
Null Hypothesis	H_0: $(\mu_1 - \mu_2) = A$	H_0: $(\mu_1 - \mu_2) = A$	H_0: $(\mu_1 - \mu_2) = A$
Alternative Hypothesis	H_1: $(\mu_1 - \mu_2) < A$	H_1: $(\mu_1 - \mu_2) \neq A$	H_1: $(\mu_1 - \mu_2) > A$
Test Statistic	$z = \dfrac{(\bar{x}_1 - \bar{x}_2) - A}{\sigma_{(\bar{x}_1 - \bar{x}_2)}}$	$z = \dfrac{(\bar{x}_1 - \bar{x}_2) - A}{\sigma_{(\bar{x}_1 - \bar{x}_2)}}$	$z = \dfrac{(\bar{x}_1 - \bar{x}_2) - A}{\sigma_{(\bar{x}_1 - \bar{x}_2)}}$
	$\approx \dfrac{(\bar{x}_1 - \bar{x}_2) - A}{\sqrt{\dfrac{s_1^2}{n_1} + \dfrac{s_2^2}{n_2}}}$	$\approx \dfrac{(\bar{x}_1 - \bar{x}_2) - A}{\sqrt{\dfrac{s_1^2}{n_1} + \dfrac{s_2^2}{n_2}}}$	$\approx \dfrac{(\bar{x}_1 - \bar{x}_2) - A}{\sqrt{\dfrac{s_1^2}{n_1} + \dfrac{s_2^2}{n_2}}}$
Rejection Region	$z < -z_\alpha$	$z < -z_{\alpha/2}$ or $z > z_{\alpha/2}$	$z > z_\alpha$

In the above chart, A is the numerical value for $(\mu_1 - \mu_2)$ as specified in the null hypothesis. In many applied problems, we are interested in testing that there is no difference between the population means. Of course, in such a situation $A = 0$.

COMMENT For these tests to work, we assume that the sample sizes, n_1 and n_2, are sufficiently large and that the two random samples are selected independently from a large population.

EXERCISES

1. Eighty-four cars coated with Brand A undercoating did not develop any rust for an average of 4.3 years. The standard deviation was 0.96 years. Sixty-nine cars coated with

Brand B undercoating did not develop any rust for an average of 5.09 years. The standard deviation was 1.98 years. Is there any significant difference between the average time for rust to develop on a car after using these brands of undercoating? (Use a 5% level of significance.)

2. **New Subway Cars**   The Transit Commission of a large city is considering purchasing new subway cars for its aging fleet from one of two manufacturers. It has road-tested 90 subway cars manufactured by Company A and 80 subway cars manufactured by Company B over a period of several months and has gathered the following information:

Subway Cars

	Company A	Company B
Average Number of Breakdowns per Month	28	32
Standard Deviation	5	2

Is there any significant difference between the average number of breakdowns per month for subway cars manufactured by each company? (Use a 1% level of significance.)

3. A baker can use one of two chemical preservatives to retard spoilage. Preservative A was added to 50 cookies. The cookies lasted an average of 9 days before spoiling. The standard deviation was 1.2 days. Preservative B was added to 75 cookies. The cookies lasted an average of 10.1 days before spoiling. The standard deviation was 2.7 days. Is there any significant difference between the two preservatives in retarding spoilage? (Use a 5% level of significance.)

4. Seventy-eight foreign cars were road-tested by professional drivers on the streets of Chicago. The cars tested at an average of 22.6 miles per gallon of gas. The standard deviation was 3.6. Fifty-three domestic cars were then road-tested by these professional drivers on the streets of Chicago. The results were an average of 23.2 miles per gallon of gas. The standard deviation was 2.8. Is the difference between the average number of miles per gallon of gas for the foreign cars and the domestic cars significant? (Use a 5% level of significance.)

5. In a random sample of 40 supermarkets in Los Angeles, the average price of a pound of lean ground beef was found to be $1.29 with a standard deviation of $0.19. In a random sample of 40 supermarkets in San Francisco, the average price of a pound of lean ground beef was found to be $1.37 with a standard deviation of $0.28. Using a 5% level of significance, test the null hypothesis that the average price of a pound of lean ground beef is the same in both cities.

6. Data gathered from 80 purchasers of a certain type of bicycle suggest that the time it takes a purchaser to fully assemble the bicycle is a random variable having a mean of

75 minutes and a standard deviation of 5.8 minutes. A group of 60 different purchasers of the bicycle is randomly selected and shown a film on how to assemble the bicycle easily. After viewing the film, each purchaser is asked to fully assemble the bicycle. It is found that they now average 68 minutes with a standard deviation of 8.6 minutes. Using a 1% level of significance, does viewing the film reduce the time required to assemble the bicycle?

7. A medical researcher divided a group of 150 volunteers into two groups of equal size. Each individual in Group A was given a new drug coated with aspirin. Each individual in Group B was given the new drug coated with buffered aspirin. The researcher then gathered the following information on the average number of hours that the new drug had an effect on the volunteers.

	Drug Coated with	
	Aspirin	*Buffered Aspirin*
Average Number of Hours Drug Had Effect	6 hr	7 hr
Standard Deviation	0.69 hr	1.22 hr

Using a 5% level of significance, test the null hypothesis that there is no difference in the lasting effect of the drug whether it is coated with aspirin or with buffered aspirin.

8. A sample of 70 families with four children in Sharon Lakes indicated that each family spends an average of $52.12 per week for food. The standard deviation was $6.92. A similar survey of 85 families in New Lakes indicated that each family spends an average of $56.43 per week for food. The standard deviation was $8.84. Is the difference between the average weekly expenditure for food by families in these two towns significant? (Use a 5% level of significance.)

9. Sixty-five police cars in a city were fitted with special radial tires manufactured by Company A. The tires lasted an average of 21,000 miles before being replaced. The standard deviation was 1012 miles. Seventy-one police cars in the same city were fitted with similar radial tires manufactured by Company B. These tires lasted an average of 27,500 miles before being replaced. The standard deviation was 1100 miles. At the 1% level of significance, is the difference between the average number of miles that these tires lasted significant?

10. Video cassette recorder tapes wear out after repeated use. A sample of 60 tapes of one brand of VCR tape had a mean lifetime of 500 hours with a standard deviation of 31 hours. A sample of 70 tapes of a different brand of VCR tape had a mean lifetime of 550 hours with a standard deviation of 56 hours. Is the difference in the average lifetimes of these tapes significant? (Use a 5% level of significance.)

10.7 TESTS CONCERNING DIFFERENCES BETWEEN MEANS FOR SMALL SAMPLES

In the last section we indicated how we test the difference between sample means. In all of the examples, the sample sizes were large enough ($n \geq 30$) to justify our use of the normal distribution. If this is not the case, we must use the t-distribution. We assume that the population from which the samples are selected have approximately normal probability distributions and that the random samples are selected independently. We then have the following small-sample-size hypothesis testing procedures.

SMALL-SAMPLE TEST OF HYPOTHESIS ABOUT THE DIFFERENCE BETWEEN MEANS

	One-tailed Test	Two-tailed Test	One-tailed Test
Null Hypothesis	H_0: $(\mu_1 - \mu_2) = A$	H_0: $(\mu_1 - \mu_2) = A$	H_0: $(\mu_1 - \mu_2) = A$
Alternative Hypothesis	H_1: $(\mu_1 - \mu_2) < A$	H_1: $(\mu_1 - \mu_2) \neq A$	H_1: $(\mu_1 - \mu_2) > A$
Test Statistic	$t = \dfrac{(\bar{x}_1 - \bar{x}_2) - A}{s_p \sqrt{\dfrac{1}{n_1} + \dfrac{1}{n_2}}}$	$t = \dfrac{(\bar{x}_1 - \bar{x}_2) - A}{s_p \sqrt{\dfrac{1}{n_1} + \dfrac{1}{n_2}}}$	$t = \dfrac{(\bar{x}_1 - \bar{x}_2) - A}{s_p \sqrt{\dfrac{1}{n_1} + \dfrac{1}{n_2}}}$

$$\text{where } s_p = \sqrt{\frac{(n_1 - 1)s_1^2 + (n_2 - 1)s_2^2}{n_1 + n_2 - 2}}$$

Rejection Region	$t < -t_\alpha$	$t < -t_{\alpha/2}$ or $t > t_{\alpha/2}$	$t > t_\alpha$

COMMENT When using the tests outlined in the above chart, it is assumed that the variances of the two populations are equal.

COMMENT When using the tests outlined in the above chart, the number of degrees of freedom for the t-distribution is $n_1 + n_2 - 2$.

We illustrate the procedure with an example.

● **Example 1**

A chemist at a paint factory claims to have developed a new oil-based paint that will dry very quickly. The manufacturer is interested in comparing this new paint with his currently best-selling paint. In order to accomplish this, he paints each of five different walls with a gallon of his best-selling paint and with a gallon of the new fast-drying paint. The number of minutes needed for each of these paints to dry thoroughly is shown as follows:

Number of Minutes Needed to Dry

Current Best-selling Paint	New Fast-drying Paint
48	42
46	43
44	45
46	43
43	44

Using a 5% level of significance, is the new paint significantly more effective in its drying time than the old paint?

Solution

Using the data in the table, we first compute the sample means and the sample standard deviation. We have

Current Best-selling Paint	New Fast-drying Paint
$n_1 = 5$	$n_2 = 5$
$\bar{x}_1 = 45.4$	$\bar{x}_2 = 43.4$
$s_1 = 1.949$	$s_2 = 1.14$

In this case, the null hypothesis is $\mu_1 = \mu_2$ and the alternative hypothesis is $\mu_1 < \mu_2$, where μ_1 is the average drying time of the current best-selling paint and μ_2 is the average drying time of the new fast-drying paint.

Based upon past experience with other paints, the manufacturer knows that the drying time of paint is approximately normally distributed and that the variances for different paints are about the same. Since the samples were randomly and independently selected we compute the test statistic. We have

$$t = \frac{\bar{x}_1 - \bar{x}_2}{s_p\sqrt{\dfrac{1}{n_1} + \dfrac{1}{n_2}}}$$

and

$$s_p = \sqrt{\frac{(n_1 - 1)s_1^2 + (n_2 - 1)s_2^2}{n_1 + n_2 - 2}}$$

so that

$$s_p = \sqrt{\frac{(5 - 1)(1.949)^2 + (5 - 1)(1.14)^2}{5 + 5 - 2}} \approx 1.597$$

Therefore,

$$t = \frac{45.4 - 43.4}{1.597\sqrt{\dfrac{1}{5} + \dfrac{1}{5}}} \approx 1.98$$

The number of degrees of freedom is $n_1 + n_2 - 2 = 5 + 5 - 2$ or 8. From Table VIII, the $t_{0.05}$ value with 8 degrees of freedom is 1.86.

The test statistic value of $t = 1.98$ falls in the critical rejection region. Hence, we reject the null hypothesis and conclude that based upon the data, the newly developed fast-drying paint does indeed dry faster than the current best-selling paint.

EXERCISES

1. A federal judge is analyzing the prison records of five convicted armed bank robbers (first offense, no prior police record) from one state and the prison records of five similar convicts from an adjoining state. The jail term for each convict is as follows:

**Length of Jail
Sentence**

State 1	State 2
5	10
4	7
6	6
8	8
3	4

Using a 5% level of significance, is the difference between the average length of jail sentences in both states significant?

2. An airline company is about to order some inflatable rubber life rafts for emergency use on its planes. Six samples of life rafts produced by Company A needed an average of 8 seconds to be fully inflated. The standard deviation was 2.1 seconds. Six samples of life rafts produced by Company B needed an average of 6 seconds to be fully inflated. The standard deviation was 2.21 seconds. Using a 5% level of significance, is the difference between the mean time required to fully inflate the life rafts produced by the different companies significant?

3. Ten patients at Brooks Hospital required an average hospital stay of 6 days for a particular surgical procedure. The standard deviation was 1.2 days. Seven patients at Hyland Hospital required an average of 7.5 days for the same surgical procedure. The standard deviation was 1.8 days. Using a 5% level of significance, is there any significant difference between the average length of stay at both hospitals for the surgical procedure?

4. A real-estate broker has determined that the average selling price for a two-family house (based on the prices of the last 15 sales) in Brighton was $121,000. The standard deviation was $7500. In neighboring Bathgate, the average selling price for a two-family house (based on the prices of the last 11 sales) was $135,000. The standard deviation was $12,400. Is there any significant difference in the average selling price for a two-family house located in Brighton and one located in Bathgate? (Use a 1% level of significance.)

5. A newspaper reporter surveyed 14 families and found that they had deducted an average of $123 for charitable contributions on their previous year's 1040 Federal Tax return. The standard deviation was $17. A second reporter surveyed 12 families and found that they had deducted an average of $152 for charitable contributions. The standard deviation was $22. Is there any significant difference between the average amount deducted by a family for charitable contributions as presented by both newspaper reporters? (Use a 1% level of significance.)

6. A consumer group accurately weighed 13 five-pound bags of name-brand sugar and found that they had an average weight of 4.967 pounds with a standard deviation of 0.23 pounds. The consumer's group then accurately weighed 10 randomly selected bags of the store brand of sugar and found that these had an average weight of 5.013 pounds with a standard deviation of 0.12 pounds. Is it true that the five-pound bags of store-brand sugar have an average weight that is significantly greater than the average weight of the other name-brand bags of sugar? (Use a 5% level of significance.)

7. In 1986, 14 randomly selected neurologists in a large Northeastern city paid an average annual premium of $37,000 for malpractice insurance. The standard deviation was $2375. In the same year, 12 randomly selected obstetricians in the city paid an average annual premium of $34,000 for malpractice insurance. The standard deviation was $2580. Is it true that the average annual premium for malpractice insurance paid by obstetricians in this city is significantly less than the average annual premium paid by neurologists? (Use a 1% level of significance.)

8. Ten randomly selected male computer science majors from one college received job offers with an average starting annual salary of $22,050. The standard deviation was $1220. Twelve randomly selected female computer science majors of the same college were offered jobs with an average starting annual salary of $20,880. The standard deviation was $1410. Is it true that the average annual starting salary for a female computer science major is significantly less than the average annual starting salary for a male computer science major of this college? (Use a 5% level of significance.)

9. In an effort to protect the toll collectors at the city tunnels, health officials frequently monitor the level of the pollutants in the air. On nine randomly selected days, the air at one of these tunnels contained an average of 32 parts per million (ppm) of a certain pollutant. The standard deviation was 2.9. At a different tunnel, on 11 randomly selected days, the air at the tunnel contained an average of 41 ppm of the same pollutant. The standard deviation was 3.7. Is there any significant difference between the average amount of the pollutant in the air at these two tunnels? (Use a 1% level of significance.)

10. Seventeen faculty members of the math department at Louis College have been teaching for an average of 14 years with a standard deviation of 2.1 years. Thirteen faculty members of the computer science department have been teaching for an average of 9 years with a standard deviation of 3.2 years. Is it true that the average number of years that the computer science faculty have been teaching is significantly less than the average number of years that the math faculty have been teaching? (Use a 5% level of significance.)

10.8 TESTS CONCERNING PROPORTIONS

Suppose a congressman claims that 60% of the voters in his district are in favor of lowering the drinking age to 16 years. If a random sample of 400 voters showed that 221 of them favored the proposal, can we reject the congressman's claim? Questions of this type occur quite often and are usually answered on the basis of observed proportions. We assume that we can use the binomial distribution and that the probability of success is the same from trial to trial. We can therefore apply Formulas 6.5 and 6.6 (see page 295), which give us the mean and standard deviation of a binomial distribution. Thus,

$$\text{Mean:} \quad \mu = np$$

$$\text{Standard deviation:} \quad \sigma = \sqrt{np(1 - p)}$$

The null hypothesis in such tests assumes that the observed proportion, p, is the same as the population proportion, π. Depending upon the situation, we have the following alternative hypotheses:

Null hypothesis: $H_0: p = \pi$

Alternative
    hypothesis: $H_1: p \neq \pi$ [which means a two-tailed test]

                  $H_2: p > \pi$ [which means a one-tailed (right-side) test]

                  $H_3: p < \pi$ [which means a one-tailed (left-side) test].

The test statistic is z, where

$$z = \frac{p - \pi}{\sqrt{\dfrac{\pi(1 - \pi)}{n}}}$$

We are assuming that the sample size is large. We reject the null hypothesis if the test statistic falls in the critical, or rejection, region.

The following examples illustrate how we test proportions:

● Example 1

The Dean of Students at a college claims that only 12% of the students commute to school by bike. To test this claim, a students' group takes a sample of 80 students. They find that 14 of these students commute by bike. Is the Dean's claim acceptable? (Use a 5% level of significance.)

Solution

In this case the population parameter being tested is π, the true proportion of students who commute by bike. The questioned value is 0.12. Since we are testing whether the true proportion is 0.12 or not, we have

H_0: $\pi = 0.12$

H_1: $\pi \neq 0.12$

We are told that 14 of the 80 sampled students commute to school by bike, so that $p = \dfrac{14}{80} = 0.175$. Thus

$$z = \frac{p - \pi}{\sqrt{\dfrac{\pi(1 - \pi)}{n}}} = \frac{0.175 - 0.12}{\sqrt{\dfrac{(0.12)(1 - 0.12)}{80}}}$$

$$= \frac{0.055}{\sqrt{0.00132}} = \frac{0.055}{0.036}$$

$$= 1.53$$

We use the two-tail rejection region of Figure 10.11 (see page 429). The value of $z = 1.53$ falls in the acceptance region. Thus we cannot reject the null hypothesis and the Dean's claim that the true proportion of students who commute to school by bike is 12%.

• Example 2

In a recent press conference a senator claimed that 55% of the American people supported the President's foreign policy. To test this claim, a newspaper editor selected a random sample of 1000 people and 490 of them said that they supported the President. Is the senator's claim justified? (Use a 1% level of significance.)

Solution

In this case the population parameter being tested is π, the true proportion of Americans who support the President. The questioned value is 0.55. Since we are testing whether the true proportion is 0.55 or not, we have

H_0: $\pi = 0.55$

H_1: $\pi \neq 0.55$

We are told that 490 of the 1000 people interviewed supported the President, so that

$$p = \frac{490}{1000} = 0.49$$

Thus

$$z = \frac{p - \pi}{\sqrt{\dfrac{\pi(1 - \pi)}{n}}} = \frac{0.49 - 0.55}{\sqrt{\dfrac{(0.55)(1 - 0.55)}{1000}}}$$

$$= \frac{-0.06}{\sqrt{0.0002475}} = \frac{-0.06}{0.016}$$

$$= -3.75$$

Since the level of significance is 1%, we use the two-tailed rejection of Figure 10.10 (see page 428). The value of $z = -3.75$ falls in the rejection region. Thus, we reject the null hypothesis and the senator's claim that 55% of the American people support the President's foreign policy.

- **Example 3**

A latest government survey indicates that 22% of the people in Camelot are illegally receiving some form of public assistance. The Mayor of Camelot believes that the figures are exaggerated. To test this claim, she carefully examines 75 cases and finds that 11 of these people are illegally receiving aid. Does this sample support the government's claim? (Use a 5% level of significance.)

Solution

In this case the population parameter being tested is π, the true proportion of people illegally receiving financial aid. The questioned value is 0.22. Since we are testing whether the true proportion is 0.22 or lower, we have

H_0: $\pi = 0.22$

H_1: $\pi < 0.22$

We are told that 11 of the 75 cases examined are illegally receiving aid, so that

$$p = \frac{11}{75} = 0.15$$

Thus

$$z = \frac{p - \pi}{\sqrt{\dfrac{\pi(1 - \pi)}{n}}} = \frac{0.15 - 0.22}{\sqrt{\dfrac{(0.22)(1 - 0.22)}{75}}}$$

$$= -1.46$$

We use the one-tail rejection region of Figure 10.12 (see page 430). The value of $z = -1.46$ falls within the acceptance region. Thus, we cannot reject the null hypothesis that 22% of the people in Camelot are illegally receiving financial aid.

● **Example 4**

Government officials claim that approximately 29% of the residents of a state are opposed to building a nuclear plant to generate electricity. Local conservation groups claim that the true percentage is much higher. To test the government's claim an independent testing group selects a random sample of 81 state residents and finds that 38 of the people are opposed to the nuclear plant. Can we conclude that the government's claim is inaccurate? (Use a 5% level of significance.)

Solution

In this case the population parameter being tested is π, the true proportion of state residents who are opposed to building the nuclear plant. The questioned value is 0.29. Since we are testing whether the true proportion is 0.29 or higher, we have

$$H_0: \quad \pi = 0.29$$

$$H_1: \quad \pi > 0.29$$

We are told that 38 of the 81 residents opposed the nuclear plant so that

$$p = \frac{38}{81} = 0.47$$

Thus

$$z = \frac{p - \pi}{\sqrt{\dfrac{\pi(1 - \pi)}{n}}} = \frac{0.47 - 0.29}{\sqrt{\dfrac{(0.29)(1 - 0.29)}{81}}}$$

$$= \frac{0.18}{\sqrt{0.002542}} = \frac{0.18}{0.0504} = 3.57$$

We use the one-tail rejection region of Figure 10.13 (see page 432). The value of $z = 3.57$ falls in the rejection region. Thus, we reject the null hypothesis that the true proportion of state residents opposed to the nuclear plant is 0.29.

EXERCISES

1. A representative from the Census Bureau claims that 23% of all American households are one-parent homes. A sociologist believes that the true percentage is much higher. In a random sample of 80 households in one city, it is found that 24 of them are one-parent homes. Should we reject the claim that 23% of all American households are one-parent homes? (Use a 5% level of significance.)

2. Consider the newspaper article on the next page. A random sample of 750 people conducted by Jones and Blakey found that 570 of them support rigorous action by the

U.S. government against terrorists. Should we reject the newspaper claim that 80% of the people support such action? (Use a 1% level of significance.)

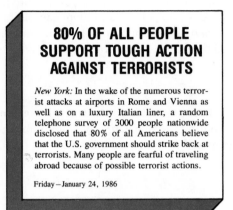

80% OF ALL PEOPLE SUPPORT TOUGH ACTION AGAINST TERRORISTS

New York: In the wake of the numerous terrorist attacks at airports in Rome and Vienna as well as on a luxury Italian liner, a random telephone survey of 3000 people nationwide disclosed that 80% of all Americans believe that the U.S. government should strike back at terrorists. Many people are fearful of traveling abroad because of possible terrorist actions.

Friday — January 24, 1986

3. *Affirmative Action*   A construction company official claims that 55% of its workers are from minority groups. In a random sample of 85 of the company's workers, it is found that 45 of the workers are from minority groups. Should we reject the claim that 55% of the company's workers are from minority groups? (Use a 5% level of significance.)

4. Industry representatives claim that 18% of all American households had a personal computer in 1985. To test this claim, an independent group randomly selects 123 households. It is found that 17 of these households have a personal computer. Should we reject the industry representative's claim? (Use a 5% level of significance.)

5. An official from the state Motor Vehicle Bureau claims that 30% of all cars on the state's highways cannot pass the new exhaust system pollution-control test. To verify this claim, an independent group randomly selects 225 cars and checks their exhaust system pollution controls. It is found that 75 of the cars cannot pass the test. Can we reject the claim of the state Motor Vehicle Bureau official? (Use a 1% level of significance.)

6. A soft-drink vending machine has been adjusted so that 90% of all cups will be filled with at least 7 ounces of soda. A random sample of 80 cups found that 65 of them were filled with at least 7 ounces of soda. Can we reject the claim that the soft-drink vending machine is adjusted properly so that 90% of all cups will be filled with at least 7 ounces of soda? (Use a 5% level of significance.)

7. A hotel manager on a Caribbean island claims that approximately 8% of all people who book reservations cancel these reservations for one reason or another. During the first three months of 1986, there were 4623 reservations and 350 cancellations of these reservations. Should we reject the hotel manager's claim? (Use a 5% level of significance.)

8. Dr. Rogers claims to have developed a new ointment which is 90% effective in treating one form of the herpes infection. To test this claim, a medical panel tests the new ointment on 65 people who are afflicted with this type of herpes infection. Forty-six of these people are cured. Should we reject the doctor's claim? (Use a 5% level of significance.)

9. An airline official claims that 84% of all passengers boarding planes have at least one piece of carry-on luggage. In one week, 4638 passengers were observed boarding planes with 3961 of these passengers carrying at least one piece of carry-on luggage. Should we reject the airline official's claim? (Use a 5% level of significance.)

10. The City Transportation Company alleges that 96% of its commuter trains arrive on time. In a random sample of 628 arriving trains, it was found that 587 of the trains arrived on time. Is the transportation company's claim justified? (Use a 5% level of significance.)

11. A Northeast utility company claims that 60% of all homes in the city are heated by gas. An oil company official disputes this claim and contends that the percentage is considerably lower. A random sample of 312 homes found that 165 of them are heated by gas. Should we reject the claim that 60% of all homes in the city are heated by gas? (Use a 1% level of significance.)

10.9 USING COMPUTER PACKAGES

We can use MINITAB to determine if the difference between sample means is significant. We illustrate the procedure by using the data given in Example 1 in Section 10.7 (page 448). We have:

```
MTB  > SET THE FOLLOWING DATA INTO C1
DATA > 48     46     44     46     43

MTB  > SET THE FOLLOWING DATA INTO C2
DATA > 42     43     45     43     44
```

We then have MINITAB determine if there is a significant difference in the means. This is accomplished as follows:

```
MTB > POOLED T FOR DATA IN C1 AND C2

TWO SAMPLE T FOR C1 VS C2
           N      MEAN      STDEV      SE MEAN
C1         5      45.40     1.95       0.87
C2         5      43.40     1.14       0.51
95 PCT CI FOR MU C1−MU C2:   (−0.33, 4.33)
T TEST MU C1=MU C2 (VS NE):   T=1.98   P=0.083   DF=8.0

MTB > STOP
```

10.10 SUMMARY

In this chapter we discussed hypothesis testing, which is a very important branch of statistical inference. Hypothesis testing is the process by which a decision is made to either reject or accept a null hypothesis about one of the parameters of the distribution. The decision to accept or reject a null hypothesis is based upon information obtained from sample data. Since any decision is subject to error, we discussed Type-I and Type-II errors.

We reject a null hypothesis when the test statistic falls in the critical, or rejection, region. The critical region is determined by two things:

1. whether we wish to perform a one-tail or two-tail test
2. the level of significance.

If a null hypothesis is not rejected, we cannot say that the sample data prove that what the null hypothesis says is necessarily true. It merely does not reject it. Some statisticians prefer to say that in this situation they *reserve judgement* rather than accept the null hypothesis.

STUDY GUIDE

You should now be able to demonstrate your knowledge of the following ideas presented in this chapter by giving definitions, descriptions, or specific examples. Page references are given for each term so that you can check your answer.

Hypothesis testing (page 419)
Hypothesis (page 419)
Critical value (page 421)
Null hypothesis (page 422)
Alternative hypothesis (page 422)
One-sided, or one-tailed, test (page 422)
Two-sided, or two-tailed, test (page 422)
Acceptance region (page 423)
Critical region (page 423)
Rejection region (page 423)
Test statistic (page 423)
Statistical tests of hypotheses (page 423)
Statistical tests of significance (page 423)
Type-I error (page 423)
Type-II error (page 423)
Level of significance (page 424)
5% significance level (page 424)
1% significance level (page 424)
Alpha, α (page 425)
Statistically significant (page 427)

FORMULAS TO REMEMBER

We discussed methods for testing means, proportions, and differences between means. All the tests studied are summarized in Table 10.1. You should be able to identify each symbol in the formulas, understand the relationships among the symbols expressed in each formula, understand the significance of each formula, and use the formulas in solving problems.

TABLE 10.1
Various Tests for Accepting or Rejecting a Null Hypothesis

Population Parameter Tested	Sample Size	Type of Test	Significance	Test Statistic	Reject Null Hypothesis if
Mean	Large	Two-tailed	0.05	$z = \dfrac{\bar{x} - \mu}{\sigma/\sqrt{n}}$	$z < -1.96$ or $z > 1.96$
Mean	Large	Two-tailed	0.01		$z < -2.58$ or $z > 2.58$
Mean	Large	One-tailed	0.05		$z < -1.64$ or $z > 1.64$
Mean	Large	One-tailed	0.01		$z < -2.33$ or $z > 2.33$
Mean	Small	Two-tailed	0.05	$t = \dfrac{\bar{x} - \mu}{s/\sqrt{n}}$	$t < -t_{0.025}$ or $t > t_{0.025}$
Mean	Small	One-tailed	0.05		$t < -t_{0.05}$ or $t > t_{0.05}$
Difference between sample means	Large	Two-tailed	0.05 or 0.01	$z = \dfrac{\bar{x}_1 - \bar{x}_2}{\sqrt{\dfrac{s_1^2}{n_1} + \dfrac{s_2^2}{n_2}}}$	Same as above
Difference between sample means	Large	One-tailed	0.05 or 0.01		
Difference between sample means	Small	Two-tailed	0.05 or 0.01	$t = \dfrac{\bar{x}_1 - \bar{x}_2}{s_p\sqrt{\dfrac{1}{n_1} + \dfrac{1}{n_2}}}$ where $s_p = \sqrt{\dfrac{(n_1 - 1)s_1^2 + (n_2 - 1)s_2^2}{n_1 + n_2 - 2}}$ $df = n_1 + n_2 - 2$	Same as above
Difference between sample means	Small	One-tailed	0.05 or 0.01		
Proportion	Large	Two-tailed	0.05 or 0.01	$z = \dfrac{p - \pi}{\sqrt{\dfrac{\pi(1 - \pi)}{n}}}$	Same as above
Proportion	Large	One-tailed	0.05 or 0.01		

The tests of the following section will be more useful if you take them after you have studied the examples and solved the exercises in this chapter.

MASTERY TESTS

Form A

1. Consider the newspaper article below. The union claims that the average number of sick days taken by an officer is 10.7 days per year. A survey of 48 randomly selected highway patrol officers found that these officers were absent an average of 12.2 days per year. The standard deviation was 5.6 days. Using a 5% level of significance, should we reject the union's claim?

STATE TO INVESTIGATE ABUSE OF SICK LEAVE PRIVILEGE

New Haven (August 2) – The governor's office announced yesterday the establishment of a commission to investigate abuses of the sick leave policy privilege afforded to highway patrol officers. Under this privilege, an officer is allowed an unlimited number of sick days per year.

The officials claim that this privilege is being abused by some officers.

Sunday, August 2, 1981

2. A sociologist believes that women from suburban areas marry at a later age than women from urban areas. Do the following results support this claim? (Use a 1% level of significance.)

	Urban Women	Suburban Women
Average Age at Which Women First Married	24.2	26.1
Standard Deviation	3.1	4.5
Number in Survey	47	52

3. Mr. Ramirez, the company payroll supervisor, claims that the 62 workers on the eighth floor take a longer coffee break on the average than do the 53 workers on the fifth floor. To support his claim, he has gathered the following information:

	Average Time Spent on Coffee Break (minutes)	Standard Deviation
Workers on the 8th Floor	25	7
Workers on the 5th Floor	18	4

Should we reject Mr. Ramirez's claim? (Use a 5% level of significance.)

4. The average hourly salary of a gas station attendant in a certain city is $4.53 with a standard deviation of 58 cents. One large chain of gas stations that employs 88 attendants pays its attendants an average hourly rate of $3.98. Can this gas station chain be accused of paying lower than the average hourly rate? (Use a 5% level of significance.)

5. A health insurance company states that the average medical claim that it paid last year was $87 with a standard deviation of $18. The union believes that the average medical claim was much less. In a random sample of 100 claims paid by the insurance company, the union finds that the average claim was $71. Should we reject the insurance company's assertion? (Use a 1% level of significance.)

6. Fifty claims representatives in the east wing of an auto insurance company processed an average of 53 claims per week. The standard deviation was 5.9. Forty claims representatives in the west wing of the auto insurance company processed an average of 62 claims per week. The standard deviation was 8.8. Is the difference between the average number of claims processed by the workers in both wings significant? (Use a 5% level of significance.)

7. Court officials in a certain city claim that any criminal case is disposed of in an average of 61 days with a standard deviation of 5 days. The local civil liberties organization claims that the average is considerably higher. An average of 65 days was required to dispose of 56 randomly selected criminal cases. Should we reject the court official's claim? (Use a 5% level of significance.)

8. An oil company executive claims in an advertisement that 88% of all people who heat their home with oil also purchase a service contract to provide service in the event of a mechanical breakdown. The Attorney General of the state believes that this percentage is inflated. To test this claim, 44 homeowners who heat their home with oil are randomly selected. It is found that 30 of these homeowners also purchase a service contract. Using a 5% level of significance should we reject the oil company executive's claim?

9. Government officials claim that 24% of all married women in a certain city work so as to provide the family with extra income even though they have young children at home. The local congressional representative disagrees and believes that the percentage is much higher. A random survey of 42 working married women found that 18 of them had young children at home. Using a 1% level of significance should we reject the government officials' claims?

10. One group of 10 cardiologists paid an average annual premium of $38,712 for malpractice insurance. The standard deviation was $5126. A second group of 12 cardiologists paid an average annual premium of $42,056 for malpractice insurance. The standard deviation was $6850. Is the difference between the average annual premium paid by both groups for malpractice insurance significant? (Use a 5% level of significance.)

MASTERY TESTS

Form B

1. The average salary of the 13 computer programmers of the Gorp Corporation is $26,041 with a standard deviation of $1469. The average salary of the 17 computer programmers of the Dowd Corporation is $31,092 with a standard deviation of $2468. Is the difference between the average annual salaries of the computer programmers at both companies significant? (Use a 5% level of significance.)

2. Company officials at Marck Corporation claim that the average age of its employees is 41.3 years. Union officials believe that this figure is inaccurate. A survey of 54 workers finds that the average age of these workers is 40.1 years with a standard deviation of 3.9 years. Using a 5% level of significance, can we accept the company official's claim?

3. Seven bags of one brand of potatoes weighed 9.96, 9.83, 10.01, 9.91, 9.97, 9.76, and 9.93 pounds. The bags were labelled 10 pounds net weight. Nine bags of a second brand of potatoes weighed 10.02, 9.86, 10.14, 10.01, 9.97, 9.89, 9.90, 9.99, and 10.04 pounds. Using a 5% level of significance, is the difference between the average weight of both brands of potatoes significant?

4. Industry representatives claim that the average retail price of one home computer is $1840. Several consumer groups claim that the average price is much higher. Seven stores are randomly selected. The average price of the computer at these stores is $1920 with a standard deviation of $183. Should we reject the industry representative's claim? (Use a 5% level of significance.)

5. Maureen is stranded on a highway awaiting the arrival of an AAA tow truck to repair her car. The local AAA club says that it responds to emergency calls in an average of 25 minutes with a standard deviation of 6.2 minutes. After speaking to 35 friends who were in similar situations, Maureen finds that the AAA responded to calls in an average of 29 minutes. Does this indicate that the average response time is more than 25 minutes? (Use a 1% level of significance.)

6. A medical researcher claims to have developed a new drug that is 92% effective for the treatment of common cold symptoms. A skeptical consumer's group randomly selects 75 volunteers and tests the new drug on them. The drug is found to be effective on 60 of these volunteers. Should we reject the medical researcher's claim? (Use a 5% level of significance.)

7. A politician claims that 85% of her constituents believe that the Social Security system will be bankrupt before the twenty-first century. A newspaper reporter claims that the percentage is higher. A survey of 56 people disclosed that 49 of them believe that the Social Security system will be bankrupt before the twenty-first century. Should we reject the politician's claim? (Use a 1% level of significance.)

8. Red Cross officials claim that potential blood donors have not been giving blood because of their fear of contracting AIDS. Furthermore, they claim that 65% of the potential blood donors expressed such a fear. A concerned citizens group believes that this figure is too high. A random sample of 58 potential blood donors found that 32 of them expressed such a fear. Do these results disagree with the Red Cross official's claim? (Use a 5% level of significance.)

9. A research scientist is interested in determining whether the addition of a special chemical compound to the soil of a growing plant affects its growth. Two groups of seven plants each are tested. The special chemical compound is added to the soil of the first group. Nothing special is added to the soil of the other group. All of the plants are then placed in a controlled environment. The heights (in centimeters) of the plants after several months are as follows:

Height of Plants Treated with Special Chemical	Height of Plants Not Treated with Special Chemical
12	11
14	13
19	15
17	16
16	14
15	17
18	16

Using a 5% level of significance, does the addition of the special chemical compound have a significant effect on the plant's growth?

10. In the Northeast, the average weekly sales of a certain type of stereo VCR in 35 cities was 246 units with a standard deviation of 14.7 units. In the

Southwest, the average weekly sales of the same stereo VCR in 35 cities was 239 units with a standard deviation of 17.8 units. Using a 5% level of significance, is there a significant difference between the average number of stereo VCRs sold in both regions?

11. *Cheating the Poor*   A consumer's group claims that many gas stations charge higher prices for a gallon of gas in poorer neighborhoods than in middle-class neighborhoods. To investigate this claim, one gallon of unleaded premium gas is purchased from each of 38 gas stations located in poorer neighborhoods and one gallon of unleaded premium gas is purchased from each of 48 gas stations located in middle-class neighborhoods. The following results were obtained:

	Poorer Neighborhoods	Middle-class Neighborhoods
Average Price	1.21	1.15
Sample Standard Deviation	0.05	0.07
Sample Size	38	48

Is there any significant difference between the average price of a gallon of unleaded gas in these neighborhoods? (Use a 5% level of significance.)

SUGGESTED FURTHER READING

1. Adler, H., and E. B. Roessler. *Introduction to Probability and Statistics*, 5th ed. San Francisco, CA: Freeman & Co., 1972.
2. Freund, J. *Statistics, A First Course*. 2nd ed. Englewood Cliffs, NJ: Prentice-Hall, 1976.
3. Johnson, N. L., and F. C. Leone. *Statistics and Experimental Design in Engineering and the Physical Sciences*. New York: Wiley, 1964.
4. Lehmann, E. L. *Testing Statistical Hypotheses*. New York: Wiley, 1959.
5. Mendenhall, William. *Introduction to Probability and Statistics*, 6th ed. North Scituate, MA: Duxbury Press, 1983.
6. Romano, A. *Applied Statistics for Science and Industry*. Boston: Allyn and Bacon, 1977.

CHAPTER 11

Linear Correlation and Regression

OBJECTIVES

- *To briefly discuss* some of the pioneers in the field of correlation and regression.

- *To learn* how to decide whether two variables are related. We draw two lines, one horizontal and one vertical, and then place dots in various places corresponding to the given data.

- *To study* linear correlation, which tells us whether there is a relationship that will cause an increase (or decrease) in the value of one variable when the other is increased (or decreased).

- *To measure* the strength of the linear relationship that exists between two variables. This is the coefficient of linear correlation.

- *To discuss* the reliability of *r*.

- *To calculate* a regression line that allows us to predict the value of one of the variables if the value of the other variable is known.

- *To analyze* the method of least squares that we use to determine the estimated regression line to be used in prediction.

- *To indicate* how the standard error of the estimate tells us how well the least squares prediction equation describes the relationship between two variables.

- *To set up* confidence intervals for our regression estimates.

- *To introduce* the concept of multiple regression where one variable depends upon many other variables.

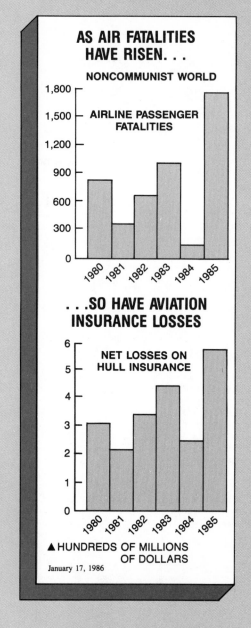

AS AIR FATALITIES HAVE RISEN. . .

NONCOMMUNIST WORLD

AIRLINE PASSENGER FATALITIES

. . .SO HAVE AVIATION INSURANCE LOSSES

NET LOSSES ON HULL INSURANCE

▲ HUNDREDS OF MILLIONS OF DOLLARS

January 17, 1986

MORE ROAD DEATHS LAID TO DRINKING

WETHERSFIELD, Conn., Sept. 10 (AP)—
Despite a police crackdown, the percentage of
Connecticut highway deaths attributed to
drunken driving has risen every year since
1982, according to the State Department of
Motor Vehicles.

A department report released this week
showed that 41.3 percent of the 435 fatal accidents that were reported last year were attributed to alcohol. The 1983 figure was 38.4
percent, and in 1982 it was 35.

The report showed that more than half the
drivers who drank and caused highway deaths
last year were between 16 and 26 years old. A
law raising the legal drinking age to 21 went
into effect Sept. 1, and beginning Oct. 1, a
driver with a blood-alcohol level of 0.1 percent
or higher will be considered legally drunk.

September 10, 1985

Consider the above newspaper clippings.
Both articles seem to imply that the variables are related. How do we determine when
variables are related? If we know the value
of one variable, can we predict the value of
the other variable? Is it true that a driver
with a high blood-alcohol level will have a
higher probability of a highway accident than
a driver with a low blood-alcohol level?
Source: U.S. Aviation Insurance Group.

11.1 INTRODUCTION

Up to this point we have been discussing the many statistical procedures for analyzing a single variable. However, when dealing with the problems of applied statistics in education, psychology, sociology, etc., we may be interested in determining whether a relationship exists between two or more variables. For example, if college officials have just administered a vocational aptitude test to 1000 entering freshmen, they may be interested in knowing whether there is any relationship between the math aptitude scores and the business aptitude scores. Do students who score well on the math section of the aptitude exam also do well on the business part? On the other hand, is it true that a student who scored poorly on the math part will necessarily score poorly on the business part? Similarly, the college officials may be interested in determining if there is a relationship between high school averages and college performance.

Questions of this nature frequently arise when we have many variables and are interested in determining relationships between these sets of scores. In this chapter we will learn how to compute a number that measures the relationship between two sets of scores. This number is called the **correlation coefficient.** The English mathematician Karl Pearson (1857–1936) studied it in great detail and to some extent so did another English mathematician, Sir Francis Galton (1822–1911).

Sir Francis Galton, a cousin of Charles Darwin, undertook a detailed study of human characteristics. He was interested in determining whether a relationship exists between the heights of fathers and the heights of their sons. Do tall parents have tall children? Do intelligent parents or successful parents have intelligent or successful children? In *Natural Inheritance* Galton introduced the idea referred to today as correlation. This mathematical idea allows us to measure the closeness of the relationship between two variables. Galton found that there exists a very close relationship between the heights of fathers and the heights of their sons. On the question of whether intelligent parents have intelligent children, it has been found that the correlation is 0.55. As we shall

---

LIKE PARENTS, LIKE CHILDREN

Children whose parents are hooked on drugs and alcohol are more likely to fall victim to these substances than other kids. *U.S. News & World Report* says: "Studies show 65 percent of those youths dependent on drugs or alcohol are from homes where at least one parent is also hooked."

Friday, January 5, 1986

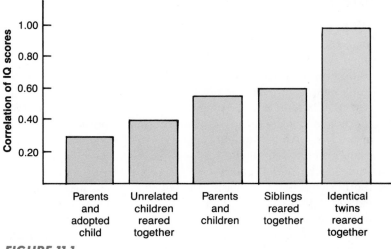

FIGURE 11.1

Correlations between intelligence of parents and that of their children.

see, this means that it is not necessarily true that intelligent parents have intelligent children. In many cases children will score higher or lower than their parents. Figure 11.1 shows how the correlations for IQ range from 0.28 between parents and adopted children to 0.97 for identical twins reared together.

The precise mathematical measure of correlation as we use it today was actually formulated by Karl Pearson.

If there is a high correlation between two variables, we may be interested in representing this correspondence by some form of an equation. So we will discuss the **method of least squares.** The statistical method of least squares was developed by Adrien-Marie Legendre (1752–1833). Although Legendre is best known for his work in geometry, he also did important work in statistics. He developed the method of least squares. This method is used when we want to find the regression equation.

Finally, we will discuss how this equation can be used to make predictions and we will discuss the reliability of these predictions.

11.2 SCATTER DIAGRAMS

To help us understand what is meant by a correlation coefficient, let us consider a guidance counselor who has just received the scores of an aptitude test administered to ten students. See Table 11.1.

	Math Aptitude	Business Aptitude	Language Aptitude	Music Aptitude
TABLE 11.1 *Different Aptitude Scores Received By Ten Students*				
Student	Math Aptitude	Business Aptitude	Language Aptitude	Music Aptitude
A	52	48	26	22
B	49	49	53	23
C	26	27	48	57
D	28	24	31	54
E	63	59	67	13
F	44	40	75	20
G	70	72	31	9
H	32	31	22	50
I	49	50	11	17
J	51	49	19	24

The counselor may be interested in determining if there is any correlation among these sets of scores. For example, the counselor may wish to know whether a student who scores well on the math aptitude part of the exam will also score well on the business aptitude part. She can analyze the situation pictorially by means of a **scatter diagram.**

To make a scatter diagram, we draw two lines, one vertical and one horizontal. On the horizontal line we indicate the math scores, and on the vertical line we indicate the business scores. Although we could put the math scores on the vertical line, we have purposely labeled the math scores on the horizontal line. This is done because we are interested in predicting the scores on the business aptitude part on the basis of the math scores. After both axes, that is, both lines, are labeled, we use one dot to represent each person's score. The dot is placed directly above the person's math score and directly to the right of the business score. Thus the dot for Student A's score is placed directly above the 52 score on the math axis and to the right of 48 on the business axis. Similarly, the dot for Student B's score is placed directly above the 49 score on the math axis and to the right of 49 on the business axis. The same procedure is used to locate all the dots of Figure 11.2.

You will notice that these dots form an approximate straight line. When this happens we say that there is a **linear correlation** between the two variables. Notice also that the higher the math score, the higher the business score. The line moves in a direction that is from lower left to upper right. When this happens, we say that there is a **positive correlation** between the math scores and the business scores. This means that a student with a higher math score will tend to have a higher business score.

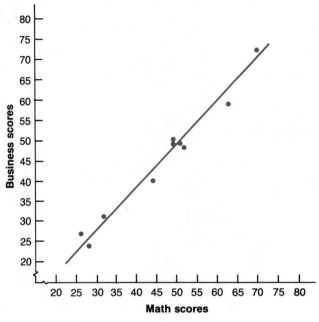

FIGURE 11.2
Scatter diagram for the math and business scores.

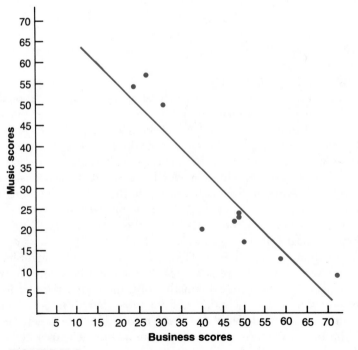

FIGURE 11.3
Scatter diagram for the business and music scores.

EMPLOYMENT STATUS OF PERSONS 16 TO 24 YEARS OF AGE AND NOT ENROLLED IN SCHOOL OCTOBER 1980

Educational Attainment	Labor Force Participation Rate	Unemployment Rate	Percent Employed
College graduates	95.2%	5.9%	89.7%
1 to 3 years of college	89.4	8.8	81.5
High school, no college	84.3	12.5	73.7
High school dropouts	67.5	25.3	50.4
TOTAL	**81.8%**	**14.0%**	**70.2%**

FIGURE 11.4

The "Labor Force Participation Rate" is the percent of the total civilian, noninstitutional population group with the indicated educational characteristic who were employed or seeking employment. The "Unemployment Rate" is the percent of those participating who were not employed. We have all heard the thought that if you want a good job, get a good education. This clipping provides documentation that there is a positive correlation between the education that one has (educational attainment) and the chances of holding or securing a job (labor force participation rate.) Furthermore, the clipping indicates that there is a negative correlation between educational attainment and unemployment. *Source: Current Population Survey* conducted by the U.S. Bureau of the Census.

Now let us draw the scatter diagram for the business aptitude scores and the music aptitude scores. It is shown in Figure 11.3. In this case you will notice that the higher the business score, the lower the music score. Again the dots arrange themselves in the form of a line, but this time the line moves in a direction that is from upper left to lower right. When this happens we say that there is a **negative correlation** between the business scores and the music scores. This means that a student with a high business score will tend to have a low music score.

Now let us draw the scatter diagram for the language scores and music scores. It is shown in Figure 11.5. In this case the dots do not form a straight line. When this happens we say that there is little or no correlation between the language scores and the music scores.

COMMENT Although we will concern ourselves with only linear, that is, a straight line, correlation, the dots may suggest different types of curves. These are studied in detail by statisticians. Several examples of such scatter diagrams are given in Figure 11.6. In this text we will analyze only linear correlation.

11.3 THE COEFFICIENT OF CORRELATION

Once we have determined that there is a linear correlation between two variables, we may be interested in determining the strength of the linear relationship. Karl Pearson developed a **coefficient of linear correlation,** which

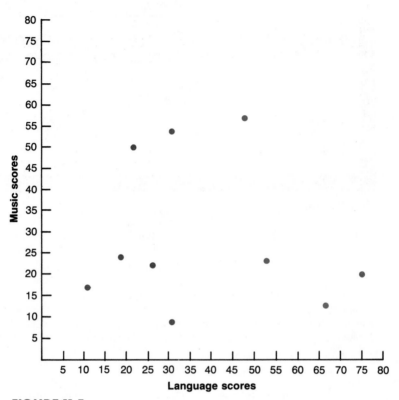

FIGURE 11.5
Scatter diagram for the language and music scores.

Language scores

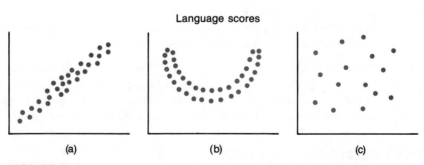

FIGURE 11.6
Scatter diagrams that suggest (a) a linear relationship, (b) a curvilinear relationship, and (c) no relationship.

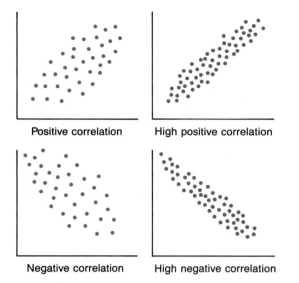

Positive correlation　　High positive correlation

Negative correlation　　High negative correlation

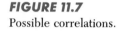

FIGURE 11.7
Possible correlations.

measures the strength of a relationship between two variables. The value of the coefficient of linear correlation is calculated by means of Formula 11.1.

FORMULA 11.1

The coefficient of linear correlation is given by

$$r = \frac{n(\Sigma xy) - (\Sigma x)(\Sigma y)}{\sqrt{n(\Sigma x^2) - (\Sigma x)^2}\sqrt{n(\Sigma y^2) - (\Sigma y)^2}}$$

where

　$x = $ *label for one of the variables*

　$y = $ *label for the other variable*

　$n = $ *number of pairs of scores*

When using Formula 11.1 the coefficient of correlation will always have a value between -1 and $+1$. A value of $+1$ means perfect positive correlation and corresponds to the situation where all the dots lie exactly on a straight line. A value of -1 means perfect negative correlation and again corresponds to the situation where all the points lie exactly on a straight line. Correlation is considered high when it is close to $+1$ or -1 and low when it is close to 0. If the coefficient of linear correlation is zero, we say that there is no linear correlation. These possibilities are indicated in Figures 11.7 and 11.8.

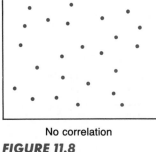

No correlation

FIGURE 11.8
No correlation.

Although Formula 11.1 looks complicated, it is rather easy to use. The only new symbol that appears is Σxy. This value is found by multiplying the corresponding values of x and y and then adding all the products. The following examples will illustrate the procedure.

• Example 1

Find the correlation coefficient for the data of Figure 11.2 (see page 470). It is repeated in the following chart:

Math Score	52	49	26	28	63	44	70	32	49	51
Business Score	48	49	27	24	59	40	72	31	50	49

Solution

We first let x represent the math score and y represent the business score. Then we arrange the data in tabular form as follows and apply Formula 11.1.

x (Math)	y (Business)	x^2	y^2	xy
52	48	2704	2304	2496
49	49	2401	2401	2401
26	27	676	729	702
28	24	784	576	672
63	59	3969	3481	3717
44	40	1936	1600	1760
70	72	4900	5184	5040
32	31	1024	961	992
49	50	2401	2500	2450
51	49	2601	2401	2499
464	449	23,396	22,137	22,729

$$\Sigma x = 464 \quad \Sigma y = 449 \quad \Sigma x^2 = 23,396 \quad \Sigma y^2 = 22,137 \quad \Sigma xy = 22,729$$

Then

$$r = \frac{n(\Sigma xy) - (\Sigma x)(\Sigma y)}{\sqrt{n(\Sigma x^2) - (\Sigma x)^2} \sqrt{n(\Sigma y^2) - (\Sigma y)^2}}$$

$$= \frac{10(22,729) - (464)(449)}{\sqrt{10(23,396) - (464)^2} \sqrt{10(22,137) - (449)^2}}$$

$$= \frac{18,954}{\sqrt{18,664} \sqrt{19,769}}$$

$$= \frac{18,954}{(136.62)(140.60)}$$

$$= \frac{18,954}{19,208.77} = 0.9867$$

Thus, the coefficient of correlation is 0.9867. Since this value is close to $+1$, we say that there is a high degree of positive correlation. Figure 11.2 also indicated the same result.

- **Example 2**

Find the coefficient of correlation for the data of Figure 11.5 (see page 472). It is repeated in the following chart:

Language Score	26	53	48	31	67	75	31	22	11	19
Music Score	22	23	57	54	13	20	9	50	17	24

Solution

We first let x represent the language score and y represent the music score. Then we arrange the data in tabular form as follows and apply Formula 11.1.

x (Language)	y (Music)	x^2	y^2	xy
26	22	676	484	572
53	23	2809	529	1219
48	57	2304	3249	2736
31	54	961	2916	1674
67	13	4489	169	871
75	20	5625	400	1500
31	9	961	81	279
22	50	484	2500	1100
11	17	121	289	187
19	24	361	576	456
383	289	18,791	11,193	10,594
$\Sigma x = 383$	$\Sigma y = 289$	$\Sigma x^2 = 18,791$	$y^2 = 11,193$	$\Sigma xy = 10,594$

Now we apply Formula 11.1. We have

$$r = \frac{n(\Sigma xy) - (\Sigma x)(\Sigma y)}{\sqrt{n(\Sigma x^2) - (\Sigma x)^2} \sqrt{n(\Sigma y^2) - (\Sigma y)^2}}$$

$$= \frac{10(10,594) - (383)(289)}{\sqrt{10(18,791) - (383)^2} \sqrt{10(11,193) - (289)^2}}$$

$$= \frac{-4747}{\sqrt{41,221} \sqrt{28,409}} = \frac{-4747}{(203.03)(168.55)}$$

$$= \frac{-4747}{34,220.71} = -0.1387$$

Thus, the coefficient of correlation is -0.1387. Since this value is close to zero, there is little correlation. Figure 11.4 indicated the same result.

How will the correlation between x and y be affected if x is coded by adding the same number to (or subtracting the same number from) each score? How is y affected?

Fortunately, it turns out that the correlation coefficient is unaffected by adding or subtracting a number to either x or y or both. Thus, x can be coded in one way—perhaps by adding or subtracting a number—and y can be coded in another way—say by multiplying by a number. In either case the value of the correlation coefficient is unaffected. Of greater importance is the fact that if we code before calculating the value of r, we do not have to uncode our results.

Let us code the data of Example 2 of this section and see how coding simplifies the computations involved.

● **Example 3**

By coding the data, find the coefficient of correlation for the data of Example 2 of this section.

Solution

We will code the data by subtracting 38 from each x value and 29 from each y value. Our new distribution of test scores then becomes

Language Score	-12	15	10	-7	29	37	-7	-16	-27	-19
Music Score	-7	-6	28	25	-16	-9	-20	21	-12	-5

Now we calculate r from the coded data. We have the following:

x (Language)	y (Music)	x^2	y^2	xy
−12	−7	144	49	84
15	−6	225	36	−90
10	28	100	784	280
−7	25	49	625	−175
29	−16	841	256	−464
37	−9	1369	81	−333
−7	−20	49	400	140
−16	21	256	441	−336
−27	−12	729	144	324
−19	−5	361	25	95
3	−1	4123	2841	−475
$\Sigma x = 3$	$\Sigma y = -1$	$\Sigma x^2 = 4123$	$\Sigma y^2 = 2841$	$\Sigma xy = -475$

Then

$$r = \frac{n(\Sigma xy) - (\Sigma x)(\Sigma y)}{\sqrt{n(\Sigma x^2) - (\Sigma x)^2} \sqrt{n(\Sigma y^2) - (\Sigma y)^2}}$$

$$= \frac{10(-475) - (3)(-1)}{\sqrt{10(4123) - (3)^2} \sqrt{10(2841) - (-1)^2}}$$

$$= \frac{-4750 + 3}{\sqrt{41,230 - 9} \sqrt{28,410 - 1}}$$

$$= \frac{-4747}{\sqrt{41,221} \sqrt{28,409}} = \frac{-4747}{(203.03)(168.55)} = -0.1387$$

Notice that the value of r obtained by coding and the value of r obtained by working with the original uncoded data is exactly the same. Coding simplifies computations if the values with which we code are chosen carefully.

EXERCISES

1. For each of the following, indicate whether you would expect a positive correlation, a negative correlation, or zero correlation.
   a. teacher's salaries and the incidence of crime
   b. the number of cigarettes smoked and the incidence of lung cancer in a city
   c. the amount of rainfall and the size of a crop
   d. the weight of an individual and the shoe size of that individual
   e. the amount of pollution in the air and the incidence of respiratory illness
   f. the number of female suicides in the United States over the past 20 years and the number of male suicides in the same period

g. overweight and the incidence of heart attacks

h. the weight of a baby at birth and the length of the baby at birth

2. *Industrial Accidents and Overtime* A large tool-manufacturing company wishes that its employees be forced to work overtime. The union claims that the more hours that an employee works, the greater the risk of an individual accident (due to fatigue). To support its claim, the union has gathered the following statistics on the average number of hours worked by an employee (per week) and the average number of accidents (per week).

Number of Hours Worked x	Number of Accidents y
35	1.6
37	3.1
39	5.8
41	7.1
42	7.3
43	7.6
45	8.1

a. Draw a scatter diagram for the data, and then compute the coefficient of correlation.

b. Does the union claim seem to be justified?

3. The following chart indicates the amount of rainfall in a certain city for the first six months of 1985 and the number of umbrellas (in thousands) sold during this period.

Month	Rainfall (inches) x	Number of Umbrellas Sold (in thousands) y
Jan.	5	23
Feb.	4.5	22
Mar.	4	19
Apr.	3	18
May	2	17
Jun.	2	16

a. Draw a scatter diagram for the data, and compute the coefficient of correlation.

b. Does an increase in rainfall mean more umbrella sales?

4. The following table gives some data on the number of highway miles obtained per gallon of gas and the engine size for several cars equipped with automatic transmission as determined by an independent testing agency.

Engine Displacement Size x	Miles per Gallon y
236	28
250	25
260	24
305	22
350	19
408	15

a. Draw a scatter diagram for the data, and then compute the coefficient of correlation.
b. Does engine size affect the number of highway miles per gallon obtained?

5. A country's gross national product (GNP) is often used as a measure of the country's standard of living. The following list presents some data on the energy consumption per capita (expressed in millions of BTUs per capita) and the GNP expressed in dollars per capita for several countries.

Country	Energy Consumption x	GNP y
Canada	131.0	$1900
India	3.4	55
Japan	30.3	550
United States	180.0	2900
USSR	69.0	800
West Germany	90.0	1410

Draw a scatter diagram for the data, and then compute the coefficient of correlation.

6. An agricultural experimenter divided a tract of land into seven plots of equal size. Each plot was treated with different levels of fertilizer to determine whether the level of fertilizer application affects yield. The results are shown below:

Plot	Level of Fertilizer Applied x	Yield (in pounds), y
A	3	53
B	3.5	57
C	4	56
D	4.5	58
E	5	62
F	5.5	59
G	6	63

Draw a scatter diagram for the data, and then compute the coefficient of correlation.

7. A local department store is interested in knowing whether a strong correlation exists between the amount of money it spends on television advertising and the total sales. It has gathered the following data:

Television Advertising Expenditure (thousands of dollars) x	Sales (thousands of dollars) y
0.75	70
0.85	83
0.92	83
1.03	84
1.55	86

Draw a scatter diagram for the data, and then compute the coefficient of correlation.

8. Professor Carmichael suspects that two of his students, Bill and Robin, have been cheating together on exams, as their grades are quite similar. They have received the following grades on the first six exams:

Bill x	Robin y
90	90
84	82
71	73
81	79
88	89
92	92

a. Draw a scatter diagram for the data, and then compute the coefficient of correlation.
b. Are Professor Carmichael's suspicions justified?

9. The number of speeding tickets issued and the number of radar traps set up by police officials on the Bayview Expressway during the past nine weekends are shown in the following table:

Number of Speed Traps x	Number of Speeding Tickets Issued y
10	76
12	84
15	93
8	61
12	79
11	70
7	55
13	84
14	85

Draw a scatter diagram for the data, and then compute the coefficient of correlation.

10. According to one supermarket manager, the number of quarts of milk returned daily by customers because the milk is sour is determined by the outside temperature as shown in the following table.

Outside Temperature x	Number of Quarts of Milk Returned y
70°	17
75°	24
80°	27
85°	31
90°	33
95°	39
100°	45

Draw a scatter diagram for the data, and then compute the coefficient of correlation.

11. According to a real-estate agent, the number of requests that she receives per day for information about a new suburban development is influenced by the number of times her commercial is broadcast on radio or television as shown in the following table:

Number of Times Commercial Is Broadcast x	Number of Requests for Information y
0	31
1	37
2	39
3	45
4	49
5	51
6	58
7	59

Draw a scatter diagram for the data, and then compute the coefficient of correlation.

11.4 THE RELIABILITY OF r

Although the coefficient of correlation is usually the first number that is calculated when we are given several sets of scores, great care must be used in how we interpret the results. It can undoubtedly be said that among all the statistical measures discussed in this book, the correlation coefficient is the one that is most misused. One reason for this misuse is the assumption that because the two variables are related, a change in one will result in a change in the other.

Many people have applied a positive correlation coefficient to prove a cause-and-effect relationship that may not even exist. To illustrate the point, it has been shown that there is a high positive correlation between teacher's salaries and the use of drugs on campus. Does this mean that reducing the teachers' salaries would reduce the use of drugs on campus or does it simply mean that the students at wealthier schools (which pay higher salaries) are more apt to use drugs?

Frequently, two variables may appear to have a high positive correlation even though they are not directly associated with each other. There may be a third variable that is highly correlated to these two variables.

There is another important consideration that is often overlooked. When r is computed on the basis of sample data, we may get a strong correlation, positive or negative, which is due purely to chance, not to some relationship that exists between x and y. For example, if x represents the amount of snowfall and y represents the number of hours that Joe studied on five consecutive days, we may have the following results:

Amount of Snow (in inches), x	1	4	2	6	3
Number of Hours Studied, y	2	6	3	4	4

The value of r in this case is 0.63. Can we conclude that if it snows more, then Joe studies more?

Fortunately, a chart has been constructed that allows us to interpret the value of the correlation coefficient correctly. This is Table VI in the Appendix. This chart allows us to determine the significance of a particular value of r. We use this table in the following way:

1. First compute the value of r using Formula 11.1.
2. Then look in the chart for the appropriate r value corresponding to some given n, where n is the number of pairs of scores.
3. The value of r is *not* statistically significant if it is between $-r_{0.025}$ and $r_{0.025}$ for a particular value of n.

The subscript, that is, the little numbers, attached to r is called the **level of significance.** If we use this chart and use the $r_{0.025}$ values of the chart as our guideline, we will be correct in saying that when there is no significant statistical correlation between x and y and we will not reject H_0 95% of the time.

Table VI also gives us the values of $r_{0.005}$. We use these chart values when we want to be correct 99% of the time. In this book we will use the $r_{0.025}$ values only. Table VI is constructed so that r can be expected to fall between $-r_{0.025}$ and $+r_{0.025}$ approximately 95% of the time and between $-r_{0.005}$ and $+r_{0.005}$ approximately 99% of the time when the true correlation between x and y is zero.

Returning to our example, we have $n = 5$ and $r = 0.63$. From Table VI we have $r_{0.025} = 0.878$. Thus, r will *not* be statistically significant if it is between -0.878 and $+0.878$. Since $r = 0.63$ is between -0.878 and $+0.878$, we conclude that the correlation may be due purely to chance. We cannot say that if it snows more, then Joe will necessarily study more.

Similarly, in Example 1 of Section 11.3 (see page 474) we found that $r = 0.9867$. There were ten scores, so $n = 10$. The chart values tell us that if r is between -0.632 and $+0.632$, there is *no* significant statistical correlation. Since the value of r that we obtained is greater than $+0.632$, we conclude that there is a *definite* positive correlation between x and y. Thus, we are justified in claiming that there is a relationship between the math aptitude scores and the business aptitude scores.

EXERCISES

1. Refer back to Exercise 2, page 478. Determine if r is significant.
2. Refer back to Exercise 3, page 478. Determine if r is significant.
3. Refer back to Exercise 4, page 478. Determine if r is significant.
4. Refer back to Exercise 5, page 479. Determine if r is significant.
5. Refer back to Exercise 6, page 479. Determine if r is significant.

6. Refer back to Exercise 7, page 480. Determine if r is significant.

7. Refer back to Exercise 8, page 480. Determine if r is significant.

8. Refer back to Exercise 9, page 480. Determine if r is significant.

9. Refer back to Exercise 10, page 481. Determine if r is significant.

10. Refer back to Exercise 11, page 481. Determine if r is significant.

11.5 LINEAR REGRESSION

Let us return to the example discussed in Section 11.2. In that example a guidance counselor was interested in determining whether a relationship existed between the different aptitudes tested. Once a relationship, in the form of an equation, can be found between two variables, the counselor can use this relationship to *predict* the value of one of the variables if the value of the other variable is known. Thus the counselor may be interested in predicting how well a student will do on the business portion of the test if she knows the student's score on the math part.

Also, the counselor might be analyzing whether any correlation exists between high school averages and college grade-point averages. Her intention would be to try to find some relationship that will *predict* a college student's academic success from a knowledge of the high school average alone.

COMMENT It should be noted that the correlation coefficient merely determines whether two variables are related, but it does not specify how. Thus the correlation coefficient cannot be used to solve prediction problems.

When given a prediction problem, we first locate all the dots on a scatter diagram as we did in Section 11.2. Then we try to fit a straight line to the data in such a way that it best represents the relationship between the two variables. Such a fitted line is called an **estimated regression line.** Once we have such a line, we try to find an equation that will determine this line. We can then use this equation to predict the value of y corresponding to a given value of x.

Fitting a line to a set of numbers is by no means an easy task. Nevertheless, methods have been designed to handle such prediction problems. These methods are known as **regression methods.** In this book we discuss **linear regression** only. This means that we will try to fit a straight line to a set of numbers.

COMMENT Occasionally the dots are so scattered that a straight line cannot be fitted to the set of numbers. The statistician may then try to fit a **curve** to the set of numbers. This is shown in Figure 11.9. We would then have **curvilinear regression.**

The following examples will illustrate how a knowledge of the regression line enables us to predict the value of y for any given value of x. It is standard notation to call the variable to be predicted the **dependent variable** and to denote it by y. The known variable is called the **independent variable** and is denoted by x.

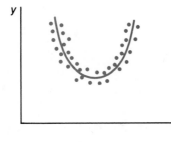

FIGURE 11.9
Curvilinear regression.

● **Example 1**

A guidance counselor notices that there is a strong positive correlation between math scores and business scores. Based upon a random sample of many students, she draws the scatter diagram shown in Figure 11.10. To this scatter diagram we have drawn a straight line that best represents the relationship between the two variables. This line enables us to predict the value of y for any given value of x. For example, if a student scored 35 on the math portion of the exam, then this line predicts that the student will score about 59 on the business part of the exam. This may be seen by first finding 35 on the horizontal axis, x, then moving straight up until you hit the estimated regression line. Finally you move directly to the left to see where you cross the vertical axis,

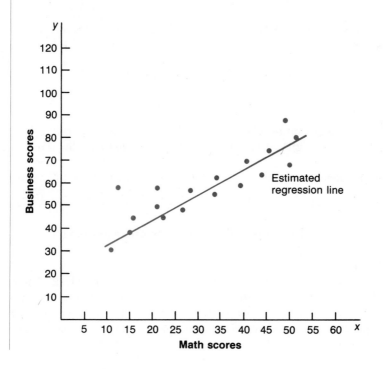

FIGURE 11.10

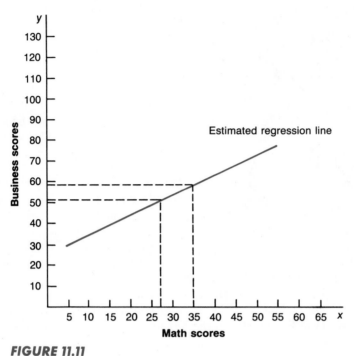

FIGURE 11.11

y. This is indicated by the dotted line of Figure 11.11. Similarly, we can predict that a student whose math score is 27 will score 51 on the business portion of the exam.

● Example 2

A certain organization claims to be able to predict a person's height if given the person's weight. It has collected the following data for ten people:

Weight (pounds), *x*	140	135	146	160	142	157	138	164	159	150
Height (inches), *y*	63	61	68	72	66	65	64	73	70	69

If a person weighs 155 pounds, what is his predicted height?

Solution

We first draw the scatter diagram as shown in Figure 11.12. Then we draw a straight line that best represents the relationship between the two variables. This line now enables us to predict a person's height if we know the person's weight. The estimated regression line predicts that a person who weighs 155 pounds will be about 70 inches tall.

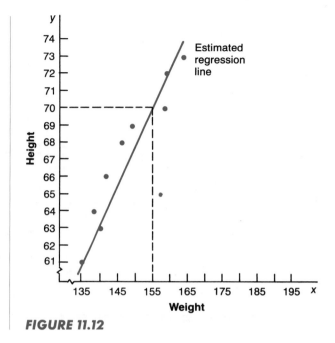

FIGURE 11.12

How does one draw an estimated regression line? Since there is no set procedure, different people are likely to draw different regression lines. So, although we may speak of finding a straight line that best fits the data, how is one to know when the best fit has been achieved? There are, in fact, several reasonable ways in which best fit can be defined. For this reason statisticians use a mathematical method for determining an equation that best describes the linear relationship between two variables. The method is known as the **least-squares method** and is discussed in detail in the next section.

11.6 THE METHOD OF LEAST SQUARES

Whenever we draw an estimated regression line, not all points will lie on the regression line. Some will be above it and some will be below. The difference between any point and the corresponding point on the regression line is called the (vertical) **deviation** from the line. It represents the difference in value between what we predicted and what actually happened. See Figure 11.13. The **least-squares method** determines the estimated regression line in such a way that the sum of the squares of these vertical deviations is as small as possible. Although a background in calculus is needed to understand how we obtain the formula for the least-square regression line, it is very easy to use the formula.

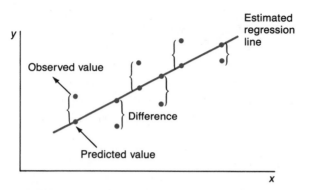

FIGURE 11.13
Difference between y-values and the estimated regression line.

FORMULA 11.2   *Regression Equation*

*The equation of the **estimated regression line** is*

$$\hat{y} = b_0 + b_1x$$

where

$$b_1 = \frac{n(\Sigma xy) - (\Sigma x)(\Sigma y)}{n(\Sigma x^2) - (\Sigma x)^2} \qquad b_0 = \frac{1}{n}(\Sigma y - b_1 \cdot \Sigma x)$$

and n is the number of pairs of scores.

COMMENT   The straight line that best fits a set of data points according to the least-squares criterion is called the **regression line** whereas the equation of the regression line is called the **regression equation.**

Let us use Formula 11.2 to find the regression equation connecting two variables.

● **Example 1**

Fifteen students were asked to indicate how many hours they studied before taking their statistics examination. Their responses were then matched with their grades on the exam, which had a maximum score of 100.

Hours, x	0.50	0.75	1.00	1.25	1.50	1.75	2.00	2.25	2.50	2.75	3.00	3.25	3.50	3.75	4.00
Scores, y	57	64	59	68	74	76	79	83	85	86	88	89	90	94	96

a. Find the regression equation that will predict a student's score if we know how many hours the student studied.

b. If a student studied 0.85 hours, what is the student's predicted grade?

Solution

a. To enable us to perform the computations, we arrange the data in the form of a chart:

x	y	x^2	xy
0.50	57	0.2500	28.5
0.75	64	0.5625	48
1.00	59	1.0000	59
1.25	68	1.5625	85
1.50	74	2.2500	111
1.75	76	3.0625	133
2.00	79	4.0000	158
2.25	83	5.0625	186.75
2.50	85	6.2500	212.5
2.75	86	7.5625	236.5
3.00	88	9.0000	264
3.25	89	10.5625	289.25
3.50	90	12.2500	315
3.75	94	14.0625	352.5
4.00	96	16.0000	384
33.75	1188	93.4375	2863

$$\Sigma x = 33.75 \quad \Sigma y = 1188 \quad \Sigma x^2 = 93.4375 \quad \Sigma xy = 2863$$

From the chart we have, $n = 15$, so that

$$b_1 = \frac{n(\Sigma xy) - (\Sigma x)(\Sigma y)}{n(\Sigma x^2) - (\Sigma x)^2}$$

$$= \frac{15(2863) - (33.75)(1188)}{15(93.4375) - (33.75)^2}$$

$$= \frac{42,945 - 40,095}{1401.5625 - 1139.0625}$$

$$= \frac{2850}{262.5}$$

$$= 10.857$$

and

$$b_0 = \frac{1}{n} (\Sigma y - b_1 \cdot \Sigma x)$$

$$= \frac{1}{15} (1188 - (10.857) \cdot (33.75))$$

$$= \frac{1}{15} (1188 - 366.424)$$

$$= \frac{1}{15}(821.576)$$

$$= 54.772$$

Thus,*

$$\hat{y} = b_0 + b_1 x$$
$$= 54.772 + 10.857x$$

The equation of the predicted regression line then is

$$\hat{y} = 54.772 + 10.857x$$

b. For $x = 0.85$, we get

$$\hat{y} = 54.772 + 10.857(0.85)$$
$$= 54.772 + 9.22845$$
$$= 64.00045$$

Thus, the predicted grade of a student who studies 0.85 hours is approximately 64.

● Example 2

A West Coast publishing company keeps accurate records of its monthly expenditure for advertising and its total monthly sales. For the first ten months of 1986, the records showed the following:

Advertising (in thousands), x	43	44	36	38	47	40	41	54	37	46
Sales (in millions), y	74	76	60	68	79	70	71	94	65	78

(Note that units are in dollars.)

a. Find the least-squares prediction equation appropriate for the data.
b. If the company plans to spend \$50,000 for advertising next month, what is its predicted sales? Assume that all other factors can be neglected.

Solution

a. We arrange the data in the form of a chart:

---

*Values for the variables throughout this chapter are calculated using computer accuracy, even though answers are often rounded to 2 or 3 decimal places.

x	y	x^2	xy
43	74	1849	3182
44	76	1936	3344
36	60	1296	2160
38	68	1444	2584
47	79	2209	3713
40	70	1600	2800
41	71	1681	2911
54	94	2916	5076
37	65	1369	2405
46	78	2116	3588
426	735	18,416	31,763
$\Sigma x = 426$	$\Sigma y = 735$	$\Sigma x^2 = 18,416$	$\Sigma xy = 31,763$

From the chart we have $n = 10$ so that

$$b_1 = \frac{n(\Sigma xy) - (\Sigma x)(\Sigma y)}{n(\Sigma x^2) - (\Sigma x)^2}$$

$$= \frac{10(31,763) - (426)(735)}{10(18,416) - (426)^2}$$

$$= \frac{317,630 - 313,110}{184,160 - 181,476}$$

$$= \frac{4520}{2684} = 1.684$$

and

$$b_0 = \frac{1}{n}(\Sigma y - b_1 \cdot \Sigma x)$$

$$= \frac{1}{10}(735 - (1.684) \cdot (426))$$

$$= \frac{1}{10}(735 - 717.384)$$

$$= \frac{1}{10}(17.616)$$

$$= 1.762$$

Thus,

$$\hat{y} = b_0 + b_1 x$$
$$= 1.762 + 1.684x$$

The equation of the predicted regression line is

$$\hat{y} = 1.762 + 1.684x$$

b. For $x = 50$, not 50,000, since x is in thousands of dollars, we get

$$\hat{y}_p = 1.762 + 1.684(50)$$
$$= 1.762 + 84.2$$
$$= 85.962$$

Thus, if the company spends \$50,000 next month for advertising, its predicted sales are \$85.962 million assuming all other factors can be neglected.

There is an alternate way to compute the equation of the regression line. This involves the sample covariance. We first compute the average of the x-values, denoted as $\bar{x}$, and the average of the y-values denoted as $\bar{y}$. Then we compute the sample standard deviation of the x-values. This is denoted as s_x. Finally, we determine the **sample covariance** of the n data points, which is defined by

$$\text{Sample covariance} = s_{xy} = \frac{\Sigma(x - \bar{x})(y - \bar{y})}{n - 1}$$

We then have the following alternate formula.

FORMULA 11.3 *Equation of Regression Line (Alternate version)*

The equation of the regression line for a set of n data points is given by

$$\hat{y} = b_0 + b_1x$$

where

$$b_1 = \frac{s_{xy}}{s_x^2}, \qquad b_0 = \bar{y} - b_1\bar{x}$$

and s_x is the sample standard deviation of the x-values

COMMENT In Exercise 12 of this section, the reader will be asked to demonstrate that Formula 11.3 is indeed equivalent to Formula 11.2, which was used earlier to calculate the equation of the regression line. We will illustrate the use of Formula 11.3 for the data presented in Example 2 of this section.

● **Example 3**
Find the equation of the regression line for the data of Example 2.

Solution
We arrange the data in the form of a chart.

x	y	x − x̄	y − ȳ	(x − x̄)(y − ȳ)	(x − x̄)²
43	74	0.4	0.5	0.2	0.16
44	76	1.4	2.5	3.5	1.96
36	60	−6.6	−13.5	89.1	43.56
38	68	−4.6	−5.5	25.3	21.16
47	79	4.4	5.5	24.2	19.36
40	70	−2.6	−3.5	9.1	6.76
41	71	−1.6	−2.5	4.0	2.56
54	94	11.4	20.5	233.7	129.96
37	65	−5.6	−8.5	47.6	31.36
46	78	3.4	4.5	15.3	11.56
426	735	0	0	452	268.4

$$\Sigma x = 426 \quad \Sigma y = 735 \qquad \Sigma(x - \bar{x})(y - \bar{y}) = 452 \quad \Sigma(x - \bar{x})^2 = 268.4$$

Thus

$$\bar{x} = \frac{\Sigma x}{n} = \frac{426}{10} = 42.6$$

$$\bar{y} = \frac{\Sigma y}{n} = \frac{735}{10} = 73.5$$

$$s_x^2 = \frac{\Sigma(x - \bar{x})^2}{n - 1} = \frac{268.4}{9} = 29.8222$$

$$s_{xy} = \frac{\Sigma(x - \bar{x})(y - \bar{y})}{n - 1} = \frac{452}{9} = 50.2222$$

so that

$$b_1 = \frac{s_{xy}}{s_x^2} = \frac{50.2222}{29.8222} = 1.684$$

and

$$b_0 = \bar{y} - b_1\bar{x} = 73.5 - (1.684)(42.6) = 1.762$$

The equation of the regression line is

$$\hat{y} = b_0 + b_1 x$$
$$= 1.762 + 1.684x$$

This is indeed the same answer that we obtained earlier.

EXERCISES

In each of the following exercises, assume that the correlation is high enough to allow for reasonable prediction.

1. Traffic department records indicate that as the posted speed limit has gradually been changed on a particular stretch of the Clearbrook Expressway, so have the number of reported accidents, as shown in the following chart:

Posted Speed Limit (mph) x	Average Number of Reported Accidents (weekly) y
55	29
50	27
45	25
40	20
30	12
25	8

a. Determine the least-squares prediction equation.

b. If the posted speed limit is changed to 35 mph, what is the predicted number of reported accidents?

2. After analyzing the salaries and batting averages of many baseball players in both leagues, a sports commentator concluded that higher-paid players will have higher batting averages. To determine if there is a relationship, ten players are randomly selected and the following statistics are obtained:

Average Salary x	Batting Average (in prior year) y
$ 70,000	0.290
62,000	0.281
98,000	0.299
125,000	0.302
400,000	0.309
900,000	0.345
200,000	0.304
350,000	0.306
85,000	0.295
500,000	0.320

a. Determine the least-squares prediction equation.

b. Assuming that the sports commentator's beliefs are accurate, what is the predicted batting average of a baseball player who is earning $175,000?

3. All yellow traffic-lane markings on the Clearbrook Expressway are applied automatically by a special paint truck. A hardening chemical is added to the paint to increase its durability. Based upon past experience, the number of months that the yellow marking will remain on the roadway before requiring a new application is determined by the amount of hardening chemical used as shown below:

Amount of Hardening Chemical Used per Truckload (units) x	Number of Months That Paint Will Remain on Highway y
19	9
22	11
25	14
28	17
35	22

a. Determine the least-squares prediction equation.

b. Because of the nature and cost of the hardening chemical, traffic department officials plan to add 30 units of the chemical. What is the predicted number of months that the paint will remain on the highway?

4. A scientist believes that there is a definite correlation between the height of a man and the height of his oldest son. Seven men have been selected at random. Their heights (in inches) and their oldest sons' heights (in inches) have been recorded:

Father x	Son y
68	66
67	65
69	67
65	61
72	71
71	70
74	74

a. Find the least-squares prediction equation.

b. If the father is 73 inches tall, what is the predicted height of his oldest son?

5. A city official believes that the number of complaints to the city's heat complaint control board (for the lack of heat) is directly related to the outdoor temperature. The following data have been collected.

Outdoor Temperature x	Number of Complaints to Heat Control Board y
30°	33
25°	42
20°	59
15°	65
10°	77
5°	88
0°	102

a. Determine the least-squares prediction equation.

b. What is the predicted number of complaints to the city's heat control board when the outdoor temperature drops to 12°?

6. The owner of an amusement park claims that the average daily attendance at the amusement park is related to the admission price charged, as shown below:

Admission Price (per adult) x	Average Daily Attendance y
$ 7.00	200,000
$ 8.00	175,000
$ 9.00	160,000
$10.00	145,000
$12.00	130,000
$15.00	120,000

a. Determine the least-squares prediction equation.

b. What is the predicted average daily attendance at the amusement park when the admission price is $13.00 per adult?

7. Giselle Markovic, president of a local chapter of La Leche Society, believes that the number of months that a mother nurses a child affects the emotional stability of the child (on a test that she has developed), as shown below:*

Number of Months That Mother Nursed Child x	Emotional Stability Score of Child y
3	17
5	19
7	23
9	29
11	35
12	42

*Giselle Markovic, *The Benefits of Nursing*. (New York, 1985.)

Assuming that her claim is valid,
a. determine the least-squares prediction equation.
b. what is the predicted emotional stability score of a child who is nursed for 10 months?

8. All visitors to a chocolate factory on the East Coast receive a free sample of the company's products. The company believes that sales at its on-premises gift shop are related to the number of free samples distributed as shown below:

Number of Free Samples Distributed x	Sales at On-site Gift Shop y
100,000	$ 600,000
200,000	$ 800,000
500,000	$1,100,000
700,000	$1,400,000
900,000	$1,800,000
1,100,000	$2,200,000
1,500,000	$3,000,000

a. Determine the least-square prediction equation.
b. If 600,000 free samples are distributed, what is the predicted sales at the on-premises gift shop?

9. A medical researcher is studying the relationship between the amount of alcohol in a person's bloodstream and the number of times that the person can safely perform a certain task. The following data are available.

Amount of Alcohol in Bloodstream (units) x	Number of Times Person Can Safely Perform Task y
10	35
15	30
20	20
30	18
40	10
50	4

a. Determine the least-squares prediction equation.
b. If a person drinks 25 units of alcohol, what is the predicted number of times that the person can safely perform the task?

10. An auto club official believes that there is a relationship between the number of hours of continuous rainfall and the average number of calls for assistance in starting a car with wet ignition wires as shown below:

Number of Hours of Continuous Rainfall x	Average Number of Calls for Assistance y
12	68
15	82
18	96
25	110
30	118
35	130
42	140
50	160

a. Determine the least-squares prediction equation.

b. What is the predicted number of calls for assistance in starting a car with wet ignition wires after 40 hours of continuous rainfall?

11. *Advertising* A chain of health-food stores is interested in determining the relationship between the number of times its commercial is broadcast on radio or television weekly and the weekly sales volume. It randomly selects nine weeks and determines the number of times that the commercial was broadcast and the weekly volume of sales as shown below.

Number of Times Commercial Is Broadcast x	Weekly Sales Volume (in thousands) y
3	40
4	50
7	70
8	90
10	120
12	150
15	190
20	230

a. Determine the least-squares prediction equation.

b. What is the predicted weekly sales volume when the commercial is broadcast 18 times weekly on radio or television?

*12. Verify that Formulas 11.2 and 11.3 are equivalent. Hence, either can be used to calculate the equation of the regression line.

13. Show that the formula for calculating the linear correlation coefficient (Formula 11.1) and the following computational formula are equivalent.

$$r = \frac{s_{xy}}{s_x s_y}$$

where

s_{xy} is the sample covariance

s_x is the sample standard deviation of the x-values

s_y is the sample standard deviation of the y-values

(*Hint:* Verify the following identity:

$$\frac{s_{xy}}{s_x s_y} = \frac{n(\Sigma xy) - (\Sigma x)(\Sigma y)}{\sqrt{n(\Sigma x^2) - (\Sigma x)^2} \sqrt{n(\Sigma y^2) - (\Sigma y)^2}})$$

11.7 STANDARD ERROR OF THE ESTIMATE

In Section 11.6 we discussed the least-squares regression line, which predicts a value of y when x has a particular value. Quite often it turns out that the predicted value of y and the observed value of y are different. If the correlation is low, these differences will be large. Only when the correlation is high can we expect the predicted values to be close to the observed values.

In general, the true population regression line is not known, so we use the data to estimate the equation of this true population regression line. For example, consider the data of Example 2 given in Section 11.6 (page 490), which is reproduced below.

x	y
43	74
44	76
36	60
38	68
47	79
40	70
41	71
54	94
37	65
46	78

The equation of the regression line is $\hat{y} = 1.762 + 1.684x$. When $x = 50$, the predicted value of y is 85.962. We cannot expect such predictions to be completely accurate. At different times, when $x = 50$, we may get different y-values. Thus, for each x, there is a corresponding population of y-values. The mean of the corresponding y-values lies on some straight line whose equation we do not know but which is of the form $y = \alpha + \beta x$. For each x-value, the distribution of the corresponding population of y-values is normally distributed. Moreover, for each x-value, the mean of the corresponding population of y-values lies on a straight line called the **population regression line** whose equation is of the form $y = \alpha + \beta x$. The population standard deviation, σ, of the population of y-values corresponding to a given x-value is the *same*, regardless of the x-value.

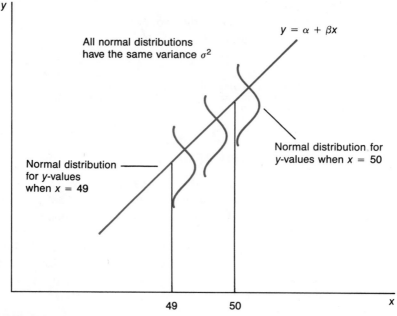

FIGURE 11.14

Normal distributions of population y-values about regression line.

Thus we assume that the kind of normal distribution occurring when $x = 50$ will also appear at any other value of x. This means that for any x-value, the distribution of the population of y-values is a normal distribution and that the variance of that normal distribution is the same for every x. This can be seen in Figure 11.14.

We can then conclude that the **error terms** (the vertical distance between the predicted y-values and the true population values) are normally distributed with mean 0 and the same standard deviation σ. How do we estimate σ, the common standard deviation of the normal distributions given in Figure 11.14? Statisticians have devised a method for measuring σ. This is the **standard error of the estimate.** However, before doing this, let us analyze our least-squares prediction equation, $\hat{y} = 1.762 + 1.684x$. Why bother computing this equation? Why not use the given data to make predictions about the sales by simply ignoring the values of x (amount of money spent for advertising) and using only the mean value of the sampled y's (the average sales) in making predictions? Thus we could use

$$\bar{y} = \frac{\Sigma y}{n} = \frac{735}{10} = 73.5$$

as our predicted sales (in all cases). How large an error have we made? In the computations below we indicate the (squared) error that is made when we predict a value of $\bar{y} = 73.5$ for the observed y-values.

y	$y - \bar{y}$	$(y - \bar{y})^2$
74	$74 - 73.5 = 0.5$	0.25
76	$76 - 73.5 = 2.5$	6.25
60	$60 - 73.5 = -13.5$	182.25
68	$68 - 73.5 = -5.5$	30.25
79	$79 - 73.5 = 5.5$	30.25
70	$70 - 73.5 = -3.5$	12.25
71	$71 - 73.5 = -2.5$	6.25
94	$94 - 73.5 = 20.5$	420.25
65	$65 - 73.5 = -8.5$	72.25
78	$78 - 73.5 = 4.5$	20.25
		780.50

$$\Sigma(y - \bar{y})^2 = 780.50$$

Thus, the total squared error that is made when we predict $\bar{y} = 73.5$ for the observed y-values is

$$\Sigma(y - \bar{y})^2 = 780.50$$

This is referred to as the **total sum of squares, SST.** Therefore

$$SST = \Sigma(y - \bar{y})^2 = 780.50$$

Instead of using $\bar{y}$ as our predicted value (in all cases), we can use the equation of the regression line values, $\hat{y}$, for our sales prediction. If we believe that our regression equation can be used to predict sales, then the squared error should be less when we use these values. The actual error obtained when we use these values is shown below:

x	y	$\hat{y}$	$y - \hat{y}$	$(y - \hat{y})^2$
43	74	74.17	-0.17	0.03
44	76	75.85	0.15	0.02
36	60	62.41	-2.41	5.81
38	68	65.77	2.23	4.97
47	79	80.89	-1.89	3.57
40	70	69.13	0.87	0.76
41	71	70.81	0.19	0.04
54	94	92.65	1.35	1.82
37	65	64.09	0.91	0.83
46	78	79.21	-1.21	1.46
				19.31

$$\Sigma(y - \hat{y})^2 = 19.31$$

In this case, the total squared error when using the equation of the regression line values $\hat{y}$ for predictions is

$$\Sigma(y - \hat{y})^2 = 19.31$$

This is called the **error sum of squares, SSE.** Thus,

$$SSE = \Sigma(y - \hat{y})^2 = 19.31$$

It should be obvious that using the regression equation for prediction reduced the total squared error considerably. The **percentage reduction** is

$$\frac{SST - SSE}{SST} = 1 - \frac{SSE}{SST} = 1 - \frac{19.31}{780.50} = 0.9753$$

or 97.53%. The percentage reduction in the total squared error obtained by using the regression equation instead of $\bar{y}$ is denoted by r^2 and is called the **coefficient of determination.** Thus, we have

---

FORMULA 11.4

$$Coefficient\ of\ determination = r^2 = 1 - \frac{SSE}{SST}$$

COMMENT The coefficient of determination, namely r^2, can be interpreted as representing the percentage of variation in the observed y-values that is explainable by the regression line.

As mentioned earlier, the population of y-values corresponding to the various x-values all have the same (usually unknown) standard deviation, σ. The value of σ can be estimated from the sample data by computing the **standard error of the estimate** also called the **residual standard deviation.**

To determine the standard error of the estimate, we first calculate the predicted value of y for each x and then compute the difference between the observed value and the predicted value. We then square these differences and divide the sum of these squares by $n - 2$. The square root of the result is called the standard error of the estimate.

FORMULA 11.5

The **standard error of the estimate** is denoted by s_e and is defined as

$$s_e = \sqrt{\frac{\Sigma(y - \hat{y})^2}{n - 2}} = \sqrt{\frac{SSE}{n - 2}}$$

where $\hat{y}$ is the predicted value, y is the observed value, and n is the number of pairs of scores.

● **Example 1**
Find the standard error of the estimate for the least-squares regression equation of Example 1 of Section 11.6 on page 488.

Solution

The least-squares regression equation was

$$\hat{y} = 54.772 + 10.857x$$

Using this equation, we find the predicted value of y corresponding to each value of x. We arrange our computations in the form of a chart:

x	y	$\hat{y}$	$y - \hat{y}$	$(y - \hat{y})^2$
0.50	57	60.2	−3.2	10.24
0.75	64	62.91	1.09	1.19
1.00	59	65.63	−6.63	43.96
1.25	68	68.34	−0.34	0.12
1.50	74	71.06	2.94	8.64
1.75	76	73.77	2.23	4.97
2.00	79	76.49	2.51	6.30
2.25	83	79.2	3.8	14.44
2.50	85	81.92	3.08	9.49
2.75	86	84.63	1.37	1.88
3.00	88	87.35	0.65	0.42
3.25	89	90.06	−1.06	1.12
3.50	90	92.78	−2.78	7.73
3.75	94	95.49	−1.49	2.22
4.00	96	98.21	−2.21	4.88
				117.60

$$\Sigma(y - \hat{y})^2 = 117.60$$

Applying Formula 11.5, we get

$$s_e = \sqrt{\frac{\Sigma(y - \hat{y})^2}{n - 2}} = \sqrt{\frac{117.60}{15 - 2}} = \sqrt{\frac{117.60}{13}} = \sqrt{9.05} = 3.01$$

Thus, the standard error of the estimate is 3.01.

• Example 2

Find the standard error of the estimate for the least-squares regression equation of Example 2 of Section 11.6 (page 490).

Solution

The least-squares regression equation was

$$\hat{y} = 1.762 + 1.684x$$

Using this equation, we find the predicted value of y corresponding to each value of x. We arrange our computations in the form of a chart:

x	y	$\hat{y}$	$y - \hat{y}$	$(y - \hat{y})^2$
43	74	74.17	−0.17	0.03
44	76	75.85	0.15	0.02
36	60	62.41	−2.41	5.81
38	68	65.77	2.23	4.97
47	79	80.89	−1.89	3.57
40	70	69.13	0.87	0.76
41	71	70.81	0.19	0.04
54	94	92.65	1.35	1.82
37	65	64.09	0.91	0.83
46	78	79.21	−1.21	1.46
				19.31

$$\Sigma(y - \hat{y})^2 = 19.31$$

Applying Formula 11.5, we get

$$s_e = \sqrt{\frac{\Sigma(y - \hat{y})^2}{n - 2}}$$

$$= \sqrt{\frac{19.31}{8}} = \sqrt{2.41} = 1.55$$

Thus, the standard error of the estimate is 1.55.

The goodness of fit of the least-squares regression line is determined by the value of the standard error of the estimate. A relatively small value of s_e indicates that the predicted and observed values of y are fairly close. This means that the regression equation is a good description of the relationship between the two variables. On the other hand, a relatively large value of s_e indicates that there is a large difference between the predicted and observed values of y. When this happens, the relationship between x and y as given by the least-squares equation is not a good indication of the relationship between the two variables. Only when the standard error of the estimate is zero can we say for sure that the least-squares regression equation is a perfect description of the relationship between x and y.

COMMENT In computing s_e, statisticians often use the following equivalent formula.

$$s_e = \sqrt{\frac{\Sigma y^2 - b_0(\Sigma y) - b_1(\Sigma xy)}{n - 2}}$$

where the values of b_0 and b_1 are the same as obtained earlier in computing the equation of the regression line. The values for the various summations should have already been obtained, thereby reducing the amount of computation needed.

EXERCISES

For each of the following, refer back to the exercise indicated and calculate the standard error of the estimate.

1. Exercise 1, page 494.

2. Exercise 2, page 494.

3. Exercise 3, page 495.

4. Exercise 4, page 495.

5. Exercise 5, page 495.

6. Exercise 6, page 496.

7. Exercise 7, page 496.

8. Exercise 8, page 497.

9. Exercise 9, page 497.

10. Exercise 10, page 497.

11. Exercise 11, page 498.

*11.8 USING STATISTICAL INFERENCE FOR REGRESSION

We can use the standard statistical inferential procedures discussed in earlier chapters for regression. Specifically, we mentioned earlier that for each x-value, the corresponding population of y-values is normally distributed with mean $y = \beta_0 + \beta_1 x$ and standard deviation σ. However, if β_1 has a value of 0, then x will be totally worthless in predicting y-values since in that case the regression equation would be

$$\hat{y} = \beta_0 + \beta_1 x$$
$$= \beta_0 + 0x$$
$$= \beta_0$$

Thus, the value of x would have absolutely nothing to do with the distribution of y-values. Hence, it is important for us to determine in advance whether x can be used as a predictor of y, that is, if x and y are linearly related. We can decide this by performing the following hypothesis test.

$H_0\colon \beta_1 = 0$

$H_1\colon \beta_1 \neq 0$

If we conclude that the null hypothesis has to be rejected, then this indicates that x and y are linearly related and that we can proceed to use the equation of the regression line for making predictions.

How do we test the null hypothesis that $\beta_1 = 0$? This can be done by using the value of b_1 which actually represents the slope of the sample regression line. We proceed as follows:

STATISTICAL INFERENCE CONCERNING β_1

To test the null hypothesis H_0 that the slope β_1 of the population regression line is zero or not (that is, whether x can be used as a predictor of y or not) do the following:

1. State the null hypothesis and the alternative hypothesis as well as the significance level α.
2. Find $n - 2$. This gives us the number of degrees of freedom for a t-distribution.
3. Find the appropriate critical values $\pm t_{\alpha/2}$ by using Table VIII in the Appendix.
4. Compute the value of the test statistic.

$$t = \frac{b_1}{s_e \bigg/ \sqrt{\Sigma x^2 - \dfrac{(\Sigma x)^2}{n}}}$$

5. If the value of the test statistic falls in the rejection region, reject H_0. Otherwise, do not reject H_0.
6. State the conclusion.

Let us illustrate the above procedure with a few examples.

● Example 1

Refer back to the data of Example 1 on page 488. Does the data indicate that the value of β_1, that is, the slope of the population regression line, is not zero, which would mean that x (number of hours studied) can be used as a predictor of y (score received)? Assume that the level of significance is 5%.

Solution

In this case, the null hypothesis is $\beta_1 = 0$, and the alternative hypothesis is $\beta_1 \neq 0$. Since $n = 15$, we will have a t-distribution with $15 - 2$ or 13 degrees of freedom. From Table VIII in the Appendix, the appropriate $t_{\alpha/2} = t_{0.05/2} = t_{0.025}$ value is 2.160. Now we compute the value of the test statistic. We have

$$t = \frac{b_1}{s_e \bigg/ \sqrt{\Sigma x^2 - \dfrac{(\Sigma x)^2}{n}}}$$

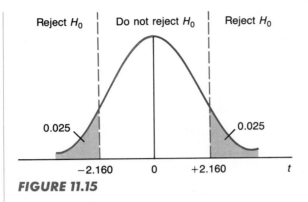

Reject H_0 | Do not reject H_0 | Reject H_0

0.025 0.025

−2.160   0   +2.160 t

FIGURE 11.15

All of the values of the variables needed to use this formula have already been found earlier. Thus we know that $b_1 = 10.857$ (computed on page 489), $\Sigma x^2 = 93.4375$ (computed on page 489), $\Sigma x = 33.75$ (computed on page 489), and $s_e = 3.01$ (computed on page 503). Also $n = 15$, so that

$$t = \frac{10.857}{3.01 \Big/ \sqrt{93.4375 - \dfrac{(33.75)^2}{15}}} = 15.089$$

Since the value of the test statistic 15.089 falls in the rejection region of Figure 11.15, we reject H_0 and conclude that the slope of the population regression line is not 0. Hence x (number of hours studied) can be used as a predictor of y.

● **Example 2**
Refer back to the data of Example 2 on page 490. Do the data indicate that the value of β_1, that is, the slope of the population regression line, is not zero, which would mean that x (amount spent on advertising) can be used as a predictor of y (sales)? Use a 5% level of significance.

Solution
The null hypothesis is $\beta_1 = 0$, and the alternative hypothesis is $\beta_1 \neq 0$. Since $n = 10$, we will have a t-distribution with $10 - 2$ or 8 degrees of freedom. From Table VIII, the $t_{0.025}$ value is 2.306. We then have the acceptance-rejection region shown in Figure 11.16. The value of the test statistic is

$$t = \frac{b_1}{s_e \Big/ \sqrt{\Sigma x^2 - \dfrac{(\Sigma x)^2}{n}}} = \frac{1.684}{1.55 \Big/ \sqrt{18{,}416 - \dfrac{(426)^2}{10}}} = 17.799$$

where the value of $b_1 = 1.684$ was computed on page 491, the value of $\Sigma x^2 = 18{,}416$ was computed on page 491, the value of $\Sigma x = 426$ was computed

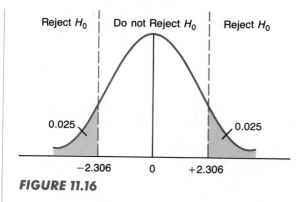

Reject H_0 Do not Reject H_0 Reject H_0

0.025   0.025

−2.306   0   +2.306

FIGURE 11.16

on page 491 and the value of $s_e = 1.55$ was computed on page 504. Since the value of the test statistic 17.799 falls in the rejection region of Figure 11.16, we reject H_0 and conclude that the slope of the population regression line is not 0. Hence x (amount spent on advertising) can be used as a predictor of y (sales).

COMMENT   Since all of the computations already have been done, we can also determine a confidence interval for the slope β_1 of the population regression line. The endpoints of the confidence interval are

$$b_1 \pm t_{\alpha/2} \cdot \frac{s_e}{\sqrt{\Sigma x^2 - \frac{(\Sigma x)^2}{n}}}$$

The interested reader should actually verify that a 95% confidence interval for β_1 of Example 1 is $10.857 \pm 2.160 \left(\dfrac{3.01}{\sqrt{17.5}} \right)$ or from 9.303 to 12.411. Thus we can be 95% confident that β_1 is somewhere between 9.303 and 12.411.

Prediction Intervals

After we have determined the least-squares prediction equation for some data, we may be interested in setting up a prediction interval for a population value of y that we are predicting, corresponding to some particular value of x. Under such circumstances we proceed as follows:

1. Find $n - 2$. This gives us the number of degrees of freedom for a t-distribution.
2. Find the appropriate $t_{\alpha/2}$ values by using Table VIII in the Appendix.
3. Find the least-squares prediction equation, $\hat{y}$, by using Formula 11.2. Use it to compute the predicted y-value, $\hat{y}$, corresponding to some particular x_p.

4. Find the standard error of the estimate, s_e, by using Formula 11.5.
5. Set up the appropriate prediction interval by using Formula 11.6, as follows.

FORMULA 11.6

A prediction interval for a particular y corresponding to some particular value of $x = x_p$ is

$$\text{Lower boundary:} \quad \hat{y}_p - t_{\alpha/2} \cdot s_e \cdot \sqrt{1 + \frac{1}{n} + \frac{n(x_p - \bar{x})^2}{n(\Sigma x^2) - (\Sigma x)^2}}$$

$$\text{Upper boundary:} \quad \hat{y}_p + t_{\alpha/2} \cdot s_e \cdot \sqrt{1 + \frac{1}{n} + \frac{n(x_p - \bar{x})^2}{n(\Sigma x^2) - (\Sigma x)^2}}$$

where $t_{\alpha/2}$ represents the t-distribution value obtained from Table VIII using $n - 2$ degrees of freedom, s_e is the standard error of the estimate, and $\hat{y}_p$ is the predicted value of y corresponding to $x = x_p$.

COMMENT The above intervals are frequently called *prediction intervals* because they give intervals for future values of y at a specified x.

We illustrate the use of Formula 11.6 with several examples.

● **Example 3**
Using the data of Example 1 on page 488, find a 95% prediction interval for the score of a student who studies for 0.85 hours.

Solution
The least-squares prediction equation was already calculated. It is

$$\hat{y} = 54.772 + 10.857x$$

Also, when $x_p = 0.85$, then the predicted value of y is $\hat{y}_p = 64$. This was calculated earlier. The standard error of the estimate was calculated on page 503. It is 3.01. In this case $n = 15$, so we will have a t-distribution with $15 - 2$, or 13, degrees of freedom. From Table VIII, the appropriate $t_{\alpha/2} = t_{0.05/2} = t_{0.025}$ value is 2.160. Now we apply Formula 11.6. We get

$$\text{Lower boundary} = \hat{y}_p - t_{\alpha/2} \cdot s_e \cdot \sqrt{1 + \frac{1}{n} + \frac{n(x_p - \bar{x})^2}{n(\Sigma x^2) - (\Sigma x)^2}}$$

$$= 64 - (2.160)(3.01) \cdot \sqrt{1 + \frac{1}{15} + \frac{15(0.85 - 2.25)^2}{15(93.4375) - (33.75)^2}}$$

$$= 64 - (2.160)(3.01)\sqrt{1 + 0.067 + 0.112}$$

$$= 64 - (2.16)(3.01)(1.086)$$

$$= 64 - 7.061 = 56.939$$

$$\text{Upper boundary} = y_p + t_{\alpha/2} \cdot s_e \cdot \sqrt{1 + \frac{1}{n} + \frac{n(x_p - \bar{x})^2}{n(\Sigma x^2) - (\Sigma x)^2}}$$

$$= 64 + (2.160)(3.01) \sqrt{1 + \frac{1}{15} + \frac{15(0.85 - 2.25)^2}{15(93.4375) - (33.75)^2}}$$

$$= 64 + (2.160)(3.01)(1.086)$$

$$= 64 + 7.061 = 71.061$$

Thus, a 95% prediction interval for the score of a student who studies for 0.85 hours is 56.939 to 71.061.

● **Example 4**

Using the data of Example 2 on page 490 find a 95% prediction interval of the predicted sales of the publishing company if it spends $50,000 for advertisement next month.

Solution

The least-squares prediction equation was already calculated. It is

$$\hat{y} = 1.762 + 1.684x$$

When $x_p = 50$, $\hat{y}_p = 85.962$. The standard error of the estimate, s_e, was calculated on page 504 and is 1.55. In this case, $n = 10$ so that we have a t-distribution with $10 - 2$, or 8, degrees of freedom. From Table VIII, the $t_{0.025}$ value is 2.306. Applying Formula 11.6 gives

$$\text{Lower boundary} = \hat{y}_p - t_{\alpha/2} \cdot s_e \cdot \sqrt{1 + \frac{1}{n} + \frac{n(x_p - \bar{x})^2}{n(\Sigma x^2) - (\Sigma x)^2}}$$

$$= 85.962 - (2.306)(1.55) \sqrt{1 + \frac{1}{10} + \frac{10(50 - 42.6)^2}{10(18,416) - (426)^2}}$$

$$= 85.962 - (2.306)(1.55)\sqrt{1 + 0.1 + 0.204}$$

$$= 85.962 - (2.306)(1.55)(1.142)$$

$$= 85.962 - 4.082 = 81.88$$

$$\text{Upper boundary} = \hat{y}_p + t_{\alpha/2} \cdot s_e \cdot \sqrt{1 + \frac{1}{n} + \frac{n(\Sigma x_p - \bar{x})^2}{n(\Sigma x^2) - (\Sigma x)^2}}$$

$$= 85.962 + (2.306)(1.55) \sqrt{1 + \frac{1}{10} + \frac{10(50 - 42.6)^2}{10(18,416) - (426)^2}}$$

$$= 85.962 + (2.306)(1.55)(1.142)$$

$$= 85.962 + 4.082 = 90.044$$

Thus, a 95% prediction interval for the predicted sales if the company spends $50,000 for advertisement next month is $81.88 million to $90.044 million (assuming that all other factors can be neglected).

EXERCISES

For each of the following, test the null hypothesis that $\beta_1 = 0$ and then find a 95% prediction interval for the indicated value.

1. The predicted number of reported accidents when the posted speed limit is 35 mph in Exercise 1 on page 494.

2. The predicted batting average of a baseball player who is earning $175,000 in Exercise 2 on page 494.

3. The predicted number of months that the paint will remain on the highway when 30 units of the hardening chemical are used in Exercise 3 on page 495.

4. The predicted height of a person's oldest son when the person is 73 inches tall in Exercise 4 on page 495.

5. The predicted number of complaints to the city's heat complaint control board when the outdoor temperature drops to 12° in Exercise 5 on page 495.

6. The predicted average daily attendance at the amusement park when the admission price is $13.00 per adult in Exercise 6 on page 496.

7. The predicted emotional stability score of a child that is nursed for 10 months in Exercise 7 on page 496.

8. The predicted sales at the on-premises gift shop when 600,000 free samples are distributed in Exercise 8 on page 497.

9. The predicted number of times a person can safely perform the task when that person drinks 25 units of alcohol in Exercise 9 on page 497.

10. The predicted number of calls for assistance in starting a car with wet ignition wires after 40 hours of continuous rainfall in Exercise 10 on page 497.

11. The predicted weekly sales volume when the commercial is broadcast 18 times weekly on radio or television in Exercise 11 on page 498.

11.9 THE RELATIONSHIP BETWEEN CORRELATION AND REGRESSION

Although it may seem to you that linear correlation and regression are very similar since many of the computations performed in both are the same, the two ideas are quite different. One uses the correlation coefficient (Formula 11.1) to determine whether two variables are linearly related. The correlation coefficient measures the strength of the linear relationship. Regression analysis, on the other hand, is used when we want to answer questions about the relationship between two variables. Just exactly how the two variables are related (that is, can we find an equation connecting the variables) requires regression analysis. The equation connecting the variables may not necessarily be linear.

Another factor that has to be considered involves **extrapolation.** Two variables may have a high positive correlation. Yet, regression analysis *cannot* be used to make predictions for new x-values that are far removed from the original data. Such uses of regression are called extrapolation. To illustrate, in Example 1 on page 488, we determined that the regression equation or the least squares prediction equation was $\hat{y} = 54.772 + 10.857x$. This equation *cannot* be used to make predictions for values of x that are far removed from the values of x between 0.50 and 4.00.

*11.10 MULTIPLE REGRESSION

Until now we have been interested in finding a linear equation connecting two variables. However, there are many practical situations where several variables may simultaneously affect a given variable. For example, the yield from an acre of land may depend upon, among other things, such variables as the amount of fertilizer used, the amount of rainfall, the amount of sunshine, and so on.

There are many formulas that can be used to express relationships between more than two variables. The most commonly used formulas are linear equations of the form

$$\hat{y} = b_0 + b_1x_1 + b_2x_2 + \cdots + b_mx_m$$

The main difficulty in deriving a linear equation in more than two variables that best describes a given set of data is that of determining the values of b_0, $b_1, b_2, \ldots, b_m$. When there are two independent variables x_1 and x_2 and the linear equation connecting them is of the form $y = b_0 + b_1x_1 + b_2x_2$, then we can apply the method of least squares. This means that we must solve the following equations simultaneously.

FORMULA 11.7 *Multiple regression formulas*

$$\Sigma y = n \cdot b_0 + b_1(\Sigma x_1) + b_2(\Sigma x_2)$$

$$\Sigma x_1 y = b_0(\Sigma x_1) + b_1(\Sigma x_1^2) + b_2(\Sigma x_1 x_2)$$

$$\Sigma x_2 y = b_0(\Sigma x_2) + b_1(\Sigma x_1 x_2) + b_2(\Sigma x_2^2)$$

Solving these equations usually involves a lot of computation. Nevertheless, we illustrate the procedure with an example.

● Example 1

The following data give the yield, y, per plot of land depending upon the quantity of fertilizer used, x_1, and the number of inches of rainfall, x_2. Compute a linear equation that will enable us to predict the average yield, y, per plot in terms of the quantity of fertilizer used, x_1, and the number of inches, x_2, of rainfall.

Yield per Plot (hundreds of bushels) y	Quantity of Fertilizer Used (units) x_1	Rainfall (inches) x_2
20	2	5
25	3	9
28	5	14
30	7	15
32	11	23

Solution

We arrange the data in the following tabular format:

x_1	x_2	y	x_1y	x_2y	x_1^2	x_1x_2	x_2^2
2	5	20	40	100	4	10	25
3	9	25	75	225	9	27	81
5	14	28	140	392	25	70	196
7	15	30	210	450	49	105	225
11	23	32	352	736	121	253	529
28	66	135	817	1903	208	465	1056

$\Sigma x_1 = 28$ $\quad \Sigma x_2 = 66$ $\quad \Sigma y = 135$ $\quad \Sigma x_1y = 817$ $\quad \Sigma x_2y = 1903$ $\quad \Sigma x_1^2 = 208$ $\quad \Sigma x_1x_2 = 465$ $\quad \Sigma x_2^2 = 1056$

We now substitute these values into the equations given in Formula 11.7. Here $n = 5$. We get

$$135 = 5b_0 + 28b_1 + 66b_2$$

$$817 = 28b_0 + 208b_1 + 465b_2$$

$$1903 = 66b_0 + 465b_1 + 1056b_2$$

This represents a system of three equations in three unknowns. If we solve these equations simultaneously, we get $b_0 = 17.4459$, $b_1 = -0.7504$, and $b_2 = 1.0422$. Thus, the least-squares prediction equation is

$$\hat{y} = 17.4459 - 0.7504x_1 + 1.0422x_2$$

When the farmer uses 4 units of fertilizer and there are 9 inches of rain, then the predicted yield per plot is

$$y = 17.4459 - 0.7504(4) + 1.0422(9)$$

or approximately 23.8241 hundreds of bushels.

EXERCISES

1. The following is a list of the grades of six students on a special math aptitude test, the IQ scores of the students, and the number of hours that each prepared for the exam.

Grade on Exam y	IQ Score x_1	Preparation for Exam (hours) x_2
84	120	18
88	130	17
81	118	15
92	142	19
78	116	10

a. Determine the least-squares prediction equation.
b. What is the predicted grade on the exam for a student who has an IQ score of 125 and who studies 12 hours preparing for the exam?

2. The systolic blood pressure of an individual is believed to be related to the person's age and weight. For eight people, the following data are available.

Systolic Blood Pressure y	Age (years) x_1	Weight (pounds) x_2
130	50	170
135	53	175
140	56	180
145	59	185
150	60	195
155	62	200
160	65	205
165	70	215

a. Determine the least-squares prediction equation.
b. What is the predicted systolic blood pressure of an individual who is 68 years old and who weighs 203 pounds?

3. A shoe salesman claims that a person's shoe size is related to his or her height and weight. The following data are available for six people.

Shoe Size y	Height (inches) x_1	Weight (pounds) x_2
$8\frac{1}{2}$	65	150
9	68	160
$9\frac{1}{2}$	70	180
10	73	190
$10\frac{1}{2}$	74	195
11	77	205

a. Determine the least-squares prediction equation.

b. What is the predicted shoe size of a man who is 69 inches tall and who weighs 200 pounds?

11.11 USING COMPUTER PACKAGES

Computer packages are ideally suited for drawing scatter diagrams, computing correlation coefficients, and deriving regression equations. The computer output for such data also provides an analysis of variance, which is a topic that we will discuss in a later chapter.

The data from Example 2 of Section 11.6 (page 490) are used to illustrate how MINITAB handles correlation and regression problems.

```
MTB > READ ADVERTISING IN C1, SALES IN C2
DATA > 43    74
DATA > 44    76
DATA > 36    60
DATA > 38    68
DATA > 47    79
DATA > 40    70
DATA > 41    71
DATA > 54    94
DATA > 37    65
DATA > 46    78

MTB > PLOT SALES IN C2 VS ADVERTISING IN C1
```

This MINITAB program produces the following scatter diagram.

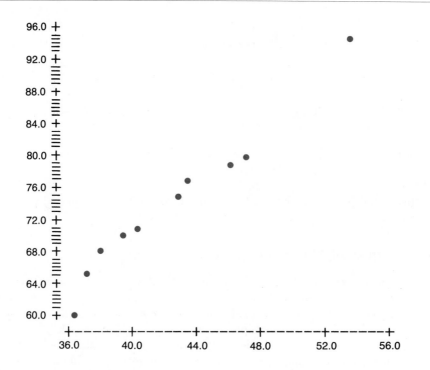

```
MTB > CORRELATION COEFFICIENT BETWEEN ADVERTISING IN C1 AND SALES IN C2
      CORRELATION OF C1 AND C2 = 0.988

MTB > REGRESS SALES IN C2 ON 1 PREDICTOR ADVERTISING IN C1
      THE REGRESSION EQUATION IS
      C2 = 1.76 + 1.68 C1
      ROW        C1         C2         PRED. Y VALUES
       1        43.0       74.0        74.17
       2        44.0       76.0        75.86
       3        36.0       60.0        62.39
       4        38.0       68.0        65.75
       5        47.0       79.0        80.91
       6        40.0       70.0        69.12
       7        41.0       71.0        70.81
       8        54.0       94.0        92.70
       9        37.0       65.0        64.07
      10        46.0       78.0        79.23
      THE ESTIMATED ST. DEV. OF Y ABOUT REGRESSION LINE
      IS 1.554 WITH (10 - 2) = 8 DEGREES OF FREEDOM
```

```
R-SQUARED = 97.5 PERCENT
R-SQUARED = 97.2 PERCENT, ADJUSTED FOR D.F.

ANALYSIS OF VARIANCE      DF        SS      MS = SS/DF
DUE TO REGRESSION          1     761.19      761.19
RESIDUAL                   8      19.31        2.41
TOTAL                      9     780.50
```

11.12 SUMMARY

In this chapter we analyzed the relationship between two variables. Scatter diagrams were drawn that help us understand this relationship. We discussed the concept of correlation coefficients, which tell us the extent to which two variables are related. Correlation coefficients vary between the values of -1 and $+1$. A value of $+1$ or -1 represents a perfect linear relationship between the two variables. A correlation coefficient of 0 means that there is no linear relationship between the two variables.

We indicated how to test whether a value of r is significant or not. Furthermore, we mentioned that even when there is an indication of positive correlation between two variables, great care must be shown in how we interpret this relationship.

Once we determine that there is a significant linear correlation between two variables, we find the least-squares equation, which expresses this relation mathematically.

We discussed the standard error of the estimate. This is a way of measuring how well the estimated least-squares regression line really fits the data. The smaller the value of s_e, the better the estimate. We also indicated how to set up prediction intervals for values of y obtained through regression analysis. Finally, we discussed multiple regression where the value of one variable depends upon several other variables.

STUDY GUIDE

You should now be able to demonstrate your knowledge of the following ideas presented in this chapter by giving definitions, descriptions, or specific examples. Page references are given for each term so that you can check your answer.

Correlation coefficient (page 467)
Scatter diagram (page 469)
Linear correlation (page 469)
Positive correlation (page 473)
Negative correlation (page 473)
Coefficient of linear correlation (page 473)

FORMULAS TO REMEMBER

You should be able to identify each symbol in the following formulas, understand the relationships among the symbols expressed in each formula, understand the significance of each formula, and use the formulas in solving problems.

1. Coefficient of linear correlation:

$$r = \frac{n(\Sigma xy) - (\Sigma x)(\Sigma y)}{\sqrt{n(\Sigma x^2) - (\Sigma x)^2} \, \sqrt{n(\Sigma y^2) - (\Sigma y)^2}}$$

2. Estimated regression line:

$$\hat{y} = b_0 + b_1 x$$

where

$$b_1 = \frac{n(\Sigma xy) - (\Sigma x)(\Sigma y)}{n(\Sigma x^2) - (\Sigma x)^2} \quad \text{and} \quad b_0 = \frac{1}{n}(\Sigma y - b_1 \cdot \Sigma x)$$

3. Sample covariance $= s_{xy} = \dfrac{\Sigma(x - \bar{x})(y - \bar{y})}{n - 1}$

4. Equation of regression line (alternate version):

$$\hat{y} = b_0 + b_1 x \qquad \text{where} \qquad b_1 = \frac{s_{xy}}{s_x^2}, \qquad b_0 = \bar{y} - b_1 \bar{x},$$

and s_x is the sample standard deviation of x-values

5. Coefficient of linear correlation (alternate version):

$$r = \frac{s_{xy}}{s_x s_y}$$

6. Total sum of squares, $SST = \Sigma(y - \bar{y})^2$

7. Error sum of squares, $SSE = \Sigma(y - \hat{y})^2$

8. Coefficient of determination or percentage reduction:

$$r^2 = \frac{SST - SSE}{SST} = 1 - \frac{SSE}{SST}$$

9. Estimate of the common standard deviation, σ, or standard error of the estimate:

$$s_e = \sqrt{\frac{\Sigma(y - \hat{y})^2}{n - 2}} = \sqrt{\frac{SSE}{n - 2}} \qquad \text{or} \qquad \sqrt{\frac{\Sigma y^2 - b_0(\Sigma y) - b_1(\Sigma y)}{n - 2}}$$

10. To test whether $\beta_1 = 0$ or not, the test statistic is

$$t = \frac{b_1}{s_e \bigg/ \sqrt{\Sigma x^2 - \frac{(\Sigma x)^2}{n}}}$$

11. Endpoints of prediction interval for β_1:

$$b_1 \pm t_{\alpha/2} \cdot \frac{s_e}{\sqrt{\Sigma x^2 - \frac{(\Sigma x)^2}{n}}}$$

12. Prediction interval for y corresponding to some given value of $x = x_p$:

$$\text{Lower boundary: } \hat{y}_p - t_{\alpha/2} \cdot s_e \sqrt{1 + \frac{1}{n} + \frac{n(x_p - \bar{x})^2}{n(\Sigma x^2) - (\Sigma x)^2}}$$

$$\text{Upper boundary: } \hat{y}_p + t_{\alpha/2} \cdot s_e \sqrt{1 + \frac{1}{n} + \frac{n(x_p - \bar{x})^2}{n(\Sigma x^2) - (\Sigma x)^2}}$$

13. Multiple regression: $\hat{y} = b_0 + b_1 x_1 + b_2 x_2$, where

$$\Sigma y = n \cdot b_0 + b_1(\Sigma x_1) + b_2(\Sigma x_2)$$

$$\Sigma x_1 y = b_0(\Sigma x_1) + b_1(\Sigma x_1^2) + b_2(\Sigma x_1 x_2)$$

$$\Sigma x_2 y = b_0(\Sigma x_2) + b_1(\Sigma x_1 x_2) + b_2(\Sigma x_2^2)$$

The tests of the following section will be more useful if you take them after you have studied the examples and solved the exercises given in this chapter.

1. What type of correlation would you expect to exist between the quantity of fatty foods consumed by a person and the blood serum cholesterol of the person?

    a. negative correlation   **b.** positive correlation

    c. zero correlation   **d.** not enough information given

    e. none of these

For Questions 2–6 refer to the following information.
During the holiday season, Mark's Department Store chain employs special plainclothes security personnel to help decrease the number of reported pickpocketing incidents. The following data are available for five of the company stores.

Number of Security Personnel Employed x	Number of Reported Pickpocketing Arrests y
9	11
5	6
4	7
6	8
7	9

2. The coefficient of correlation for these data is

    a. 0.84   **b.** 0.93   **c.** 0.98   **d.** 0.91   **e.** none of these

3. The least-squares prediction equation for these data is

    a. $\hat{y} = 2.419 + 0.9324x$ **b.** $\hat{y} = 0.9324 + 2.419x$

    c. $\hat{y} = 12.0955 + 0.9324x$ **d.** $\hat{y} = 0.9324 + 12.0955x$

    e. none of these

4. What is the predicted number of reported arrests for pickpocketing in a sixth company store, if that store employs eight plainclothes security personnel?

    a. 6.01   **b.** 2.73   **c.** 9.88   **d.** 52.03   **e.** none of these

5. What is the value of s_e, the standard error of the estimate?

6. Compute a 95% prediction interval for the number of reported arrests for pickpocketing by a store that employs eight security personnel.

7. What type of correlation would you think exists between high blood pressure and the incidence of left-handedness?
 a. positive correlation **b.** zero correlation
 c. negative correlation **d.** depends upon the value
 e. none of these

8. A correlation coefficient of −0.12
 a. indicates a strong negative correlation
 b. indicates a weak negative correlation
 c. is insignificant
 d. is impossible
 e. none of these

9. Consider the following correlations. Are any of the correlations between IQ and achievement in specific school subjects or skills significant?

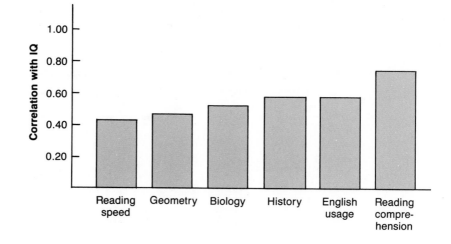

10. *At What Age Are We Smartest?* The bar graph on the following page shows the average scores achieved on a skills test by men when they were 20 years old and again when the same men were 60 years old. Is there a correlation between age and skills? Explain your answer.

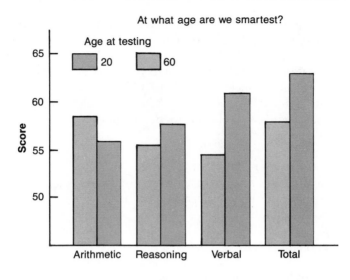

At what age are we smartest?

MASTERY TESTS

Form B

1. An auto mechanic believes that the estimated gas mileage, y, depends upon the car engine's size and upon the number of cylinders in the engine. The following data are available for seven cars:*

Engine Size (liters) x_1	Number of Cylinders x_2	Estimated Gas Mileage (mpg) y
2.5	4	22
4.2	6	16
4.4	8	18
1.6	4	28
5.8	8	16
3.8	6	20
5.0	8	17

Determine the least-squares prediction equation.

2. Refer back to Question 1. What is the estimated gas mileage for a car which has 4 cylinders and whose engine size is 2.3?

*Source: *Gas Mileage Guide*, U.S. Department of Energy.

3. A forest ranger compared the number of reported fires in a certain wooded area of a national preserve with the annual number of inches of rainfall for that region. The following data were collected for several years:

Rainfall (inches) x	Number of Reported Fires y
50	50
60	35
20	78
40	60
55	41
38	62

a. Compute the coefficient of correlation for the data.
b. Is the relationship between the number of inches of rainfall and the number of reported fires significant?

4. Refer back to Question 3. Find a 95% prediction interval for the predicted number of fires when 45 inches of rain fall during the year.

5. A professor in a business school suspects that there is a linear relationship between the height of a college graduate and the starting salary offered to the college graduate. The professor believes that taller people tend to be offered higher starting salaries than shorter people. The following data have been collected for seven recent graduates:

Height (inches) x	Starting Salary (per year) y
64	$21,000
67	24,000
66	23,000
68	26,000
69	26,000
71	27,000
65	25,000

a. Compute the coefficient of correlation for the data.
b. Are the professor's suspicions justified?

6. The ages and salaries of seven executives at seven companies are as follows:

Age (years) x	Annual Salary y
50	$ 89,000
48	85,000
42	82,000
55	98,000
62	109,000
53	110,000
65	120,000

Determine the least-squares prediction equation.

7. A greeting cards company has done extensive market research and believes that the number of Valentine's Day cards sold is related to the price of each card as shown below.

Price per Card x	Cards Sold (thousands) y
$1.00	62
$1.25	53
$1.50	40
$1.75	35
$2.00	25
$2.50	20

If a card is priced at $2.25, what is the predicted number of those cards to be sold?

8. Seven randomly selected families had the following annual income and accumulated savings in a bank account.

Family	Income, x	Accumulated Savings, y
A	$21,000	$15,000
B	28,000	20,000
C	39,000	25,000
D	71,000	37,000
E	53,000	29,000
F	62,000	28,000
G	40,000	30,000

Compute the equation of estimated regression line.

9. A Wall Street broker analyzed the expenditures for research and the profits of several companies and came up with the following data:

Company	Expenditure for Research (in thousands of dollars) x	Profit (in thousands of dollars) y
A	80	100
B	80	120
C	60	80
D	100	100

a. Determine the least-squares prediction equation.

b. Does this equation show how "research expenditure generates profits"? Explain your answer.

10. A patient in a medical laboratory was given various doses of a certain drug to determine its effect on the patient's heart rate. The following data summarize the results of this experiment:

Units of Drug Administered x	Heart Beats per Minute y
9	70
6	60
3	50
9	80
15	100
12	90

a. Determine the least-squares prediction equation for the data.

b. If 10 units of the drug are administered, will the patient's heart rate be stabilized at approximately 80 beats per minute?

c. If it is desired to stabilize the patient's heart rate at 75 beats per minute, how many units of the drug should be administered?

SUGGESTED FURTHER READING

1. Draper, N., and H. Smith. *Applied Regression Analysis.* New York: Wiley, 1986.
2. Jencks, C. *Inequality: A Reassessment of the Effect of Family and Schooling in America.* New York: Basic Books, 1972.
3. Johnson, N. L., and F. C. Leone. *Statistics and Experimental Design in Engineering and the Physical Sciences.* Belmont, CA: Wadsworth, 1968.
4. Mendenhall, W., and J. T. McClave. *A Second Course in Business Statistics: Regression Analysis.* San Francisco: Dellen, 1981. (Chapter 3)
5. Neter, J., and W. Wasserman. Applied Linear Statistical Models. Homewood, IL: Richard Irwin, 1974. (Chapters 2–6)
6. Tanur, J. M., F. Mosteller, W. H. Kruskal, R. F. Link, R. S. Pieters, and G. R. Rising, eds. *Statistics: A Guide to the Unknown.* San Francisco: Holden-Day, 1978.

CHAPTER 12

Analyzing Count Data: The Chi-Square Distribution

OBJECTIVES

- *To present* chi-square tests which provide the basis for testing whether more than two population proportions can be considered as equal.

- *To study* contingency tables. These are tabular arrangements of data into a two-way classification. The chi-square test statistic tells us whether the two ways of classifying the data are independent.

- *To analyze* expected frequencies. These are the numbers, that is, the frequencies, that should appear in each of the boxes of a contingency table.

- *To discuss* a chi-square test statistic which can be applied to determine whether an observed frequency is in agreement with the expected mathematical distribution. This is known as the goodness of fit.

The Army's New Physical Fitness Standards

Age Groups	Push-ups		Sit-ups		Two-mile run	
	Min.	Max.	Min.	Max.	Min.	Max.
17-21	42	82	52	92	15:54	11:45
	18	58	50	90	18:45	14:45
22-26	40	80	47	87	16:36	12:36
	16	56	45	85	19:36	15:36
27-31	38	78	42	82	17:18	13:18
	15	54	40	80	21:00	17:00
32-36	33	73	38	78	18:00	14:00
	14	52	35	75	22:36	18:36
37-41	32	72	33	73	18:42	14:42
	13	48	30	70	22:36	19:36
42-46	26	66	29	69	19:06	15:12
	12	45	27	67	24:00	20:00
47-51	22	62	27	67	19:36	15:36
	10	41	24	64	24:30	20:30
52 & over	16	56	26	66	20:00	16:00
	09	40	22	62	25:00	21:00

■ MALE   ■ FEMALE

Source: Office of Deputy Chief of Staff for Operations, Individual Training Branch of the U.S Army.

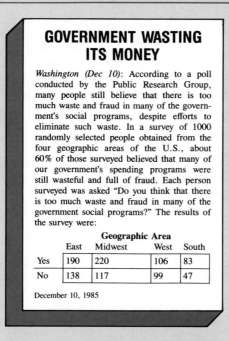

GOVERNMENT WASTING ITS MONEY

Washington (Dec 10): According to a poll conducted by the Public Research Group, many people still believe that there is too much waste and fraud in many of the government's social programs, despite efforts to eliminate such waste. In a survey of 1000 randomly selected people obtained from the four geographic areas of the U.S., about 60% of those surveyed believed that many of our government's spending programs were still wasteful and full of fraud. Each person surveyed was asked "Do you think that there is too much waste and fraud in many of the government social programs?" The results of the survey were:

| | **Geographic Area** | | | |
	East	Midwest	West	South
Yes	190	220	106	83
No	138	117	99	47

December 10, 1985

The first newspaper article gives the Army's new physical fitness standards for both men and women in different age groups. By looking at the data, can we conclude that people in different age groups are capable of performing different numbers of push-ups, sit-ups, and so on, or is the age of the person independent of these standards? Such information is very valuable since it enables an individual to measure his or her performance against that of others.

Now consider the second article. Is the geographical area in which a person lives a factor in determining a person's opinion as to whether the government is wasting its money? Are there ways of comparing the proportions of responses?

12.1 INTRODUCTION

In Chapter 10 we discussed methods for testing whether the observed difference between two sample means is significant. In this chapter we will analyze whether differences among two or more sample proportions are significant or whether they are due purely to chance. For example, suppose a college professor distributes a faculty-evaluation form to the 150 students of his Psychology 12 classes. The following are two examples of the multiple-choice questions appearing on the form:

1. What is your grade point average? (Assume A = 4 points, B = 3 points, C = 2 points, D = 1 point, and F = 0 points.)
   a. 3.0 to 4.0   b. 2.0 to 2.99   c. below 2.0
2. Would you be willing to take another course with this teacher?
   a. Yes   b. No

The results for these two questions are summarized in the following chart:

		Grade Point Average		
		3.0–4.0	*2.0–2.99*	*Below 2.0*
Would You	Yes	28	36	11
Take This	No	22	44	9
Teacher Again?	Total	50	80	20

The teacher may be interested in knowing whether these ratings are influenced by the student's grade point average. In the 3.0–4.0 category the proportion of students who said that they would take this teacher for another course is $\frac{28}{50}$. In the 2.0–2.99 category the proportion is $\frac{36}{80}$, and in the below 2.0 category the proportion is $\frac{11}{20}$. Is it true that students with a higher grade point average tend to rate the teacher differently than students with a lower grade point average?

Now consider the magazine clipping (Fig. 12.1), which gives the victimization rates by type of crime and the age of the victim. Is the type of crime committed related to the age of the intended victim? Such information is vital to law-enforcement officials. Questions of this type occur quite often.

In this chapter we will study the chi-square distribution, which is of great help in studying differences between proportions.

PERSONAL CRIMES: VICTIMIZATION RATES FOR PERSONS AGE 12 AND OVER, BY TYPE OF CRIME AND AGE OF VICTIMS, 1980

(Rate per 1,000 population in each age group)

Type of crime	12–15 (16,527,000)	16–19 (15,792,000)	20–24 (17,609,000)	25–34 (29,211,000)	35–49 (33,783,000)	50–64 (30,847,000)	65 and over (20,792,000)
Crimes of violence	52.6	67.9	61.1	38.6	20.8	11.8	9.0
Rape	1.5	2.5	2.1	1.4	0.2[1]	0.3	0.2[1]
Robbery	12.7	11.3	10.7	7.0	5.5	4.1	3.9
Robbery with injury	3.3	3.5	3.3	2.1	2.1	1.5	1.9
From serious assault	1.4	1.9	1.7	1.4	1.3	0.9	1.0
From minor assault	2.0	1.6	1.6	0.6	0.8	0.6	0.9
Robbery without injury	9.4	7.8	7.4	5.0	3.4	2.6	2.0
Assault	38.5	54.1	48.3	30.2	15.2	7.3	4.9
Aggravated assault	12.9	23.7	22.0	12.6	7.0	2.7	1.6
With injury	5.7	7.7	6.9	3.7	2.1	0.7	0.4[1]
Attempted assault with weapon	7.2	16.0	15.2	8.9	4.9	2.0	1.2
Simple assault	25.6	30.4	26.3	17.7	8.2	4.6	3.4
With injury	7.9	8.3	6.5	3.7	1.7	1.0	0.5
Attempted assault without weapon	17.8	22.1	19.7	13.9	6.5	3.6	2.9
Crimes of theft	166.7	159.8	146.3	106.2	79.2	49.4	21.9
Personal larceny with contact	3.1	3.7	3.4	2.6	2.6	3.5	3.4
Purse snatching	0.4[1]	0.6[1]	0.9	0.6	0.7	1.5	1.4
Pocket picking	2.7	3.1	2.4	2.0	1.9	1.9	2.0
Personal larceny without contact	163.6	156.1	143.0	103.5	76.7	45.9	18.5

NOTE: Detail may not add to total shown because of rounding. Numbers in parentheses refer to population in the group.
[1]Estimate, based on about 10 or fewer sample cases, is statistically unreliable.

FIGURE 12.1

Source: Criminal victimization in the United States, U.S. Department of Justice—Law Enforcement Assistance Administration, Washington, D.C.

12.2 THE CHI-SQUARE DISTRIBUTION

To illustrate the method that is used when analyzing several sample proportions, let us return to the first example discussed in the introduction. Let p_1 be the true proportion of students in the 3.0–4.0 category who will take another course with this teacher. Similarly let p_2 and p_3 represent the proportion of students in the 2.0–2.99 and below 2.0 categories who will take another course with this teacher. The null hypothesis that we wish to test is

$$H_0: \quad p_1 = p_2 = p_3$$

This means that a student's grade point average does not affect the student's decision to take this teacher again. The alternative hypothesis is that at least one of p_1, p_2, and p_3 are different. This means that a student's grade point average does affect the student's decision.

If the null hypothesis is true, the observed difference between the proportions in each of the grade point average categories is due purely to chance. Under this assumption we combine all the samples into one and consider it as one large sample. We then obtain the following estimate of the true proportion of *all* students in the school who are willing to take another course with the teacher. We get

$$\frac{28 + 36 + 11}{50 + 80 + 20} = \frac{75}{150} = 0.5$$

Thus our estimate of the true proportion of students who are willing to take another course with this teacher is 0.5.

There are 50 students in the 3.0–4.0 category. We would therefore expect 50(0.5), or 25, of these students to indicate yes they would take another course with this teacher, and we would expect 50(0.5), or 25, to indicate no. Similarly, in the 2.0–2.99 category there are 80 students so we would expect 80(0.5), or 40, yes answers and 40 no answers. Also, in the below 2.0 category there are 20 students, so that we would expect 20(0.5), or 10, yes answers and 10 no answers. The numbers that should appear are called **expected frequencies.** In the following table we have indicated these numbers in parentheses below the ones that were actually observed. We call the numbers that were actually observed **observed frequencies.**

	Grade Point Average		
	3.0–4.0	2.0–2.99	Below 2.0
Yes	28 (25)	36 (40)	11 (10)
No	22 (25)	44 (40)	9 (10)
Total	50	80	20

Notice that the expected frequencies and the observed frequencies are not the same. If the null hypothesis that $p_1 = p_2 = p_3$ is true, the observed frequencies should be fairly close to the expected frequencies. Since this rarely will happen, we need some way of determining when these differences are significant.

When is the difference between the observed frequencies and the expected frequencies significant? To answer this question we calculate a test statistic called the **chi-square statistic.**

FORMULA 12.1

Let E represent the expected frequency and let O represent the observed frequency. Then the **chi-square test statistic,** denoted as χ^2 is defined as

$$\chi^2 = \Sigma \frac{(O - E)^2}{E}$$

COMMENT   In using Formula 12.1 we must calculate the square of the difference for each box, that is, cell, of the table. Then we divide the squares of the difference for each cell by the expected frequency for that box. Finally, we add these results together.

Returning to our example we have

$$\chi^2 = \frac{(28 - 25)^2}{25} + \frac{(36 - 40)^2}{40} + \frac{(11 - 10)^2}{10} + \frac{(22 - 25)^2}{25} + \frac{(44 - 40)^2}{40} + \frac{(9 - 10)^2}{10}$$

$$= \quad 0.36 \quad + \quad 0.40 \quad + \quad 0.10 \quad + \quad 0.36 \quad + \quad 0.40 \quad + \quad 0.10$$

$$= 1.72$$

The value of the χ^2 statistic is 1.72.

It should be obvious from Formula 12.1 that the value of χ^2 will be 0 when there is perfect agreement between the observed frequencies and the expected frequencies, since in this case $O - E = 0$. Generally speaking, if the value of χ^2 is small, the observed frequencies and the expected frequencies will be pretty close to each other. On the other hand, if the value of χ^2 is large, this indicates that there is a considerable difference between the observed frequency and the expected frequency.

To determine when the value of the χ^2 statistic is significant, we use the **chi-square distribution.** This is pictured in Figure 12.2. We reject the null hypothesis when the value of the chi-square statistic falls in the rejection region of Figure 12.2. Table IX in the Appendix gives us the critical values, that is, the dividing line, depending upon the number of degrees of freedom. Thus, $\chi^2_{0.05}$ represents the dividing line that cuts off 5% of the right tail of the distribution. *The number of degrees of freedom is always 1 less than the number of sample proportions that we are testing.*

In our example we are comparing three proportions so that the number of degrees of freedom is $3 - 1$, or 2. Now we look at Table IX in the Appendix to find the χ^2 value that corresponds to two degrees of freedom. We have $\chi^2_{0.05} = 5.991$. Since the test statistic value that we obtained, $\chi^2 = 1.72$, is much less than the table value of 5.991, we do not reject the null hypothesis. The difference between what was expected and what actually happened can be attributed to chance.

COMMENT   Although we will usually use the 5% level of significance, Table IX in the Appendix also gives us the χ^2 values for the 1% level of significance.

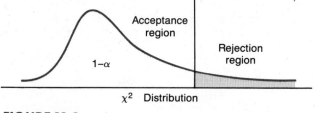

FIGURE 12.2

We use these values when we wish to find the dividing line that cuts off 1% of the right tail of the distribution.

Let us further illustrate the χ^2 test with several examples.

● Example 1

There are 10,000 students at a college. Twenty-seven hundred are freshmen, 2300 are sophomores, 3000 are juniors, and 2000 are seniors. Recently a new president was appointed. Two thousand students attended the reception party for the president. The attendance breakdown is shown in the following table:

		Freshmen	Sophomores	Juniors	Seniors
Attended Reception?	Yes	300	700	650	350
	No	2400	1600	2350	1650
	Total	2700	2300	3000	2000

Test the null hypothesis that the proportion of freshmen, sophomores, juniors, and seniors that attended the reception is the same. (Use a 5% level of significance.)

Solution

In order to compute the χ^2 test statistic, we must first compute the expected frequency for each box, or cell. To do this we obtain an estimate of the true proportion of students who attended the reception. We have

$$\frac{300 + 700 + 650 + 350}{2700 + 2300 + 3000 + 2000} = \frac{2000}{10,000} = 0.20$$

Out of 2700 freshmen we would expect 2700(0.20), or 540, to attend and 2700 − 540, or 2160, not to attend. Similarly, out of 2300 sophomores we would expect 2300(0.20), or 460, to attend and 2300 − 460, or 1840, not to attend. Also out of 3000 juniors we would expect 3000(0.20), or 600, to attend and 3000 − 600, or 2400, not to attend. Finally, out of 2000 seniors, we would expect 2000(0.20), or 400, to attend and 2000 − 400, or 1600, not to attend. We have indicated these expected frequencies just below the observed values in the following chart:

		Freshmen	Sophomores	Juniors	Seniors
Attended Reception?	Yes	300 (540)	700 (460)	650 (600)	350 (400)
	No	2400 (2160)	1600 (1840)	2350 (2400)	1650 (1600)
	Total	2700	2300	3000	2000

Now we calculate the value of the χ^2 statistic. We have

$$\chi^2 = \Sigma \frac{(O - E)^2}{E}$$

$$= \frac{(300 - 540)^2}{540} + \frac{(700 - 460)^2}{460} + \frac{(650 - 600)^2}{600}$$

$$+ \frac{(350 - 400)^2}{400} + \frac{(2400 - 2160)^2}{2160} + \frac{(1600 - 1840)^2}{1840}$$

$$+ \frac{(2350 - 2400)^2}{2400} + \frac{(1650 - 1600)^2}{1600}$$

$$= 106.67 + 125.22 + 4.17 + 6.25 + 26.67$$

$$+ 31.30 + 1.04 + 1.56$$

$$= 302.88$$

There are four proportions that we are testing so that there are $4 - 1$, or 3, degrees of freedom. From Table IX in the Appendix we find that the $\chi^2_{0.05}$ value with 3 degrees of freedom is 7.815. The value of the χ^2 test statistic ($\chi^2 = 302.88$) is definitely greater than 7.815. Hence we reject the null hypothesis. The proportions of freshmen, sophomores, juniors, and seniors that attended the reception are not the same.

● **Example 2**
A survey of the marital status of the members of three health clubs was taken. The following table indicates the results of the survey.

		Club 1	Club 2	Club 3
	Yes	11	17	8
Married?	No	29	33	22
	Total	40	50	30

Test the null hypothesis that the proportion of members that are married in each of these health clubs is the same. (Use a 5% level of significance.)

Solution
We must first compute the expected frequency for each cell. To do this we obtain an estimate of the true proportion of members who are married. We have

$$\frac{11 + 17 + 8}{40 + 50 + 30} = \frac{36}{120} = 0.3$$

Thus, the estimate of the true proportion is 0.3. Out of the 40 members in Club 1 we would expect 40(0.3), or 12, members to be married and 40 − 12, or 28, not to be married. In Club 2 we would expect 50(0.3), or 15, members to be married and 50 − 15, or 35, members not to be married. In Club 3 we would expect 30(0.3), or 9, members to be married and 30 − 9, or 21, not to be married. We have indicated these expected frequencies in parentheses in the following table:

		Club 1	Club 2	Club 3
	Yes	11 (12)	17 (15)	8 (9)
Married?	No	29 (28)	33 (35)	22 (21)
	Total	40	50	30

Now we calculate the value of the χ^2 statistic. We have

$$\chi^2 = \Sigma \frac{(O - E)^2}{E}$$

$$= \frac{(11 - 12)^2}{12} + \frac{(17 - 15)^2}{15} + \frac{(8 - 9)^2}{9} + \frac{(29 - 28)^2}{28}$$

$$+ \frac{(33 - 35)^2}{35} + \frac{(22 - 21)^2}{21}$$

$$= 0.08 + 0.27 + 0.11 + 0.04 + 0.11 + 0.05$$

$$= 0.66$$

There are three proportions that we are testing so that there are 3 − 1, or 2, degrees of freedom. From Table IX in the Appendix we find that the $\chi^2_{0.05}$ value with two degrees of freedom is 5.991. Since the value of the test statistic, 0.66, is less than 5.991, we do not reject the null hypothesis.

COMMENT   Experience has shown us that the χ^2 test can only be used when the expected frequency in each cell is at least 5. If the expected frequency of a cell is not larger than 5, this cell should be combined with other cells until the expected frequency is at least 5. We will not, however, concern ourselves with this situation.

EXERCISES

1. A sociologist is interested in determining if a woman's age is a factor in whether she smokes or not. A random survey of 1200 working women in New York produced the following results:

		Age group (in years)			
		Between 20 and 29	Between 30 and 39	Between 40 and 49	Over 50
Smoker?	Yes	167	131	49	25
	No	320	278	148	82

Using a 5% level of significance, test the null hypothesis that there is no significant difference between the corresponding proportion of women in the various age groups who smoke.

2. A psychologist reported that in a study of 240 people boarding an airplane it was found that 41 of the 158 male passengers were left-handed and 25 of the 82 female passengers were left-handed. Using a 1% level of significance, test the null hypothesis that there is no significant difference between the corresponding proportion of male or female passengers who are left-handed.

3. A random survey of 800 families in St. Louis, Miami, and Los Angeles was taken to determine the proportion of families who own a home computer system.* The following table indicates the results of the survey.

		City		
		St. Louis	Miami	Los Angeles
Own a Home Computer System?	Yes	40	55	70
	No	210	145	280

Using a 5% level of significance, test the null hypothesis that the proportion of families in all three cities that own a home computer system is the same.

4. A nationwide grocery store chain purchases cookies from five different suppliers located in different parts of the country. Due to the numerous complaints received about the number of broken cookies in each box of its most-popular-selling cookie, management samples 1000 boxes of cookies produced by these suppliers and determines the number of broken cookies in each box. The results are shown below:

		Suppliers				
		A	B	C	D	E
Acceptable Number of Broken Cookies in Box?	Yes	202	186	142	189	179
	No	18	22	33	12	17

*Public Interest Research Group, "Who Owns A Home Computer System," 1985.

Using a 1% level of significance, test the null hypothesis that the proportion of boxes with an acceptable number of broken cookies in each is the same for all the suppliers.

5. A survey of 1816 people located throughout the state was taken to see how many of them were in favor of building a new nuclear electrical generating station to meet the future demands of the people living within the state. The following table indicates the results of the survey according to the geographic location of the person interviewed:

		Geographic Location Within State			
		North	South	East	West
In Favor of Building a New Nuclear Station?	Yes	129	137	164	158
	No	317	288	302	321

Using a 1% level of significance, test the null hypothesis that the proportion of residents in favor of building a new nuclear electrical generating station is the same in each of the geographic locations surveyed.

6. A study was conducted to determine whether an individual's social class has any effect on the individual's belief that the quality of his or her life has changed since 1980. Each of 812 randomly selected individuals was asked the same question: "Do you believe that the quality of your life has changed since 1980?" A summary of their responses is shown in the accompanying table:

		Social Class			
		Upper	Upper-Middle	Middle-Lower	Lower
Has the Quality of Your Life Changed?	Yes	69	84	81	69
	No	111	138	131	129

Using a 5% level of significance, test the null hypotheses that the percentage of people who believe that the quality of their life has changed is the same for all social classes.

7. A legislator polled 450 of her constituents to determine their views regarding a new piece of proposed legislation requiring restaurant owners to set up special sections for smokers and nonsmokers. A summary of their responses is shown below.

		Frequency of Restaurant Visits			
		Almost Every Day	Once a Week	Once a Month	Almost Never
In Favor of Proposed Legislation?	Yes	98	84	69	27
	No	76	53	32	11

Using a 1% level of significance, test the null hypothesis that the percentage of constituents in favor of the proposed legislation is the same for *all* the patrons of the restaurants, regardless of frequency of visits.

8. Consider the newspaper article below. The commission surveyed homes on six streets and obtained the following data:

		Homes Located on					
Dangerous Level of Radon Gas Present?		Bay Street	Colby Square	Morris Ave	Francis Blvd	Clarence Road	Hylan Ave
	Yes	7	9	3	8	6	10
	No	31	29	37	26	39	40

Using a 5% level of significance, test the null hypothesis that the percentage of homes containing excessive levels of radon gas is the same for homes located on all the streets.

MORE HOMES CONTAMINATED WITH RADON

Bergen (Sept. 12): A new study released today by the state's environmental commission reveals that the number of homes containing excessive amounts of deadly radon gas is larger than expected. The radon gas is entering into many homes on the East Coast (particularly those in the states of New Jersey and Pennsylvania) through holes or cracks in the foundation. It is believed that the gas is coming from deep within the earth.

The Commission urged the governor to appropriate additional funds to enable it to expand its monitoring activities.

September 12, 1985

9. A textbook publisher is interested in knowing whether the percentage of books that it has published over the years that contain misprints or mathematical errors is the same for all math or math-related books. The following data have been obtained from the production department of the company.

		Type of Book Published			
	Statistics	Algebra or Trigonometry	Geometry	Calculus and Analysis	Advanced Theoretical Books
Does the Book Have 5 or Fewer Errors? Yes	13	19	10	15	22
No	8	11	4	7	10

Using a 5% level of significance, test the null hypothesis that the percentage of books with five or fewer errors is the same for all types of math or math-related books.

10. It is often claimed that the percentage of students from one-parent homes that attend the public high schools of a large city is the same for all the high schools in the city. A recent study by Hodges and Giliksen (Los Angeles, 1985) revealed the following information about students at five high schools.

		High-School			
	A	B	C	D	E
Student from One-parent Home? Yes	193	203	128	309	512
No	646	716	517	1462	784

Using a 1% level of significance, test the null hypothesis that the percentage of students from one-parent homes is the same for these high schools.

11. A newspaper reporter conducted a random survey of 3127 people nationwide to determine their views on the fiscal soundness of the Social Security system. The results are summarized below.

	Geographic Location of Respondent				
	Far West	West	South	Northeast	Southeast
Do You Believe in the Fiscal Soundness of the Social Security System? Yes	288	390	333	262	220
No	312	414	375	280	253

Using a 1% level of significance, test the null hypothesis that the percentage of people who believe in the fiscal soundness of the Social Security system is the same for all geographical regions.

12.3 GOODNESS OF FIT

In addition to the applications mentioned in the previous section, the chi-square test statistic can also be used to determine whether an observed frequency distribution is in agreement with the expected mathematical distribution. For example, when a die is rolled, we assume that the probability of any one face coming up is $\frac{1}{6}$. Thus, if a die is rolled 120 times, we would expect

each face to come up approximately 20 times since $\mu = np = 120\left(\frac{1}{6}\right) = 20$.

Suppose we actually rolled a die 120 times and obtained the results shown in Table 12.1. In this table we have also indicated the expected frequencies. Are these observed frequencies reasonable? Do we actually have an honest die?

TABLE 12.1
Expected and Observed Frequencies
When a Die Was Tossed 120 Times

Die Shows	Expected Frequency	Observed Frequency
1	20	18
2	20	21
3	20	17
4	20	21
5	20	19
6	20	24
	Total = 120	Total = 120

To check whether the differences between the observed frequencies and the expected frequencies are due purely to chance or are significant, we use the chi-square test statistic of Formula 12.1. We reject the null hypothesis that the observed differences are not significant only when the test statistic falls in the rejection region.

In our case the value of the test statistic is

$$\chi^2 = \Sigma \frac{(O - E)^2}{E}$$

$$= \frac{(18 - 20)^2}{20} + \frac{(21 - 20)^2}{20} + \frac{(17 - 20)^2}{20} + \frac{(21 - 20)^2}{20}$$

$$+ \frac{(19 - 20)^2}{20} + \frac{(24 - 20)^2}{20}$$

$$= 1.60$$

There are $6 - 1$, or 5, degrees of freedom. From Table IX in the Appendix we find that the $\chi^2_{0.05}$ value with five degrees of freedom is 11.070. Since the test statistic has a value of only 1.60, which is considerably less than 11.070, we do not reject the null hypothesis. Any differences between the observed frequencies and the expected frequencies are due purely to chance.

The following examples will further illustrate how the chi-square test statistic can be used to test **goodness of fit,** that is, to determine whether the observed frequencies fit with what was expected.

● Example 1

The number of phone calls received per day by a local chapter of Alcoholics Anonymous is as follows:

	M	T	W	T	F
Number of Calls Received	173	153	146	182	193

Using a 5% level of significance, test the null hypothesis that the number of calls received is independent of the day of the week.

Solution

We first calculate the number of expected calls per day. If the number of calls received is independent of the day of the week, we would expect to receive

$$\frac{173 + 153 + 146 + 182 + 193}{5} = 169.4$$

calls per day. We can then set up the following table:

	M	T	W	T	F
Observed Number of Calls	173	153	146	182	193
Expected Number of Calls	169.4	169.4	169.4	169.4	169.4

Now we calculate the value of the chi-square test statistic. We have

$$\chi^2 = \Sigma \frac{(O - E)^2}{E}$$

$$= \frac{(173 - 169.4)^2}{169.4} + \frac{(153 - 169.4)^2}{169.4} + \frac{(146 - 169.4)^2}{169.4}$$

$$+ \frac{(182 - 169.4)^2}{169.4} + \frac{(193 - 169.4)^2}{169.4}$$

$$= 0.0765 + 1.5877 + 3.2323 + 0.9372 + 3.2878$$

$$= 9.1215$$

There are $5 - 1$, or 4, degrees of freedom. From Table IX in the Appendix we find that the $\chi^2_{0.05}$ value with four degrees of freedom is 9.488. Since the test statistic has a value of 9.1215, which is less than 9.488, we do not reject the null hypothesis and the claim that the number of calls received is independent of the day of the week.

● Example 2

A scientist has been experimenting with rats. As a result of certain injections, the scientist claims that when two black rats are mated, the offspring will be black, white, and gray in the proportion $5:4:3$. (This means that the probability of a black offspring is $\frac{5}{12}$, the probability of a white offspring is $\frac{4}{12}$, and the probability of a gray rat is $\frac{3}{12}$.) Many rats were mated after being injected with the chemical. Of 180 newborn rats 71 were black, 69 were white, and 40 were gray. Can we accept the scientist's claim that the true proportion is $5:4:3$? Use a 5% level of significance.

Solution

We first calculate the expected frequencies. Out of 180 rats we would expect $180\left(\frac{5}{12}\right)$, or 75, of them to be black. Similarly, out of 180 rats we would expect $180\left(\frac{4}{12}\right)$, or 60, of them to be white, and we would expect $180\left(\frac{3}{12}\right)$, or 45, of them to be gray. We now set up the following table:

Color of Rat	Expected Frequency	Observed Frequency
Black	75	71
White	60	69
Gray	45	40

The value of the χ^2 test statistic is

$$\chi^2 = \Sigma \frac{(O - E)^2}{E}$$

$$= \frac{(71 - 75)^2}{75} + \frac{(69 - 60)^2}{60} + \frac{(40 - 45)^2}{45}$$

$$= 2.12$$

There are $3 - 1$, or 2, degrees of freedom. From Table IX in the Appendix we find that the $\chi^2_{0.05}$ value with two degrees of freedom is 5.991. Since the test statistic has a value of 2.12, which is less than 5.991, we do not reject the null hypothesis and the scientist's claim.

EXERCISES

1. There were 1421 people treated in the emergency room at Lincoln Hospital during one week. These were as follows:

Day of Week	Number of People Treated
Sun.	237
Mon.	188
Tues.	201
Wed.	179
Thurs.	210
Fri.	200
Sat.	206

Using a 5% level of significance, test the null hypothesis that the proportion of people treated daily is independent of the day of the week.

2. The incomes of the 212 families who applied to adopt a child in Willington are indicated in the following table.

Income	Number of Families
Under $20,000	48
$20,000–29,999	45
$30,000–39,999	48
$40,000–49,999	36
Above $50,000	35

Using a 5% level of significance, test the null hypothesis that the proportion of families applying to adopt a child from each of the income groups is the same.

3. There are five entrance ramps to the main parking lot at an airport. The number of cars entering the parking lot through these ramps during one period is as follows:

Entrance Ramp on	Number of Cars Entering Through This Ramp
Northeast	238
Northwest	216
South	302
West	294
Southeast	199

Using a 5% level of significance, test the null hypothesis that all the entrances to the parking lot are used by cars with the same frequency.

4. According to one theory, when a particular plant is crossbred with another plant, the offspring should be red, pink, yellow, and orange in the ratio $1:2:3:5$. Of 220 new plants 18 were red, 45 were pink, 66 were yellow, and 91 were orange. Using a 5% level of significance, can we accept the theory that the true proportion is $1:2:3:5$?

5. An advertising company conducted a taste test at a shopping mall. Each of 240 randomly selected volunteers was given five samples of a soft drink. The volunteers were then asked to select the one with the greatest taste appeal. The results of the survey were then tallied and presented to the manufacturers. After analyzing the accompanying data, one of the manufacturers believes that the volunteers expressed no significant difference in their opinions of the taste appeal of the products and that any observed difference was due merely to random selection.

Drink with Greatest Appeal	Frequency of Response
A	48
B	39
C	53
D	60
E	40

Using a 1% level of significance, test the null hypothesis that the manufacturer's assumptions are correct, and that customers prefer each of these brands equally.

6. An automatic salad vending machine dispenses units of pepper, cucumber, tomatoes, and lettuce in the ratio $2:3:7:11$ when functioning properly. A food manager notices that in 552 salads prepared by the machine, the following was dispensed:

Item	Number of Units Dispensed
Pepper	42
Cucumber	70
Tomatoes	180
Lettuce	260

Using a 5% level of significance, test the null hypothesis that the machine is functioning properly.

7. An obstetrician delivered 38 babies in the fall, 69 babies in the winter, 41 babies in the spring, and 17 babies in the summer. Using a 5% level of significance, test the null hypothesis that the proportion of babies delivered by the obstetrician during the various seasons of the year is independent of the season of the year.

8. A college guidance counselor is analyzing the majors of this year's 1424 graduates. The counselor has compiled the following data:

Major	Number of Students With This Major
Computer Science	282
Engineering	258
Nursing	276
Business	309
Other	299

Using a 5% level of significance, test the null hypothesis that the proportion of students majoring in the various categories mentioned is not significantly different.

9. Using gene-splicing techniques, a researcher claims to have developed a new strain of cattle with many desirable characteristics. Pending further investigation, the researcher coded these characteristics as A, B, C, and D. The following results were obtained in several gene-splicing experiments:

717 had characteristic A
200 had characteristic B
210 had characteristic C
 80 had characteristic D

Using a 5% level of significance, test the null hypothesis that the frequencies of these types of characteristics obtained by gene-splicing are in the ratio 9:3:3:2.

10. "Fear of heights" is a feeling demonstrated by many people. A psychologist is interested in knowing whether frequency of expression of such a fear varies by age group. The following data have been collected for 279 people who are afraid of heights.

Age of Person (in years)	10–19	20–29	30–39	40–49	50–59
Number of People Expressing Fear of Heights	58	42	53	61	65

Using a 5% level of significance, does the above data indicate that people in some age groups are more likely to be afraid of heights than people in other age groups?

11. The Austrian monk Gregor Mendel performed many experiments with garden peas. In one such experiment the following results were obtained:

322 round and yellow peas
104 wrinkled and yellow peas
108 round and green peas
 37 wrinkled and green peas

Using a 5% level of significance, test the null hypothesis that the frequencies of these types of peas should be in the ratio 9:3:3:1.

12.4 CONTINGENCY TABLES

A very useful application of the χ^2 test discussed in Section 12.2 occurs in connection with contingency tables. Contingency tables are used when we wish to determine whether two variables of classification are related or dependent one upon the other. For example, consider the following chart, which indicates the eye color and hair color of 100 randomly selected girls:

	Brown Eyes	Blue Eyes
Light Hair	10	33
Dark Hair	44	13

Is eye color independent of hair color or is there a significant relationship between hair color and eye color?

Contingency tables are especially useful in the social sciences where data are collected and often classified into two main groups. We might be interested in determining whether a relationship exists between these two ways of classifying the data or whether they are independent. We have the following definition:

DEFINITION 12.1

A **contingency table** is an arrangement of data into a two-way classification. One of the classifications is entered in rows and the other in columns.

When dealing with contingency tables, remember that the null hypothesis always assumes that the two ways of classifying the data are independent. We use the χ^2 test statistic as discussed in Section 12.2. The only difference is that we compute the expected frequency for each cell by using the following formula.

FORMULA 12.2

The expected frequency of any cell in a contingency table is found by multiplying the total of the row with the total of the column to which the cell belongs. The product is then divided by the total sample size.

The following examples will illustrate how we apply the χ^2 test to contingency tables.

● Example 1

Let us consider the contingency table given at the beginning of this section. Is eye color independent of hair color?

Solution

In order to compute the χ^2 test statistic, we must first compute the row total, column total, and the total sample size. We have

$$
\begin{aligned}
\textit{Row Totals:}& \quad 10 + 33 = 43 \\
& \quad 44 + 13 = 57 \\
\textit{Column Totals:}& \quad 10 + 44 = 54 \\
& \quad 33 + 13 = 46 \\
\textit{Total Sample Size:}& \quad 10 + 33 + 44 + 13 = 100
\end{aligned}
$$

We indicate these values in the following table:

	Brown Eyes	Blue Eyes	Row Total
Light Hair	10 (23.22)	33 (19.78)	43
Dark Hair	44 (30.78)	13 (26.22)	57
Column Total	54	46	

The expected value for the cell in the first row first column is obtained by multiplying the first row total with the first column total and then dividing the product by the total sample size. We get

$$\frac{54 \times 43}{100} = 23.22$$

For the first row second column we have

$$\frac{46 \times 43}{100} = 19.78$$

For the second row first column we have

$$\frac{54 \times 57}{100} = 30.78$$

For the second row second column we have

$$\frac{46 \times 57}{100} = 26.22$$

These values are entered in parentheses in the appropriate cell. We now use Formula 12.1 of Section 12.2 and calculate the χ^2 statistic. We have

$$\chi^2 = \Sigma \frac{(O - E)^2}{E}$$

$$= \frac{(10 - 23.22)^2}{23.22} + \frac{(33 - 19.78)^2}{19.78} + \frac{(44 - 30.78)^2}{30.78} + \frac{(13 - 26.22)^2}{26.22}$$

$$= 7.53 + 8.84 + 5.68 + 6.67$$
$$= 28.72$$

The χ^2 test statistic has a value of 28.72. *If the contingency table has r rows and c columns, then the number of degrees of freedom is $(r - 1)(c - 1)$.* In this example there are two rows and two columns, so there are $(2 - 1) \cdot (2 - 1)$, or $1 \cdot 1$, which is 1 degree of freedom. From Table IX in the Appendix we find that the $\chi^2_{0.05}$ value with 1 degree of freedom is 3.841. Since we obtained a value of 28.72, we reject the null hypothesis and conclude that hair color and eye color are *not* independent.

- **Example 2** *Criminal Analysis*
A sociologist is interested in determining whether the occurrence of different types of crimes varies from city to city. An analysis of 1100 reported crimes produced the following results:

Type of Crime

	Rape	Auto Theft	Robbery and Burglary	Other	Total
City A	76 (61.35)	112 (146.34)	87 (72.66)	102 (96.65)	377
City B	64 (68.83)	184 (164.20)	77 (81.52)	98 (108.44)	423
City C	39 (48.82)	131 (116.45)	48 (57.82)	82 (76.91)	300
Total	179	427	212	282	1100

Does this data indicate that the occurrence of a type of crime is dependent upon the location of the city? (Use a 5% level of significance.)

Solution

We first calculate the expected frequency for each cell. We have

$$\textit{First Row:} \quad \frac{(179)(377)}{1100} = 61.35 \quad\quad \frac{(427)(377)}{1100} = 146.34$$

$$\frac{(212)(377)}{1100} = 72.66 \quad\quad \frac{(282)(377)}{1100} = 96.65$$

$$\textit{Second Row:} \quad \frac{(179)(423)}{1100} = 68.83 \quad\quad \frac{(427)(423)}{1100} = 164.20$$

$$\frac{(212)(423)}{1100} = 81.52 \quad\quad \frac{(282)(423)}{1100} = 108.44$$

$$\textit{Third Row:} \quad \frac{(179)(300)}{1100} = 48.82 \quad\quad \frac{(427)(300)}{1100} = 116.45$$

$$\frac{(212)(300)}{1100} = 57.82 \quad\quad \frac{(282)(300)}{1100} = 76.91$$

These numbers appear in parentheses in the above chart. Now we calculate the χ^2 test statistic, getting

$$\chi^2 = \Sigma \frac{(O - E)^2}{E}$$

$$= \frac{(76 - 61.35)^2}{61.35} + \frac{(112 - 146.34)^2}{146.34} + \cdots + \frac{(82 - 76.91)^2}{76.91}$$

$$= 24.46$$

There are three rows and four columns so that there are six degrees of freedom since

$$(3 - 1)(4 - 1) = 2 \cdot 3 = 6$$

From Table IX in the Appendix we find that the $\chi^2_{0.05}$ value with six degrees of freedom is 12.592. Since we obtained a χ^2 value of 24.46, we reject the null hypothesis and conclude that the type of crime and the location of the city are not independent.

EXERCISES

1. A hospital administrator wishes to know if there is a relationship between a patient's length of stay in the hospital for a specified illness and the extent of the person's hospitalization insurance. Of people who were patients in the hospital over the past six months for the specified illness, 1440 are randomly selected. The extent of hospital insurance reimbursement is determined and is shown below.

Extent of Insurance Reimbursement

	Between 50% and 60%	Between 60% and 80%	Between 80% and 90%	Above 90%
1–5	39	55	69	84
6–8	60	71	88	103
9–12	79	92	104	127
Above 12	81	101	132	155

Length of Hospital Stay (in days)

Using a 5% level of significance, test the null hypothesis that the length of hospital stay is independent of the extent of hospital insurance reimbursement.

2. Recently, a nationwide survey by the Brown Associates of used-car buyers in different age groups was conducted to determine which feature in a used car was of utmost concern to the buyer. The following results were obtained:

| | **Car Feature** | | | |
	Economy of Operation	Styling	Size	Color
Under 30	262	392	301	361
30–50	427	309	296	203
Over 50	522	283	371	152

Age (in years) of Respondent

Using a 5% level of significance, test the null hypothesis that the age of the respondent is independent of the car feature selected.

TIGHTER SECURITY AT AIRPORTS

Washington (July 5): As a direct result of the recent TWA hijacking, government officials have announced a series of security checks at airports of all international travelers and their luggage to thwart any possible terrorist attacks. Yesterday, the *Tribune* conducted a random survey of travellers at the nation's airports to determine whether they were satisfied with the new security arrangements and the resulting delays. The results are summarized below:

	Support the checks enthusiastically despite delays	Support the checks moderately	Are opposed to the checks because of delays
Male Frequent traveler	43	67	21
Infrequent traveler	79	99	32
Female Frequent traveler	48	58	17
Infrequent traveler	69	95	27

Monday, July 5, 1985

3. Consider the newspaper article given above. Using a 5% level of significance, test the null hypothesis that the type of traveler is independent of the degree of satisfaction with the new security arrangement and the resulting delays.

4. In a recent survey conducted by the Acme Insurance Company of 300 cars equipped with some anti-theft device, the following information was obtained.

		Type of Anti-theft Device		
		Ignition Shut-off	Steering Wheel Lock	Burglar Alarm
	Compact	48	27	53
Size of Car	Intermediate	32	19	46
	Large	17	22	36

Using a 5% level of significance, test the null hypothesis that the type of anti-theft device used is independent of the size of the car.

5. An analysis of the medical records of the Bakst Corporation reveals the following information about 632 of its employees who applied for coverage under the company's extended health-coverage plans.

	Heavy Smoker and Drinker	Heavy Smoker and Nondrinker	Nonsmoker but Heavy Drinker	Nonsmoker and Nondrinker
Male	162	98	114	6
Female	64	136	42	10

Using a 5% level of significance, test the null hypothesis that the sex of the employee is independent of whether the employee is a heavy smoker or a drinker.

6. In an effort to generate sales and influence the shopper, many businesses advertise heavily in magazines or media directed primarily toward some particular segment of the population. An advertising agency recently conducted a random survey of shoppers at five different sales outlets to determine who usually buys shirts for a man—the man himself or a woman. The results of the survey are presented in the accompanying chart.

		Sales Location				
		I	II	III	IV	V
Who Bought	Male	42	49	51	68	81
the Shirt?	Female	48	57	50	83	92

Using a 1% level of significance, test the null hypothesis that the number of males or females who will buy a shirt for a man is not significantly different at each of these locations.

7. Consider the newspaper article below. A competing drug company conducted a survey to determine the preference of its customers for various forms of medicines. Their responses as well as their age are indicated in the accompanying table.

NO MORE TYLENOL CAPSULES

New Jersey (Jan. 19): As a direct result of the recent drug tampering, Johnson and Johnson, makers of Tylenol, and one of the nation's large drug manufacturers, announced yesterday that it would no longer manufacture any over-the-counter drugs in capsule form.

Wednesday, Jan. 19, 1986

Medicine Preference Form for Pills

	Caplet	Capsule	Tablet
Age of User (in years) Between 20 and 30	35	38	44
Between 30 and 50	69	87	49
Over 50	57	78	32

Using a 1% level of significance, test the null hypothesis that the preference in medicine form is the same for all age groups.

8. A quality-control engineer at General Electronics Company samples parts from each of the company's production lines on a daily basis. The following data are available for one week's production. (The production lines are closed on Friday.)

Number of Defective Items Produced on Production Line

Day of Week	1	2	3	4	5
Mon.	18	12	16	12	11
Tues.	15	17	19	21	13
Wed.	14	18	15	16	10
Thurs.	11	13	15	16	14

Using a 5% level of significance, test the null hypothesis that the number of defective items produced on the various production lines is independent of the day of the week.

9. The number of arrests by U.S. Border Patrol agents of aliens attempting to enter the United States illegally at several locations during a four-month period is as follows:

		Location			
		A	*B*	*C*	*D*
	Jan.	17	15	16	10
Month	Feb.	13	12	18	11
	Mar.	21	24	19	16

Using a 5% level of significance, test the null hypothesis that the number of arrests of illegal aliens attempting to enter the United States at the various locations is independent of the month of the year.

10. A politician obtained the following data in a random sample of 2485 voters in a district.

		Political Preference			
		Democrat	Republican	Conservative	Other
	Under 25	351	287	158	78
Age of Voter (*in years*)	Between 25 and 50	297	317	142	98
	Over 50	225	210	179	143

Using a 5% level of significance, test the null hypothesis that the political preference of a voter is independent of the age of the voter.

12.5 USING COMPUTER PACKAGES

MINITAB is well-suited to perform all calculations needed to test for independence. One need only enter the cell frequencies and obtain the results. MINITAB is especially useful when the number of rows and columns of a contingency table is large. Let us apply the MINITAB program to the data given in Example 2 of Section 12.4 (page 549).

```
MTB  > READ THE TABLE INTO C1, C2, C3, C4
DATA >    76    112    87    102
DATA >    64    184    77     98
DATA >    39    131    48     82


MTB  > CHISQUARE ANALYSIS ON TABLE IN C1, C2, C3, C4
EXPECTED FREQUENCIES ARE PRINTED BELOW OBSERVED FREQUENCIES
```

```
            !   C1   !   C2   !   C3   !   C4   ! TOTALS
  ----------!--------!--------!--------!--------!--------
            !   76   !  112   !   87   !  102   !
      1     !        !        !        !        !   377
            !   61.3 !  146.3 !   72.7 !   96.6 !
  ----------!--------!--------!--------!--------!--------
            !   64   !  184   !   77   !   98   !
      2     !        !        !        !        !   423
            !   68.8 !  164.2 !   81.5 !  108.4 !
  ----------!--------!--------!--------!--------!--------
            !   39   !  131   !   48   !   82   !
      3     !        !        !        !        !   300
            !   48.8 !  116.5 !   57.8 !   76.9 !
  ----------!--------!--------!--------!--------!--------
  TOTALS  !  179   !  427   !  212   !  282   !  1100

  TOTAL CHI SQUARE =
        3.50 + 8.06 + 2.83 + 0.30 + 0.34 + 2.39 +
        0.25 + 1.01 + 1.97 + 1.82 + 1.67 + 0.34 = 24.48
  DEGREES OF FREEDOM = (3-1)  X  (4-1) = 6

  MTB > STOP
```

12.6 SUMMARY

In this chapter we discussed the chi-square distribution and how it can be used to test hypotheses that differences between expected frequencies and observed frequencies are due purely to chance.

We applied the chi-square test statistic to test whether observed frequency distributions are in agreement with expected mathematical frequencies.

The chi-square test statistic can also be used to analyze whether the two factors of a contingency table are independent. This is very useful, especially in the social sciences, where the data are often grouped according to two factors.

When using the χ^2 test statistic, we must take great care in determining the number of degrees of freedom. Also, as we pointed out in a comment on page 536, each expected cell frequency must be at least 5 for the χ^2 test statistic to be applied.

STUDY GUIDE

You should now be able to demonstrate your knowledge of the following ideas presented in this chapter by giving definitions, descriptions, or specific examples. Page references are given for each term so that you can check your answer.

Expected frequency (page 532)
Observed frequency (page 532)
Chi-square test statistic (page 532)

The chi-square distribution (page 533)
Goodness of fit (page 540)
Contingency table (page 547)

FORMULAS TO REMEMBER

You should be able to identify each symbol in the following formulas, understand the relationships among the symbols expressed in each formula, understand the significance of each formula, and use the formulas in solving problems.

1. $\chi^2 = \Sigma \dfrac{(O - E)^2}{E}$

where O = observed frequency and E = expected frequency.

2. The expected frequency for any cell of a contingency table:

$$\dfrac{\text{(total of row to which cell belongs)} \cdot \text{(total of column to which cell belongs)}}{\text{total sample size}}$$

3. The number of degrees of freedom for a contingency table:

$(r - 1)(c - 1)$

where c = number of columns and r = number of rows

The tests in the following section will be more useful if you take them after
you have studied the examples and solved the exercises in this chapter.

MASTERY TESTS

Form A

1. One hundred computer science majors were asked to indicate what they
used their microcomputers for. The results of the survey are shown below:
(The null hypothesis is that there is no significant difference between uses.)

	Word Processing	Games	Spread Sheet Analysis	Data Base
Observed	28	31	18	23
Expected	25	25	25	25

The number of degrees of freedom for this example is
a. 4   **b.** 3   **c.** 2   **d.** 1   **e.** none of these

2. Refer back to Question 1. When we calculate the χ^2 (chi-square) test statistic we get
a. 25 **b.** 5.04 **c.** 3.92 **d.** 4.12 **e.** none of these

3. A recent survey of many residents of a city contained the following question: "Should the city administration expand the operation of the child day-care centers?" The responses of the 600 "yes" votes were grouped as follows.

		Gender of Respondent	
		Male	Female
	Under 20 years	77	136
Age of Respondent	20–30 years	78	94
	Over 30 years	33	182

What is the expected number of females 20–30 years of age who are in favor of expanding the operations of the day-care centers? (Round to the nearest whole number.)
a. 94 **b.** 118 **c.** 172 **d.** 412 **e.** none of these

4. The number of degrees of freedom for the data given in Question 3 is
a. 6 **b.** 3 **c.** 2 **d.** 1 **e.** none of these

5. Refer back to Question 3. What is the value of χ^2?

6. Refer back to Question 3. Using a 5% level of significance, test the null hypothesis that the gender of the respondent is independent of the age of the respondent.

7. A recent survey of 1470 smokers and nonsmokers contained the following question: "Do you believe that smoking is harmful to your health?"* The responses are shown below:

	Smokers	Nonsmokers
Harmful	198	712
Not Harmful	452	108

What is the expected number of smokers who believe that smoking is harmful?
a. 402.4 **b.** 242.1 **c.** 398.3 **d.** 424.72 **e.** none of these

8. The Florida State Attorney General's office receives a number of telephone complaints daily from senior citizens dealing with deceptive business practices. During one week, the number of calls received was as follows:

*Source: Giligman, "The Effects of Smoking," (1985).

Day of Week	Number of Calls Received
Mon.	29
Tue.	34
Wed.	32
Thurs.	27
Fri.	31

Using a 5% level of significance, test the null hypothesis that the number of calls received is independent of the day of the week.

9. A housing sales agency interviewed 931 potential buyers of apartments in a new housing complex to determine which factor was important in deciding whether or not to buy. The results are presented in the accompanying table.

		Factor				
		Crime-free Neighborhood	Many Playgrounds	Schools Nearby	Shopping Nearby	Parking
Gender of Respondent	Male	168	77	68	47	102
	Female	186	91	61	58	73

Using a 5% level of significance, test the null hypothesis that the gender of the respondent is independent of the factor considered important by a potential buyer.

10. The Atlas Travel Agency arranges many tour packages. It recently surveyed 586 of its customers to obtain some information needed to arrange such tour packages. Each customer was asked the same question: "How should the tour packages be arranged?" The answers of the customers are presented in the accompanying table:

		How Should Tour Be Arranged?		
		Every Day Planned Out	Several Days Planned. The Remainder Free Time	All Free Time. Only Arrange Air Transportation
Age of Customer (in years)	20–30 years	31	82	102
	Between 30 and 50 years	46	71	65
	Over 50 years	57	93	39

Using a 5% level of significance, test the null hypothesis that the age of the customer is independent of the customer's views as to how the tour should be arranged.

MASTERY TESTS

Form B

1. An analysis of the cause of death of 717 patients at the Morgantown Hospital and Convalescent Home reveals the following information:

		Cause of Death				
		Disease of Heart and Blood Vessels	Cancer	Pneumonia and Influenza	Accidents	All Other Causes
Age of Person (in years)	Under 50	42	29	12	58	74
	Between 50 and 65	59	34	43	19	68
	Over 65	67	42	81	7	82

Using a 5% level of significance, test the null hypothesis that the cause of death is not related to a person's age.

2. The number of marriages ending in divorce each year has been steadily increasing. A sociologist conducted a survey to determine if the highest level of education attained by at least one of the partners is independent of or affects the number of years that a marriage will last (before ending in divorce). A summary of that survey is given below.

		Number of Years Marriage Lasted Before Ending in Divorce			
		0–1	2–5	6–15	16–20
Highest Education Level Attained by at Least One of the Partners	Elementary School	81	72	64	30
	High School	99	81	69	47
	College	123	101	86	58

Using a 5% level of significance, test the null hypothesis that the highest educational level attained by at least one of the partners is independent of the number of years that a marriage will last.

3. An airline company has five check-in counters at an airport. An observer notices that 450 randomly selected passengers used these check-in counters with the following frequency:

Check-in Counter Number	Number of Passengers Using This Check-in Counter
1	88
2	64
3	99
4	73
5	126

Using a 5% level of significance, test the null hypothesis that all the check-in counters are used with the same frequency.

4. A vending-machine operator believes that the number of cans of Coca-Cola, Pepsi-Cola, Seven-Up, Orange Soda, and Sprite sold per day from his vending machines is in the ratio 7:5:4:2:3. In a random sample of 630 purchases of cans of soda, the following sodas were purchased.

Soda Purchased	Frequency
Coca-Cola	298
Pepsi-Cola	127
Seven-Up	117
Orange Soda	71
Sprite	17

Using a 1% level of significance, test the null hypothesis that the vending-machine operator's claim is correct.

5. A politician surveyed 550 people to determine whether they were in favor of legalizing gambling and the setting up of gambling casinos. The following table indicates their responses according to their socio-economic standing:

		Socio-economic Standing		
		Lower Class	Middle Class	Upper Class
In Favor of Legalizing Gambling Casinos?	Yes	154	117	82
	No	66	63	68

Using a 5% level of significance, test the null hypothesis that the proportion of people in favor of legalizing gambling casinos is the same for all socio-economic classes.

6. *When is an Auto Accident More Likely to Occur?* An insurance company statistician is analyzing claims involving car accidents that occurred over a period of many months at different times of the day, as shown in the accompanying chart.

	Frequency for the Number of Reported Accidents			
	0–10 Accidents	11–20 Accidents	21–40 Accidents	Over 40 Accidents
Time of Day Accident Occurred Between Midnight and 6 A.M.	12	5	3	0
Between 6 A.M. and 9 A.M.	28	19	17	4
Between 9 A.M. and 3 P.M.	22	16	10	3
Between 3 P.M. and 7 P.M.	32	19	11	2
Between 7 P.M. and Midnight	15	10	6	1

Using a 5% level of significance, test the null hypothesis that the frequency with which the different number of accidents occurring that were reported is independent of the time of the day when the accident occurs.

7. Do Obese Parents Have Obese Children? A panel of researchers conducted a survey of 210 families. The results of the survey are shown below:

		Obese Parent(s)	
		Yes	No
Obese Child?	Yes	98	52
	No	37	23

Using a 1% level of significance, test the null hypothesis that the obesity of a parent is not a factor in the obesity of the child.

8. Teenage drug use is a growing problem facing our society. A psychologist is interested in knowing whether there is a relationship between the use of drugs by teenagers and the financial status of the parents. The following data have been collected.

		Financial Status of Parents		
		Low Income	Middle Income	Upper Income
Frequency of Drug Use by Teenagers	Never	10	11	13
	Occasionally	61	47	53
	Frequently	37	29	42

Using a 5% level of significance, test the null hypothesis that the financial status of parents is independent of the frequency of the use of drugs by teenagers.

9. A sociologist is interested in knowing whether women that come from large families also have large families themselves. To investigate this question, a random survey is taken of 938 women from different-sized families. The data are presented below:

		Number of Children in Family from Which Mother Came			
		1	2–3	4–5	6 and Over
	1	90	85	76	33
Number of Children	2–3	43	160	82	23
Mother Now Has	4–5	28	35	46	27
	6 and Over	39	42	53	76

Using a 5% level of significance, test the null hypothesis that the number of children that a woman has is independent of the size of the family from which she comes.

10. A school principal is interested in knowing whether the grade level of a child determines which parent will come (assuming only one parent comes) on open school day to discuss the child's academic progress with the teacher. The following randomly selected data are available.

		Grade Level of Child in School		
		Elementary School	Junior High School	Senior High School
Which Parent Came	Father	32	47	96
on Open School Day?	Mother	78	53	29

At the 5% level of significance, is grade level a factor in determining which parent will come to school to discuss the child's academic progress with the teacher on open school day?

SUGGESTED FURTHER READING

1. Anderson, T. W., and S. L. Sclove. *An Introduction to the Statistical Analysis of Data.* Boston: Houghton Mifflin, 1978. (Chapter 12)

2. Chapman, D. G., and R. A. Schaufele. *Elementary Probability Models and Statistical Inference.* Lexington, MA: Xerox College Publishing, 1970. (Chapter 11)

3. Cochran, W. G. "The χ^2 Test of Goodness of Fit" in *Annals of Mathematical Statistics* 23 (1952), pp. 315–345.

4. Dixon, W. J., and F. J. Massey Jr. *Introduction to Statistical Analysis*, 3rd ed. New York: McGraw-Hill, 1969.

5. Nie, N., D. H. Bent, and C. H. Hull. *Statistical Package for the Social Sciences*, New York: McGraw-Hill, 1970.

CHAPTER 13

Analysis of Variance

OBJECTIVES

- *To analyze* a technique that is used when comparing several sample means. This technique is called Single Factor Analysis of Variance.

- *To discuss* how to set up Analysis of Variance (ANOVA) charts and develop formulas to use with ANOVA charts.

- *To study* a distribution that we use when comparing variances. This is the F-distribution.

- *To apply* the F-distribution to help us determine if differences in sample means are significant.

- *To learn* how to analyze two-way ANOVA tables where two factors may affect the sample means.

COMPARISON OF NEW YORK STATE AND NATIONAL AVERAGES ON COLLEGE BOARD ACHIEVEMENT TEST, 1985

	Average Score	
Achievement Test	NYS	National
English Composition	517	512
Mathematics Level 1	572	539
Biology	590	546
Chemistry	602	571
American History	538	508
Spanish	547	529
Mathematics Level 2	675	654
Physics	613	595
French	559	546
Literature	522	517
European History	587	544
German	582	551
Hebrew	629	602
Latin	574	548
Russian	708	642
All Tests Combined	552	532

Source: New York State Department of Education

An analysis of the above article indicates that the average scores of students on many different tests are given. How do we determine if there is any significant difference between the average scores of New York students and the national average on the various tests? A test-by-test comparison is very time-consuming. This is true even if we know the sample sizes.

13.1 INTRODUCTION

Suppose a company is interested in determining whether changing the lighting conditions of its factory will have any effect on the number of items produced by a worker. It can arrange the lighting conditions in four different ways: A, B, C, and D. To determine how lighting affects production, the factory supervisor decides to randomly arrange the lights under each of the four possible conditions for an equal number of days. She will then measure worker productivity under each of these conditions.

After all the data are collected, the manager calculates the number of items produced under each of these lighting conditions. She now wishes to test whether there is any significant difference between these sample means. How does she proceed?

The null hypothesis is

$$H_0: \quad \mu_A = \mu_B = \mu_C = \mu_D$$

This tells us that the mean number of items produced is the same for each lighting condition. The alternative hypothesis is that not all the means are the same. Thus the lighting condition does affect production. The supervisor will reject the null hypothesis if one (or more) of the means is different from the others.

To determine if lighting affects production, she could use the techniques of Section 10.5 for testing differences between means. However, she would have to apply those techniques many times since each time she would be able to test only two means. Thus she would have to test the following null hypotheses:

$$H_0: \quad \mu_A = \mu_B \qquad H_0: \quad \mu_A = \mu_D \qquad H_0: \quad \mu_B = \mu_D$$

$$H_0: \quad \mu_A = \mu_C \qquad H_0: \quad \mu_B = \mu_C \qquad H_0: \quad \mu_C = \mu_D$$

In order to conclude that there is no significant difference between the sample means, she would have to accept each of the six separate null hypotheses previously listed. Performing these tests involves a tremendous amount of computation. Furthermore, in doing it this way, the probability of making a Type-I error is quite large.

Since many of the problems that occur in applied statistics involve testing whether there is any significant difference between several means, statisticians use a special analysis of variance technique. This is abbreviated as **ANOVA.** The reason why these types of problems occur so often is because many companies often hire engineers to design new techniques or processes for producing products. The company must then compare the sample means of these new processes with the sample mean of the old process. Using the ANOVA technique, we have to test only one hypothesis in order to determine when to reject or to accept a null hypothesis.

ANOVA techniques can be applied to many different types of problems. In this chapter we first discuss one simple application of ANOVA techniques. This is the case in which the data are classified into groups on the basis of one single property. Then we discuss two-way ANOVA techniques, where the data are classified into groups on the basis of two properties.

13.2 DIFFERENCES AMONG *r* SAMPLE MEANS

Suppose that a chemical researcher is interested in comparing the average number of months that four different paints will last on an exterior wall before beginning to blister. The researcher has available the following data on the lasting time of three paint samples for each of the four different brands of paint.

Lasting Time (months)

Paint A	Paint B	Paint C	Paint D
13.4	14.8	12.9	13.7
11.8	13.9	14.3	14.1
14.4	12.4	13.3	12.1
Average: 13.2	13.7	13.5	13.3

The average lasting time for these paints is 13.2, 13.7, 13.5, and 13.3 months, respectively. Since not all of the paints had the same average lasting time, the chemical researcher is interested in knowing whether the observed differences between the sample means is significant or whether they can be attributed purely to chance. Is there a rule available for determining when observed differences between sample means are significant?

Before proceeding further, we wish to emphasize the fact that our discussion will be based upon certain assumptions. These are the following:

1. We will always assume that the populations from which the samples are obtained are normally distributed and that the samples are obtained independently of one another.
2. We will also assume that the populations from which the samples are obtained all have the same (often unknown) variance, σ^2.

When the above assumptions are satisfied, we can obtain an estimate of σ^2 in two different ways. First, we note that if the paints have the same mean, then the average lasting time of the different paints (in our example) is

$$\bar{x} = \frac{13.2 + 13.7 + 13.5 + 13.3}{4} = 13.425$$

The averages will be normally distributed with a variance of σ^2. This can be estimated by the following procedure:

$$s_{\bar{x}}^2 = \frac{(13.2 - 13.425)^2 + (13.7 - 13.425)^2 + (13.5 - 13.425)^2 + (13.3 - 13.425)^2}{4 - 1}$$

$$= \frac{0.1475}{3} = 0.0491667$$

From our earlier work, we already know that

$$s_{\bar{x}}^2 \approx \left(\frac{\sigma}{\sqrt{n}} \right)^2 = \frac{\sigma^2}{n}$$

Here n = the number of samples of each brand of paint. Thus, $n \cdot s_{\bar{x}}^2$ can be used as an estimate of σ^2. In our case we get $3(0.0491667) = 0.1475$. Our first estimate of σ^2, which is based upon the variation among the sample means (assuming that all the groups have the same mean), has a value of 0.1475.

Since under our second assumption the populations from which the samples are obtained all have the same (often unknown) variance, σ^2, we could obtain another estimate of σ^2 by selecting any one of the sample variances. For each brand of paint we can obtain an estimate of σ^2 based upon two degrees of freedom. We can then pool the estimates of the σ from each of the brands of paint to obtain an estimate of σ. This estimate of σ is not affected by variation of the population means among the paints. We have the following computations:

Paint Brand	Variance
A	$s_1^2 = \dfrac{(13.4 - 13.2)^2 + (11.8 - 13.2)^2 + (14.4 - 13.2)^2}{3 - 1} = \dfrac{3.44}{2} = 1.72$
B	$s_2^2 = \dfrac{(14.8 - 13.7)^2 + (13.9 - 13.7)^2 + (12.4 - 13.7)^2}{3 - 1} = \dfrac{2.94}{2} = 1.47$
C	$s_3^2 = \dfrac{(12.9 - 13.5)^2 + (14.3 - 13.5)^2 + (13.3 - 13.5)^2}{3 - 1} = \dfrac{1.04}{2} = 0.52$
D	$s_4^2 = \dfrac{(13.7 - 13.3)^2 + (14.1 - 13.3)^2 + (12.1 - 13.3)^2}{3 - 1} = \dfrac{2.24}{2} = 1.12$

Taking the average of these gives

$$\frac{s_1^2 + s_2^2 + s_3^2 + s_4^2}{4} = \frac{1.72 + 1.47 + 0.52 + 1.12}{4}$$

$$= \frac{4.83}{4} = 1.2075$$

Thus, our second estimate of σ^2, which is based on the variation within the samples, is 1.2075.

How do we compare these two estimates of σ^2? Our first estimate of σ^2, which is based upon the variation among the sample means, had a value of 0.1475, whereas our second estimate, which is based on the variation within the samples (or on the fact that the variation is due purely to chance), had a value of 1.2075. It seems reasonable that the variation among the sample means should be larger than that which is due purely to chance.

Fortunately, statisticians use a special F-distribution that facilitates such comparisons. It tells us under what conditions to reject the null hypothesis assuming the variation of the sample means is simply too great to be attributed to chance. This, of course, implies that the differences among the sample means is significant. The exact technique will be discussed in the next section after we introduce a convenient tabular arrangement for displaying our computations.

COMMENT   The straight averaging technique discussed until now works since the example given is a case with equal sample sizes.

13.3 ONE-WAY OR SINGLE FACTOR ANOVA

Let us return to the factory supervisor problem discussed in the introduction. The manager repeated each of the lighting conditions on five different days. Table 13.1 indicates the number of items produced under each of the conditions.

TABLE 13.1
Number of Items Produced Under Different Lighting Conditions

		Day					
		1	*2*	*3*	*4*	*5*	Average
	A	12	10	15	12	13	12.4
	B	16	14	9	10	15	12.8
Lighting Conditions	C	11	15	8	12	10	11.2
	D	15	14	12	11	13	13

The means for each of these lighting conditions are 12.4, 12.8, 11.2, and 13. Since the average number of items produced is not the same for all the lighting conditions, the question becomes "Is the variation among individual sample means due purely to chance or are these differences due to the different lighting conditions?" The null hypothesis is H_0: $\mu_A = \mu_B = \mu_C = \mu_D$. If the null hypothesis is true, then the lighting condition does not affect production. We can then consider our results as a listing of the number of items produced on 20 randomly selected days under one of the lighting conditions.

As mentioned earlier, when working with such problems we will assume that the number of items produced is normally distributed and that the variance is the same for each of the lighting conditions. We will also assume that the experiments with the different lighting conditions are independent of each other. (Usually the experiments are conducted in random order so that we have independence.)

Let us now apply the ANOVA techniques to this problem. Notice that we have included the row totals in Table 13.2. In applying the ANOVA technique, we analyze the reasons for the difference among the means. What is the source of the variance?

TABLE 13.2
Number of Items Produced Under Different Lighting Conditions

		Day					Row Total
		1	2	3	4	5	
	A	12	10	15	12	13	62
	B	16	14	9	10	15	64
Lighting Conditions	C	11	15	8	12	10	56
	D	15	14	12	11	13	65
							247 Grand Total

The first number that we calculate is called the **Total Sum of Squares** and is abbreviated as SS (total). To obtain the SS (total) we first square each of the numbers in the table and add these squares together. Then we divide the square of the grand total (total of all the rows or the total of all the columns) by the number in the sample. (In our case there are 20 in the sample.) Subtracting this result from the sum of the squares, we get, in our case,

$$[12^2 + 10^2 + 15^2 + 12^2 + 13^2 + 16^2 + 14^2 + 9^2 + 10^2$$
$$+ 15^2 + 11^2 + 15^2 + 8^2 + 12^2 + 10^2 + 15^2 + 14^2 + 12^2 + 11^2 + 13^2]$$

$$- \left[\frac{(247)^2}{20} \right]$$

$$= 3149 - 3050.45$$

$$= 98.55$$

*Thus SS (total) = 98.55. (An * will precede important results.)

To find the next important result we first square each of the row totals and divide the sum of these squares by the number in each row, which is 5. Then we divide the square of the grand total by the number in the sample. We subtract this result from the sum of squares and get

$$\left(\frac{62^2 + 64^2 + 56^2 + 65^2}{5}\right) - \frac{(247)^2}{20} = 3060.2 - 3050.45 = 9.75$$

This result is called the **Sum of Squares Due to the Factor.** The factor in our example is the lighting conditions. Thus

* SS(lighting) = 9.75

Now we calculate a number that is called the **Sum of Squares of the Error** abbreviated as SS(error). We first square each entry of the table and add these squares together. Then we square each row total and divide the sum of these squares by the number in each row. Subtracting this result from the sum of squares, we get

$$\left(12^2 + 10^2 + 15^2 + 12^2 + \cdots + 11^2 + 13^2\right) - \frac{62^2 + 64^2 + 56^2 + 65^2}{5}$$

$$= 3149 - 3060.2$$

$$= 88.8$$

Thus

* SS(error) = 88.8

COMMENT Actually, SS(error) = SS(total) − SS(factor).

We enter these results along with the appropriate number of degrees of freedom, which we will determine shortly, in a table known as an **ANOVA table.** The general form of an ANOVA table is shown in Table 13.3.

TABLE 13.3
ANOVA Table

Source of Variation	Sum of Squares (SS)	Degrees of Freedom (df)	Mean Square (MS)	F-Ratio
Factors of Experiment				
Error				
Total				

In our case we have the ANOVA table shown in Table 13.4. We have entered the important results that we obtained in the preceding paragraphs in the

TABLE 13.4
ANOVA Table

Source of Variation	Sum of Squares	Degrees of Freedom	Mean Square	F-Ratio
Factors (lighting)				$\dfrac{3.25}{5.55} = 0.59$
Error	88.8	16	5.55	
Total				

appropriate space of the table. The degrees of freedom are obtained according to the following rules:

RULES

1. The number of degrees of freedom for the factor tested is one less than the number of possible levels at which the factor is tested. If there are r levels of the factor, the number of degrees of freedom is $r - 1$.
2. The number of degrees of freedom for the error is one less than the number of repetitions of each condition multiplied by the number of possible levels of the factor. If each condition is repeated c times, then the number of degrees of freedom is $r(c - 1)$.
3. Finally, the number of degrees of freedom for the total is one less than the total number in the sample. If the total sample consists of n things, the number of degrees of freedom is $n - 1$.

In our case there are 4 experimental conditions and each condition is repeated 5 times so that $r = 4$, $c = 5$, and $n = 20$.

Thus

$$df(\text{factor}) = r - 1 = 4 - 1 = 3$$

$$df(\text{error}) = r(c - 1) = 4(5 - 1) = 4 \cdot 4 = 16$$

$$df(\text{total}) = n - 1 = 20 - 1 = 19$$

We enter these values in an ANOVA table as shown in Table 13.4.

Now we calculate the values that belong in the **Mean Square** column. We divide the sum of squares for the row by the number of degrees of freedom for that row. Thus,

$$MS(\text{lighting}) = \frac{SS(\text{lighting})}{df(\text{lighting})} = \frac{9.75}{3} = 3.25$$

$$MS(\text{error}) = \frac{SS(\text{error})}{df(\text{error})} = \frac{88.8}{16} = 5.55$$

Although we are comparing the sample means for the different lighting conditions, the ANOVA technique compares variances under the assumption that the variance among all the levels is 0. If the variance is 0, then all the means are the same.

Suppose we are given two samples of size n_1 and n_2 for two populations that have roughly the shape of a normal distribution. Then we base tests on the equality of the two population standard deviations (or variances) on the ratio $\frac{s_1^2}{s_2^2}$ or $\frac{s_2^2}{s_1^2}$ where s_1 and s_2 are the standard deviations of both samples. The distribution of such a ratio is a continuous distribution called the **F-distribution.** This distribution depends upon the number of degrees of freedom of both sample estimates, $n_1 - 1$ and $n_2 - 1$. When using the distribution, we reject the null hypotheses of equal variances ($\sigma_1 = \sigma_2$) when the test statistic value of F exceeds $F_{\alpha/2}$, where α is the level of significance.

When comparing variances we use the **F-distribution.** The test statistic is

$$F = \frac{MS(\text{lighting})}{MS(\text{error})}$$

COMMENT When using the F-distribution we are testing whether or not the variances among the factors is zero (in which case the means are equal). These situations are pictured in Figure 13.1 for the case involving three means.

As mentioned earlier, the values of the F-distribution depend on the number of degrees of freedom of the numerator, that of the denominator, and upon the level of significance. Table X in the Appendix gives us the different values corresponding to different degrees of freedom. From Table X in the Appendix we find that the F-value with 3 degrees of freedom for the numerator and 16 degrees of freedom for the denominator, at the 5% level of significance, is 3.24. We reject the null hypothesis if the F-value is greater than 3.24.

In our case the value of the F statistic is

$$\frac{MS(\text{lighting})}{MS(\text{error})} = \frac{3.25}{5.55} = 0.59$$

Since 0.59 is less than 3.24, we do not reject the null hypothesis. Thus the data do not indicate that the lighting condition affects production.

Let us summarize the procedure to be used when testing several sample means by the ANOVA technique. First draw an ANOVA table such as shown in Table 13.5. In this table we have placed numbers in parentheses in the

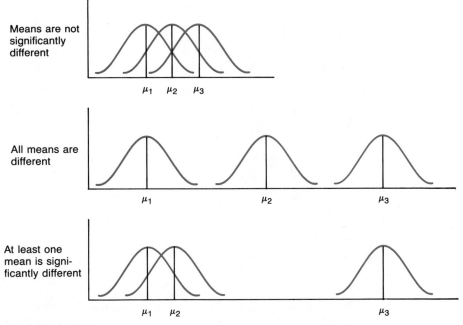

FIGURE 13.1

various cells. The values that belong in each of these cells are obtained by using the following formulas:

$$\text{cell (1):} \quad \frac{\Sigma(\text{each row total})^2}{\text{number in each row}} - \frac{(\text{grand total})^2}{\text{total sample size}}$$

$$\text{cell (2):} \quad \Sigma(\text{each original number})^2 - \frac{\Sigma(\text{each row total})^2}{\text{number in each row}}$$

TABLE 13.5
ANOVA Table

Source of Variation	Sum of Squares	Degrees of Freedom	Mean Square	F-Ratio
Factor of the Experiment	(1)	(4)	(7)	(9)
Error	(2)	(5)	(8)	
Total	(3)	(6)		

cell (3): $\quad \Sigma(\text{each original number})^2 - \dfrac{(\text{grand total})^2}{\text{total sample size}}$

cell (4): $\quad$ If there are r levels of the factor, $r - 1$

cell (5): $\quad$ If there are c repetitions of each of the r levels, $r(c - 1)$

cell (6): $\quad$ If the total sample consists of n things, $n - 1$

cell (7): $\quad \dfrac{\text{cell (1) value}}{\text{cell (4) value}}$

cell (8): $\quad \dfrac{\text{cell (2) value}}{\text{cell (5) value}}$

cell (9): $\quad \dfrac{\text{cell (7) value}}{\text{cell (8) value}}$

Although these formulas seem complicated, they are easy to use as the following examples illustrate.

● **Example 1** *Testing the Tar Content of Cigarettes*
Three brands of cigarettes, six cigarettes from each brand, were tested for tar content. Do the data shown below indicate that there is a significant difference in the average tar content for these three brands of cigarettes? (Use a 5% level of significance.)

		X							
	X	14	16	12	18	11	13	84	
Brand	Y	10	11	22	19	9	18	89	
	Z	24	22	19	18	20	19	122	
								295	Grand Total

Solution
The null hypothesis is that the sample mean tar content is the same for the three brands of cigarettes. To solve this problem we set up the following ANOVA table:

Source of Variation	Sum of Squares	Degrees of Freedom	Mean Square	F-Ratio
Cigarettes	(1)	(4)	(7)	(9)
Error	(2)	(5)	(8)	
Total	(3)	(6)		

Now we calculate the values that belong in each of the cells by using the formulas given on pages 575–576. We have

cell (1): $\dfrac{\Sigma(\text{each row total})^2}{\text{number in each row}} - \dfrac{(\text{grand total})^2}{\text{total sample size}}$

$= \dfrac{84^2 + 89^2 + 122^2}{6} - \dfrac{(295)^2}{18}$

$= 4976.83 - 4834.72$

$= 142.11$

cell (2): $\Sigma(\text{each original number})^2 - \dfrac{\Sigma(\text{each row total})^2}{\text{number in each row}}$

$= \left(14^2 + 16^2 + 12^2 + \cdots + 19^2\right) - \dfrac{(84^2 + 89^2 + 122^2)}{6}$

$= 5187 - 4976.83$

$= 210.17$

cell (3): $\Sigma(\text{each original number})^2 - \dfrac{(\text{grand total})^2}{\text{total sample size}}$

$= \left(14^2 + 16^2 + 12^2 + \cdots + 19^2\right) - \dfrac{(295)^2}{18}$

$= 5187 - 4834.72$

$= 352.28$

cell (4): $df = r - 1$

$= 3 - 1 = 2$

cell (5): $df = r(c - 1)$

$= 3(6 - 1) = 3(5) = 15$

cell (6): $df = n - 1$

$= 18 - 1 = 17$

cell (7): $\dfrac{\text{cell (1) value}}{\text{cell (4) value}} = \dfrac{142.11}{2} = 71.06$

cell (8): $\dfrac{\text{cell (2) value}}{\text{cell (5) value}} = \dfrac{210.17}{15} = 14.01$

cell (9): $\dfrac{\text{cell (7) value}}{\text{cell (8) value}} = \dfrac{71.06}{14.01} = 5.07$

Now we enter these numbers on the ANOVA table:

Source of Variation	Sum of Squares	Degrees of Freedom	Mean Square	F-Ratio
Cigarettes	142.11 (1)	2 (4)	71.06 (7)	5.07 (9)
Error	210.17 (2)	15 (5)	14.01 (8)	
Total	352.28 (3)	17 (6)		

From Table X in the Appendix we find that the F-value with 2 degrees of freedom for the numerator and 15 degrees of freedom for the denominator at the 5% level of significance is 3.68. Since we obtained an F-value of 5.07, which is larger than 3.68, we reject the null hypothesis that the sample mean tar content is the same for the three brands of cigarettes.

COMMENT   Rejection of the null hypothesis H_0: $\mu_1 = \mu_2 = \mu_3$ does not tell us where the difference lies.

● **Example 2**

Four groups of five students each were taught a skill by four different teaching techniques. At the end of a specified time the students were tested and their scores recorded. Does the following data indicate that there is a significant difference in the mean achievement for the four teaching techniques? (Use a 1% level of significance.)

						Row Total
A	64	73	69	75	78	359
B	73	82	71	69	74	369
C	61	79	71	73	66	350
D	63	69	68	74	75	349
					1427	Grand Total

Teaching Technique labels the rows A, B, C, D.

Solution
The null hypothesis is that the mean achievement for the four teaching techniques is the same. To solve this problem we set up the following ANOVA table:

Source of Variation	Sum of Squares	Degrees of Freedom	Mean Square	F-Ratio
Teaching Method	(1) 52.15	(4) 3	(7) 17.38	(9) 0.56
Error	500.4 (2)	16 (5)	31.28 (8)	
Total	552.55 (3)	19 (6)		

Now we calculate the values that belong in each of the cells by using the formulas given on pages 575–576. We have

cell (1): $\dfrac{\Sigma(\text{each row total})^2}{\text{number in each row}} - \dfrac{(\text{grand total})^2}{\text{total sample size}}$

$= \dfrac{359^2 + 369^2 + 350^2 + 349^2}{5} - \dfrac{(1427)^2}{20}$

$= 101{,}868.6 - 101{,}816.45$

$= 52.15$

cell (2): $\Sigma(\text{each original number})^2 - \dfrac{\Sigma(\text{each row total})^2}{\text{number in each row}}$

$= (64^2 + 73^2 + 69^2 + 75^2 + \cdots + 74^2 + 75^2)$

$\quad - \left(\dfrac{359^2 + 369^2 + 350^2 + 349^2}{5}\right)$

$= 102{,}369 - 101{,}868.6$

$= 500.4$

cell (3): $\Sigma(\text{each original number})^2 - \dfrac{(\text{grand total})^2}{\text{total sample size}}$

$= \left(64^2 + 73^2 + 69^2 + 75^2 + \cdots + 74^2 + 75^2\right) - \dfrac{(1427)^2}{20}$

$= 102{,}369 - 101{,}816.45$

$= 552.55$

cell (4): $df = r - 1 = 4 - 1 = 3$

cell (5): $df = r(c - 1) = 4(5 - 1) = 4(4) = 16$

cell (6): $df = n - 1 = 20 - 1 = 19$

cell (7): $\dfrac{\text{cell (1) value}}{\text{cell (4) value}} = \dfrac{52.15}{3} = 17.38$

cell (8): $\dfrac{\text{cell (2) value}}{\text{cell (5) value}} = \dfrac{500.4}{16} = 31.28$

cell (9): $\dfrac{\text{cell (7) value}}{\text{cell (8) value}} = \dfrac{17.38}{31.28} = 0.56$

We enter all these values on the ANOVA table as shown.

From Table X in the Appendix we find that the *F*-value with 3 degrees of freedom for the numerator and 16 degrees of freedom for the denominator, at the 1% level of significance, is 5.29. Since we obtained an *F*-value of 0.56 we do not reject the null hypothesis that the mean achievement for the four teaching techniques is the same.

*13.4 TWO-WAY ANOVA

Let us return to the example given in Section 13.2 and construct an analysis-of-variance table for the data that are reproduced below except that we have now rewritten the chart in a slightly different manner. (The reason for this will be apparent shortly.)

	Lasting Time (months)			Row Total
Paint A	13.4	11.8	14.4	39.6
Paint B	14.8	13.9	12.4	41.1
Paint C	12.9	14.3	13.3	40.5
Paint D	13.7	14.1	12.1	39.9
Column Total	54.8	54.1	52.2	161.1 Grand Total

We set up the following ANOVA table to which we have added the values that belong in each of the cells by using the formulas given on pages 575–576.

Source of Variation	Sum of Squares		Degrees of Freedom		Mean Square		F-Ratio
Different Paints	0.4425	(1)	3	(4)	0.1475	(7)	0.1222 (9)
Error	9.66	(2)	8	(5)	1.2075	(8)	
Total	10.1025	(3)	11	(6)			

Thus, we have

cell (1): $\left(\dfrac{(39.6)^2 + (41.1)^2 + (40.5)^2 + (39.9)^2}{3} \right) - \dfrac{(161.1)^2}{12}$

$$= 2163.21 - 2162.7675 = 0.4425$$

cell (2): $[(13.4)^2 + (11.8)^2 + \cdots + (12.1)^2]$

$$- \left[\frac{(39.6)^2 + (41.1)^2 + (40.5)^2 + (39.9)^2}{3} \right]$$

$$= 2172.87 - 2163.21 = 9.66$$

cell (3): $\left[(13.4)^2 + (11.8)^2 + \cdots + (12.1)^2 \right] - \dfrac{(161.1)^2}{4 \cdot 3}$

$$= 2172.87 - 2162.7675 = 10.1025$$

cell (4): $4 - 1 = 3$

cell (5): $4(3 - 1) = 8$

cell (6): $12 - 1 = 11$

cell (7): $\dfrac{0.4425}{3} = 0.1475$

cell (8): $\dfrac{9.66}{8} = 1.2075$

cell (9): $\dfrac{0.1475}{1.2075} = 0.1222$

From Table X in the Appendix we find that the F-value with three degrees of freedom for the numerator and eight degrees of freedom for the denominator at the 5% level of significance is 4.07. Therefore, since we obtained an F-value of 0.1222, we do not reject the null hypothesis that there is a significant difference in the average lasting times of the four brands of paint.

After analyzing the data carefully, the chemical engineer discovers that the test conditions for each of the brands of paint were not the same and that the samples for each brand of paint were subject to different temperature and humidity conditions. This factor, that was not considered earlier, definitely has to be considered. Thus, instead of just having to determine if the average lasting time for the four different brands of paint is the same, we also have to consider the various test conditions as a possible cause of the variation as well as pure chance. Under these new circumstances the data must be arranged as follows:

Test Conditions

	Test Condition I	Test Condition II	Test Condition III	Row Total	
Paint A	13.4	11.8	14.4	39.6	
Paint B	14.8	13.9	12.4	41.1	
Paint C	12.9	14.3	13.3	40.5	
Paint D	13.7	14.1	12.1	39.9	
Column Total	54.8	54.1	52.2	161.1	*Grand Total*

When discussing this second factor, it is customary to refer to it as **blocks** as opposed to the original factor, which is referred to as the **treatments**. We then have the following ANOVA table for such a **two-way analysis** of variance.

Source of Variation	Sum of Squares	Degrees of Freedom	Mean Square	F-Ratio
Treatments	(1)	(5)	(9)	(12)
Blocks	(2)	(6)	(10)	(13)
Error	(3)	(7)	(11)	
Total	(4)	(8)		

The values that belong in each of these cells for a two-way analysis of variance table are obtained by using the following formulas.

$$\text{cell (1):} \quad \frac{\Sigma(\text{each row total})^2}{\text{number in each row}} - \frac{(\text{grand total})^2}{r \cdot c}$$

where r = number of levels of one factor (row)

c = number of levels of second factor (column)

$$\text{cell (2):} \quad \frac{\Sigma(\text{each column total})^2}{\text{number in each column}} - \frac{(\text{grand total})^2}{r \cdot c}$$

$$\text{cell (3):} \quad \Sigma(\text{each original number})^2 - \frac{\Sigma(\text{each row total})^2}{\text{number in each row}}$$

$$- \frac{\Sigma(\text{each column total})^2}{\text{number in each column}} + \frac{(\text{grand total})^2}{r \cdot c}$$

$$\text{cell (4):} \quad \Sigma(\text{each original number})^2 - \frac{(\text{grand total})^2}{r \cdot c}$$

cell (5): $r - 1$

cell (6): $c - 1$

cell (7): $(r - 1)(c - 1)$

cell (8): $rc - 1$

cell (9): $\dfrac{\text{cell (1) value}}{\text{cell (5) value}}$

cell (10): $\dfrac{\text{cell (2) value}}{\text{cell (6) value}}$

cell (11): $\dfrac{\text{cell (3) value}}{\text{cell (7) value}}$

cell (12): $\dfrac{\text{cell (9) value}}{\text{cell (11) value}}$

cell (13): $\dfrac{\text{cell (10) value}}{\text{cell (11) value}}$

Let us apply these formulas to our example. We have the following two-way ANOVA table to which we have added the values that belong in each of the cells by using the formulas given previously.

Source of Variation	Sum of Squares		Degrees of Freedom		Mean Square		F-Ratio	
Paints	0.4425	(1)	3	(5)	0.1475	(9)	0.1011	(12)
Condition	0.905	(2)	2	(6)	0.4525	(10)	0.3101	(13)
Error	8.755	(3)	6	(7)	1.4592	(11)		
Total	10.1025	(4)	11	(8)				

We have

cell (1) value: $\left[\dfrac{(39.6)^2 + (41.1)^2 + (40.5)^2 + (39.9)^2}{3} \right] - \dfrac{(161.1)^2}{4 \cdot 3}$

$= 2163.21 - 2162.7675 = 0.4425$

cell (2) value: $\left[\dfrac{(54.8)^2 + (54.1)^2 + (52.2)^2}{4} \right] - \dfrac{(161.1)^2}{4 \cdot 3}$

$= 2163.6725 - 2162.7675 = 0.905$

cell (3) value: $[(13.4)^2 + (11.8)^2 + \cdots + (12.1)^2]$

$$- \left[\frac{(39.6)^2 + (41.1)^2 + (40.5)^2 + (39.9)^2}{3} \right]$$

$$- \left[\frac{(54.8)^2 + (54.1)^2 + (52.2)^2}{4} \right] + \frac{(161.1)^2}{4 \cdot 3}$$

$$= 2172.87 - 2163.21 - 2163.6725 + 2162.7675$$

$$= 8.755$$

cell (4) value: $= \left[(13.4)^2 + (11.8)^2 + \cdots + (12.1)^2 \right] - \dfrac{(161.1)^2}{4 \cdot 3}$

$$= 2172.87 - 2162.7675 = 10.1025$$

cell (5) value: $\quad 4 - 1 = 3$

cell (6) value: $\quad 3 - 1 = 2$

cell (7) value: $\quad (4 - 1)(3 - 1) = 6$

cell (8) value: $\quad 4 \cdot 3 - 1 = 12 - 1 = 11$

cell (9) value: $\quad \dfrac{0.4425}{3} = 0.1475$

cell (10) value: $\quad \dfrac{0.905}{2} = 0.4525$

cell (11) value: $\quad \dfrac{8.755}{6} = 1.4592$

cell (12) value: $\quad \dfrac{0.1475}{1.4592} = 0.1011$

cell (13) value: $\quad \dfrac{0.4525}{1.4592} = 0.3101$

From Table X in the Appendix, we find that the F-value with three degrees of freedom for the numerator and six degrees of freedom for the denominator is 4.76 at the 5% level of significance. Similarly the F-value with two degrees of freedom for the numerator and six degrees of freedom for the denominator is 5.14. In both cases we obtained values of F equal to 0.1011 and 0.3101, respectively, that are less than 4.76 and 5.14. Thus, we do not reject the null hypothesis that the average drying time for the several brands of paint is significantly different nor do we reject the null hypothesis that the different testing conditions was a factor in determining average drying time.

COMMENT   Our analysis of two-way ANOVA is by no means a complete discussion of the topic. A thorough two-way ANOVA problem and solution depends upon how the treatments were assigned to the experimental units (i.e., how the randomization was conducted) as well as the number of replications of each experimental condition. Our intention was merely to introduce you to the idea of two-way ANOVA. A more detailed analysis can be found by consulting any of the references listed in the *Suggested Further Reading* for this chapter.

EXERCISES

1. An independent fire prevention agency is testing smoke alarms produced by four different companies to determine if there is any significant difference between the average amount of time elapsed before the smoke alarm emits an audible siren in response to a certain stimulus. The agency randomly purchases three smoke alarms produced by each of the four companies and subjects them to the same test conditions. The number of seconds that elapses before the smoke alarms emit an audible siren is as follows:

Time Elapsed (seconds)

Brand A	22	14	17
Brand B	16	18	11
Brand C	23	15	18
Brand D	19	13	16

Do the data indicate that there is a significant difference in the average time elapsed before the smoke alarms emit an audible sound among the different brands? (Use a 5% level of significance.)

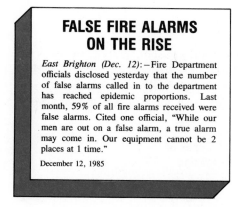

FALSE FIRE ALARMS ON THE RISE

East Brighton (Dec. 12): – Fire Department officials disclosed yesterday that the number of false alarms called in to the department has reached epidemic proportions. Last month, 59% of all fire alarms received were false alarms. Cited one official, "While our men are out on a false alarm, a true alarm may come in. Our equipment cannot be 2 places at 1 time."

December 12, 1985

2. Consider the newspaper article above. After reading it, the mayor orders the fire commissioner to submit a detailed report on the number of false alarms for one week covering five different sections of the city. The following report was submitted to the mayor.

**Number of False
Alarms Reported
Daily Over A Five-Day Period**

	A	23	39	32	22	42
	B	13	17	24	11	19
Section of City	C	37	31	26	30	34
	D	20	11	13	17	10
	E	16	12	11	18	21

Using a 1% level of significance, do the data indicate that the average number of false alarms reported is significantly different for the various sections of the city?

3. The Commissioner of Education of a state is analyzing the average math SAT scores of students from high schools located throughout the state. The commissioner believes that the average math SAT scores of students is significantly different for the various high schools and is related to the geographic region of the school district in which the high school is located. In support of this claim the following data are available for students from four high schools located in different geographic regions within the state.

SAT Scores

Geographic Location of High School Within Part of the State	Northern	458	473	496	446
	Southern	512	501	498	422
	Eastern	416	440	432	423
	Western	508	488	469	510

At the 5% level of significance, is the Commissioner's claim justified?

4. Accidental breakage of glass items by customers as they push their shopping carts through the aisles of a supermarket can be dangerous to other customers as well as costly to the management. A large supermarket chain reported the following number of reported cases of the breakage of glass items on a weekly basis for a period of four weeks, and the store location.

**Number of Cases of
Glass Breakage**

Store Location	Brooks Mall	45	32	36	40
	Weaver Mall	37	34	46	34
	Tanner Mall	36	38	36	45
	Acres Mall	42	39	31	40

Using a 5% level of significance, test the null hypothesis that the average number of accidental glass breakage cases is the same for all the stores, irrespective of location.

5. A consumer's group wishes to verify the claim made by several paint manufacturers that the average gallon of a certain type of paint will cover a 420-square-foot wall when applied according to specifications. Fifteen walls are divided into groups of five each. Each wall in a subgroup is then painted with one of three brands of paint. The coverage (in square feet) by each of these brands of paint is as follows:

Paint Coverage (square feet)

Brand A	400	390	402	407	410
Brand B	422	416	421	413	420
Brand C	415	422	425	410	411

Using a 5% level of significance, test the null hypothesis that the average coverage by each of these brands of paint is not significantly different.

6. To compare the effectiveness of four new pills in lowering blood serum cholesterol levels, 20 people with high cholesterol levels were carefully selected so as to make them as comparable as possible. The 20 people were divided into 4 groups of 5 each. Each person in a subgroup was then given a pill. The decrease in the blood cholesterol level of each person taking the pill is indicated below:

Decrease in Blood Cholesterol Level (mg)

Pill A	12	16	18	13	15
Pill B	10	15	22	11	17
Pill C	16	3	9	12	10
Pill D	17	19	11	14	15

Using a 1% level of significance, test the null hypothesis that the average decrease in the blood serum cholesterol levels of people taking each of these types of pills is the same.

7. A chicken farmer is interested in testing which of four different vitamin supplements will increase the average weight gain of her chickens. She gives each vitamin supplement to four different chickens over a period of a month and records the following average weight gains (in pounds) for these chickens.

		Average Weight Gain (pounds)			
Vitamin Supplement	*I*	2.3	1.7	2.1	1.8
	II	3.6	1.4	2.8	3.1
	III	2.5	3.5	1.7	2.4
	IV	3.1	1.8	1.9	2.3

At the 5% level of significance, is there any significant difference in the average weight gain produced by these vitamin supplements?

* **8.** After a home computer has been purchased and removed from the packing material, the various components have to be assembled and connected for the machine to become functional. Many home computer stores provide this service at no charge. One large midwestern computer store employs five people who perform this function. The number of minutes required by each to complete the assembly task for 26 identical machines is shown below:

		Time to Complete Task (minutes)						
Employee	*Bill*	65	47	51	72	62	81	
	Mary	53	58	62	49	53		
	Bob	61	57	49	58	57	64	72
	Sue	65	63	59	56			
	Sandra	49	58	51	45			

Using a 1% level of confidence, test the null hypothesis that the average time required to assemble the computer is the same for all of the employees.

9. Twenty moviegoers were asked to rate five different movies (on a certain scale). However, the air-conditioning system in the theater at the time the movies were shown was not functioning properly. As a result, it is suspected that the room temperature may have been an additional factor in the moviegoer's ratings of the movies, which are given below.

		Temperature in Theater			
		70–74°	*75–79°*	*80–84°*	*85–90°*
Movie Viewed	A	35	25	21	18
	B	32	22	25	28
	C	29	26	23	21
	D	29	28	31	34
	E	24	26	27	31

Using a two-way analysis of variance, test the null hypothesis that the average ratings for these movies is the same and that the temperature in the theater is not a factor. (Use a 5% level of significance.)

10. An experimenter is interested in knowing which method yields the greatest amount of usable oil from shale rock. The experimenter gathers shale rock from five different parts of the country and applies one of four conversion techniques. The quantity of oil obtained (on a certain scale) is given below:

Quantity of Oil Obtained

		Region A	Region B	Region C	Region D	Region E
Conversion Technique Used	I	37	33	41	39	38
	II	29	32	36	44	53
	III	41	32	21	30	36
	IV	21	24	20	23	22

Using a two-way analysis of variance, test the null hypothesis that the average quantity of oil obtained is the same for all the conversion techniques and is also the same for shale rock obtained from each of the regions. (Use a 1% level of significance.)

11. Consider the newspaper article below. After taking samples of the air on different days at the various tunnels and bridges, the following levels of pollutants were obtained:

TOLL COLLECTORS PROTEST UNHEALTHY WORKING CONDITIONS

New Dorp (May 17): — Toll collectors staged a 3-hour protest yesterday against the unhealthy working conditions occurring at different times of the day at the toll plazas of the city's bridges and tunnels. A spokesperson for the toll collectors indicated that the level of sulfur oxide pollutants in the air at the plazas was far in excess of the recommended level. Between the hours of 10 A.M. and 1 P.M., motorists were able to use the tunnel without paying any toll. In an effort to resolve the dispute, Bill Sigowsky has been appointed as mediator.

May 17, 1985

	Day of the Week				
	Mon.	Tues.	Wed.	Thurs.	Fri.
Bridge 1	2.8	3.3	2.6	2.9	3.5
Bridge 2	2.1	2.3	2.8	3.7	2.7
Bridge 3	3.8	3.4	3.5	3.1	2.9
Tunnel 1	2.8	2.7	3.1	3.2	3.3
Tunnel 2	2.8	3.3	2.9	3.0	3.1

Location Where Sample was Obtained

Using a two-way analysis of variance, test the null hypothesis that the average amount of sulfur oxide pollutants is the same at all the bridges and tunnels and is the same for each day of the week. (Use a 5% level of significance.)

12. A lawyer has gathered the following statistics from several states for the length of a jail term for people with the same prior record convicted of different types of crimes.

Jail Term by Crime (years)

State	Burglary	Armed Robbery	Assault	Murder
1	4	7	2	35
2	3	6	3	37
3	5	8	1	36
4	4	9	2	38

Using a two-way analysis of variance, test the null hypothesis that the average length of a jail term is the same for the various crimes mentioned and that this average is not significantly different for the several states. (Use a 1% level of significance.)

13.5 USING COMPUTER PACKAGES

MINITAB greatly simplifies the computations involved when working with ANOVA problems. We illustrate the output from MINITAB when applied to the data of Example 2 of Section 13.3 (page 578).

```
MTB  > READ THE FOLLOWING DATA INTO C1, C2, C3, C4
DATA > 64    73    61    63
DATA > 73    82    79    69
DATA > 69    71    71    68
DATA > 75    69    73    74
DATA > 78    74    66    75

MTB  > AOVONE WAY ON C1-C4
```

```
ANALYSIS OF VARIANCE
SOURCE   DF     SS     MS       F
FACTOR    3   52.2   17.4    0.56
ERROR    16  500.4   31.3
TOTAL    19  552.6        INDIVIDUAL 95 PCT CI'S FOR MEAN
                         BASED ON POOLED STDEV
LEVEL     N   MEAN  STDEV  --------+---------+---------+-------
C1        5  71.80   5.45      (-------------*-----------)
C2        5  73.80   4.97        (-------------*-----------)
C3        5  70.00   6.86   (-----------*-----------)
C4        5  69.80   4.87  (------------*-----------)
                         --------+---------+---------+-------
POOLED STDEV = 5.59           68.0      72.0      76.0
MTB > STOP
```

COMMENT Although the MINITAB analysis of variance for this problem will produce additional output (involving confidence intervals), we will not analyze such information here, as our main objective was merely to indicate how MINITAB handles simple ANOVA problems.

13.6 SUMMARY

In this chapter we briefly introduced the important statistical technique known as analysis of variance or ANOVA. This technique is used when we wish to test a hypothesis about the equality of several means. We limited our discussion to normal populations with equal variances.

When applying the ANOVA technique, we do not test the differences between the means directly. Instead we test the variances. If the variances are zero, there is no difference between the means. We therefore analyze the source of the variation. Is it due purely to chance or is the difference in the variation significant? When analyzing variation by means of analysis of variance techniques, we use an ANOVA table.

Although we discussed only single-factor (one-way) and two-way ANOVA, it is worth noting that ANOVA techniques can be applied to more complicated situations or when we have replications for each test condition. However, these are beyond the scope of this book.

STUDY GUIDE

You should now be able to demonstrate your knowledge of the following ideas presented in this chapter by giving definitions, descriptions, or specific examples. Page references are given for each term so that you can check your answer.

FORMULAS TO REMEMBER

The most important thing to remember when testing several means is how to set up an ANOVA table and how to find the appropriate values for each cell of the ANOVA table. Both of these ideas are summarized on pages 575–576 for single-factor ANOVA and on pages 582–583 for a two-way ANOVA.

The tests of the following section will be more useful if you take them after you have studied the examples and solved the exercises in this chapter.

MASTERY TESTS

Form A

For Questions 1–9 use the following information. A large utility company in the Northeast receives numerous complaints daily dealing with billing discrepancies. A record on the number of customer complaints that were resolved in a mutually acceptable way by five company employees for one week is shown below:

		Number of Customer Complaints Resolved				
	Fran	15	14	19	32	18
	Arlene	27	35	17	14	19
Employee	Sherry	33	24	21	25	14
	Barry	11	31	27	33	34
	Al	25	28	21	19	27

In an effort to determine whether there is a significant difference in the average daily processing ability of these workers, the following ANOVA table has been set up.

ANOVA Table

Source of Variation	Sum of Squares	Degrees of Freedom	Mean Square	F-ratio
Different workers	(1)	(4)	(7)	(9)
Error	(2)	(5)	(8)	
Total	(3)	(6)		

1. The appropriate entry for cell (1) is _____.

2. The appropriate entry for cell (2) is _____.

3. The appropriate entry for cell (3) is _____.

4. The appropriate entry for cell (4) is _____.

5. The appropriate entry for cell (5) is _____.

6. The appropriate entry for cell (6) is _____.

7. The appropriate entry for cell (7) is _____.

8. The appropriate entry for cell (8) is _____.

9. The appropriate entry for cell (9) is _____.

10. It is appropriate to use the F-distribution when comparing
    a. differences between sample means.
    b. differences between variances.
    c. differences resulting from the use of different sample sizes.
    d. variances from non–normally distributed populations.
    e. none of these.

11. Which of the following assumptions are made when we use the ANOVA techniques?
    a. The populations from which the samples are selected are normally distributed.
    b. The sample sizes are large.
    c. The Central Limit Theorem is not applicable.
    d. There is a correlation between the factors of the experiment.
    e. None of these.

12. The total number of degrees of freedom when using a two-way ANOVA, where there are r levels of one factor and c levels of another factor, is
    a. $r - 1$ b. $c - 1$ c. $rc - 1$ d. $r(c - 1)$ e. none of these

1. Many drug companies claim that their products are superior in their ability to bring quick relief to suffering patients. A consumer's group decides to test the drugs manufactured by four different companies. The average time (in minutes) required for 20 patients to feel relief is shown in the accompanying chart:

	Average Time to Pain Relief (minutes)				
	Company A	11	19	22	21	26
Drug Manufactured by	Company B	21	23	22	26	19
	Company C	10	9	12	8	17
	Company D	11	14	13	16	22

Do the data indicate that there is a significant difference between the average amount of time that elapses when a patient takes the drugs produced by the different companies before relief is felt? (Use a 5% level of significance.)

2. Environmentalists took four water samples at each of three different locations of a river in an effort to measure whether the quantity of dissolved oxygen in the water varied from one location to another and from season to season. The quantity of dissolved oxygen (on an appropriate scale) in the water is used to determine the extent of the water pollution. The results of the survey are as follows:

		Quantity of Dissolved Oxygen		
		Location I	Location II	Location III
	Autumn	18.4	15.8	15.9
Season	Winter	17.6	17.3	16.8
	Spring	17.9	16.5	18.9
	Summer	14.3	16.6	17.8

Using a two-way analysis of variance, do the data indicate that the average quantity of dissolved oxygen is the same at all three locations and is the same for all the seasons of the year? (Use a 1% level of significance.)

3. A perfume manufacturer is interested in determining whether different package wrappings for perfume have any effect on sales. She decides to package the perfume in one of four possible package wrappings and to sell the perfume with each of the possible wrappings at five large department stores. The price for the perfume at each store is the same, irrespective of the packaging, and each package contains the same quantity of perfume. The following are the sales reported for the perfume:

Package Wrapping

		A	B	C	D
	I	77	88	71	93
	II	61	53	42	76
Store	III	64	72	58	84
	IV	42	31	54	62
	V	57	63	59	68

Using a two-way analysis of variance, do the data indicate that the average sales reported for the perfume is the same for all four possible wrappings and is the same for all the stores? (Use a 5% level of significance.)

4. A medical researcher is interested in comparing the effects of four different diets on hypertension (high blood pressure). Twenty men with high blood pressure were randomly assigned (five each) to each of the four diet groups. After several months the blood pressure of the participants was measured and is given below:

Blood Pressure

Diet A	148	160	151	170	160
Diet B	152	158	142	148	145
Diet C	162	139	135	152	144
Diet D	153	155	145	158	162

Do the above data indicate that the type of diet used affects the average blood pressure level? (Use a 1% level of significance.)

* 5. There are three hospitals with emergency rooms in Adams City. Thirty-three patients who had been treated at these emergency rooms are randomly selected and asked how long they had to wait before seeing a doctor. Their answers are as follows:

Length of Wait (minutes)

Lincoln Hospital	19	17	14	31	23	17	25	27	12	8	21	12
Mt. Sinai Hospital	9	16	32	21	22	17	12	18	19	24	17	
Columbia Hospital	13	12	16	17	13	12	24	11	18	25		

Do the data indicate that there is a significant difference between the average waiting time in the emergency rooms at these hospitals before seeing a doctor? (Use a 5% level of significance.)

6. The number of arrests for illegal drug use on a daily basis at the three city beaches during the five-day July 4th weekend is as follows:

Number of Arrests

Beach 1	26	31	35	32	21
Beach 2	39	32	25	38	30
Beach 3	19	27	32	31	37

Do the data indicate that the average number of arrests for illegal drug use is not significantly different at these beaches? (Use a 1% level of significance.)

7. Mary is interested in buying a washing machine and is considering three equally rated brands. She has obtained the following prices for the cost of a washing machine for each of these brands.

Washing Machine Prices

Brand A	$429	$436	$453	$402	$444
Brand B	409	428	419	437	403
Brand C	399	437	388	419	423

Do the data indicate that there is a significant difference in the average cost of a washing machine for each of these brands? (Use a 1% level of significance.)

8. The number of unemployment insurance claims handled by each of five claim examiners each day of a particular week is as follows:

Number of Claims Handled

	Bill	32	27	42	26	29
	Joyce	29	46	34	28	41
Claims Examiner	Cassandra	35	39	24	40	36
	Louis	30	43	26	31	27
	Heather	35	27	32	44	37

Do the data indicate that there is a significant difference in the average number of unemployment insurance claims handled by these examiners? (Use a 5% level of significance.)

9. The following table gives the gains in weight of four different types of hogs fed three different rations.

Weight Gain (pounds)

		Hog Type I	Hog Type II	Hog Type III	Hog Type IV
	A	8	15	11	14
Ration	B	17	18	24	21
	C	7	15	8	13

Using a two-way analysis of variance, test the null hypothesis that there is no significant difference in the average weight gain of the different types of hogs and in the different rations used. (Use a 1% level of significance.)

10. Sixteen music students were divided into four groups of four students each. Each group was taught how to play a new song on a particular instrument, each by a different method. The following chart indicates the number of minutes needed by each student to learn to play the song.

Time to Learn New Song (minutes)

	A	50	43	44	53
Method	B	37	34	54	45
	C	41	27	46	43
	D	33	42	49	34

Do the data indicate that there is a significant difference in the average time needed by each student to learn to play the song by the different methods? (Use a 5% level of significance.)

SUGGESTED FURTHER READING

1. Brownlee, K. A. *Statistical Theory and Methodology in Science and Engineering*, 2nd ed. New York: Wiley, 1965.
2. Guenther, W. C. *Analysis of Variance*. Englewood Cliffs, NJ: Prentice-Hall, 1964.
3. Lapin, Lawrence. *Statistics*. New York: Harcourt Brace Jovanovich, 1975.
4. Li, J. C. R. *Introduction to Statistical Inference*. Ann Arbor, MI: J. W. Edwards Press, 1961.
5. Neter, J., and W. Wasserman. *Applied Linear Statistics Models*. Homewood, IL: Richard D. Irwin, 1974.
6. Snedecor, George W., and William G. Cochran. *Statistical Methods*, 6th ed. Ames, IA: The Iowa State University Press, 1967.
7. Steel, Robert G., and James H. Torrie. *Principles and Procedures of Statistics*. New York: McGraw-Hill, 1960.
8. Winer, B. J. *Statistical Principles in Experimental Design*, 2nd ed. New York: McGraw-Hill, 1971.
9. Yamane, Taro. *Statistics: An Introductory Analysis*, 2nd ed. New York: Harper & Row, 1967.

CHAPTER 14

Nonparametric Statistics

OBJECTIVES

- *To analyze* several tests that can be used when assumptions about normally distributed populations or sample size cannot be satisfied. These are called nonparametric statistics or distribution-free methods.

- *To study* the sign test that is used in the "before and after" type study. We test whether or not $\mu_1 = \mu_2$ when we know that the samples are not independent. We can also use the Wilcoxon signed-rank test.

- *To discuss* an alternative to the standard significance tests for the difference between two sample means that is used when we have non-normally distributed populations. This is the rank-sum test. We can also use this test when the population variances are not equal.

- *To present* the Spearman rank correlation test.

- *To point out* how we use the runs test when we wish to test for randomness.

- *To introduce* the Kruskal-Wallis H-test as a nonparametric test that can be used to test whether the difference between numerous sample means is significant.

INCIDENCE OF CAR THEFT BEGINNING TO DECLINE

Albany (Dec 12): Police department records, as well as insurance company statistics, indicate that the number of reported claims of car theft has been decreasing in the state ever since the state instituted computer checks of all reported car thefts and vigorous prosecution of all false car theft reports. This represents the first time that the reported number of car thefts showed any decline. Over the past years car theft had become a burgeoning crime.

December 12, 1985

Referring to the article above, one can conclude that a technique is needed for measuring the effects of a new policy like the one described here. Such "before and after" types of situations often occur when new products are introduced or when a new technique is tried.

Are such statistical tests available?

14.1 INTRODUCTION

In previous chapters we discussed procedures for testing various hypotheses involving means, proportions, variances, and the like. In almost all of the cases discussed we assumed that the populations from which the samples were taken were approximately normally distributed. Only when we applied the chi-square distribution in comparing observed frequencies with expected frequencies did we not specify the normal distribution.

Since there are many situations where this requirement cannot be satisfied, statisticians have developed techniques to be used in such cases. These techniques are known as **nonparametric statistics** or **distribution-free methods.** As the names imply, these methods are not dependent upon the distribution or parameters involved.

There are advantages and disadvantages associated with using nonparametric statistics. The advantages in using these methods as opposed to the **standard methods** are as follows:

1. They are easier to understand.
2. They often involve much less computation.
3. They are less demanding in their assumptions about the nature of the sampled populations.

For these reasons many people often refer to nonparametric statistics as shortcut statistics. The disadvantages associated with nonparametric statistics are that they usually waste information, as we will see shortly, and that they tend to result in the acceptance of null hypotheses more often than they should.

Nonparametric statistical methods are often used when samples are small, since most of the **standard tests** require that the sample sizes be reasonably large.

In this chapter we will discuss only briefly some of the more commonly used nonparametric tests. A complete discussion of all these methods would require many chapters or perhaps even several volumes.

14.2 THE ONE-SAMPLE SIGN TEST

All of the standard sample tests involving means that we have discussed so far in this book are based upon the assumption that the populations are approximately normally distributed. This often may not be the case.

When the above assumption is not necessarily true, then we can replace the standard tests by one of the numerous nonparametric tests. By far, the simplest of all the nonparametric tests is the one-sample sign test. It is used as an alternative to the t-test with one mean that was discussed in Section 10.5. We apply this test when the hypothesis we are testing concerns the value of the mean or median of the population and we are sampling a continuous population in the vicinity of the unknown mean or median M. Under these assumptions, the probability that a sample value is less than the mean or median and the probability that a sample value is greater than the mean or median are both $\frac{1}{2}$.

If the sample size is small, we can perform the sign test by referring to a table of binomial probabilities (Table IV in the Appendix.) When the sample size is large we can perform the sign test by using the normal curve approximation to the binomial. We illustrate the techniques with several examples.

● Example 1

A trucking industry spokesperson claims that the median weight of a load carried by a truck traveling on State Highway No. 7 is 15,000 pounds. A transportation official believes that this figure is much too low. To verify the trucking industry claim, a random survey of 15 trucks is taken and the weight of each truck's load is determined. It is found that 5 trucks have loads with weights below 15,000 pounds, 2 trucks have loads equal to 15,000 pounds, and 8 trucks have loads above 15,000 pounds. Using the one-sample sign test, test the null hypothesis that the median weight of a load carried by a truck is 15,000 pounds against the alternative hypothesis that the median weight is more than 15,000 pounds. (Use a 5% level of significance.)

Solution

We replace each truck's load weight above 15,000 pounds with a plus sign and each truck's load weight below 15,000 pounds with a minus sign. We discard those trucks whose load weight equals 15,000 pounds. We then test the null hypothesis that the plus signs and minus signs are values of a random variable having the binomial distribution with $p = \frac{1}{2}$. In our case, we have 5 minus signs and 8 plus signs. Here $n = 13$ and not 15 since we disregard those trucks whose load weight is exactly 15,000 pounds. We then determine whether 8 plus signs in 13 observed trials agrees with the null hypothesis that $p = \frac{1}{2}$ or with the alternative hypothesis that $p > \frac{1}{2}$. Using Table IV in the Appendix, we find that for $n = 13$ and $p = \frac{1}{2}$ the probability of obtaining 8 or more successes is

$$\text{Prob (8 successes)} = 0.157$$

$$\text{Prob (9 successes)} = 0.087$$

$$\text{Prob (10 successes)} = 0.035$$

$$\text{Prob (11 successes)} = 0.010$$

$$\text{Prob (12 successes)} = 0.002$$

$$\text{Prob (13 successes)} = \text{-------}$$

$$\text{Prob (8 or more successes)} = \overline{0.291}$$

Since this value is more than $\alpha = 0.05$, we do not reject the null hypothesis. The sample data do not contradict the trucking industry's claim.

The previous example was a one-sided hypothesis problem. However, the same procedure can be used when testing two-sided alternative hypotheses problems. In essence, what we do is calculate the number of plus signs and the number of minus signs. We then determine whether a random variable having a binomial distribution can have the calculated number of plus signs or minus signs and still have $p = \dfrac{1}{2}$. By referring to the binomial probability chart (Table IV in the Appendix), the critical rejection region is then specified.

Since applications using the sign test occur often, statisticians have constructed a chart which enables us to determine whether the number of plus signs or minus signs is significant. This eliminates the need to use the binomial distribution probability chart and add probabilities. One such chart is given in Table 14.1.

RULE

When using Table 14.1 for a two-sided test, the following applies:

1. n represents the total number of plus signs and negative signs, disregarding any zeros.
2. The test statistic is the number of the less frequent sign. This means that we first determine the number of plus signs and the number of minus signs. We select the smaller number between the number of plus signs or minus signs. This represents the test statistic.
3. We reject the null hypothesis if the test statistic value is less than or equal to the chart value. If the test statistic is larger than the chart value, we do not reject the null hypothesis.

COMMENT   Table 14.1 gives us the critical values for a two-sided test. When working with a one-sided test, double the value of α specified in the problem.

Use of the above rule is illustrated in the following example.

● Example 2

A doctor suspects that the median annual cost for malpractice insurance in her specialty is approximately $18,000. Her nurse believes that the median annual

TABLE 14.1
Critical Values for Sign Test

n	Level of Significance ($\alpha =$)				n	Level of Significance ($\alpha =$)			
	0.01	0.05	0.10	0.25		0.01	0.05	0.10	0.25
1					21	4	5	6	7
2					22	4	5	6	7
3				0	23	4	6	7	8
4				0	24	5	6	7	8
5			0	0	25	5	7	7	9
6		0	0	1	26	6	7	8	9
7		0	0	1	27	6	7	8	10
8	0	0	1	1	28	6	8	9	10
9	0	1	1	2	29	7	8	9	10
10	0	1	1	2	30	7	9	10	11
11	0	1	2	3	31	7	9	10	11
12	1	2	2	3	32	8	9	10	12
13	1	2	3	3	33	8	10	11	12
14	1	2	3	4	34	9	10	11	13
15	2	3	3	4	35	9	11	12	13
16	2	3	4	5	36	9	11	12	14
17	2	4	4	5	37	10	12	13	14
18	3	4	5	6	38	10	12	13	14
19	3	4	5	6	39	11	12	13	15
20	3	5	5	6	40	11	13	14	15

cost is not equal to $18,000. She samples 13 insurance companies and obtains the following quotes for the identical cost of malpractice insurance.

$14,350 $17,010 $13,936 $17,073 $17,985 $18,840 $17,240
$18,000 $19,420 $17,840 $16,090 $17,360 $17,053

Using the one-sample sign test, test the null hypothesis that the median cost for malpractice insurance is $18,000 against the alternative hypothesis that the median cost is not $18,000. (Use a 5% level of significance.)

Solution

We replace each price quote with a minus sign if it is less than $18,000 and with a plus sign if it is above $18,000. We neglect those quotes that equal $18,000. We then have

$$- \ - \ - \ - \ + \ - \ + \ - \ - \ - \ -$$

or a total of 10 minus signs and 2 plus signs in 12 trials. Here we are testing whether the number of plus signs (two in our case) supports the null hypothesis that $p = \frac{1}{2}$ or the alternative hypothesis that $p \neq \frac{1}{2}$. The number of plus signs is 2 and the number of minus signs is 10. The smaller of these two numbers is 2. This represents the test statistic. Now we look in Table 14.1 to find the appropriate critical value for $n = 12$ and $\alpha = 0.05$. The chart value is 2. Since the test statistic value that we obtained is less than or equal to the chart value of 2 (in our case, it equals 2), we reject the null hypothesis and conclude that the median cost of malpractice insurance is not $18,000.

14.3 THE PAIRED-SAMPLE SIGN TEST

The sign test can also be used when working with **paired data** that occur when we deal with **two dependent samples.** This often happens when we measure the same sample twice, as is done in the **before and after type study.** We proceed in a manner similar to what was done in Section 14.2.

To illustrate, suppose a college administrator is interested in knowing how a particular three-week math mini-course affects a student's grade. Twenty students are selected and are given a math test. Then these students attend the mini-course and are retested. The administration would like to use the results of these two tests to determine whether the mini-course actually improves a student's score.

Table 14.2 contains the scores of the twenty students on the precourse test and the postcourse test. In this table we have taken each student's precourse test score and subtracted it from the student's postcourse test score to obtain the change score. An increase in score is assigned a plus ($+$) sign and a decrease in score is assigned a minus ($-$) sign. No sign is indicated when the two scores are identical.

> **TABLE 14.2**
> Scores of Twenty Students on Precourse and Postcourse Tests

Student	Precourse Score	Postcourse Score	Sign of Difference
1	68	71	+
2	63	65	+
3	82	88	+
4	70	79	+
5	65	57	−
6	66	77	+
7	64	62	−
8	69	73	+
9	72	70	−
10	74	76	+
11	71	68	−
12	80	80	
13	59	71	+
14	85	80	−
15	57	65	+
16	83	87	+
17	43	48	+
18	94	94	
19	82	93	+
20	91	94	+

The null hypothesis in this case is $\mu_1 = \mu_2$. This means that the mini-course does not significantly affect a student's score. Under this assumption we would expect an equal number of plus signs and minus signs. Thus if p is the proportion of plus signs, we would expect p to be around 0.5 (subject only to chance error).

Since we think that the mini-course does increase a student's score, the alternative hypothesis would be $\mu_2 > \mu_1$. Thus

H_0: $\mu_1 = \mu_2$

H_1: $\mu_2 > \mu_1$

We can now apply the methods discussed in Section 10.8 (pages 452–455) for testing a proportion. Recall that for testing a proportion we use the test statistic

$$z = \frac{p - \pi}{\sqrt{\dfrac{\pi(1 - \pi)}{n}}}$$

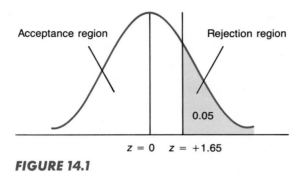

Acceptance region Rejection region

0.05

$z = 0$ $z = +1.65$

FIGURE 14.1

In applying this test statistic we let π be the true proportion of plus signs as specified in the null hypothesis. Thus if the mini-course does not affect a student's scores, we would expect as many plus signs as minus signs. There should be no more students obtaining higher scores than students obtaining lower scores as a result of this mini-course. Therefore, $\pi = 0.50$. We now count the number of plus signs. There are 13 plus signs out of a possible 18 sign changes. We ignore the cases that involve no change. So, $p = \dfrac{13}{18}$, or 0.72, $n = 18$, and $\pi = 0.50$. Applying the test statistic, we have

$$z = \frac{p - \pi}{\sqrt{\dfrac{\pi(1 - \pi)}{n}}}$$

$$= \frac{0.72 - 0.50}{\sqrt{\dfrac{(0.5)(1 - 0.5)}{18}}}$$

$$= \frac{0.22}{\sqrt{0.013889}} = \frac{0.22}{0.118}$$

$$= 1.86$$

We use the one-tail rejection region of Figure 14.1. Since the value of $z = 1.86$ falls in the critical region, we reject the null hypothesis at the 5% level of significance. Thus the mini-course seems to have improved a student's score.

The following example will further illustrate the paired-sign test technique.

● **Example 1**

A new weight-reducing pill is given to 15 people once a week for three months to determine its effectiveness in reducing weight. The following data indicate the before and after weights (in pounds) of these 15 people:

Weight Before Taking Pill	Weight After Taking Pill
131	125
127	128
116	118
153	155
178	179
202	200
192	195
183	180
171	180
182	180
169	174
155	150
163	169
171	172
208	200

Using a 5% level of significance, test the null hypothesis that the pill is not effective in reducing weight.

Solution

We arrange the data as follows:

Weight Before	Weight After	Sign of Difference
131	125	−
127	128	+
116	118	+
153	155	+
178	179	+
202	200	−
192	195	+
183	180	−
171	180	+
182	180	−
169	174	+
155	150	−
163	169	+
171	172	+
208	200	−

For each person we determine the change in weight. A plus sign indicates a gain and a minus sign indicates a loss. If the weight reducing pill is not effective, the average weight should be the same before and after taking this pill. Since we are testing whether a person's weight remains the same or is reduced, we have

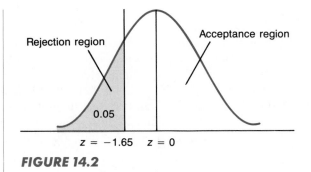

FIGURE 14.2

H_0: $\mu_1 = \mu_2$

H_1: $\mu_2 < \mu_1$

Out of the 15 sign changes 6 are minus so that

$$p = \frac{6}{15} = 0.4 \quad \text{and} \quad n = 15$$

Also $\pi = 0.50$. Applying the test statistic, we get

$$z = \frac{p - \pi}{\sqrt{\dfrac{\pi(1 - \pi)}{n}}}$$

$$= \frac{0.4 - 0.5}{\sqrt{\dfrac{(0.5)(1 - 0.5)}{15}}}$$

$$= \frac{-0.1}{\sqrt{0.01667}} = \frac{-0.1}{0.129}$$

$$= -0.78$$

We use the one-tail rejection region of Figure 14.2. Since the value of $z = -0.78$ falls in the acceptance region, we cannot reject the null hypothesis. The weight-reducing pill does not seem to be effective in reducing weight. It may even cause an increase in weight.

COMMENT We mentioned earlier that nonparametric methods are wasteful. A plus or minus sign merely tells us that a person gained weight or lost weight. It does not specify whether the gain was 1 pound, 10 pounds, or even 100 pounds. The same is true for the minus signs.

COMMENT Since the sign test is so easy to use, many people use it even when the *standard* tests can be used.

COMMENT The paired-sample sign test is actually a nonparametric alternative to the paired difference *t*-test.

EXERCISES

1. An economist believes that the median starting salary for a computer programmer in a certain city is $21,000. To verify this claim, a computer club randomly selects 12 computer programmers with similar backgrounds who recently obtained jobs. Their starting salaries were $19,100, $17,638, $24,360, $20,350, $21,430, $22,000, $21,000, $22,500, $20,000, $21,850, $20,800, $22,400. Using the one-sample sign test, test the null hypothesis that the median starting salary for a computer programmer in this city is $21,000 against the alternative hypothesis that the median starting salary is not $21,000. Use a 5% level of significance.

2. A medical researcher believes that the side effects of a particular drug will begin to disappear approximately 10 hours after a patient has ingested the drug. A doctor administers this drug to 15 patients. These patients report that the side effects of the drug began to disappear 8, 12, 11, 16, 9, 10, 7, 11, 15, 10, 14, 13, 8, 14, and 15 hours, respectively, after ingestion of the drug. Using the one-sample sign test, test the null hypothesis that the median time for the side effects of the particular drug to begin to disappear is 10 hours against the alternative hypothesis that the median time is more than 10 hours. Use a 5% level of significance.

3. A drug enforcement agent believes that the median age of a person arrested for illegal use of drugs in a particular city is 19 years. A politician disagrees with this claim and believes that the median age is higher. A random sample of 13 people arrested for illegal use of drugs is taken. The ages of these people at the time of arrest is found to be 15, 23, 18, 17, 19, 34, 28, 26, 20, 24, 22, 18, and 16 years. Use the one-sample sign test to test the null hypothesis that the drug enforcement agent is correct against the alternative hypothesis that the politician is correct. Use a 5% level of significance.

4. A merchant has been accused of price gouging by charging higher prices for a quart of milk in a poorer neighborhood than the price charged by competitors. The merchant claims that the median cost of a container of milk in the neighborhood is 64 cents. A random survey of the prices charged by 10 competitors for a quart of milk disclosed prices of 58, 59, 64, 62, 65, 60, 62, 69, 63, and 61 cents respectively. Use the sample sign test to test the null hypothesis that the merchant's claim is correct against the alternative hypothesis that the merchant is price-gouging. Use a 5% level of significance.

5. A large department-store chain has received numerous complaints about pickpocketing incidents in the store. The manager decides to hire more security personnel (both uniformed and nonuniformed). The following table gives the number of arrests per store for pickpocketing in 12 stores of this department-store chain, before and after more security personnel were added.

Number of Weekly Arrests for Pickpocketing at Various Stores

		Before New Personnel Hired	After New Personnel Hired
	1	10	9
	2	9	8
	3	8	8
	4	6	7
	5	5	3
Store	6	9	8
	7	9	7
	8	9	9
	9	5	4
	10	8	6
	11	4	3
	12	7	2

Using a 5% level of significance, test the null hypothesis that the new security personnel hired have no effect on the number of weekly arrests for pickpocketing against the alternative hypothesis that the number of weekly arrests has decreased.

6. An industrial engineer is analyzing the number of products completed per day by 11 workers before and after the installation of a new air-conditioning system and the introduction of piped-in music to the work area. The following data are available.

Number of Products Completed Daily by Each Worker

		Before New Equipment Installed	After New Equipment Installed
	A	38	47
	B	34	39
	C	36	36
	D	28	28
	E	49	47
Worker	F	27	37
	G	17	21
	H	18	24
	I	16	19
	J	10	11
	K	14	15

Using a 5% level of significance, test the null hypothesis that the number of products completed daily is not affected by the change in working conditions against the alternative hypothesis that the number of products completed daily has increased.

7. Several homeowners in the Northeast who recently converted from oil heat to gas heat in order to save money are comparing their current heating bills with the heating bills of years prior to converting. All of the homes are insulated in a comparable manner. The heating bills for a full season are as follows:

		Heating Bill	
		Oil	Gas
	A	$2200	$2400
	B	$2000	$1900
	C	$2700	$2800
	D	$2600	$2480
Homeowner	E	$3000	$3100
	F	$2500	$2540
	G	$2300	$2380
	H	$2100	$2000
	I	$2375	$2395

Using a 5% level of significance test the null hypothesis that there is no significant difference in the annual heating bills for homes heated by oil or by gas against the alternative hypothesis that gas heating is cheaper.

8. The number of different times that dizziness was reported per month by patients prior to and after taking a certain medicine is shown below:

		Reported Dizziness	
		Before Taking Medicine	After Taking Medicine
	1	19	17
	2	18	24
	3	9	12
	4	8	4
Patient	5	7	7
	6	12	15
	7	16	19
	8	22	25
	9	19	16
	10	18	24

Using a 5% level of significance, test the null hypothesis that the medicine has no effect on the number of times that dizziness is reported against the alternative hypothesis that the medicine increases the number of times that dizziness is reported.

9. A large Wall Street brokerage firm has instituted a new policy whereby it randomly "listens in" and records all broker conversations with clients. The number of complaints about incorrect orders executed by ten sales brokers both before and after the new policy was instituted is shown in the accompanying table.

		Before New Policy	After New Policy
	1	47	38
	2	39	34
	3	28	16
	4	31	26
Broker	5	30	30
	6	53	47
	7	23	39
	8	42	38
	9	23	21
	10	14	10

Using a 5% level of significance, test the null hypothesis that the new policy has no effect on the number of incorrect orders executed against the alternative hypothesis that the number of complaints has decreased.

ONE OFFICER PATROL CARS A FIASCO

New York (May 10): Police department officials have refused to comment on surveys conducted by civic groups which indicate that the number of police arrests for various crimes including robberies, felonies, burglaries, etc. has dropped sharply ever since the department's new one police officer per patrol car policy went into effect. Although this new policy was started as an economy measure, its effectiveness is being questioned.

The department officials steadfastly refuse to admit that this new policy has had any effect on the number of police arrests.

May 10, 1986

10. Refer to the newspaper article given above. The number of felony arrests over a comparable period for ten precincts both before and after the one-police-officer-per-patrol-car policy was instituted is as follows:

Number of Felony Arrests

	Before One-Police-Officer-per-Car Policy	After One-Police-Officer-per-Car Policy
8	10	6
13	4	2
17	8	7
22	6	6
25	8	4
26	12	7
28	9	3
31	8	4
32	7	5
44	11	10

Precinct (label at left spanning rows 25, 26)

Using a 5% level of significance, test the null hypothesis that the daily number of felony arrests is not affected by the number of police officers in the patrol car against the alternative hypothesis that the number of felony arrests has decreased.

14.4 THE WILCOXON SIGNED-RANK TEST

As we indicated in the last section, the **paired-sample sign test** merely utilizes information concerning whether the differences between pairs of numbers is positive or negative. Quite often the sign test will accept a null hypothesis simply because too much information is "thrown away." The **Wilcoxon signed-rank test** is less likely to accept a null hypothesis since it considers both the *magnitude* as well as the *direction* (positive or negative) of the differences between pairs.

To illustrate the use of the Wilcoxon signed-rank test, let us consider the following exam scores of 12 students who were tested both before and after receiving special instruction.

Student	Pre-instruction Score X_B	Post-instruction Score X_A	Difference $D = X_B - X_A$	Absolute Value of Difference $|D|$	Rank of $|D|$	Signed Rank
1	46	81	$46 - 81 = -35$	35	9	-9
2	58	73	$58 - 73 = -15$	15	8	-8
3	69	72	$69 - 72 = -3$	3	2	-2
4	72	77	$72 - 77 = -5$	5	4	-4
5	82	82	$82 - 82 = 0$	---	---	---
6	65	72	$65 - 72 = -7$	7	6	-6
7	69	63	$69 - 63 = 6$	6	5	$+5$
8	72	68	$72 - 68 = 4$	4	3	$+3$
9	73	85	$73 - 85 = -12$	12	7	-7
10	87	88	$87 - 88 = -1$	1	1	-1

Notice that we have added several new columns.

To apply the Wilcoxon signed-rank test we proceed as follows:

1. Find the entry for the difference column by subtracting the new value from the corresponding old value. These differences may be positive, negative, or zero.
2. Form the absolute value of these differences.
3. Rank these absolute values in order from lowest (1) to highest.
4. Give each rank a plus ($+$) sign or a minus ($-$) sign, which is the same sign as in the column for D.
5. The test statistic is the sum of the ranks with the smaller sum. If the null hypothesis is correct, then we would expect the sum of the positive ranks and the sum of the negative ranks to balance each other. If the sum of the ranks is considerably more positive or considerably more negative, then we would be more likely to reject the null hypothesis.

In our case, the sum of the ranks is

Positive sign ranks $= (+5) + (+3) = +8$

Negative sign ranks $= (-9) + (-8) + (-2) + (-4)$
$$+ (-6) + (-7) + (-1) = -37$$

We select $+8$, since this is the sum of the ranks with the smaller sum. We now compare this test statistic value with the critical value given in Table 14.3 using $n = 9$ and $\alpha = 0.05$. The chart value for a two-tailed test is 5. Since our test statistic value of $+8$ is larger than the chart value, we do not reject the null hypothesis. If, on the other hand, the test statistic value is less than or equal to the chart value, then we reject the null hypothesis.

COMMENT In performing the calculations in the previous example, we did not use the fifth student's scores since, as with the sign test, a difference of zero is not considered as positive or negative for our purposes.

COMMENT All entries given in Table 14.3 are for absolute values of the test statistic.

COMMENT Occasionally, two differences will have the same rank. For example, if two differences are tied for the fourth place, then each is assigned a rank of 4.5. We then assign the next value a rank of 6. The same procedure is followed when there is a tie for any rank.

We illustrate the use of the Wilcoxon signed-rank test with another example.

● **Example 1**

A new cholesterol-lowering pill is given to 15 people once a day for two months to determine its effectiveness in lowering blood serum cholesterol levels. The following data indicate the before and after cholesterol levels (in mg) of these 15 people.

TABLE 14.3
Critical Values for Wilcoxon Signed-rank Test

n	One-tailed Test Level of Significance ($\alpha =$)				n	Two-tailed Test Level of Significance ($\alpha =$)			
	0.005	0.01	0.025	0.05		0.01	0.02	0.05	0.10
5	----	----	----	0	5	----	----	----	0
6	----	----	0	2	6	----	----	0	2
7	----	0	2	3	7	----	0	2	3
8	0	1	3	5	8	0	1	3	5
9	1	3	5	8	9	1	3	5	8
10	3	5	8	10	10	3	5	8	10
11	5	7	10	13	11	5	7	10	13
12	7	9	13	17	12	7	9	13	17
13	9	12	17	21	13	9	12	17	21
14	12	15	21	25	14	12	15	21	25
15	15	19	25	30	15	15	19	25	30
16	19	23	29	35	16	19	23	29	35
17	23	27	34	41	17	23	27	34	41
18	27	32	40	47	18	27	32	40	47
19	32	37	46	53	19	32	37	46	53
20	37	43	52	60	20	37	43	52	60
21	42	49	58	67	21	42	49	58	67
22	48	55	65	75	22	48	55	65	75
23	54	62	73	83	23	54	62	73	83
24	61	69	81	91	24	61	69	81	91
25	68	76	89	100	25	68	76	89	100
26	75	84	98	110	26	75	84	98	110
27	83	92	107	119	27	83	92	107	119
28	91	101	116	130	28	91	101	116	130
29	100	110	126	140	29	100	110	126	140
30	109	120	137	151	30	109	120	137	151

Cholesterol Level Before Taking Pill (mg)	Cholesterol Level After Taking Pill (mg)
240	227
261	238
283	257
276	276
220	208
186	193
195	198
198	199
247	233
238	227
220	210
250	241
263	255
298	276
317	269

Using a 5% level of significance, test the null hypothesis that the pill has no effect on a person's blood serum cholesterol level; it neither lowers nor raises it.

Solution

We arrange the data as follows:

Cholesterol Level Before X_B	Cholesterol Level After X_A	Difference $D = X_B - X_A$	Absolute Value of Difference $|D|$	Rank of $|D|$	Signed Rank
240	227	$240 - 227 = 13$	13	9	$+9$
261	238	$261 - 238 = 23$	23	12	$+12$
283	257	$283 - 257 = 26$	26	13	$+13$
276	276	$276 - 276 = 0$	---	---	---
220	208	$220 - 208 = 12$	12	8	$+8$
186	193	$186 - 193 = -7$	7	3	-3
195	198	$195 - 198 = -3$	3	2	-2
198	199	$198 - 199 = -1$	1	1	-1
247	233	$247 - 233 = 14$	14	10	$+10$
238	227	$238 - 227 = 11$	11	7	$+7$
220	210	$220 - 210 = 10$	10	6	$+6$
250	241	$250 - 241 = 9$	9	5	$+5$
263	255	$263 - 255 = 8$	8	4	$+4$
298	276	$298 - 276 = 22$	22	11	$+11$
317	269	$317 - 269 = 48$	48	14	$+14$

For each person, we determine the difference in the cholesterol level, and the absolute value of the difference. Then we assign ranks from lowest to highest and assign a plus sign or a minus sign to each of these ranks. The sign is the same as the sign in the column for D. Now we calculate the sum of the ranks. We have

$$\text{Positive sign ranks} = (+9) + (+12) + (+13) + (+8) + (+10)$$
$$+ (+7) + (+6) + (+5) + (+4) + (+11) + (+14) = +99$$

$$\text{Negative sign ranks} = (-3) + (-2) + (-1) = -6$$

We select -6 since this is the sum of the ranks with the smaller sum. The absolute value of -6 is 6. We now compare this test statistic value with the critical value given in Table 14.3 using $n = 14$ and $\alpha = 0.05$. The chart value for a two-tailed test is 21. Since our test statistic value of 6 is less than the chart value, we reject the null hypothesis and conclude that the pill does affect the blood serum cholesterol level of an individual.

COMMENT When using the Wilcoxon signed-rank test, it is assumed that the sample data are continuous and that the sampled population is symmetric. Furthermore, the data results can be arranged into relationships of "greater than" or "less than." In an applied example, it is often difficult to verify that the sampled population is symmetric.

EXERCISES

1. Many schools claim that a lot of food is wasted simply because students do not eat the lunches that have been prepared for them. At one school, a new caterer has been hired. The caterer is supposed to make the food served more visually appealing. After a ten-week trial period the following data are obtained.

Number of Students Completing Their Lunch

Week	Before New Caterer Began Preparing Lunches	After New Caterer Began Preparing Lunches
1	240	253
2	230	247
3	260	255
4	253	261
5	270	275
6	290	283
7	310	307
8	260	313
9	265	299
10	280	332

Using the Wilcoxon signed-rank test, test the null hypothesis that the number of students completing their lunch has not changed even after the new caterer was hired. Use a 5% level of significance.

2. A radio talk show has been experimenting with having guest celebrities on the air. The estimated audience (in hundreds of thousands of people) both before and after instituting this policy for eight sections of a large city over a given time period is shown below.

		Radio Audience (in hundred-thousands)	
		Before New Policy	After New Policy
	Brighton	23	27
	Queens	22	26
	Boro Hall	19	21
Section of City	Gravesend	25	24
	Togo	18	25
	Braverly	24	24
	Beaver	17	25
	Kensington	26	26

Using the Wilcoxon signed-rank test at a 5% level of significance, test the null hypothesis that the new policy has no effect on radio audience.

3. Refer back to the newspaper article given at the beginning of this chapter. The number of reported cases of car theft for a specific time period both before and after the use of the extensive computer tracking is shown below:

		Number of Reported Cases of Car Theft	
		Before New Policy	After New Policy
	A	35	31
	B	39	33
	C	43	40
	D	41	31
	E	38	39
City/Town	F	37	37
	G	53	45
	H	47	46
	I	55	53
	J	46	43

Using the Wilcoxon signed-rank test at a 5% level of significance, test the null hypothesis that the new policy has no effect on reducing car theft.

4. Twelve people were randomly selected and asked to rate a toothpaste (on a certain scale) both before and after a certain ingredient was added. Their ratings are shown below:

Toothpaste Rating

Subject	Before Ingredient Added	After Ingredient Added
1	110	107
2	118	119
3	137	139
4	115	114
5	119	119
6	136	139
7	127	135
8	116	128
9	119	103
10	117	128
11	135	123
12	126	148

Using the Wilcoxon signed-rank test, do the data indicate that there is no significant difference in the ratings of the toothpaste before and after the ingredient was added? (Use a 5% level of significance.)

14.5 THE MANN-WHITNEY TEST

An important nonparametric test that is used as an alternative to the standard significance tests for the difference between two sample means is the **Wilcoxon rank-sum test** or **the Mann-Whitney Test.** We can use this test when the assumption about normality is not satisfied.

To illustrate how this test is used, we consider the following data on the number of minutes needed by two independent groups of music students to learn to play a particular song. Group A received special instruction whereas Group B did not.

										Average
Group A	35	39	51	63	48	31	29	41	55	43.56
Group B	85	28	42	37	61	54	36	57		50

The means of these two samples are 43.56 and 50. In this case we wish to decide whether the difference between the means is significant.

The two samples are arranged jointly, as if they were one sample, in order of increasing time. We get

Time	Group	Rank
28	B	1
29	A	2
31	A	3
35	A	4
36	B	5
37	B	6
39	A	7
41	A	8
42	B	9
48	A	10
51	A	11
54	B	12
55	A	13
57	B	14
61	B	15
63	A	16
85	B	17

We indicate each value, whether it belongs to Group A or to Group B. Then we assign the ranks 1, 2, 3, 4, . . . , 17 to the scores, in this order, as shown.

Notice that the Group A scores occupy the ranks of 2, 3, 4, 7, 8, 10, 11, 13, and 16. The Group B scores occupy the ranks of 1, 5, 6, 9, 12, 14, 15, and 17. Now we sum the ranks of the group with the *smaller* sample size, in this case Group B, getting

$$1 + 5 + 6 + 9 + 12 + 14 + 15 + 17 = 79$$

The sum of the ranks is denoted by R. In this case $R = 79$.

We always let n_1 and n_2 denote the sizes of the two samples where n_1 represents the smaller of the two sample sizes. Thus R represents the sum of the ranks of this smaller group. (If both groups are of equal sizes, then either one is called n_1, and R represents the sum of the ranks of this group.) Statistical theory tells us that if both n_1 and n_2 are large enough, each equal to eight or more, then the distribution of R can be approximated by a normal distribution. The test statistic is given by Formula 14.1.

FORMULA 14.1

$$z = \frac{R - \mu_R}{\sigma_R}$$

where

$$\mu_R = \frac{n_1(n_1 + n_2 + 1)}{2}$$

$$\sigma_R = \sqrt{\frac{n_1 n_2 (n_1 + n_2 + 1)}{12}}$$

Using a 5% level of significance, we reject the null hypothesis of equal means if $z > 1.96$ or $z < -1.96$.

In our case $R = 79$, $n_1 = 8$, and $n_2 = 9$ so that

$$\mu_R = \frac{n_1(n_1 + n_2 + 1)}{2}$$

$$= \frac{8(8 + 9 + 1)}{2}$$

$$= 72$$

and

$$\sigma_R = \sqrt{\frac{n_1 n_2 (n_1 + n_2 + 1)}{12}}$$

$$= \sqrt{\frac{8(9)(8 + 9 + 1)}{12}} = \sqrt{108}$$

$$= 10.39$$

The test statistic then becomes $z = \dfrac{R - \mu_R}{\sigma_R} = \dfrac{79 - 72}{10.39} = 0.67$

Since the value of $z = 0.67$ falls in the acceptance region of Figure 14.3, we *do not* reject the null hypothesis. There is no significant difference between the means of these two groups.

COMMENT The test that we have just described is the Wilcoxon rank-sum test with the normal approximation of the test statistic. **Mann-Whitney's test is**

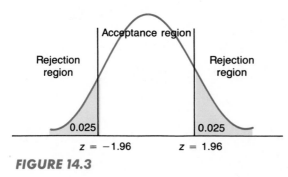

FIGURE 14.3

equivalent, but the test statistic is calculated in a slightly different way. Statisticians have constructed tables that give the appropriate critical values when both sample sizes, n_1 and n_2, are smaller than 8. The interested reader can find such tables in many books on nonparametric statistics. The corresponding exact statistic is called the **Mann-Whitney U test.**

● **Example 1**

An animal trainer in a circus is teaching 20 lions to perform a special trick. The lions have been divided into two groups, A and B. Group A gets positive reinforcement of food and favorable comments during the learning session whereas Group B does not. The following table indicates the number of days needed by each lion to learn the trick:

Group A	78	95	82	69	111	65	73	84	92	110
Group B	121	132	101	79	94	88	102	93	98	127

Using a 5% level of significance, test the null hypothesis that the mean time for both groups is the same.

Solution

The two samples are first arranged jointly, as if they were one large sample, in order of increasing size. We get

Days	Group	Rank
65	A	1
69	A	2
73	A	3
78	A	4
79	B	5
82	A	6
84	A	7
88	B	8
92	A	9
93	B	10
94	B	11
95	A	12
98	B	13
101	B	14
102	B	15
110	A	16
111	A	17
121	B	18
127	B	19
132	B	20

Since both groups are of equal size, we will work with Group A. The sum of the ranks of Group A is

$$1 + 2 + 3 + 4 + 6 + 7 + 9 + 12 + 16 + 17 = 77$$

Thus $R = 77$. Now we apply Formula 14.1. We have $R = 77$, $n_1 = 10$, and $n_2 = 10$ so that

$$\mu_R = \frac{n_1(n_1 + n_2 + 1)}{2}$$

$$= \frac{10(10 + 10 + 1)}{2}$$

$$= 105$$

and

$$\sigma_R = \sqrt{\frac{n_1 n_2 (n_1 + n_2 + 1)}{12}}$$

$$= \sqrt{\frac{10 \cdot 10(10 + 10 + 1)}{12}}$$

$$= \sqrt{175}$$

$$\approx 13.23$$

The test statistic then becomes

$$z = \frac{R - \mu_R}{\sigma_R}$$

$$= \frac{77 - 105}{13.23}$$

$$= -2.12$$

We use the two-tail rejection of Figure 14.3. Since the value of $z = -2.12$ falls in the rejection region, we reject the null hypothesis and conclude that the number of minutes needed by each group is not the same. Positive reinforcement affects learning time.

● Example 2

Many airlines are forced to cancel scheduled flights for a variety of reasons. Records submitted to aviation officials indicate that the weekly number of cancelled flights reported by two large airline companies over a consecutive period of weeks is as follows:

Airline A	Airline B
7	13
9	5
12	4
14	8
8	11
6	3
11	17
	15

Using a 5% level of significance, test the null hypothesis that the average number of cancelled flights is the same for both airlines.

Solution

The null hypothesis is that the average number of cancelled flights is the same for both airlines. We arrange both samples jointly, as if they were one large sample, in order of increasing size. We get

Number of Cancelled Flights	Airline	Rank
3	B	1
4	B	2
5	B	3
6	A	4
7	A	5
8	A	6.5
8	B	6.5
9	A	8
11	A	9.5
11	B	9.5
12	A	11
13	B	12
14	A	13
15	B	14
17	B	15

You will notice that there are some ties for several ranks. Whenever a tie comes up, each of the tied observations is assigned the mean of the ranks they occupy. Thus, we assign a rank of 6.5 when the number of cancelled flights was 8, and a rank of 9.5 when the number of cancelled flights was 11. For Airline A we have 7 weeks of data and for Airline B we have 8 weeks of data. Thus, we will work with the smaller group, which is A. The sum of the ranks for Airline A is

$$4 + 5 + 6.5 + 8 + 11 + 13 = 47.5$$

Thus, $R = 47.5$. Applying Formula 14.1 with $n_1 = 7$ and $n_2 = 8$ gives

$$\mu_R = \frac{n_1(n_1 + n_2 + 1)}{2}$$

$$= \frac{7(7 + 8 + 1)}{2}$$

$$= 56$$

and

$$\sigma_R = \sqrt{\frac{n_1 n_2(n_1 + n_2 + 1)}{12}}$$

$$= \sqrt{\frac{7(8)(7 + 8 + 1)}{12}}$$

$$= \sqrt{448}$$

$$\approx 21.166$$

The test statistic then becomes

$$z = \frac{R - \mu_R}{\sigma_R}$$

$$= \frac{47.5 - 56}{21.166}$$

$$= -0.402$$

The value of $z = -0.402$ falls in the acceptance region of Figure 14.2. Thus, we cannot reject the null hypothesis. There are not sufficient data to conclude that the average number of cancellations of scheduled flights is significantly different for both airlines.

EXERCISES

1. Two groups of students have been given a special math competency exam. Group-A students are from schools located in the northern part of the city, an area that is predominantly middle class. Group B students are from schools located in the southern part of the city, an area that is predominantly working class. The accompanying data indicate the results of the exam:

Group A	Group B
45	52
50	41
48	30
54	42
47	43
58	38
53	41
44	43
49	37
55	42
57	44
	45
	41
	40
	46

Using the rank-sum test, test the null hypothesis that the mean score for both groups is the same. (Use a 5% level of significance.)

2. The following data indicate the number of minutes of practice needed by two groups of people at a health club to learn a particular aerobic exercise.

Group A	55	60	52	49	63	64	57	48	58	53		
Group B	45	65	61	53	57	56	62	69	54	59	50	55

Using the rank-sum test, test the null hypothesis that the mean time needed by both groups to learn the aerobic exercise is the same. (Use a 1% level of significance.)

3. *The Effects of Alcohol*   Nine younger volunteers aged 20–30 years and ten older volunteers aged 40–50 years were each given the same amount of alcohol and then asked to perform a special task correctly. The number of trials required by each volunteer before completing the task is shown below:

Younger Volunteer	Older Volunteer
20	34
28	36
32	33
17	28
24	19
19	29
25	24
21	27
23	23
	29

Using the rank-sum test, test the null hypothesis that the average number of trials required by a member of each group to complete the task after being given alcohol is the same. (Use a 5% level of significance.)

4. The number of patients treated in the emergency room of Spovik Hospital for the first eight days of January and in the emergency room of Mercy Hospital for the first nine days of January are as follows:

Spovik Hospital	157	160	172	149	155	182	169	132	
Mercy Hospital	161	146	183	195	103	113	188	170	153

Using a 5% level of significance, test the null hypothesis that the average number of patients treated at both hospitals is the same.

5. The number of airplanes arriving at least 15 minutes after their scheduled arrival time daily for two large carriers over a period of time is as follows:

Airline A	6	17	8	19	15	16	11	9	12	3	7	29	
Airline B	18	5	12	22	17	14	5	9	13	4	8	16	10

Using a 5% level of significance, test the null hypothesis that the average number of late arrivals is the same for both airlines.

6. Ten male and 13 female drivers were asked to drive over a road which had many obstacles. The number of minutes needed by each to complete the trip is as follows:

Female	57	60	58	64	65	72	68	44	55	61	53	68	62
Male	46	70	62	61	58	55	69	73	53	48			

Using the rank-sum test, test the hypothesis that the average time needed by both groups is the same. (Use a 5% level of significance.)

14.6 THE SPEARMAN RANK CORRELATION TEST

For many years the most widely used of the nonparametric statistical tests was the rank correlation test developed by C. Spearman in the early 1900s. Although originally devised as a shortcut method for computing the coefficient of correlation discussed in Chapter 11, it has the advantage that it uses rankings only, it makes no assumptions about the distribution of the underlying populations, and it does not assume normality. We simply arrange some data in rank order and then apply the following formula.

> FORMULA 14.2   The Spearman Rank Correlation Coefficient
>
> $$R = 1 - \frac{6\ \Sigma(x - y)^2}{n(n^2 - 1)}$$
>
> where $\Sigma(x - y)^2$ represents the sum of the squares of the difference in ranks and n stands for the number of individuals who have been ranked.

COMMENT   When using Formula 14.2, the value of R will be between -1 and $+1$. It is used in much the same way that we used the correlation coefficient in Chapter 11.

COMMENT   When using Formula 14.2, the null hypothesis to be tested is that there is no significant correlation between the two rankings as opposed to the alternative hypothesis, which assumes that there is a significant correlation between the rankings.

We illustrate the use of this formula with several examples.

● Example 1   Beer Tasting

Two judges are testing five different brands of beer for their taste appeal, and the judges rate the beers as follows:

	Judge 1	Judge 2
Brand A	3	4
Brand B	4	3
Brand C	5	1
Brand D	2	2
Brand E	1	5

Using a 5% level of significance, test the null hypothesis that there is no significant (linear) correlation between the two judge's ratings:

Solution

We first rewrite the above rankings by letting x represent Judge 1's rankings and by letting y represent Judge 2's rankings. We then have the following:

x	y	x − y	(x − y)²
3	4	−1	1
4	3	1	1
5	1	4	16
2	2	0	0
1	5	−4	16
			34

$$\Sigma(x - y)^2 = 34$$

Now we apply Formula 14.2. We have

$$R = 1 - \frac{6 \Sigma(x - y)^2}{n(n^2 - 1)}$$

$$= 1 - \frac{6(34)}{5(25 - 1)}$$

$$= 1 - 1.7 = -0.7$$

Table XII in the Appendix allows us to interpret the value of the Spearman Rank Coefficient correctly. We use this table in the following way:

1. First compute the value of R using Formula 14.2.
2. Then look in the chart for the appropriate R value corresponding to some given value of n where n is the number of pairs of scores.
3. The value of R is *not* statistically significant if it is between $-R_{0.025}$ and $R_{0.025}$ for a particular value of n at the 5% level of significance.

Returning to the previous example, we conclude that since the value of R is -0.7, it is not significant at the 5% level of significance. We cannot conclude that there is a significant correlation between the two rankings.

● **Example 2**

Several 9-year old children recently competed in the Brown Bowling League playoffs and the Southview Bowling League games. The scores of these contestants were as follows:

<table>
<tr><th></th><th colspan="2">Score</th></tr>
<tr><th>Contestant</th><th>Brown League Playoff Game</th><th>Southview Bowling League Game</th></tr>
<tr><td>Kim</td><td>180</td><td>162</td></tr>
<tr><td>Heather</td><td>176</td><td>157</td></tr>
<tr><td>Jason</td><td>198</td><td>176</td></tr>
<tr><td>Cassandra</td><td>197</td><td>183</td></tr>
<tr><td>Wilfredo</td><td>171</td><td>188</td></tr>
</table>

Using a 5% level of significance, test the null hypothesis that there is no significant correlation between the Brown League playoff game scores and the Southview scores.

Solution

We first rewrite each of the scores in terms of rankings from highest score, which is assigned a ranking of 1 to lowest score, which is assigned a ranking of 5. Now we let x represent the Brown score rankings and we let y represent the Southview score rankings. We then have the following:

x	y	x − y	$(x - y)^2$
3	4	−1	1
4	5	−1	1
1	3	−2	4
2	2	0	0
5	1	4	$\underline{16}$
			22

$$\Sigma(x - y)^2 = 22$$

Applying Formula 14.2 gives

$$R = 1 - \frac{6\,\Sigma(x - y)^2}{n(n^2 - 1)}$$

$$= 1 - \frac{6(22)}{5(25 - 1)}$$

$$= 1 - 1.1 = -0.1$$

By referring to Table XII in the Appendix, we conclude that since our value of R is -0.1, it is not significant at the 5% level of significance. We cannot conclude that there is a significant correlation between the two bowling league scores.

14.7 THE RUNS TEST

All the samples discussed so far in this book were assumed to be random samples. How does one test for randomness?

In recent years mathematicians have developed a **runs test** for determining the randomness of samples. This test is based on the order in which the observations are made. For example, suppose 25 people are waiting in line for admission to a theater and they are arranged as follows where m denotes male and f denotes female:

f, f, f, f, m, f, m, m, m, m, f, f, f, f, f, f, m, m, m, m, f, m, m, f, m

Is this a random arrangement of the m's and f's?

In order to answer this question, statisticians use the **theory of runs**. We first have the following definition.

DEFINITION 14.1

*A **run** is a succession of identical letters or symbols that is followed and preceded by a different letter or by no letter at all.*

There are ten runs in the above sequence of m's and f's. These are

Run	Letters
1	ffff
2	m
3	f
4	mmmm
5	ffffff
6	mmmm
7	f
8	mm
9	f
10	m

Many runs would indicate that the data occur in definite cycles according to some pattern. The same is true for data with few runs. In either case we do not have a random sample. We still need some way of determining when the number of runs is reasonable.

When using the runs test, note that the length of each individual run is not important. What is important is the number of times that each letter appears in the entire sequence of letters. Thus, in our example there are 25 people waiting in line; 13 are female and 12 are male, so that f appears 13 times and m appears 12 times. We have n_1 samples of one kind and n_2 samples of another kind. We now wish to test whether this sample is random.

Table XI in the Appendix gives us the critical values for the total number of runs. To use this table we first determine the larger of n_1 and n_2. In our case there are 13 f's and 12 m's so that the larger is 13 and the smaller is 12. We now move across the top of the chart until we reach 13. Then we move down until we get to the 12 row. Notice that there are two numbers in the box corresponding to larger 13, smaller 12. These are the numbers 8 and 19. These are also the critical values. If the number of runs is between 8 and 19, we do not reject the null hypothesis. This would mean that we have a random sample. If the number of runs is less than 8 or more than 19, we no longer have a random sample. In our case since we had 10 runs, we do not reject the null hypotheses of randomness.

Let us further illustrate the runs test with several examples.

● Example 1

Twenty people are waiting on line in a bank. These people will either deposit or withdraw money. Let d represent a customer who makes a deposit and let w represent a customer who is making a withdrawal. If the people are arranged in the following order, test for randomness. (Use a 5% level of significance.)

d, d, d, w, w, w, w, w, d, d, d, d, d, d, d, w, w, w, d, d

Solution

There are five runs as shown in the following chart:

Run	Letters
1	ddd
2	wwwww
3	ddddddd
4	www
5	dd

There are also 12 d's and 8 w's in this succession of letters, so that the larger is 12 and the smaller is 8. From Table XI we find that the critical values, where the larger number is 12 and the smaller number is 8, are 6 and 16 runs. Since we obtained only 5 runs, we reject the null hypotheses and conclude that these people are not arranged in random order.

● **Example 2**

Thirty dresses are arranged on a rack as follows, where r represents a red dress and b represents a blue dress. Using a 5% level of significance, determine if the dresses are arranged in a random order.

r, b, r, b, r, r, b, r, b, b, r, r, r, r, r, r, b, r, b, b, r, b, r, b, r, b, b, b, b, r, r

Solution

There are 16 r's and 14 b's so that the larger number is 16 and the smaller number is 14. There are also 19 runs as shown.

Run	Letters
1	r
2	b
3	r
4	b
5	rr
6	b
7	r
8	bb
9	rrrrr
10	b
11	r
12	bb
13	r
14	b
15	r
16	b
17	r
18	bbbb
19	rr

From Table XI we note that the critical values, where the larger number is 16 and the smaller number is 14, are 10 and 22. Since we obtained 19 runs, we do not reject the null hypothesis of randomness.

The theory of runs can also be applied to any set of numbers to determine whether or not these numbers appear in a random order. In such cases we first calculate the median of these numbers (see page 96), since approximately one half of the numbers are below the median and one half of the numbers are above the median. We then go through the sequence of numbers and replace each number with the letter a if it is above the median and with the letter b if it is below the median. We omit any values that equal the median. Once we have a sequence of a's and b's, we proceed in exactly the same way as we did in Examples 1 and 2 of this section.

● **Example 3**

The number of defective items produced by a machine per day over a period of a month is:

13, 17, 14, 20, 18, 16, 14, 19, 21, 20, 14, 17, 12, 14, 19, 20, 17, 18, 14, 20, 17, 19, 17, 14, 19, 21, 16, 12, 15, 22

Using a 5% level of significance, test for randomness.

Solution

We first arrange the numbers in order, from smallest to largest, to determine the median. We get 12, 12, 13, 14, 14, 14, 14, 14, 14, 15, 16, 16, 17, 17, 17, 17, 17, 18, 18, 19, 19, 19, 19, 20, 20, 20, 20, 21, 21, 22. The median of these numbers is 17. We now replace all the numbers of the original sequence with the letter b if the number is below 17 and with the letter a if the number is above 17. We do not replace the 17 with any letter. The new sequence then becomes

b, b, a, a, b, b, a, a, a, b, b, b, a, a, a, b, a, a, b, a, a, b, b, b, a

There are now 13 a's and 12 b's, and there are 12 runs as follows:

Run	Letters
1	bb
2	aa
3	bb
4	aaa
5	bbb
6	aaa
7	b
8	aa
9	b
10	aa
11	bbb
12	a

From Table XI we note that the critical values, where the larger number is 13 and the smaller number is 12, are 8 and 19. Since we obtained 12 runs, we do not reject the null hypothesis.

COMMENT   Table XI can be used only when n_1 and n_2 are not greater than 20. If either is larger than 20, we use the normal curve approximation. In this case the test statistic is

$$z = \frac{X - \mu_R}{\sigma_R}$$

where

$$\mu_R = \frac{2n_1 n_2}{n_1 + n_2} + 1$$

and

$$\sigma_R = \sqrt{\frac{2n_1 n_2 (2n_1 n_2 - n_1 - n_2)}{(n_1 + n_2)^2 (n_1 + n_2 - 1)}}$$

EXERCISES

1. Four talent scouts have rated five potential prospects (Jonathan, Richard, Thomas, Donald, and Mark) in terms of the prospect's potential according to the following rankings:

Prospect Rankings

Talent Scout 1	Talent Scout 2	Talent Scout 3	Talent Scout 4
Jonathan	Jonathan	Richard	Richard
Mark	Richard	Thomas	Mark
Thomas	Donald	Donald	Donald
Donald	Mark	Mark	Thomas
Richard	Thomas	Jonathan	Jonathan

Using a 5% level of significance, test the null hypothesis that there is no correlation between the rankings of
a. talent scouts 1 and 2.   **b.** talent scouts 1 and 3.   **c.** talent scouts 2 and 4.

2. All entering freshmen at Unity College must take the SAT (Scholastic Aptitude Test) as well as the college's own math placement exam. The following list gives the test scores of ten entering freshmen on both of these tests.

Student	Math Score on SAT Test	College Math Placement Exam
Margaret	582	50
Gina	579	90
Lorraine	562	74
Donna	574	75
Lisa	570	70
Joseph	578	68
Gail	586	66
Christine	558	74
Renée	566	79
Maria	596	94

Using a 5% level of significance, test the null hypothesis that there is no correlation between the SAT test scores and the college test scores, that is, that they are independent.

3. The ages of the 19 people who applied for unemployment insurance benefits on Monday (in the order that they applied) are as follows:

19, 24, 58, 63, 42, 39, 31, 28, 22, 67, 44, 37, 53, 32, 29, 18, 22, 34, 46.

Test for randomness.

4. The number of child abuse cases per month which were classified as serious by the welfare department of a city over the past two years is as follows:

75, 62, 94, 108, 83, 95, 62, 78, 84, 80, 95, 83, 60,
72, 123, 111, 85, 83, 61, 78, 63, 76, 93, 100.

Test for randomness.

5. The number of homeless individuals in the United States has been a continuing problem. A newspaper reporter assigned to investigate the situation observed 20 homeless people with the following sequence of males (M) and females (F).

M, F, M, M, F, M, M, F, F, F, M, M, F, F, F, M, M, F, F, M

Test for randomness.

6. The first 25 cars that enter a parking lot are classified as either foreign-made (F) or American-made (A) as follows:

F, F, F, F, A, A, A, F, F, F, F, F, A, F, A, F, A, F, F, A, A, A, A, F, F

Test for randomness.

14.8 THE KRUSKAL-WALLIS H-TEST

In Chapter 13, we discussed ANOVA techniques for determining whether the sample means of several populations are equal. In order to apply the ANOVA techniques and the F-distribution we assumed that the samples were randomly selected from independent populations that were approximately normally distributed.

A nonparametric statistical test that does not require that these assumptions be satisfied is the Kruskal-Wallis H-test. This rank-sum test is used when we wish to test the null hypothesis that r independent random samples were obtained from identical (not necessarily normal) populations. Of course, the alternative hypothesis is that the means of these populations are not the same. The only requirement is that each sample have at least five observations.

When using the Kruskal-Wallis H-test we combine all the data and rank them jointly as if they would represent one single sample. If the null hypothesis is true, then the sampling distribution of these numbers can be approximated by a chi-square distribution with $r - 1$ degrees of freedom. We accept the null hypothesis that the sample means are equal whenever the test statistic value is less than the χ^2 value for $r - 1$ degrees of freedom at a level of significance of α. Otherwise we reject the null hypothesis. The test statistic value is calculated using the following formula.

TEST STATISTIC WHEN USING THE KRUSKAL-WALLIS H-TEST

$$\frac{12}{n(n + 1)} \sum_{i=1}^{r} \frac{R_i^2}{n_i} - 3(n + 1)$$

When using the above test statistic,

$n =$ the total number in the entire sample (when the data are joined together.)

that is,

$n = n_1 + n_2 + \cdots + n_r$

$R_i =$ the sum of the ranks assigned to n_i values of the ith sample.

We illustrate the use of this formula with several examples.

● **Example 1**

A consumer's group tested numerous cans of paint from three different companies to determine whether there is any significant difference between these brands in average drying time. The number of minutes needed for each of these brands of paint to dry when applied to identical walls was as follows:

Brand A	Brand B	Brand C
38	52	47
32	48	30
27	39	37
29	42	41
23	46	44
43	53	49

Use the Kruskal-Wallis H-test to determine if there is any significant difference in the average drying time for these brands of paint. (Use a 5% level of significance.)

Solution

The null hypothesis is that the means of these different brands are the same. The three samples are first arranged jointly, as if they were one large sample, in order of increasing size. We get

Drying Time	Brand	Rank
23	A	1
27	A	2
29	A	3
30	C	4
32	A	5
37	C	6
38	A	7
39	B	8
41	C	9
42	B	10
43	A	11
44	C	12
46	B	13
47	C	14
48	B	15
49	C	16
52	B	17
53	B	18

The sum of the rankings for Brand A is

$$R_1 = 1 + 2 + 3 + 5 + 7 + 11 = 29$$

The sum of the rankings for Brand B is

$$R_2 = 8 + 10 + 13 + 15 + 17 + 18 = 81$$

The sum of the rankings for Brand C is

$$R_3 = 4 + 6 + 9 + 12 + 14 + 16 = 61$$

In this case, the number of cans of paint tested for each brand is 6, so that $n_1 = n_2 = n_3 = 6$. Also, the total number in the entire joined sample is $n = 18$. Substituting these values into the test statistic formula gives

$$\text{Test statistic} = \frac{12}{18(18 + 1)} \left(\frac{29^2}{6} + \frac{81^2}{6} + \frac{61^2}{6} \right) - 3(18 + 1)$$

$$= 8.048$$

From Table IX in the Appendix, the χ^2 value with $3 - 1$ or 2 degrees of freedom at the 5% level of significance is 5.991. Since our test statistic value of 8.048 exceeds this value, we reject the null hypothesis and conclude that the average drying time for these brands of paint is not the same.

● Example 2

An independent agency is interested in determining how long (in minutes) it takes a teller to complete the identical transaction at each of three banks. The following results were obtained when numerous tellers at the three banks were surveyed:

Time to Transaction Completion (minutes)

Bank 1	Bank 2	Bank 3
12	23	15
14	8	20
11	17	21
16	9	7
10	13	22
18	24	
19		

Do the data indicate that there is a significant difference in the average time required to complete the transaction at these banks? (Use the Kruskal-Wallis H-test at the 5% level of significance.)

Solution

The null hypothesis is that the means for the three banks are the same. The three samples are first arranged jointly, as if they constitute one large sample, in order of increasing size. We get

Time	Bank	Rank
7	3	1
8	2	2
9	2	3
10	1	4
11	1	5
12	1	6
13	2	7
14	1	8
15	3	9
16	1	10
17	2	11
18	1	12
19	1	13
20	3	14
21	3	15
22	3	16
23	2	17
24	2	18

The sum of the rankings for Bank 1 is

$$R_1 = 4 + 5 + 6 + 8 + 10 + 12 + 13 = 58$$

The sum of the rankings for Bank 2 is

$$R_2 = 2 + 3 + 7 + 11 + 17 + 18 = 58$$

The sum of the rankings for Bank 3 is

$$R_3 = 1 + 9 + 14 + 15 + 16 = 55$$

In this case, the number of tellers from Bank 1 is seven, so $n_1 = 7$; the number of tellers from Bank 2 is six, so $n_2 = 6$; and the number of tellers from Bank 3 is five, so $n_3 = 5$. Also, the total number in the entire joined sample is $n = 18$. Substituting these values into the test statistic formula gives

$$\text{Test statistic} = \frac{12}{18(18 + 1)} \left(\frac{58^2}{7} + \frac{58^2}{6} + \frac{55^2}{5} \right) - 3(18 + 1)$$

$$= 0.7627$$

From Table IX in the Appendix, the χ^2 value with $3 - 1$ or 2 degrees of freedom at the 5% level of significance is 5.991. Since our test statistic value of 0.7627 is less than this value, we do not reject the null hypothesis. We conclude that there is no significant difference in the average time required to complete the transaction at these banks.

EXERCISES

1. The number of summonses issued on a daily basis by each of five police officers during a five-day period is as follows:

Number of Summonses by Police Officer

Smith	Eskey	Kien	Walsh	Nuzzo
18	20	17	21	14
30	16	22	31	37
26	29	27	23	24
15	33	34	12	28
19	36	32	35	25

Do the above data indicate that there is a significant difference in the average number of summonses issued by each of these police officers? (Use a 5% level of significance.)

2. **Water Pollution**   Environmentalists have accused a chemical company of polluting the waters of a particular river by dumping untreated chemical wastes in it. To test this charge, a judge orders that water samples from four different locations on the river be taken and the quantity of dissolved oxygen contained in the river at each location be determined. (The quantity of dissolved oxygen contained in water is often used to determine the extent of water pollution. The lower the dissolved oxygen content in the water, the higher the level of water pollution.) The locations to be used are:

1. upstream above the chemical plant.
2. adjacent to the chemical plant's discharge pipe.
3. one-half mile downstream from the chemical plant.
4. a considerable distance downstream from the chemical plant.

It is decided that at least five samples will be taken at each location. The results of the experiment are as follows:

Average Quantity of Oxygen in Water

	1	4.7	4.2	6.4	4.1	5.1	6.6		
Location	2	6.2	6.3	3.8	5.7	6.1	5.4	3.9	4.6
	3	5.2	5.8	5.9	5.5	5.6	7.3		
	4	6.5	6.0	5.3	4.8	3.7			

Do the above data indicate that there is a significant difference in the average dissolved oxygen content at these four locations? (Use a 1% level of significance.)

14.9 COMPARISON OF PARAMETRIC AND NONPARAMETRIC STATISTICS

In this chapter we have discussed only some of the nonparametric statistical techniques. Actually, there are many other such techniques that can be used. The question that is often asked is: Why should one bother using the standard statistical techniques discussed in the remainder of this book, when the nonparametric statistical techniques discussed in this chapter are easier to use? As a matter of fact many statisticians actually recommend the use of such nonparametric techniques in many different situations.

The decision as to which statistical technique to use depends upon the particular situation. Generally speaking, if you are sure that the data come from a population that is approximately normally distributed, you should use a parametric test. On the other hand, if you are not sure, then use the appropriate nonparametric test. Other factors, such as the possible error generated by using one test as opposed to another, have to be considered. While nonparametric calculations are easier, the widespread availability and use of the computer today for parametric calculations should make such techniques easier to use.

For the benefit of the reader we summarize some of the nonparametric and parametric statistical techniques discussed and the cases in which each is used.

When Performing Tests Involving	Parametric Test to Use	Nonparametric Test to Use
One mean (median)	*t*-test (page 436)	One-sample or paired-sample sign test (pages 601 and 605)
Two means (independent samples)	*t*-test (page 448)	Mann-Whitney test (page 620)
Two means (paired samples)	Paired *t*-test	Sign test or Wilcoxon signed-rank test (page 614)
More than two means (independent samples)	ANOVA (page 570)	Kruskal-Wallis H-test (page 637)
Correlation	Linear correlation (page 472)	Spearman's rank correlation test (page 628)
Randomness		Runs test (page 631)

14.10 USING COMPUTER PACKAGES

The MINITAB statistical package is well-suited to perform many of the nonparametric statistical tests discussed in this chapter. We illustrate the use of the MINITAB for one such nonparametric test. Let us return to Example 1 of Section 14.6 (page 629). In that example, two judges were asked to rate five different brands of beer on the basis of taste appeal. The actual test results were:

	Judge 1	Judge 2
Brand A	72	68
Brand B	68	70
Brand C	62	79
Brand D	81	77
Brand E	87	63

Although we deleted the actual test results and presented only the judge's rankings, when using the MINITAB we do not have to do this. MINITAB does the ranking and then computes the Spearman rank coefficient. This is shown below:

```
MTB > READ JUDGE 1 INTO C1, JUDGE 2 INTO C3
DATA > 72    68
DATA > 68    70
DATA > 62    79
DATA > 81    77
DATA > 87    63

DATA > RANK THE VALUES IN C1, PUT RANKS INTO C2

     5 ROWS READ

DATA > RANK THE VALUES IN C3, PUT THE RANKS INTO C4

MTB > PRINT C1-C4

    ROW    C1    C2    C3    C4
     1     72     3    68     4
     2     68     4    70     3
     3     62     5    79     1
     4     81     2    77     2
     5     87     1    63     5

MTB > CORRELATION COEFFICIENT BETWEEN RANKS IN C2 AND C4

     CORRELATION OF C2 AND C4 = -0.7

MTB > STOP
```

14.11 SUMMARY

In this chapter we discussed several of the nonparametric statistical methods that are often used when we cannot use the standard tests. By far the easiest and most popular of these methods is the sign test. This test is used when we wish to compare two sample means and we know that the samples are not independent. Because of its simplicity the sign test is used by many people even when a standard test can be used. However, this method is very wasteful of information.

Another important nonparametric test is the rank-sum test (Mann-Whitney test), which is used when the normality assumption is not satisfied or when the variances are not equal. The sum of the ranks is normally distributed when the sample size is large enough, in which case we can use an appropriate z statistic.

We also mentioned the Spearman rank coefficient test, which was originally devised as a shortcut method for computing the coefficient of correlation.

Then we discussed the runs test, which is used to test for randomness or a lack of it. In determining whether or not we have a random sample, we use Table XI in the Appendix to find the appropriate critical values.

Finally, we presented the Kruskal-Wallis H-test, which we can use instead of the ANOVA techniques discussed in Chapter 13.

STUDY GUIDE

You should now be able to demonstrate your knowledge of the following ideas presented in this chapter by giving definitions, descriptions, or specific examples. Page references are given for each term so that you can check your answer.

Nonparametric statistics (page 601)
Distribution-free methods (page 601)
Standard tests (page 601)
Sign test (page 601)
One-sample sign test (page 601)
Paired data (page 605)
Two-dependent samples (page 605)
Before and after type study (page 605)
Paired-sample sign test (page 614)
Wilcoxon signed-rank sum test (page 614)
Rank-sum test (page 620)
Mann-Whitney test (page 620)
Mann-Whitney U test (page 623)
Spearman rank correlation test (page 629)
Runs test (page 631)
Run (page 631)
Testing randomness (page 631)
Kruskal-Wallis H-test (page 637)
Comparison of parametric and nonparametric statistics (page 642)

FORMULAS TO REMEMBER

You should be able to identify each symbol in the following formulas, understand the relationship among the symbols expressed in the formulas, understand the significance of the formulas, and use the formulas in solving problems.

1. When using the Mann-Whitney test (rank-sum test):

$$z = \frac{R - \mu_R}{\sigma_R}$$

where

$$\mu_R = \frac{n_1(n_1 + n_2 + 1)}{2}$$

and

$$\sigma_R = \sqrt{\frac{n_1 n_2(n_1 + n_2 + 1)}{12}}$$

and n_1 is the smaller sample size.

2. Spearman rank coefficient test:

$$R = 1 - \frac{6 \, \Sigma(x - y)^2}{n(n^2 - 1)}$$

3. Normal curve approximation for the runs test:

$$z = \frac{X - \mu_R}{\sigma_R}$$

$$\mu_R = \frac{2n_1 n_2}{n_1 + n_2} + 1$$

$$\sigma_R = \sqrt{\frac{2n_1 n_2(2n_1 n_2 - n_1 - n_2)}{(n_1 + n_2)^2(n_1 + n_2 - 1)}}$$

4. Kruskal-Wallis H-test statistic:

$$\frac{12}{n(n + 1)} \sum_{i=1}^{r} \frac{R_i^2}{n_i} - 3(n + 1)$$

where n = the total number in the entire sample (when the data are joined together) and R_i = the sum of the ranks assigned to n_i values of the ith sample.

The tests of the following section will be more useful if you take them after you have studied the examples and solved the exercises in this chapter.

MASTERY TESTS

Form A

1. The weights of ten people before they stopped smoking and four months after they stopped smoking are shown below:

		Before	After
	A	150	145
	B	142	148
	C	170	166
	D	160	155
Subject	E	164	156
	F	177	170
	G	157	153
	H	168	163
	I	175	170
	J	172	165

Using the paired-sample sign test, test the null hypothesis that stopping smoking has no effect on a person's weight against the alternative hypothesis that stopping smoking helps reduce a person's weight. (Use a 5% level of significance.)

2. Melissa and Rhoda were asked to rate five different daytime "soap opera" television shows for their overall appeal. Their ratings are as follows:

		Melissa	Rhoda
	A	1	5
	B	4	4
Soap Opera	C	3	3
	D	2	2
	E	5	1

Using a 5% level of significance, test the null hypothesis that both ratings are independent. (Use the Spearman rank coefficient test.)

3. The amount of pollution in several lakes both before and after being treated with special chemicals by environmentalists is shown in the accompanying chart:

Before Treatment	After Treatment
21	17
16	19
18	17
18	18
11	10
13	9
24	16
15	13
14	11

Using a 5% level of significance, test the null hypothesis that the amount of pollution in the lakes is not affected by the addition of the special chemicals against the alternative hypothesis that the amount of pollution in the lakes is less as a result of the treatments. (Use the paired-sign test.)

4. The daily number of people who saw a particular movie showing at a drive-in theater during the first 21 days of July was as follows: 371, 421, 602, 533, 191, 504, 256, 337, 289, 691, 376, 369, 279, 601, 694, 457, 307, 403, 326, 492, and 257. Test for randomness.

5. A credit-card company has compiled the following list of the number of billing inquiries received daily and the day of the week. The following list is available:

Day	Number of Billing Inquiries Received
Mon.	89
Tues.	91
Wed.	48
Thurs.	93
Fri.	56
Mon.	88
Tues.	72
Wed.	67
Thurs.	75
Fri.	84

By using Spearman's rank coefficient test, test the null hypothesis that there is no correlation between the day of the week on which the billing inquiry is received and the number of billing inquiries received. (Use a 5% level of significance.) *Hint:* Assume that the two weeks represent rankings by two different people, where Monday = 1, Tuesday = 2, . . . , in terms of the number of inquiries received.

6. What is the test statistic when using the Kruskal-Wallis H-test?

7. When performing tests involving more than two means (independent samples), which nonparametric test would be most appropriate?
 a. Mann-Whitney test   **b.** Wilcoxon signed-rank test
 c. Runs test   **d.** Kruskal-Wallis H-test
 e. none of these

8. When using the rank-sum test, what happens when a tie occurs?

9. When using the one-sample sign test, we assume that the number of plus signs (or minus signs) is a random variable having a binomial distribution with $p = ?$

a. 1   **b.** 0   **c.** $\frac{1}{2}$ **d.** 3   **e.** none of these

10. Which of the following nonparametric tests considers both the magnitude as well as the direction (positive or negative) of the difference between pairs of numbers?
   a. Sign test
   b. Paired-sample sign test
   c. Spearman's rank correlation test
   d. Wilcoxon signed-rank test
   e. none of these

MASTERY TESTS

Form B

1. The Marvo Medical Group consists of six doctors each of whom sees patients on a rotating basis. Recently, two patients were asked to rank these doctors in terms of courteousness and bedside manner. The following rankings were obtained:

		Patient 1	Patient 2
	Kok	3	2
	Sanjurai	2	3
Doctor	Ghandi	4	1
	Shabsi	1	4
	Nova	5	6
	Speyer	6	5

   Using a 5% level of significance, test the null hypothesis that the rankings are not related. (Use the Spearman rank coefficient test.)

2. A computer-literacy test was given to students in a non-air-conditioned room on a humid day. The same test was then given to these students in an air-conditioned room. The following list gives the results of the two tests.

	Result in Non-air-conditioned Room	Result in Air-conditioned Room
A	79	82
B	38	60
C	81	75
D	83	81
E	32	45
F	10	45
Subject G	63	60
H	80	80
I	51	81
J	77	81
K	87	76
L	81	68
M	76	81

Using the Wilcoxon signed-rank test, test the null hypothesis that an air-conditioned room on a humid day has no effect on the performance on this literacy test against the alternative hypothesis that an air-conditioned room on a humid day increases performance on the literacy test.

3. The student government at State University subscribes to eight different magazines, which are available for students to read in the lounge. Two students are randomly selected and asked to rank the magazines from most preferred to least preferred. Their rankings are as follows:

		Rankings by Student I	Rankings by Student II
	Time	1	3
	Newsweek	2	5
	Business Week	6	4
	U.S. News and		
Magazine	World Report	3	2
	Reader's Digest	4	1
	Consumer's Reports	5	7
	Sports Illustrated	8	8
	New Yorker	7	6

Using a 5% level of significance, test the null hypothesis that there is no correlation between the students' rankings of the magazines. (Use Spearman's rank correlation test.)

4. An interviewer asks people as they are getting off an elevator whether they are in favor of concessions by our government in order to achieve nuclear disarmament. They answer yes (Y) or no (N) as follows:

N,N,N,Y,Y,Y,N,N,Y,N,Y,Y,N,N,N,N,N,Y,Y,Y,N,Y,N,Y,N,Y,N

Test for randomness.

5. The number of daily calls to the AAA for emergency road assistance over the last few weeks is as follows:

65, 58, 30, 35, 65, 77, 79, 63, 58, 20, 87, 55, 80, 79,
55, 87, 88, 70, 73, 86, 84

Test for randomness.

6. Bill Sadowsky finds it very difficult to arrive at work on time. Over the past 20 working days Bill has been late the following number of minutes:

13, 20, 6, 22, 18, 19, 17, 8, 26, 20, 21, 19, 22, 31, 23, 21, 19, 17, 15, 22

Test for randomness.

7. The number of defective items produced per day on two different production lines is as follows. (One production line operates on a seven-day schedule.)

Production line A	9	3	11	2	2	6	10	4	9	5			
Production line B	7	10	5	3	8	11	1	12	8	4	12	0	7	6

Using a 5% level of significance, test the null hypothesis that the average number of defective items produced by both production lines is the same. (Use the rank-sum test.)

8. *Do Doctors Order Many Unnecessary Blood Tests?* To answer this question, a random survey is taken of the number of blood tests ordered daily by four doctors over a one-week period. The results are as follows:

Number of Blood Tests Ordered

Dr. Rogers	Dr. DeMaria	Dr. Scarslow	Dr. Koch
36	32	49	43
47	30	28	24
45	35	33	37
41	21	29	27
31	42	22	40

Do the data indicate that there is a significant difference between the average number of blood tests ordered by these doctors? (Use a 5% level of significance.) Use the Kruskal-Wallis H-test.

9. The number of similar audits completed weekly by three different IRS agents over a five-week period is as follows:

Number of Audits Completed

IRS Agent A	IRS Agent B	IRS Agent C
34	20	39
32	37	26
21	22	29
24	36	35
27	23	33

Do the data indicate that there is a significant difference between the average number of audits completed by these agents? (Use a 1% level of significance.) Use the Kruskal-Wallis H-test.

10. The number of pounds of newspaper collected for recycling during one month from houses on the north and south sides of Broadway is as follows:

North Side	362	328	504	272	438	304	440		
South Side	320	391	424	362	504	496	512	480	410

Test the null hypothesis that the average number of pounds of newspapers collected from houses on both sides of Broadway is the same. (Use a 5% level of significance.) Use the rank-sum test.

11. A consumer believes that the median cost of a VCR (video cassette recorder) is approximately $220. Another consumer believes that the median cost is not equal to $220. She samples 11 stores and obtains the following price quotes for the identical model VCR:

$203, $256, $230, $249, $257, $225, $229, $207, $220, $227, $251

Using the one-sample sign test, test the null hypothesis that the median cost for the VCR is $220 against the alternative hypothesis that the median cost is not $220. (Use a 5% level of significance.)

SUGGESTED FURTHER READING

1. Daniel, Wayne. *Applied Non-parametric Statistics*. Boston: Houghton Mifflin, 1978.
2. Gibbons, J. D. *Non-parametric Statistical Inference*. New York: McGraw-Hill, 1971.
3. Lehmann, E. L. *Non-parametrics: Statistical Methods Based on Ranks*. San Francisco: Holden-Day, 1975.
4. Mosteller, F., and R. Rourke. *Sturdy Statistics*. Reading, MA: Addison-Wesley, 1973.
5. Noether, G. E. *Introduction to Statistics: A Nonparametric Approach*. Boston: Houghton Mifflin, 1976.

APPENDIX

Statistical Tables

TABLE I: SQUARES, SQUARE ROOTS, AND RECIPROCALS

Suggestions for Using the Square Root Table

To find the square root of any number between 1.00 and 10.00, we use the third column. For example, $\sqrt{3.21}$ is 1.79165 and $\sqrt{6.58}$ is 2.56515. The fourth column enables us to find the square root of all numbers between 10.0 and 100.0, since $\sqrt{10N}$ gives us 10(1.00), and 10(10.00), or 10.0 to 100.0. For example, from the fourth column we find that $\sqrt{32.1}$ is 5.66569 and $\sqrt{65.8}$ is 8.11172. Thus, using the third and fourth columns, we can find the square root of all numbers between 1.00 and 100.0.

Suppose we want $\sqrt{321}$. We can find $\sqrt{3.21}$ and $\sqrt{32.1}$. Which should we use? We locate the decimal point. (If no decimal point is indicated, it is understood to be at the end of the number.) Then move the decimal point an even number of places to the right or to the left until a number greater than or equal to 1 but less than 100 is reached. If the resulting number is less than 10, go to the $\sqrt{N}$ column; if it is between 10 and 100, go to the $\sqrt{10N}$ column. The next question is, where do we put the decimal? Since we moved the decimal point an even number of places to the left or to the right to get a number between 1 and 100, the decimal point of the answer obtained from Table I is moved half as many places in the *opposite* direction.

Thus, to find $\sqrt{321}$, we move the decimal two places to the left, to $\sqrt{3.21}$. From Table I, we get 1.79165. Since we moved the decimal two places to the left to arrive at this answer, we now move the decimal half as many places in the opposite direction. Therefore $\sqrt{321} = 17.9165$. In the same way, $\sqrt{3210}$ becomes $\sqrt{32.10}$, which is 5.66569, so $\sqrt{3210} = 56.6569$. $\sqrt{0.000321}$ becomes $\sqrt{0003.21}$, which is 1.79165, so $\sqrt{0.000321} = 0.0179165$.

TABLE I
Squares, Square Roots, and Reciprocals

N	N^2	$\sqrt{N}$	$\sqrt{10N}$	N	N^2	$\sqrt{N}$	$\sqrt{10N}$
1.00	1.0000	1.00000	3.16228	**1.50**	2.2500	1.22474	3.87298
1.01	1.0201	1.00499	3.17805	1.51	2.2801	1.22882	3.88587
1.02	1.0404	1.00995	3.19374	1.52	2.3104	1.23288	3.89872
1.03	1.0609	1.01489	3.20936	1.53	2.3409	1.23693	3.91152
1.04	1.0816	1.01980	3.22490	1.54	2.3716	1.24097	3.92428
1.05	1.1025	1.02470	3.24037	1.55	2.4025	1.24499	3.93700
1.06	1.1236	1.02956	3.25576	1.56	2.4336	1.24900	3.94968
1.07	1.1449	1.03441	3.27109	1.57	2.4649	1.25300	3.96232
1.08	1.1664	1.03923	3.28634	1.58	2.4964	1.25698	3.97492
1.09	1.1881	1.04403	3.30151	1.59	2.5281	1.26095	3.98748
1.10	1.2100	1.04881	3.31662	**1.60**	2.5600	1.26491	4.00000
1.11	1.2321	1.05357	3.33167	1.61	2.5921	1.26886	4.01248
1.12	1.2544	1.05830	3.34664	1.62	2.6244	1.27279	4.02492
1.13	1.2769	1.06301	3.36155	1.63	2.6569	1.27671	4.03733
1.14	1.2996	1.06771	3.37639	1.64	2.6896	1.28062	4.04969
1.15	1.3225	1.07238	3.39116	1.65	2.7225	1.28452	4.06202
1.16	1.3456	1.07703	3.40588	1.66	2.7556	1.28841	4.07431
1.17	1.3689	1.08167	3.42053	1.67	2.7889	1.29228	4.08656
1.18	1.3924	1.08628	3.43511	1.68	2.8224	1.29615	4.09878
1.19	1.4161	1.09087	3.44964	1.69	2.8561	1.30000	4.11096
1.20	1.4400	1.09545	3.46410	**1.70**	2.8900	1.30384	4.12311
1.21	1.4641	1.10000	3.47851	1.71	2.9241	1.30767	4.13521
1.22	1.4884	1.10454	3.49285	1.72	2.9584	1.31149	4.14729
1.23	1.5129	1.10905	3.50714	1.73	2.9929	1.31529	4.15933
1.24	1.5376	1.11355	3.52136	1.74	3.0276	1.31909	4.17133
1.25	1.5625	1.11803	3.53553	1.75	3.0625	1.32288	4.18330
1.26	1.5876	1.12250	3.54965	1.76	3.0976	1.32665	4.19524
1.27	1.6129	1.12694	3.56371	1.77	3.1329	1.33041	4.20714
1.28	1.6384	1.13137	3.57771	1.78	3.1684	1.33417	4.21900
1.29	1.6641	1.13578	3.59166	1.79	3.2041	1.33791	4.23084
1.30	1.6900	1.14018	3.60555	**1.80**	3.2400	1.34164	4.24264
1.31	1.7161	1.14455	3.61939	1.81	3.2761	1.34536	4.25441
1.32	1.7424	1.14891	3.63318	1.82	3.3124	1.34907	4.26615
1.33	1.7689	1.15326	3.64692	1.83	3.3489	1.35277	4.27785
1.34	1.7956	1.15758	3.66060	1.84	3.3856	1.35647	4.28952
1.35	1.8225	1.16190	3.67423	1.85	3.4225	1.36015	4.30116
1.36	1.8496	1.16619	3.68782	1.86	3.4596	1.36382	4.31277
1.37	1.8769	1.17047	3.70135	1.87	3.4969	1.36748	4.32435
1.38	1.9044	1.17473	3.71484	1.88	3.5344	1.37113	4.33590
1.39	1.9321	1.17898	3.72827	1.89	3.5721	1.37477	4.34741
1.40	1.9600	1.18322	3.74166	**1.90**	3.6100	1.37840	4.35890
1.41	1.9881	1.18743	3.75500	1.91	3.6481	1.38203	4.37035
1.42	2.0164	1.19164	3.76829	1.92	3.6864	1.38564	4.38178
1.43	2.0449	1.19583	3.78153	1.93	3.7249	1.38924	4.39318
1.44	2.0736	1.20000	3.79473	1.94	3.7636	1.39284	4.40454
1.45	2.1025	1.20416	3.80789	1.95	3.8025	1.39642	4.41588
1.46	2.1316	1.20830	3.82099	1.96	3.8416	1.40000	4.42719
1.47	2.1609	1.21244	3.83406	1.97	3.8809	1.40357	4.43847
1.48	2.1904	1.21655	3.84708	1.98	3.9204	1.40712	4.44972
1.49	2.2201	1.22066	3.86005	1.99	3.9601	1.41067	4.46094
1.50	2.2500	1.22474	3.87298	**2.00**	4.0000	1.41421	4.47214
N	N^2	$\sqrt{N}$	$\sqrt{10N}$	N	N^2	$\sqrt{N}$	$\sqrt{10N}$

TABLE I
Squares, Square Roots, and Reciprocals (continued)

N	N^2	$\sqrt{N}$	$\sqrt{10N}$	N	N^2	$\sqrt{N}$	$\sqrt{10N}$
2.00	4.0000	1.41421	4.47214	**2.50**	6.2500	1.58114	5.00000
2.01	4.0401	1.41774	4.48330	2.51	6.3001	1.58430	5.00999
2.02	4.0804	1.42127	4.49444	2.52	6.3504	1.58745	5.01996
2.03	4.1209	1.42478	4.50555	2.53	6.4009	1.59060	5.02991
2.04	4.1616	1.42829	4.51664	2.54	6.4516	1.59374	5.03984
2.05	4.2025	1.43178	4.52769	2.55	6.5025	1.59687	5.04975
2.06	4.2436	1.43527	4.53872	2.56	6.5536	1.60000	5.05964
2.07	4.2849	1.43875	4.54973	2.57	6.6049	1.60312	5.06952
2.08	4.3264	1.44222	4.56070	2.58	6.6564	1.60624	5.07937
2.09	4.3681	1.44568	4.57165	2.59	6.7081	1.60935	5.08920
2.10	4.4100	1.44914	4.58258	**2.60**	6.7600	1.61245	5.09902
2.11	4.4521	1.45258	4.59347	2.61	6.8121	1.61555	5.10882
2.12	4.4944	1.45602	4.60435	2.62	6.8644	1.61864	5.11859
2.13	4.5369	1.45945	4.61519	2.63	6.9169	1.62173	5.12835
2.14	4.5796	1.46287	4.62601	2.64	6.9696	1.62481	5.13809
2.15	4.6225	1.46629	4.63681	2.65	7.0225	1.62788	5.14782
2.16	4.6656	1.46969	4.64758	2.66	7.0756	1.63095	5.15752
2.17	4.7089	1.47309	4.65833	2.67	7.1289	1.63401	5.16720
2.18	4.7524	1.47648	4.66905	2.68	7.1824	1.63707	5.17687
2.19	4.7961	1.47986	4.67974	2.69	7.2361	1.64012	5.18652
2.20	4.8400	1.48324	4.69042	**2.70**	7.2900	1.64317	5.19615
2.21	4.8841	1.48661	4.70106	2.71	7.3441	1.64621	5.20577
2.22	5.9284	1.48997	4.71169	2.72	7.3984	1.64924	5.21536
2.23	4.9729	1.49332	4.72229	2.73	7.4529	1.65227	5.22494
2.24	5.0176	1.49666	4.73286	2.74	7.5076	1.65529	5.23450
2.25	5.0625	1.50000	4.74342	2.75	7.5625	1.65831	5.24404
2.26	5.1076	1.50333	4.75395	2.76	7.6176	1.66132	5.25357
2.27	5.1529	1.50665	4.76445	2.77	7.6729	1.66433	5.26308
2.28	5.1984	1.50997	4.77493	2.78	7.7284	1.66733	5.27257
2.29	5.2441	1.51327	4.78539	2.79	7.7841	1.67033	5.28205
2.30	5.2900	1.51658	4.79583	**2.80**	7.8400	1.67332	5.29150
2.31	5.3361	1.51987	4.80625	2.81	7.8961	1.67631	5.30094
2.32	5.3824	1.52315	4.81664	2.82	7.9524	1.67929	5.31037
2.33	5.4289	1.52643	4.82701	2.83	8.0089	1.68226	5.31977
2.34	5.4756	1.52971	4.83735	2.84	8.0656	1.68523	5.32917
2.35	5.5225	1.53297	4.84768	2.85	8.1225	1.68819	5.33854
2.36	5.5696	1.53623	4.85798	2.86	8.1796	1.69115	5.34790
2.37	5.6169	1.53948	4.86626	2.87	8.2369	1.69411	5.35724
2.38	5.6644	1.54272	4.87852	2.88	8.2944	1.69706	5.36656
2.39	5.7121	1.54596	4.88876	2.89	8.3521	1.70000	5.37587
2.40	5.7600	1.54919	4.89898	**2.90**	8.4100	1.70294	5.38516
2.41	5.8081	1.55252	4.90918	2.91	8.4681	1.70587	5.39444
2.42	5.8564	1.55563	4.91935	2.92	8.5264	1.70880	5.40370
2.43	5.9049	1.55885	4.92950	2.93	8.5849	1.71172	5.41295
2.44	5.9536	1.56205	4.93964	2.94	8.6436	1.71464	5.42218
2.45	6.0025	1.56525	4.94975	2.95	8.7025	1.71756	5.43139
2.46	6.0516	1.56844	4.95984	2.96	8.7616	1.72047	5.44059
2.47	6.1109	1.57162	4.96991	2.97	8.8209	1.72337	5.44977
2.48	6.1054	1.57480	4.97996	2.98	8.8804	1.72627	5.45894
2.49	6.2001	1.57797	4.98999	2.99	8.9401	1.72916	5.46809
2.50	6.2500	1.58114	5.00000	**3.00**	9.0000	1.73205	5.47723
N	N^2	$\sqrt{N}$		N	N^2	$\sqrt{N}$	$\sqrt{10N}$

TABLE I
Squares, Square Roots, and Reciprocals (continued)

N	N^2	$\sqrt{N}$	$\sqrt{10N}$	N	N^2	$\sqrt{N}$	$\sqrt{10N}$
3.00	9.0000	1.73205	5.47723	**3.50**	12.2500	1.87083	5.91608
3.01	9.0601	1.73494	5.48635	3.51	12.3201	1.87350	5.92453
3.02	9.1204	1.73781	5.49545	3.52	12.3904	1.87617	5.93296
3.03	9.1809	1.74069	5.50454	3.53	12.4609	1.87883	5.94138
3.04	9.2416	1.74356	5.51362	3.54	12.5316	1.88149	5.94979
3.05	9.3025	1.74642	5.52268	3.55	12.6025	1.88414	5.95819
3.06	9.3636	1.74929	5.53173	3.56	12.6736	1.88680	5.96657
3.07	9.4249	1.75214	5.54076	3.57	12.7449	1.88944	5.97495
3.08	9.4864	1.75499	5.54977	3.58	12.8164	1.89209	5.98331
3.09	9.5481	1.75784	5.55878	3.59	12.8881	1.89473	5.99166
3.10	9.6100	1.76068	5.56776	**3.60**	12.9600	1.89737	6.00000
3.11	9.6721	1.76352	5.57674	3.61	13.0321	1.90000	6.00833
3.12	9.7344	1.76635	5.58570	3.62	13.1044	1.90263	6.01664
3.13	9.7969	1.76918	5.59464	3.63	13.1769	1.90526	6.02495
3.14	9.8596	1.77200	5.60357	3.64	13.2496	1.90788	6.03324
3.15	9.9225	1.77482	5.61249	3.65	13.3225	1.91050	6.04152
3.16	9.9856	1.77764	5.62139	3.66	13.3956	1.91311	6.04949
3.17	10.0489	1.78045	5.63028	3.67	13.4689	1.91572	6.05805
3.18	10.1124	1.78326	5.63915	3.68	13.5424	1.91833	6.06630
3.19	10.1761	1.78606	5.64801	3.69	13.6161	1.92094	6.07454
3.20	10.2400	1.78885	5.65685	**3.70**	13.6900	1.92354	6.08276
3.21	10.3041	1.79165	5.66569	3.71	13.7641	1.92614	6.09098
3.22	10.3684	1.79444	5.67450	3.72	13.8384	1.92873	6.09918
3.23	10.4329	1.79722	5.68331	3.73	13.9129	1.93132	6.10737
3.24	10.4976	1.80000	5.69210	3.74	13.9876	1.93391	6.11555
3.25	10.5625	1.80278	5.70088	3.75	14.0625	1.93649	6.12372
3.26	10.6276	1.80555	5.70964	3.76	14.1376	1.93907	6.13188
3.27	10.6929	1.80831	5.71839	3.77	14.2129	1.94165	6.14003
3.28	10.7584	1.81108	5.72713	3.78	14.2884	1.94422	6.14817
3.29	10.8241	1.81384	5.73585	3.79	14.3641	1.94679	6.15630
3.30	10.8900	1.81659	5.74456	**3.80**	14.4400	1.94936	6.16441
3.31	10.9561	1.81934	5.75326	3.81	14.5161	1.95192	6.17252
3.32	10.0224	1.82209	5.76194	3.82	14.5924	1.95448	6.18061
3.33	11.0889	1.82483	5.77062	3.83	14.6689	1.95704	6.18870
3.34	11.1556	1.82757	5.77927	3.84	14.7456	1.95959	6.19677
3.35	11.2225	1.83030	5.78792	3.85	14.8225	1.96214	6.20484
3.36	11.2896	1.83303	5.79655	3.86	14.8996	1.96469	6.21289
3.37	11.3569	1.83576	5.80517	3.87	14.9769	1.96723	6.22093
3.38	11.4244	1.83848	5.81378	3.88	15.0544	1.96977	6.22896
3.39	11.4921	1.84120	5.82237	3.89	15.1321	1.97231	6.23699
3.40	11.5600	1.84391	5.83095	**3.90**	51.2100	1.97484	6.24500
3.41	11.6281	1.84662	5.83952	3.91	15.2881	1.97737	6.25300
3.42	11.6964	1.84932	5.84808	3.92	15.3664	1.97990	6.26099
3.43	11.7649	1.85203	5.85662	3.93	15.4449	1.98242	6.26897
3.44	11.8336	1.85472	5.86515	3.94	15.5236	1.98494	6.27694
3.45	11.9025	1.85742	5.87367	3.95	15.6025	1.98746	6.28490
3.46	11.9716	1.86011	5.88218	3.96	15.6816	1.98997	6.29285
3.47	12.0409	1.86279	5.89067	3.97	15.7609	1.99249	6.30079
3.48	12.1104	1.86548	5.89915	3.98	15.8404	1.99499	6.30872
3.49	12.1801	1.86815	5.90762	3.99	15.9201	1.99750	6.31644
3.50	12.2500	1.87083	5.91608	**4.00**	16.0000	2.00000	6.32456
N	N^2	$\sqrt{N}$	$\sqrt{10N}$	N	N^2	$\sqrt{N}$	$\sqrt{10N}$

TABLE I
Squares, Square Roots, and Reciprocals (continued)

N	N^2	$\sqrt{N}$	$\sqrt{10N}$	N	N^2	$\sqrt{N}$	$\sqrt{10N}$
4.00	16.0000	2.00000	6.32456	**4.50**	20.2500	2.12132	6.70820
4.01	16.0801	2.00250	6.33246	4.51	20.3401	2.12368	6.71565
4.02	16.1604	2.00499	6.34035	4.52	20.4304	2.12603	6.72309
4.03	16.2409	2.00749	6.34823	4.53	20.5209	2.12838	6.73053
4.04	16.3216	2.00998	6.35610	4.54	20.6116	2.13073	6.73795
4.05	16.4025	2.01246	6.36396	4.55	20.7025	2.13307	6.74537
4.06	16.4836	2.01494	6.37181	4.56	20.7936	2.13542	6.75278
4.07	16.5649	2.01742	6.37966	4.57	20.8849	2.13776	6.76018
4.08	16.6464	2.01990	6.38749	4.58	20.9764	2.14009	6.76757
4.09	16.7281	2.02237	6.39531	4.59	21.0681	2.14243	6.77495
4.10	16.8100	2.02485	6.40312	**4.60**	21.1600	2.14476	6.78233
4.11	16.8921	2.02731	6.41093	4.61	21.2521	2.14709	6.78970
4.12	16.9744	2.02978	6.41872	4.62	21.3444	2.14942	6.79706
4.13	17.0569	2.03224	6.42651	4.63	21.4369	2.15174	6.80441
4.14	17.1396	2.03470	6.43428	4.64	21.5296	2.15407	6.81175
4.15	17.2225	2.03715	6.44205	4.65	21.6225	2.15639	6.81909
4.16	17.3056	2.03961	6.44981	4.66	21.7156	2.15870	6.82642
4.17	17.3889	2.04206	6.45755	4.67	21.8089	2.16102	6.83374
4.18	17.4724	2.04450	6.46529	4.68	21.9024	2.16333	6.84105
4.19	17.5561	2.04695	6.47302	4.69	21.9961	2.16564	6.84836
4.20	17.6400	2.04939	6.48074	**4.70**	22.0900	2.16795	6.85565
4.21	17.7241	2.05183	6.48845	4.71	22.1841	2.17025	6.86294
4.22	17.8084	2.05426	6.49615	4.72	22.2784	2.17256	6.87023
4.23	17.8929	2.05670	6.50384	4.73	22.3729	2.17486	6.87750
4.24	17.9776	2.05913	6.51153	4.74	22.4676	2.17715	6.88477
4.25	18.0625	2.06155	6.51920	4.75	22.5625	2.17945	6.89202
4.26	18.1476	2.06398	6.52687	4.76	22.6576	2.18174	6.89928
4.27	18.2329	2.06640	6.53452	4.77	22.7529	2.18403	6.90652
4.28	18.3184	2.06882	6.54217	4.78	22.8484	2.18632	6.91375
4.29	18.4041	2.07123	6.54981	4.79	22.9441	2.18861	6.92098
4.30	18.4900	2.07364	6.55744	**4.80**	23.0400	2.19089	6.92820
4.31	18.5761	2.07605	6.66506	4.81	23.1361	2.19317	6.93542
4.32	18.6624	2.07846	6.57267	4.82	23.2324	2.19545	6.94262
4.33	18.7489	2.08087	6.58027	4.83	23.3289	2.19773	6.94982
4.34	18.8356	2.08327	6.58787	4.84	23.4256	2.20000	6.95701
4.35	18.9225	2.08567	6.59545	4.85	23.5225	2.20227	6.96419
4.36	19.0096	2.08806	6.60303	4.86	23.6196	2.20454	6.97137
4.37	19.0969	2.09045	6.61060	4.87	23.7169	2.20681	6.97854
4.38	19.1844	2.09284	6.61816	4.88	23.8144	2.20907	6.98570
4.39	19.2721	2.09523	6.62571	4.89	23.9121	2.21133	6.99285
4.40	19.3600	2.09762	6.63325	**4.90**	24.0100	2.21359	7.00000
4.41	19.4481	2.10000	6.64078	4.91	24.1081	2.21585	7.00714
4.42	19.5364	2.10238	6.64831	4.92	24.2064	2.21811	7.01427
4.43	19.6249	2.10476	6.65582	4.93	24.3649	2.22036	7.02140
4.44	19.7136	2.10713	6.66333	4.94	24.4036	2.22261	7.02851
4.45	19.8025	2.10950	6.67083	4.95	24.5025	2.22486	7.03562
4.46	19.8916	2.11187	6.67832	4.96	24.6016	2.22711	7.04273
4.47	19.9809	2.11424	6.68581	4.97	24.7009	2.22935	7.04982
4.48	20.0704	2.11660	6.69328	4.98	24.8004	2.23159	7.05691
4.49	20.1601	2.11896	6.70075	4.99	24.9001	2.23383	7.06399
4.50	20.2500	2.12132	6.70820	**5.00**	25.0000	2.23607	7.07107
N	N^2	$\sqrt{N}$	$\sqrt{10N}$	N	N^2	$\sqrt{N}$	$\sqrt{10N}$

TABLE I
Squares, Square Roots, and Reciprocals (continued)

N	N^2	$\sqrt{N}$	$\sqrt{10N}$	N	N^2	$\sqrt{N}$	$\sqrt{10N}$
5.00	25.0000	2.23607	7.07107	**5.50**	30.2500	2.34521	7.41620
5.01	25.1001	2.23830	7.07814	5.51	30.3601	2.34734	7.42294
5.02	25.2004	2.24054	7.08520	5.52	30.4704	2.34947	7.42967
5.03	25.3009	2.24277	7.09225	5.53	30.5809	2.35160	7.43640
5.04	25.4016	2.24499	7.09930	5.54	30.6916	2.35372	7.44312
5.05	25.5025	2.24722	7.10634	5.55	30.8025	2.35584	7.44983
5.06	25.6036	2.24944	7.11337	5.56	30.9136	2.35797	7.45654
5.07	25.7049	2.25167	7.12039	5.57	31.0249	2.36008	7.46324
5.08	25.8064	2.25389	7.12741	5.58	31.1364	2.36220	7.46994
5.09	25.9081	2.25610	7.13442	5.59	31.2481	2.36432	7.47663
5.10	26.0100	2.25832	7.14143	**5.60**	31.3600	2.36643	7.48331
5.11	26.1121	2.26053	7.14843	5.61	31.4721	2.36854	7.48999
5.12	26.2144	2.26274	7.15542	5.62	31.5844	2.37065	7.49667
5.13	26.3169	2.26495	7.16240	5.63	31.6969	2.37276	7.50333
5.14	26.4196	2.26716	7.16938	5.64	31.8096	2.37487	7.50999
5.15	26.5225	2.26936	7.17635	5.65	31.9225	2.37697	7.51665
5.16	26.6256	2.27156	7.18331	5.66	32.0356	2.37908	7.52330
5.17	26.7289	2.27376	7.19027	5.67	32.1489	2.38118	7.52994
5.18	26.8324	2.27596	7.19722	5.68	32.2624	2.38328	7.53658
5.19	26.9361	2.27816	7.20417	5.69	32.3761	2.38537	7.54321
5.20	27.0400	2.28035	7.21110	**5.70**	32.4900	2.38747	7.54983
5.21	27.1441	2.28254	7.21803	5.71	32.6041	2.38956	7.55645
5.22	27.2484	2.28473	7.22496	5.72	32.7184	2.39165	7.56307
5.23	27.3529	2.28692	7.23187	5.73	32.8329	2.39374	7.56968
5.24	27.4576	2.28910	7.23838	5.74	32.9476	2.39583	7.57628
5.25	27.5625	2.29129	7.24569	5.75	33.0625	2.39792	7.58288
5.26	27.6676	2.29347	7.25259	5.76	33.1776	2.40000	7.58947
5.27	27.7729	2.29565	7.25948	5.77	33.2929	2.40208	7.59605
5.28	27.8784	2.29783	7.26636	5.78	33.4084	2.40416	7.60263
5.29	27.9841	2.30000	7.27324	5.79	33.5241	2.40624	7.60920
5.30	28.0900	2.30217	7.28011	**5.80**	33.6400	2.40832	7.61577
5.31	28.1961	2.30434	7.28697	5.81	33.7561	2.41039	7.62234
5.32	28.3024	2.30651	7.29383	5.82	33.8724	2.41247	7.62889
5.33	28.4089	2.30868	7.30068	5.83	33.9889	2.41454	7.63544
5.34	28.5156	2.31084	7.30753	5.84	34.1056	2.41661	7.64199
5.35	28.6225	2.31301	7.31437	5.85	34.2225	2.41868	7.64853
5.36	28.7296	2.31517	7.32120	5.86	34.3396	2.42074	7.65506
5.37	28.8369	2.31733	7.32803	5.87	34.4569	2.42281	7.66159
5.38	28.9444	2.31948	7.33485	5.88	34.5744	2.42487	7.66812
5.39	29.0521	2.32164	7.34166	5.89	34.6921	2.42693	7.67463
5.40	29.1600	2.32379	7.34847	**5.90**	34.8100	2.42899	7.68115
5.41	29.2681	2.32594	7.35527	5.91	34.9281	2.43105	7.68765
5.42	29.3764	2.32809	7.36206	5.92	35.0464	2.43311	7.69415
5.43	29.4849	2.33024	7.36885	5.93	35.1649	2.43516	7.70065
5.44	29.5936	2.33238	7.37564	5.94	35.2836	2.43721	7.70714
5.45	29.7025	2.33452	7.38241	5.95	35.4025	2.43926	7.71362
5.46	29.8116	2.33666	7.38918	5.96	35.5216	2.44131	7.72010
5.47	29.9209	2.33880	7.39594	5.97	35.6409	2.44336	7.72658
5.48	30.0304	2.34094	7.40270	5.98	35.7604	2.44540	7.73305
5.49	30.1401	2.34307	7.40945	5.99	35.8801	2.44745	7.73951
5.50	30.2500	2.34521	7.41620	**6.00**	36.0000	2.44949	7.74597
N	N^2	$\sqrt{N}$	$\sqrt{10N}$	N	N^2	$\sqrt{N}$	$\sqrt{10N}$

TABLE I
Squares, Square Roots, and Reciprocals (continued)

N	N^2	$\sqrt{N}$	$\sqrt{10N}$	N	N^2	$\sqrt{N}$	$\sqrt{10N}$
6.00	36.0000	2.44949	7.74597	**6.50**	42.2500	2.54951	8.06226
6.01	36.1201	2.45153	7.75242	6.51	42.3801	2.55147	8.06846
6.02	36.2404	2.45357	7.75887	6.52	42.5104	2.55343	8.07465
6.03	36.3609	2.45561	7.76531	6.53	42.6409	2.55539	8.08084
6.04	36.4816	2.45764	7.77174	6.54	42.7716	2.55734	8.08703
6.05	36.6025	2.45967	7.77817	6.55	42.9025	2.55930	8.09321
6.06	36.7236	2.46171	7.78460	6.56	43.0336	2.56125	8.09938
6.07	36.8449	2.46374	7.79102	6.57	43.1649	2.56320	8.10555
6.08	36.9664	2.46577	7.79744	6.58	43.2964	2.56515	8.11172
6.09	37.0881	2.46779	7.80385	6.59	43.4281	2.56710	8.11788
6.10	37.2100	2.46982	7.81025	**6.60**	43.5600	2.56905	8.12404
6.11	37.3321	2.47184	7.81665	6.61	43.6921	2.57099	8.13019
6.12	37.4544	2.47386	7.82304	6.62	43.8244	2.57294	8.13634
6.13	37.5769	2.47588	7.82943	6.63	43.9569	2.57488	8.14248
6.14	37.6996	2.47790	7.83582	6.64	44.0896	2.57682	8.14862
6.15	37.8225	2.47992	7.84219	6.65	44.2225	2.57876	8.15475
6.16	37.9456	2.48193	7.84857	6.66	44.3556	2.58070	8.16088
6.17	38.0689	2.48395	7.85493	6.67	44.4889	2.58263	8.16701
6.18	38.1924	2.48596	7.86130	6.68	44.6224	2.58457	8.17313
6.19	38.3161	2.48797	7.86766	6.69	44.7561	2.58650	8.17924
6.20	38.4400	2.48998	7.87401	**6.70**	44.8900	2.58844	8.18535
6.21	38.5641	2.49199	7.88036	6.71	45.0241	2.59037	8.19146
6.22	38.6884	2.49399	7.88670	6.72	45.1584	2.59230	8.19756
6.23	38.8129	2.49600	7.89303	6.73	45.2929	2.59422	8.20366
6.24	38.9376	2.49800	7.89937	6.74	45.4276	2.59615	8.20975
6.25	39.0625	2.50000	7.90569	6.75	45.5625	2.59808	8.21584
6.26	39.1876	2.50200	7.91202	6.76	45.6976	2.60000	8.22192
6.27	39.3129	2.50400	7.91833	6.77	45.8329	2.60192	8.22800
6.28	39.4384	2.50599	7.92465	6.78	45.9684	2.60384	8.23408
6.29	39.5641	2.50799	7.93095	6.79	46.1041	2.60576	8.24015
6.30	39.6900	2.50998	7.93725	**6.80**	46.2400	2.60768	8.24621
6.31	39.8161	2.51197	7.94355	6.81	46.3761	2.60960	8.25227
6.32	39.9424	2.51396	7.94984	6.82	46.5124	2.61151	8.25833
6.33	40.0689	2.51595	7.95613	6.83	46.6489	2.61343	8.26438
6.34	40.1956	2.51794	7.96241	6.84	46.7856	2.61534	8.27043
6.35	40.3225	2.51992	7.96869	6.85	46.9225	2.61725	8.27647
6.36	40.4496	2.52190	7.97496	6.86	47.0596	2.61916	8.28251
6.37	40.5769	2.52389	7.98123	6.87	47.1969	2.62107	8.28855
6.38	40.7044	2.52587	7.98749	6.88	47.3344	2.62298	8.29458
6.39	40.8321	2.52784	7.99375	6.89	47.4721	2.62488	8.30060
6.40	40.9600	2.52982	8.00000	**6.90**	47.6100	2.62679	8.30662
6.41	41.0881	2.53180	8.00625	6.91	47.7481	2.62869	8.31264
6.42	41.2164	2.53377	8.01249	6.92	47.8864	2.63059	8.31865
6.43	41.3449	2.53574	8.01873	6.93	48.0249	2.63249	8.32466
6.44	41.4736	2.53772	8.02496	6.94	48.1636	2.63439	8.33067
6.45	41.6025	2.53969	8.03119	6.95	48.3025	2.63629	8.33667
6.46	41.7316	2.54165	8.03741	6.96	48.4416	2.63818	8.34266
6.47	41.8609	2.54362	8.04363	6.97	48.5809	2.64008	8.34865
6.48	41.9904	2.54558	8.04984	6.98	48.7204	2.64197	8.35464
6.49	42.1201	2.54755	8.05605	6.99	48.8601	2.64386	8.36062
6.50	42.2500	2.54951	8.06226	**7.00**	49.0000	2.64575	8.36660
N	N^2	$\sqrt{N}$	$\sqrt{10N}$	N	N^2	$\sqrt{N}$	$\sqrt{10N}$

TABLE I
Squares, Square Roots, and Reciprocals (continued)

N	N²	√N	√10N		N	N²	√N	√10N
7.00	49.0000	2.64575	8.36660		**7.50**	56.2500	2.73861	8.66025
7.01	49.1401	2.64764	8.37257		7.51	56.4001	2.74044	8.66603
7.02	49.2804	2.64953	8.37854		7.52	56.5504	2.74226	8.67179
7.03	49.4209	2.65141	8.38451		7.53	56.7009	2.74408	8.67756
7.04	49.5616	2.65330	8.39047		7.54	56.8516	2.74591	8.68332
7.05	49.7025	2.65518	8.39643		7.55	57.0025	2.74773	8.68907
7.06	49.8436	2.65707	8.40238		7.56	57.1536	2.74955	8.69483
7.07	49.9849	2.65895	8.40833		7.57	57.3049	2.75136	8.70057
7.08	50.1264	2.66083	8.41427		7.58	57.4564	2.75318	8.70632
7.09	50.2681	2.66271	8.42021		7.59	57.6081	2.75500	8.71206
7.10	50.4100	2.66458	8.42615		**7.60**	57.7600	2.75681	8.71780
7.11	50.5521	2.66646	8.43208		7.61	57.9121	2.75862	8.72353
7.12	50.6944	2.66833	8.43801		7.62	58.0644	2.76043	8.72926
7.13	50.8369	2.67021	8.44393		7.63	58.2169	2.76225	8.73499
7.14	50.9796	2.67208	8.44985		7.64	58.3696	2.76405	8.74071
7.15	51.1225	2.67395	8.45577		7.65	58.5225	2.76586	8.74643
7.16	51.2656	2.67582	8.46168		7.66	58.6756	2.76767	8.75214
7.17	51.4089	2.67769	8.46759		7.67	58.8289	2.76948	8.75785
7.18	51.5524	2.67955	8.47349		7.68	58.9824	2.77128	8.76356
7.19	51.6961	2.68142	8.47939		7.69	59.1361	2.77308	8.76926
7.20	51.8400	2.68328	8.48528		**7.70**	59.2900	2.77489	8.77496
7.21	51.9841	2.68514	8.49117		7.71	59.4441	2.77669	8.78066
7.22	52.1284	2.68701	8.49706		7.72	59.5984	2.77849	8.78635
7.23	52.2729	2.68887	8.50294		7.73	59.7529	2.78029	8.79204
7.24	52.4176	2.69072	8.50882		7.74	59.9076	2.78209	8.79773
7.25	52.5625	2.69258	8.51469		7.75	60.0625	2.78388	8.80341
7.26	52.7076	2.69444	8.52056		7.76	60.2176	2.78568	8.80909
7.27	52.8529	2.69629	8.52643		7.77	60.3729	2.78747	8.81476
7.28	52.9984	2.69815	8.53229		7.78	60.5284	2.78927	8.82043
7.29	53.1441	2.70000	8.53815		7.79	60.6841	2.79106	8.82610
7.30	53.2900	2.70185	8.54400		**7.80**	60.8400	2.79285	8.83176
7.31	53.4361	2.70370	8.54985		7.81	60.9961	2.79464	8.83742
7.32	53.5824	2.70555	8.55570		7.82	61.1524	2.79643	8.84308
7.33	53.7289	2.70740	8.56154		7.83	61.3089	2.79821	8.84873
7.34	53.8756	2.70924	8.56738		7.84	61.4656	2.80000	8.85438
7.35	54.0225	2.71109	8.57321		7.85	61.6225	2.80179	8.86002
7.36	54.1696	2.71293	8.57904		7.86	61.7796	2.80357	8.86566
7.37	54.3169	2.71477	8.58487		7.87	61.9369	2.80535	8.87130
7.38	54.4644	2.71662	8.59069		7.88	62.0944	2.80713	8.87694
7.39	54.6121	2.71846	8.59651		7.89	62.2521	2.80891	8.88257
7.40	54.7600	2.72029	8.60233		**7.90**	62.4100	2.81069	8.88819
7.41	54.9081	2.72213	8.60814		7.91	62.5681	2.81247	8.89382
7.42	55.0564	2.72397	8.61394		7.92	62.7264	2.81425	8.89944
7.43	55.2049	2.72580	8.61974		7.93	62.8849	2.81603	8.90505
7.44	55.3536	2.72764	8.62554		7.94	63.0436	2.81780	8.91067
7.45	55.5025	2.72947	8.63134		7.95	63.2025	2.81957	8.91628
7.46	55.6516	2.73130	8.63713		7.96	63.3616	2.82135	8.92188
7.47	55.8009	2.73313	8.64292		7.97	63.5209	2.82312	8.92749
7.48	55.9504	2.73496	8.64870		7.98	63.6804	2.82489	8.93308
7.49	56.1001	2.73679	8.65448		7.99	63.8401	2.82666	8.93868
7.50	56.2500	2.73861	8.66025		**8.00**	64.0000	2.82843	8.94427
N	N²	√N	√10N		N	N²	√N	√10N

TABLE I
Squares, Square Roots, and Reciprocals (continued)

N	N^2	$\sqrt{N}$	$\sqrt{10N}$	N	N^2	$\sqrt{N}$	$\sqrt{10N}$
8.00	64.0000	2.82843	8.94427	**8.50**	72.2500	2.91548	9.21954
8.01	64.1601	2.83019	8.94986	8.51	72.4201	2.91719	9.22497
8.02	64.3204	2.83196	8.95545	8.52	72.5904	2.91890	9.23038
8.03	64.4809	2.83373	8.96103	8.53	72.7609	2.92062	9.23580
8.04	64.6416	2.83549	8.96660	8.54	72.9316	2.92233	9.24121
8.05	64.8025	2.83725	8.97218	8.55	73.1025	2.92404	9.24662
8.06	64.9636	2.83901	8.97775	8.56	73.2736	2.92575	9.25203
8.07	65.1249	2.84077	8.98332	8.57	73.4449	2.92746	9.25743
8.08	65.2864	2.84253	8.98888	8.58	73.6164	2.92916	9.26283
8.09	65.4481	2.84429	8.99444	8.59	73.7881	2.93087	9.26823
8.10	65.6100	2.84605	9.00000	**8.60**	73.9600	2.93258	9.27362
8.11	65.7721	2.84781	9.00555	8.61	74.1321	2.93428	9.27901
8.12	65.9344	2.84956	9.01110	8.62	74.3044	2.93598	9.28440
8.13	66.0969	2.85132	9.01665	8.63	74.4769	2.93769	9.28978
8.14	66.2596	2.85307	9.02219	8.64	74.6496	2.93939	9.29516
8.15	66.4225	2.85482	9.02774	8.65	74.8225	2.94109	9.30054
8.16	66.5856	2.85657	9.03327	8.66	74.9956	2.94279	9.30591
8.17	66.7489	2.85832	9.03881	8.67	75.1689	2.94449	9.31128
8.18	66.9124	2.86007	9.04434	8.68	75.3424	2.94618	9.31665
8.19	67.0761	2.86182	9.04986	8.69	75.5161	2.94788	9.32202
8.20	67.2400	2.86356	9.05539	**8.70**	75.6900	2.94958	9.32738
8.21	67.4041	2.86531	9.06091	8.71	75.8641	2.95127	9.33274
8.22	67.5684	2.86705	9.06642	8.72	76.0384	2.95296	9.33809
8.23	67.7329	2.86880	9.07193	8.73	76.2129	2.95466	9.34345
8.24	67.8976	2.87054	9.07744	8.74	76.3876	2.95635	9.34880
8.25	68.0625	2.87228	9.08295	8.75	76.5625	2.95804	9.35414
8.26	68.2276	2.87402	9.08845	8.76	76.7376	2.95973	9.35949
8.27	68.3929	2.87576	9.09395	8.77	76.9129	2.96142	9.36483
8.28	68.5584	2.87750	9.09945	8.78	77.0884	2.96311	9.37017
8.29	68.7241	2.87924	9.10494	8.79	77.2641	2.96479	9.37550
8.30	68.8900	2.88097	9.11045	**8.80**	77.4400	2.96648	9.38083
8.31	69.0561	2.88271	9.11592	8.81	77.6161	2.96816	9.38616
8.32	69.2224	2.88444	9.12140	8.82	77.7924	2.96985	9.39149
8.33	69.3889	2.88617	9.12688	8.83	77.9689	2.97153	9.39681
8.34	69.5556	2.88791	9.13236	8.84	78.1456	2.97321	9.40213
8.35	69.7225	2.88964	9.13783	8.85	78.3225	2.97489	9.40744
8.36	69.8896	2.89137	9.14330	8.86	78.4996	2.97658	9.41276
8.37	70.0569	2.89310	9.14877	8.87	78.6769	2.97825	9.41807
8.38	70.2244	2.89482	9.15423	8.88	78.8544	2.97993	9.42338
8.39	70.3921	2.89655	9.15969	8.89	79.0321	2.98161	9.42868
8.40	70.5600	2.89828	9.16515	**8.90**	79.2100	2.98329	9.43398
8.41	70.7281	2.90000	9.17061	8.91	79.3881	2.98496	9.43928
8.42	70.8964	2.90172	9.17606	8.92	79.5664	2.98664	9.44458
8.43	71.0649	2.90345	9.18150	8.93	79.7449	2.98831	9.44987
8.44	71.2336	2.90517	9.18695	8.94	79.9236	2.98998	9.45516
8.45	71.4025	2.90689	9.19239	8.85	80.1025	2.99166	9.46044
8.46	71.5716	2.90861	9.19783	8.96	80.2816	2.99333	9.46573
8.47	71.7409	2.91033	9.20326	8.97	80.4609	2.99500	9.47101
8.48	71.9104	2.91204	9.20869	8.98	80.6404	2.99666	9.47629
8.49	72.0801	2.91376	9.21412	8.99	80.8201	2.99833	9.48156
8.50	72.2500	2.91548	9.21954	**9.00**	81.0000	3.00000	9.48683
N	N^2	$\sqrt{N}$	$\sqrt{10N}$	N	N^2	$\sqrt{N}$	$\sqrt{10N}$

TABLE I
Squares, Square Roots, and Reciprocals (continued)

N	N^2	$\sqrt{N}$	$\sqrt{10N}$	N	N^2	$\sqrt{N}$	$\sqrt{10N}$
9.00	81.0000	3.00000	9.48683	**9.50**	90.2500	3.08221	9.74679
9.01	81.1801	3.00167	9.49210	9.51	90.4401	3.08383	9.75192
9.02	81.3604	3.00333	9.49737	9.52	90.6304	3.08545	9.75705
9.03	81.5409	3.00500	9.50263	9.53	90.8209	3.08707	9.76217
9.04	81.7216	3.00666	9.50789	9.54	91.0116	3.08869	9.76729
9.05	81.9025	3.00832	9.51315	9.55	91.2025	3.09031	9.77241
9.06	82.0836	3.00998	9.51840	9.56	91.3936	3.09192	9.77753
9.07	82.2649	3.01164	9.52365	9.57	91.5849	3.09354	9.78264
9.08	82.4464	3.01330	9.52890	9.58	91.7764	3.09516	9.78775
9.09	82.6281	3.01496	9.53415	9.59	91.9681	3.09677	9.79285
9.10	82.8100	3.01662	9.53939	**9.60**	92.1600	3.09839	9.79796
9.11	82.9921	3.01828	9.54463	9.61	92.3521	3.10000	9.80306
9.12	83.1744	3.01993	9.54987	9.62	92.5444	3.10161	9.80816
9.13	83.3569	3.02159	9.55510	9.63	92.7369	3.10322	9.81326
9.14	83.5396	3.02324	9.56033	9.64	92.9296	3.10483	9.81835
9.15	83.7225	3.02490	9.56556	9.65	93.1225	3.10644	9.82344
9.16	83.9056	3.02655	9.57079	9.66	93.3156	3.10805	9.82853
9.17	84.0889	3.02820	9.57601	9.67	93.5089	3.10966	9.83362
9.18	84.2724	3.02985	9.58123	9.68	93.7024	3.11127	9.83870
9.19	84.4561	3.03150	9.58645	9.69	93.8961	3.11288	9.84378
9.20	84.6400	3.03315	9.59166	**9.70**	94.0900	3.11448	9.84886
9.21	84.8241	3.03480	9.59687	9.71	94.2841	3.11609	9.85393
9.22	85.0084	3.03645	9.60208	9.72	94.4784	3.11769	9.85901
9.23	85.1929	3.03809	9.60729	9.73	94.6729	3.11929	9.86408
9.24	85.3776	3.03974	9.61249	9.74	94.8676	3.12090	9.86914
9.25	85.5625	3.04138	9.61769	9.75	95.0625	3.12250	9.87421
9.26	85.7476	3.04302	9.62289	9.76	95.2576	3.12410	9.87927
9.27	85.9329	3.04467	9.62808	9.77	95.4529	3.12570	9.88433
9.28	86.1184	3.04631	9.63328	9.78	95.6484	3.12730	9.88939
9.29	86.3041	3.04795	9.63846	9.79	95.8441	3.12890	9.89444
9.30	86.4900	3.04959	9.64365	**9.80**	96.0400	3.13050	9.89949
9.31	86.6761	3.05123	9.64883	9.81	96.2361	3.13209	9.90454
9.32	86.8624	3.05287	9.65401	9.82	96.4324	3.13369	9.90959
9.33	86.0489	3.05450	9.65919	9.83	96.6289	3.13528	9.91464
9.34	87.2356	3.05614	9.66437	9.84	96.8256	3.13688	9.91968
9.35	87.4225	3.05778	9.66954	9.85	97.0225	3.13847	9.92472
9.36	87.6096	3.05941	9.67471	9.86	97.2196	3.14006	9.92974
9.37	87.7969	3.06105	9.67988	9.87	97.4169	3.14166	9.93479
9.38	87.9844	3.06268	9.68504	9.88	97.6144	3.14325	9.93982
9.39	88.1721	3.06431	9.69020	9.89	97.8121	3.14484	9.94485
9.40	88.3600	3.06594	9.69536	**9.90**	98.0100	3.14643	9.94987
9.41	88.5481	3.06757	9.70052	9.91	98.2081	3.14802	9.95490
9.42	88.7364	3.06920	9.70567	9.92	98.4064	3.14960	9.95992
9.43	88.9249	3.07083	9.71082	9.93	98.6049	3.15119	9.96494
9.44	89.1136	3.07246	9.71597	9.94	98.8036	3.15278	9.96995
9.45	89.3025	3.07409	9.72111	9.95	99.0025	3.15436	9.97497
9.46	89.4916	3.07571	9.72625	9.96	99.2016	3.15595	9.97998
9.47	89.6809	3.07734	9.73139	9.97	99.4009	3.15753	9.98499
9.48	89.8704	3.07896	9.73653	9.98	99.6004	3.15911	9.98999
9.49	90.0601	3.08058	9.74166	9.99	99.8001	3.16070	9.99500
9.50	90.2500	3.08221	9.74679	**10.0**	100.000	3.16228	10.0000
N	N^2	$\sqrt{N}$	$\sqrt{10N}$	N	N^2	$\sqrt{N}$	$\sqrt{10N}$

TABLE II
Factorials

n	n!
0	1
1	1
2	2
3	6
4	24
5	120
6	720
7	5,040
8	40,320
9	362,880
10	3,628,800
11	39,916,800
12	479,001,600
13	6,227,020,800
14	87,178,291,200
15	1,307,674,368,000
16	20,922,789,888,000
17	355,687,428,096,000
18	6,402,373,705,728,000
19	121,645,100,408,832,000
20	2,432,902,008,176,640,000

Table III Binomial Coefficient n!/x!(n − x!) **A.13**

TABLE III

Binomial Coefficients $\dfrac{n!}{x!(n - x!)}$

n \ x	2	3	4	5	6	7	8	9	10
2	1								
3	3	1							
4	6	4	1						
5	10	10	5	1					
6	15	20	15	6	1				
7	21	35	35	21	7	1			
8	28	56	70	56	28	8	1		
9	36	84	126	126	84	36	9	1	
10	45	120	210	252	210	120	45	10	1
11	55	165	330	462	462	330	165	55	11
12	66	220	495	792	924	792	495	220	66
13	78	286	715	1,287	1,716	1,716	1,287	715	286
14	91	364	1,001	2,002	3,003	3,432	3,003	2,002	1,001
15	105	455	1,365	3,003	5,005	6,435	6,435	5,005	3,003
16	120	560	1,820	4,368	8,008	11,440	12,870	11,440	8,008
17	136	680	2,380	6,188	12,376	19,448	24,310	24,310	19,448
18	153	816	3,060	8,568	18,564	31,824	43,758	48,620	43,758
19	171	969	3,876	11,628	27,132	50,388	75,582	92,378	92,378
20	190	1,140	4,845	15,504	38,760	77,520	125,970	167,960	184,756

TABLE IV
Binomial Probabilities

n	x	0.05	0.1	0.2	0.3	0.4	0.5	0.6	0.7	0.8	0.9	0.95
2	0	0.902	0.810	0.640	0.490	0.360	0.250	0.160	0.090	0.040	0.010	0.002
	1	0.095	0.180	0.320	0.420	0.480	0.500	0.480	0.420	0.320	0.180	0.095
	2	0.002	0.010	0.040	0.090	0.160	0.250	0.360	0.490	0.640	0.810	0.902
3	0	0.857	0.729	0.512	0.343	0.216	0.125	0.064	0.027	0.008	0.001	
	1	0.135	0.243	0.384	0.441	0.432	0.375	0.288	0.189	0.096	0.027	0.007
	2	0.007	0.027	0.096	0.189	0.288	0.375	0.432	0.441	0.384	0.243	0.135
	3		0.001	0.008	0.027	0.064	0.125	0.216	0.343	0.512	0.729	0.857
4	0	0.815	0.656	0.410	0.240	0.130	0.062	0.026	0.008	0.002		
	1	0.171	0.292	0.410	0.412	0.346	0.250	0.154	0.076	0.026	0.004	
	2	0.014	0.049	0.154	0.265	0.346	0.375	0.346	0.265	0.154	0.049	0.014
	3		0.004	0.026	0.076	0.154	0.250	0.346	0.412	0.410	0.292	0.171
	4			0.002	0.008	0.026	0.062	0.130	0.240	0.410	0.656	0.815
5	0	0.774	0.590	0.328	0.168	0.078	0.031	0.010	0.002			
	1	0.204	0.328	0.410	0.360	0.259	0.156	0.077	0.028	0.006		
	2	0.021	0.073	0.205	0.309	0.346	0.312	0.230	0.132	0.051	0.008	0.001
	3	0.001	0.008	0.051	0.132	0.230	0.312	0.346	0.309	0.205	0.073	0.021
	4			0.006	0.028	0.077	0.156	0.259	0.360	0.410	0.328	0.204
	5				0.002	0.010	0.031	0.078	0.168	0.328	0.590	0.774
6	0	0.735	0.531	0.262	0.118	0.047	0.016	0.004	0.001			
	1	0.232	0.354	0.393	0.303	0.187	0.094	0.037	0.010	0.002		
	2	0.031	0.098	0.246	0.324	0.311	0.234	0.138	0.060	0.015	0.001	
	3	0.002	0.015	0.082	0.185	0.276	0.312	0.276	0.185	0.082	0.015	0.002
	4		0.001	0.015	0.060	0.138	0.234	0.311	0.324	0.246	0.098	0.031
	5			0.002	0.010	0.037	0.094	0.187	0.303	0.393	0.354	0.232
	6				0.001	0.004	0.016	0.047	0.118	0.262	0.531	0.735
7	0	0.698	0.478	0.210	0.082	0.028	0.008	0.002				
	1	0.257	0.372	0.367	0.247	0.131	0.055	0.017	0.004			
	2	0.041	0.124	0.275	0.318	0.261	0.164	0.077	0.025	0.004		
	3	0.004	0.023	0.115	0.227	0.290	0.273	0.194	0.097	0.029	0.003	
	4		0.003	0.029	0.097	0.194	0.273	0.290	0.227	0.115	0.023	0.004
	5			0.004	0.025	0.077	0.164	0.261	0.318	0.275	0.124	0.041
	6				0.004	0.017	0.055	0.131	0.247	0.367	0.372	0.257
	7					0.002	0.008	0.028	0.082	0.210	0.478	0.698
8	0	0.663	0.430	0.168	0.058	0.017	0.004	0.001				
	1	0.279	0.383	0.336	0.198	0.090	0.031	0.008	0.001			
	2	0.051	0.149	0.294	0.296	0.209	0.109	0.041	0.010	0.001		
	3	0.005	0.033	0.147	0.254	0.279	0.219	0.124	0.047	0.009		
	4		0.005	0.046	0.136	0.232	0.273	0.232	0.136	0.046	0.005	
	5			0.009	0.047	0.124	0.219	0.279	0.254	0.147	0.033	0.005
	6			0.001	0.010	0.041	0.109	0.209	0.296	0.294	0.149	0.051
	7				0.001	0.008	0.031	0.090	0.198	0.336	0.383	0.279
	8					0.001	0.004	0.017	0.058	0.168	0.430	0.663
9	0	0.630	0.387	0.134	0.040	0.010	0.002					
	1	0.299	0.387	0.302	0.156	0.060	0.018	0.004				
	2	0.063	0.172	0.302	0.267	0.161	0.070	0.021	0.004			
	3	0.008	0.045	0.176	0.267	0.251	0.164	0.074	0.021	0.003		
	4	0.001	0.007	0.066	0.172	0.251	0.246	0.167	0.074	0.017	0.001	
	5		0.001	0.017	0.074	0.167	0.246	0.251	0.172	0.066	0.007	0.001
	6			0.003	0.021	0.074	0.164	0.251	0.267	0.176	0.045	0.008
	7				0.004	0.021	0.070	0.161	0.267	0.302	0.172	0.063
	8					0.004	0.018	0.060	0.156	0.302	0.387	0.299
	9						0.002	0.010	0.040	0.134	0.387	0.630

Table IV Binomial Probabilities   **A.15**

TABLE IV
Binomial Probabilities (continued)

n	x	p 0.05	0.1	0.2	0.3	0.4	0.5	0.6	0.7	0.8	0.9	0.95
10	0	0.599	0.349	0.107	0.028	0.006	0.001					
	1	0.315	0.387	0.268	0.121	0.040	0.010	0.002				
	2	0.075	0.194	0.302	0.233	0.121	0.044	0.011	0.001			
	3	0.010	0.057	0.201	0.267	0.215	0.117	0.042	0.009	0.001		
	4	0.001	0.011	0.088	0.200	0.251	0.205	0.111	0.037	0.006		
	5		0.001	0.026	0.103	0.201	0.246	0.201	0.103	0.026	0.001	
	6			0.006	0.037	0.111	0.205	0.251	0.200	0.088	0.011	0.001
	7			0.001	0.009	0.042	0.117	0.215	0.267	0.201	0.057	0.010
	8				0.001	0.011	0.044	0.121	0.233	0.302	0.194	0.075
	9					0.002	0.010	0.040	0.121	0.268	0.387	0.315
	10						0.001	0.006	0.028	0.107	0.349	0.599
11	0	0.569	0.314	0.086	0.020	0.004						
	1	0.329	0.384	0.236	0.093	0.027	0.005	0.001				
	2	0.087	0.213	0.295	0.200	0.089	0.027	0.005	0.001			
	3	0.014	0.071	0.221	0.257	0.177	0.081	0.023	0.004			
	4	0.001	0.016	0.111	0.220	0.236	0.161	0.070	0.017	0.002		
	5		0.002	0.039	0.132	0.221	0.226	0.147	0.057	0.010		
	6			0.010	0.057	0.147	0.226	0.221	0.132	0.039	0.002	
	7			0.002	0.017	0.070	0.161	0.236	0.220	0.111	0.016	0.001
	8				0.004	0.023	0.081	0.177	0.257	0.221	0.071	0.014
	9				0.001	0.005	0.027	0.089	0.200	0.295	0.213	0.087
	10					0.001	0.005	0.027	0.093	0.236	0.384	0.329
	11							0.004	0.020	0.086	0.314	0.569
12	0	0.540	0.282	0.069	0.014	0.002						
	1	0.341	0.377	0.206	0.071	0.017	0.003					
	2	0.099	0.230	0.283	0.168	0.064	0.016	0.002				
	3	0.017	0.085	0.236	0.240	0.142	0.054	0.012	0.001			
	4	0.002	0.021	0.133	0.231	0.213	0.121	0.042	0.008	0.001		
	5		0.004	0.053	0.158	0.227	0.193	0.101	0.029	0.003		
	6			0.016	0.079	0.177	0.226	0.177	0.079	0.016		
	7			0.003	0.029	0.101	0.193	0.227	0.158	0.053	0.004	
	8			0.001	0.008	0.042	0.121	0.213	0.231	0.133	0.021	0.002
	9				0.001	0.012	0.054	0.142	0.240	0.236	0.085	0.017
	10					0.002	0.016	0.064	0.168	0.283	0.230	0.099
	11						0.003	0.017	0.071	0.206	0.377	0.341
	12						0.002	0.014	0.069	0.282	0.540	
13	0	0.513	0.254	0.055	0.010	0.001						
	1	0.351	0.367	0.179	0.054	0.011	0.002					
	2	0.111	0.245	0.268	0.139	0.045	0.010	0.001				
	3	0.021	0.100	0.246	0.218	0.111	0.035	0.006	0.001			
	4	0.003	0.028	0.154	0.234	0.184	0.087	0.024	0.003			
	5		0.006	0.069	0.180	0.221	0.157	0.066	0.014	0.001		
	6		0.001	0.023	0.103	0.197	0.209	0.131	0.044	0.006		
	7			0.006	0.044	0.131	0.209	0.197	0.103	0.023	0.001	
	8			0.001	0.014	0.066	0.157	0.221	0.180	0.069	0.006	
	9				0.003	0.024	0.087	0.184	0.234	0.154	0.028	0.003
	10				0.001	0.006	0.035	0.111	0.218	0.246	0.100	0.021
	11					0.001	0.010	0.045	0.139	0.268	0.245	0.111
	12						0.002	0.011	0.054	0.179	0.367	0.351
	13							0.001	0.010	0.055	0.254	0.513

TABLE IV
Binomial Probabilities (continued)

n	x	0.05	0.1	0.2	0.3	0.4	0.5	0.6	0.7	0.8	0.9	0.95
14	0	0.488	0.229	0.044	0.007	0.001						
	1	0.359	0.356	0.154	0.041	0.007	0.001					
	2	0.123	0.257	0.250	0.113	0.032	0.006	0.001				
	3	0.026	0.114	0.250	0.194	0.085	0.022	0.003				
	4	0.004	0.035	0.172	0.229	0.155	0.061	0.014	0.001			
	5		0.008	0.086	0.196	0.207	0.122	0.041	0.007			
	6		0.001	0.032	0.126	0.207	0.183	0.092	0.023	0.002		
	7			0.009	0.062	0.157	0.209	0.157	0.062	0.009		
	8			0.002	0.023	0.092	0.183	0.207	0.126	0.032	0.001	
	9				0.007	0.041	0.122	0.207	0.196	0.086	0.008	
	10				0.001	0.014	0.061	0.155	0.229	0.172	0.035	0.004
	11					0.003	0.022	0.085	0.194	0.250	0.114	0.026
	12					0.001	0.006	0.032	0.113	0.250	0.257	0.123
	13						0.001	0.007	0.041	0.154	0.356	0.359
	14							0.001	0.007	0.044	0.229	0.488
15	0	0.463	0.206	0.035	0.005							
	1	0.366	0.343	0.132	0.031	0.005						
	2	0.135	0.267	0.231	0.092	0.022	0.003					
	3	0.031	0.129	0.250	0.170	0.063	0.014	0.002				
	4	0.005	0.043	0.188	0.219	0.127	0.042	0.007	0.001			
	5	0.001	0.010	0.103	0.206	0.186	0.092	0.024	0.003			
	6		0.002	0.043	0.147	0.207	0.153	0.061	0.012	0.001		
	7			0.014	0.081	0.177	0.196	0.118	0.035	0.003		
	8			0.003	0.035	0.118	0.196	0.177	0.081	0.014		
	9			0.001	0.012	0.061	0.153	0.207	0.147	0.043	0.002	
	10				0.003	0.024	0.092	0.186	0.206	0.103	0.010	0.001
	11				0.001	0.007	0.042	0.127	0.219	0.188	0.043	0.005
	12					0.002	0.014	0.063	0.170	0.250	0.129	0.031
	13						0.003	0.022	0.092	0.231	0.267	0.135
	14							0.005	0.031	0.132	0.343	0.366
	15								0.005	0.035	0.206	0.463

Table V The Standard Normal Distribu

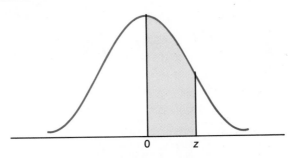

The entries in Table V are the probabilities that a random variable having the standard normal distribution takes on a value between 0 and z; they are given by the area under the curve shaded in the diagram.

TABLE V
The Standard Normal Distribution

z	.00	.01	.02	.03	.04	.05	.06	.07	.08	.09
0.0	.0000	.0040	.0080	.0120	.0160	.0199	.0239	.0279	.0319	.0359
0.1	.0398	.0438	.0478	.0517	.0557	.0596	.0636	.0675	.0714	.0753
0.2	.0793	.0832	.0871	.0910	.0948	.0987	.1026	.1064	.1103	.1141
0.3	.1179	.1217	.1255	.1293	.1331	.1368	.1406	.1443	.1480	.1517
0.4	.1554	.1591	.1628	.1664	.1700	.1736	.1772	.1808	.1844	.1879
0.5	.1915	.1950	.1985	.2019	.2054	.2088	.2123	.2157	.2190	.2224
0.6	.2257	.2291	.2324	.2357	.2389	.2422	.2454	.2486	.2517	.2549
0.7	.2580	.2611	.2642	.2673	.2704	.2734	.2764	.2794	.2823	.2852
0.8	.2881	.2910	.2939	.2967	.2995	.3023	.3051	.3078	.3106	.3133
0.9	.3159	.3186	.3212	.3238	.3264	.3289	.3315	.3340	.3365	.3389
1.0	.3413	.3438	.3461	.3485	.3508	.3531	.3554	.3577	.3599	.3621
1.1	.3643	.3665	.3686	.3708	.3729	.3749	.3770	.3790	.3810	.3830
1.2	.3849	.3869	.3888	.3907	.3925	.3944	.3962	.3980	.3997	.4015
1.3	.4032	.4049	.4066	.4082	.4099	.4115	.4131	.4147	.4162	.4177
1.4	.4192	.4207	.4222	.4236	.4251	.4265	.4279	.4292	.4306	.4319
1.5	.4332	.4345	.4357	.4370	.4382	.4394	.4406	.4418	.4429	.4441
1.6	.4452	.4463	.4474	.4484	.4495	.4505	.4515	.4525	.4535	.4545
1.7	.4554	.4564	.4573	.4582	.4591	.4599	.4608	.4616	.4625	.4633
1.8	.4641	.4649	.4656	.4664	.4671	.4678	.4686	.4693	.4699	.4706
1.9	.4713	.4719	.4726	.4732	.4738	.4744	.4750	.4756	.4761	.4767
2.0	.4772	.4778	.4783	.4788	.4793	.4798	.4803	.4808	.4812	.4817
2.1	.4821	.4826	.4830	.4834	.4838	.4842	.4846	.4850	.4854	.4857
2.2	.4861	.4864	.4868	.4871	.4875	.4878	.4881	.4884	.4887	.4890
2.3	.4893	.4896	.4898	.4901	.4904	.4906	.4909	.4911	.4913	.4916
2.4	.4918	.4920	.4922	.4925	.4927	.4929	.4931	.4932	.4934	.4936
2.5	.4938	.4940	.4941	.4943	.4945	.4946	.4948	.4949	.4951	.4952
2.6	.4953	.4955	.4956	.4957	.4959	.4960	.4961	.4962	.4963	.4964
2.7	.4965	.4966	.4967	.4968	.4969	.4970	.4971	.4972	.4973	.4974
2.8	.4974	.4975	.4976	.4977	.4977	.4978	.4979	.4979	.4980	.4981
2.9	.4981	.4982	.4982	.4983	.4984	.4984	.4985	.4985	.4986	.4986
3.0	.4987	.4987	.4987	.4988	.4988	.4989	.4989	.4989	.4990	.4990

TABLE VI
Critical Values of r

n	$r_{0.025}$	$r_{0.005}$	n	$r_{0.025}$	$r_{0.005}$
3	0.997		18	0.468	0.590
4	0.950	0.999	19	0.456	0.575
5	0.878	0.959	20	0.444	0.561
6	0.811	0.917	21	0.433	0.549
7	0.754	0.875	22	0.423	0.537
8	0.707	0.834	27	0.381	0.487
9	0.666	0.798	32	0.349	0.449
10	0.632	0.765	37	0.325	0.418
11	0.602	0.735	42	0.304	0.393
12	0.576	0.708	47	0.288	0.372
13	0.553	0.684	52	0.273	0.354
14	0.532	0.661	62	0.250	0.325
15	0.514	0.641	72	0.232	0.302
16	0.497	0.623	82	0.217	0.283
17	0.482	0.606	92	0.205	0.267

This table is abridged from Table VI of R. A. Fisher and F. Yates: *Statistical Tables for Biological, Agricultural, and Medical Research*, published by Longman Group, Ltd., London (previously published by Oliver & Boyd, Edinburgh), by permission of the authors and publishers.

TABLE VII
Table of Random Digits

LINE / COL	(1)	(2)	(3)	(4)	(5)	(6)	(7)	(8)	(9)	(10)	(11)	(12)	(13)	(14)
1	10480	15011	01536	02011	81647	91646	69179	14194	62590	36207	20969	99570	91291	90700
2	22368	46573	25595	85393	30995	89198	27982	53402	93965	34095	52666	19174	39615	99505
3	24130	48360	22527	97265	76393	64809	15179	24830	49340	32081	30680	19655	63348	58629
4	42167	93093	06243	61680	07856	16376	39440	53537	71341	57004	00849	74917	97758	16379
5	37570	39975	81837	16656	06121	91782	60468	81305	49684	60672	14110	06927	01263	54613
6	77921	06907	11008	42751	27756	53498	18602	70659	90655	15053	21916	81825	44394	42880
7	99562	72905	56420	69994	98872	31016	71194	18738	44013	48840	63213	21069	10634	12952
8	96301	91977	05463	07972	18876	20922	94595	56869	69014	60045	18425	84903	42508	32307
9	89579	14342	63661	10281	17453	18103	57740	84378	25331	12566	58678	44947	05585	56941
10	85475	36857	53342	53988	53060	59533	38867	62300	08158	17983	16439	11458	18593	64952
11	28918	69578	88231	33276	70997	79936	56865	05859	90106	31595	01547	85590	91610	78188
12	63553	40961	48235	03427	49626	69445	18663	72695	52180	20847	12234	90511	33703	90322
13	09429	93969	52636	92737	88974	33488	36320	17617	30015	08272	84115	27156	30613	74952
14	10365	61129	87529	85689	48237	52267	67689	93394	01511	26358	85104	20285	29975	89868
15	07119	97336	71048	08178	77233	13916	47564	81056	97735	85977	29372	74461	28551	90707
16	51085	12765	51821	51259	77452	16308	60756	92144	49442	53900	70960	63990	75601	40719
17	02368	21382	52404	60268	89368	19885	55322	44819	01188	65255	64835	44919	05944	55157
18	01011	54092	33362	94904	31273	04146	18594	29852	71585	85030	51132	01915	92747	64951
19	52162	53916	46369	58586	23216	14513	83149	98736	23495	64350	94738	17752	35156	35749
20	07056	97628	33787	09998	42698	06691	76988	13602	51851	46104	88916	19509	25625	58104
21	48663	91245	85828	14346	09172	30168	90229	04734	59193	22178	30421	61666	99904	32812
22	54164	58492	22421	74103	47070	25306	76468	26384	58151	06646	21524	15227	96909	44592
23	32639	32363	05597	24200	13363	38005	94342	28728	35806	06912	17012	64161	18296	22851
24	29334	27001	87637	87308	58731	00256	45834	15398	46557	41135	10367	07684	36188	18510
25	02488	33062	28834	07351	19731	92420	60952	61280	50001	67658	32586	86679	50720	94953
26	81525	72295	04839	96423	24878	82651	66566	14778	76797	14780	13300	87074	79666	95725
27	29676	20591	68086	26432	46901	20849	89768	81536	86645	12659	92259	57102	80428	25280
28	00742	57392	39064	66432	84673	40027	32832	61362	98947	96067	64760	64584	96096	98253
29	05366	04213	25669	26422	44407	44048	37937	63904	45766	66134	75470	66520	34693	90449
30	91921	26418	64117	94305	26766	25940	39972	22209	71500	64568	91402	42416	07844	69618
31	00582	04711	87917	77341	42206	35126	74087	99547	81817	42607	43808	76655	62028	76630
32	00725	69884	62797	56170	86324	88072	76222	36086	84637	93161	76038	65855	77919	88006
33	69011	65795	95876	55293	18988	27354	26575	08625	40801	59920	29841	80150	12777	48501
34	25976	57948	29888	88604	67917	48708	18912	82271	65424	69774	33611	54262	85963	03547
35	09763	83473	73577	12908	30883	18317	28290	35797	05998	41688	34952	37888	38917	88050
36	91567	42595	27958	30134	04024	86385	29880	99730	55536	84855	29080	09250	79656	73211
37	17955	56349	90999	49127	20044	59931	06115	20542	18059	02008	73708	83517	36103	42791
38	46503	18584	18845	49618	02304	51038	20655	58727	28168	15475	56942	53389	20562	87338
39	92157	89634	94824	78171	84610	82834	09922	25417	44137	48413	25555	21246	35509	20468
40	14577	62765	35605	81263	39667	47358	56873	56307	61607	49518	89656	20103	77490	18062
41	98427	07523	33362	64270	01638	92477	66969	98420	04880	45585	46565	04102	46880	45709
42	34914	63976	88720	82765	34476	17032	87589	40836	32427	70002	70663	88863	77775	69348
43	70060	28277	39475	46473	23219	53416	94970	25832	69975	94884	19661	72828	00102	66794
44	53976	54914	06990	67245	68350	82948	11398	42878	80287	88267	47363	46634	06541	97809
45	76072	29515	40980	07391	58745	25774	22987	80059	39911	96189	41151	14222	60697	59583
46	90725	52210	83974	29992	65831	38857	50490	83765	55657	14361	31720	57375	56228	41546
47	64364	67412	33339	31926	14883	24413	59744	92351	97473	89286	35931	04110	23726	51900
48	08962	00358	31662	25388	61642	34072	81249	35648	56891	69352	48373	45578	78547	81788
49	95012	68379	93526	70765	10592	04542	76463	54328	02349	17247	28865	14777	62730	92277
50	15664	10493	20492	38391	91132	21999	59516	81652	27195	48223	46751	22923	32261	85653

Page 1 of *Table of 105,000 Random Decimal Digits*, Statement No. 4914, May 1949, File No. 261-A-1, Interstate Commerce Commission, Washington, D.C.

TABLE VIII
The t-distribution

df	$t_{0.050}$	$t_{0.025}$	$t_{0.010}$	$t_{0.005}$	df
1	6.314	12.706	31.821	63.657	1
2	2.920	4.303	6.965	9.925	2
3	2.353	3.182	4.541	5.841	3
4	2.132	2.776	3.747	4.604	4
5	2.015	2.571	3.365	4.032	5
6	1.943	2.447	3.143	3.707	6
7	1.895	2.365	2.998	3.499	7
8	1.860	2.306	2.896	3.355	8
9	1.833	2.262	2.821	3.250	9
10	1.812	2.228	2.764	3.169	10
11	1.796	2.201	2.718	3.106	11
12	1.782	2.179	2.681	3.055	12
13	1.771	2.160	2.650	3.012	13
14	1.761	2.145	2.624	2.977	14
15	1.753	2.131	2.602	2.947	15
16	1.746	2.120	2.583	2.921	16
17	1.740	2.110	2.567	2.898	17
18	1.734	2.101	2.552	2.878	18
19	1.729	2.093	2.539	2.861	19
20	1.725	2.086	2.528	2.845	20
21	1.721	2.080	2.518	2.831	21
22	1.717	2.074	2.508	2.819	22
23	1.714	2.069	2.500	2.807	23
24	1.711	2.064	2.492	2.797	24
25	1.708	2.060	2.485	2.787	25
26	1.706	2.056	2.479	2.779	26
27	1.703	2.052	2.473	2.771	27
28	1.701	2.048	2.467	2.763	28
29	1.699	2.045	2.462	2.756	29
inf.	1.645	1.960	2.326	2.576	inf.

This table is abridged from Table IV of R. A. Fisher and F. Yates: *Statistical Tables for Biological, Agricultural, and Medical Research,* published by Longman Group, Ltd., London (previously published by Oliver & Boyd, Edinburgh), by permission of the authors and publishers.

Table IX The χ^2 Distribution **A.21**

TABLE IX
The χ^2 distribution

df	$\chi^2_{0.05}$	$\chi^2_{0.01}$	df
1	3.841	6.635	1
2	5.991	9.210	2
3	7.815	11.345	3
4	9.488	13.277	4
5	11.070	15.086	5
6	12.592	16.812	6
7	14.067	18.475	7
8	15.507	20.090	8
9	16.919	21.666	9
10	18.307	23.209	10
11	19.675	24.725	11
12	21.026	26.217	12
13	22.362	27.688	13
14	23.685	29.141	14
15	24.996	30.578	15
16	26.296	32.000	16
17	27.587	33.409	17
18	28.869	34.805	18
19	30.144	36.191	19
20	31.410	37.566	20
21	32.671	38.932	21
22	33.924	40.289	22
23	35.172	41.638	23
24	36.415	42.980	24
25	37.652	44.314	25
26	38.885	45.642	26
27	40.113	46.963	27
28	41.337	48.278	28
29	42.557	49.588	29
30	43.773	50.892	30

TABLE X
Critical Values of the F-distribution ($\alpha = 0.05$)

		Degrees of Freedom for Numerator								
	1	2	3	4	5	6	7	8	9	10
1	161	200	216	225	230	234	237	239	241	242
2	18.5	19.0	19.2	19.2	19.3	19.3	19.4	19.4	19.4	19.4
3	10.1	9.55	9.28	9.12	9.01	8.94	8.89	8.85	8.81	8.79
4	7.71	6.94	6.59	6.39	6.26	6.16	6.09	6.04	6.00	5.96
5	6.61	5.79	5.41	5.19	5.05	4.95	4.88	4.82	4.77	4.74
6	5.99	5.14	4.76	4.53	4.39	4.28	4.21	4.15	4.10	4.06
7	5.59	4.74	4.35	4.12	3.97	3.87	3.79	3.73	3.68	3.64
8	5.32	4.46	4.07	3.84	3.69	3.58	3.50	3.44	3.39	3.35
9	5.12	4.26	3.86	3.63	3.48	3.37	3.29	3.23	3.18	3.14
10	4.96	4.10	3.71	3.48	3.33	3.22	3.14	3.07	3.02	2.98
11	4.84	3.98	3.59	3.36	3.20	3.09	3.01	2.95	2.90	2.85
12	4.75	3.89	3.49	3.26	3.11	3.00	2.91	2.85	2.80	2.75
13	4.67	3.81	3.41	3.18	3.03	2.92	2.83	2.77	2.71	2.67
14	4.60	3.74	3.34	3.11	2.96	2.85	2.76	2.70	2.65	2.60
15	4.54	3.68	3.29	3.06	2.90	2.79	2.71	2.64	2.59	2.54
16	4.49	3.63	3.24	3.01	2.85	2.74	2.66	2.59	2.54	2.49
17	4.45	3.59	3.20	2.96	2.81	2.70	2.61	2.55	2.49	2.45
18	4.41	3.55	3.16	2.93	2.77	2.66	2.58	2.51	2.46	2.41
19	4.38	3.52	3.13	2.90	2.74	2.63	2.54	2.48	2.42	2.38
20	4.35	3.49	3.10	2.87	2.71	2.60	2.51	2.45	2.39	2.35
21	4.32	3.47	3.07	2.84	2.68	2.57	2.49	2.42	2.37	2.32
22	4.30	3.44	3.05	2.82	2.66	2.55	2.46	2.40	2.34	2.30
23	4.28	3.42	3.03	2.80	2.64	2.53	2.44	2.37	2.32	2.27
24	4.26	3.40	3.01	2.78	2.62	2.51	2.42	2.36	2.30	2.25
25	4.24	3.39	2.99	2.76	2.60	2.49	2.40	2.34	2.28	2.24
30	4.17	3.32	2.92	2.69	2.53	2.42	2.33	2.27	2.21	2.16
40	4.08	3.23	2.84	2.61	2.45	2.34	2.25	2.18	2.12	2.08
60	4.00	3.15	2.76	2.53	2.37	2.25	2.17	2.10	2.04	1.99
120	3.92	3.07	2.68	2.45	2.29	2.18	2.09	2.02	1.96	1.91
∞	3.84	3.00	2.60	2.37	2.21	2.10	2.01	1.94	1.88	1.83

Degrees of Freedom for Denominator

Table X Critical Values of the F-distribution ($\alpha = 0.05$) **A.23**

TABLE X
Critical Values of the F-distribution ($\alpha = 0.05$)

		Degrees of Freedom for Numerator								
		12	15	20	24	30	40	60	120	∞
	1	6,106	6,157	6,209	6,235	6,261	6,287	6,313	6,339	6,366
	2	99.4	99.4	99.4	99.5	99.5	99.5	99.5	99.5	99.5
	3	27.1	26.9	26.7	26.6	26.5	26.4	26.3	26.2	26.1
	4	14.4	14.2	14.0	13.9	13.8	13.7	13.7	13.6	13.5
	5	9.89	9.72	9.55	9.47	9.38	9.29	9.20	9.11	9.02
	6	7.72	7.56	7.40	7.31	7.23	7.14	7.06	6.97	6.88
	7	6.47	6.31	6.16	6.07	5.99	5.91	5.82	5.74	5.65
	8	5.67	5.52	5.36	5.28	5.20	5.12	5.03	4.95	4.86
	9	5.11	4.96	4.81	4.73	4.65	4.57	4.48	4.40	4.31
	10	4.71	4.56	4.41	4.33	4.25	4.17	4.08	4.00	3.91
	11	4.40	4.25	4.10	4.02	3.94	3.86	3.78	3.69	3.60
	12	4.16	4.01	3.86	3.78	3.70	3.62	3.54	3.45	3.36
Degrees	13	3.96	3.82	3.66	3.59	3.51	3.43	3.34	3.25	3.17
of	14	3.80	3.66	3.51	3.43	3.35	3.27	3.18	3.09	3.00
Freedom	15	3.67	3.52	3.37	3.29	3.21	3.13	3.05	2.96	2.87
for	16	3.55	3.41	3.26	3.18	3.10	3.02	2.93	2.84	2.75
Denominator	17	3.46	3.31	3.16	3.08	3.00	2.92	2.83	2.75	2.65
	18	3.37	3.23	3.08	3.00	2.92	2.84	2.75	2.66	2.57
	19	3.30	3.15	3.00	2.92	2.84	2.76	2.67	2.58	2.49
	20	3.23	3.09	2.94	2.86	2.78	2.69	2.61	2.52	2.42
	21	3.17	3.03	2.88	2.80	2.72	2.64	2.55	2.46	2.36
	22	3.12	2.98	2.83	2.75	2.67	2.58	2.50	2.40	2.31
	23	3.07	2.93	2.78	2.70	2.62	2.54	2.45	2.35	2.26
	24	3.03	2.89	2.74	2.66	2.58	2.49	2.40	2.31	2.21
	25	2.99	2.85	2.70	2.62	2.53	2.45	2.36	2.27	2.17
	30	2.84	2.70	2.55	2.47	2.39	2.30	2.21	2.11	2.01
	40	2.66	2.52	2.37	2.29	2.20	2.11	2.02	1.92	1.80
	60	2.50	2.35	2.20	2.12	2.03	1.94	1.84	1.73	1.60
	120	2.34	2.19	2.03	1.95	1.86	1.76	1.66	1.53	1.38
	∞	2.18	2.04	1.88	1.79	1.70	1.59	1.47	1.32	1.00

TABLE X
Critical Values of the F-distribution ($\alpha = 0.01$)

		1	2	3	4	5	6	7	8	9	10
						Degrees of Freedom for Numerator					
	1	4,052	5,000	5,403	5,625	5,764	5,859	5,928	5,982	6,023	6,056
	2	98.5	99.0	99.2	99.2	99.3	99.3	99.4	99.4	99.4	99.4
	3	34.1	30.8	29.5	28.7	28.2	27.9	27.7	27.5	27.3	27.2
	4	21.2	18.0	16.7	16.0	15.5	15.2	15.0	14.8	14.7	14.5
	5	16.3	13.3	12.1	11.4	11.0	10.7	10.5	10.3	10.2	10.1
	6	13.7	10.9	9.78	9.15	8.75	8.47	8.26	8.10	7.98	7.87
	7	12.2	9.55	8.45	7.85	7.46	7.19	6.99	6.84	6.72	6.62
	8	11.3	8.65	7.59	7.01	6.63	6.37	6.18	6.03	5.91	5.81
	9	10.6	8.02	6.99	6.42	6.06	5.80	5.61	5.47	5.35	5.26
	10	10.0	7.56	6.55	5.99	5.64	5.39	5.20	5.06	4.94	4.85
	11	9.65	7.21	6.22	5.67	5.32	5.07	4.89	4.74	4.63	4.54
	12	9.33	6.93	5.95	5.41	5.06	4.82	4.64	4.50	4.39	4.30
Degrees	13	9.07	6.70	5.74	5.21	4.86	4.62	4.44	4.30	4.19	4.10
of	14	8.86	6.51	5.56	5.04	4.70	4.46	4.28	4.14	4.03	3.94
Freedom	15	8.68	6.36	5.42	4.89	4.56	4.32	4.14	4.00	3.89	3.80
for	16	8.53	6.23	5.29	4.77	4.44	4.20	4.03	3.89	3.78	3.69
Denominator	17	8.40	6.11	5.19	4.67	4.34	4.10	3.93	3.79	3.68	3.59
	18	8.29	6.01	5.09	4.58	4.25	4.01	3.84	3.71	3.60	3.51
	19	8.19	5.93	5.01	4.50	4.17	3.94	3.77	3.63	3.52	3.43
	20	8.10	5.85	4.94	4.43	4.10	3.87	3.70	3.56	3.46	3.37
	21	8.02	5.78	4.87	4.37	4.04	3.81	3.64	3.51	3.40	3.31
	22	7.95	5.72	4.82	4.31	3.99	3.76	3.59	3.45	3.35	3.26
	23	7.88	5.66	4.76	4.26	3.94	3.71	3.54	3.41	3.30	3.21
	24	7.82	5.61	4.72	4.22	3.90	3.67	3.50	3.36	3.26	3.17
	25	7.77	5.57	4.68	4.18	3.86	3.63	3.46	3.32	3.22	3.13
	30	7.56	5.39	4.51	4.02	3.70	3.47	3.30	3.17	3.07	2.98
	40	7.31	5.18	4.31	3.83	3.51	3.29	3.12	2.99	2.89	2.80
	60	7.08	4.98	4.13	3.65	3.34	3.12	2.95	2.82	2.72	2.63
	120	6.85	4.79	3.95	3.48	3.17	2.96	2.79	2.66	2.56	2.47
	∞	6.63	4.61	3.78	3.32	3.02	2.80	2.64	2.51	2.41	2.32

Table X Critical Values of the F-distribution $(\alpha = 0.01)$ **A.25**

TABLE X
Critical Values of the F-distribution $(\alpha = 0.01)$

		Degrees of Freedom for Numerator								
		12	15	20	24	30	40	60	120	∞
	1	244	246	248	249	250	251	252	253	254
	2	19.4	19.4	19.4	19.5	19.5	19.5	19.5	19.5	19.5
	3	8.74	8.70	8.66	8.64	8.62	8.59	8.57	8.55	8.53
	4	5.91	5.86	5.80	5.77	5.75	5.72	5.69	5.66	5.63
	5	4.68	4.62	4.56	4.53	4.50	4.46	4.43	4.40	4.37
	6	4.00	3.94	3.87	3.84	3.81	3.77	3.74	3.70	3.67
	7	3.57	3.51	3.44	3.41	3.38	3.34	3.30	3.27	3.23
	8	3.28	3.22	3.15	3.12	3.08	3.04	3.01	2.97	2.93
	9	3.07	3.01	2.94	2.90	2.86	2.83	2.79	2.75	2.71
	10	2.91	2.85	2.77	2.74	2.70	2.66	2.62	2.58	2.54
	11	2.79	2.72	2.65	2.61	2.57	2.53	2.49	2.45	2.40
	12	2.69	2.62	2.54	2.51	2.47	2.43	2.38	2.34	2.30
	13	2.60	2.53	2.46	2.42	2.38	2.34	2.30	2.25	2.21
Degrees	14	2.53	2.46	2.39	2.35	2.31	2.27	2.22	2.18	2.13
of	15	2.48	2.40	2.33	2.29	2.25	2.20	2.16	2.11	2.07
Freedom										
for	16	2.42	2.35	2.28	2.24	2.19	2.15	2.11	2.06	2.01
Denominator	17	2.38	2.31	2.23	2.19	2.15	2.10	2.06	2.01	1.96
	18	2.34	2.27	2.19	2.15	2.11	2.06	2.02	1.97	1.92
	19	2.31	2.23	2.16	2.11	2.07	2.03	1.98	1.93	1.88
	20	2.28	2.20	2.12	2.08	2.04	1.99	1.95	1.90	1.84
	21	2.25	2.18	2.10	2.05	2.01	1.96	1.92	1.87	1.81
	22	2.23	2.15	2.07	2.03	1.98	1.94	1.89	1.84	1.78
	23	2.20	2.13	2.05	2.01	1.96	1.91	1.86	1.81	1.76
	24	2.18	2.11	2.03	1.98	1.94	1.89	1.84	1.79	1.73
	25	2.16	2.09	2.01	1.96	1.92	1.87	1.82	1.77	1.71
	30	2.09	2.01	1.93	1.89	1.84	1.79	1.74	1.68	1.62
	40	2.00	1.92	1.84	1.79	1.74	1.69	1.64	1.58	1.51
	60	1.92	1.84	1.75	1.70	1.65	1.59	1.53	1.47	1.39
	120	1.83	1.75	1.66	1.61	1.55	1.50	1.43	1.35	1.25
	∞	1.75	1.67	1.57	1.52	1.46	1.39	1.32	1.22	1.00

From E. S. Pearson and H. O. Hartley, *Biometrika Tables for Statisticians*, 1 (1958), 159–163. Reprinted by permission of the Biometrika Trustees.

TABLE XI
Critical Values for Total Number of Runs (Table Shows Critical Values for Two-tailed Test at $\alpha = 0.05$)

The Larger of n_1 and n_2

Each cell shows the lower critical value (top) over the upper critical value (bottom).

The Smaller of n_1 and n_2	5	6	7	8	9	10	11	12	13	14	15	16	17	18	19	20
2								2/6	2/6	2/6	2/6	2/6	2/6	2/6	2/6	2/6
3		2/8	2/8	2/8	2/8	2/8	2/8	2/8	2/8	2/8	3/8	3/8	3/8	3/8	3/8	3/8
4	2/9	2/9	2/10	3/10	3/10	3/10	3/10	3/10	3/10	3/10	3/10	4/10	4/10	4/10	4/10	4/10
5	2/10	3/10	3/11	3/11	3/12	3/12	4/12	4/12	4/12	4/12	4/12	4/12	4/12	5/12	5/12	5/12
6		3/11	3/12	3/12	4/13	4/13	4/13	4/13	5/14	5/14	5/14	5/14	5/14	5/14	6/14	6/14
7			3/13	4/13	4/14	5/14	5/14	5/14	5/15	5/15	5/15	6/16	6/16	6/16	6/16	6/16
8				4/14	5/14	5/15	5/15	6/16	6/16	6/16	6/16	6/17	7/17	7/17	7/17	7/17
9					5/15	5/16	6/16	6/16	6/17	7/17	7/18	7/18	7/18	8/18	8/18	8/18
10						6/16	6/17	7/17	7/18	7/18	7/18	8/19	8/19	8/19	8/20	9/20
11							7/17	7/18	7/19	8/19	8/19	8/20	9/20	9/20	9/21	9/21
12								7/19	8/19	8/20	8/20	9/21	9/21	9/21	10/22	10/22
13									8/20	9/20	9/21	9/21	10/22	10/22	10/23	10/23
14										9/21	9/22	10/22	10/23	10/23	11/23	11/24
15											10/22	10/23	11/23	11/24	11/24	12/25
16												11/23	11/24	11/25	12/25	12/25
17													11/25	12/25	12/26	13/26
18														12/26	13/26	13/27
19															13/27	13/27
20																14/28

From C. Eisenhart and F. Swed, "Tables for testing randomness of grouping in a sequence of alternatives," *The Annals of Statistics*, 14(1943), 66–87. Reprinted by permission.

TABLE XII
Critical Value of Spearman's Rank
Correlation Coefficient

	Level of Significance for One-tailed Test			
	0.05	0.025	0.01	0.005
	Level of Significance for Two-tailed Test			
n	0.10	0.05	0.02	0.01
5	0.900	1.000	1.000	——
6	0.829	0.886	0.943	1.000
7	0.714	0.786	0.893	0.929
8	0.643	0.738	0.833	0.881
9	0.600	0.683	0.783	0.833
10	0.564	0.648	0.745	0.794
11	0.523	0.623	0.736	0.818
12	0.497	0.591	0.703	0.780
13	0.475	0.566	0.673	0.745
14	0.457	0.545	0.646	0.716
15	0.441	0.525	0.623	0.689
16	0.425	0.507	0.601	0.666
17	0.412	0.490	0.582	0.645
18	0.399	0.476	0.564	0.625
19	0.388	0.462	0.549	0.608
20	0.377	0.450	0.534	0.591
21	0.368	0.438	0.521	0.576
22	0.359	0.428	0.508	0.562
23	0.351	0.418	0.496	0.549
24	0.343	0.409	0.485	0.537
25	0.336	0.400	0.475	0.526
26	0.329	0.392	0.465	0.515
27	0.323	0.385	0.456	0.505
28	0.317	0.377	0.448	0.496
29	0.311	0.370	0.440	0.487
30	0.305	0.364	0.432	0.478

TABLE XIII
The Exponential Function

x	e^{-x}	x	e^{-x}	x	e^{-x}	x	e^{-x}
.00	1.00000	**.40**	.67032	**.80**	.44933	**1.20**	.30119
.01	.99005	.41	.66365	.81	.44486	1.21	.29820
.02	.98020	.42	.65705	.82	.44043	1.22	.29523
.03	.97045	.43	.65051	.83	.43605	1.23	.29229
.04	.96079	.44	.64404	.84	.43171	1.24	.28938
.05	.95123	.45	.63763	.85	.42741	1.25	.28650
.06	.94176	.46	.63128	.86	.42316	1.26	.28365
.07	.93239	.47	.62500	.87	.41895	1.27	.28083
.08	.92312	.48	.61878	.88	.41478	1.28	.27804
.09	.91393	.49	.61263	.89	.41066	1.29	.25727
.10	.90484	**.50**	.60653	**.90**	.40657	**1.30**	.27253
.11	.89583	.51	.60050	.91	.40252	1.31	.26982
.12	.88692	.52	.59452	.92	.39852	1.32	.26714
.13	.87810	.53	.58860	.93	.39455	1.33	.26448
.14	.86936	.54	.58275	.94	.39063	1.34	.26185
.15	.86071	.55	.57695	.95	.38674	1.35	.25924
.16	.85214	.56	.57121	.96	.38289	1.36	.25666
.17	.84366	.57	.56553	.97	.37908	1.37	.25411
.18	.83527	.58	.55990	.98	.37531	1.38	.25158
.19	.82696	.59	.55433	.99	.37158	1.39	.24908
.20	.81873	**.60**	.54881	**1.00**	.36788	**1.40**	.24660
.21	.81058	.61	.54335	1.01	.36422	1.41	.24414
.22	.80252	.62	.53794	1.02	.36059	1.42	.24171
.23	.79453	.63	.53259	1.03	.35701	1.43	.23931
.24	.78663	.64	.52729	1.04	.35345	1.44	.23693
.25	.77880	.65	.52205	1.05	.34994	1.45	.23457
.26	.77105	.66	.51685	1.06	.34646	1.46	.23224
.27	.76338	.67	.51171	1.07	.34301	1.47	.22993
.28	.75578	.68	.50662	1.08	.33960	1.48	.22764
.29	.74826	.69	.51058	1.09	.33622	1.49	.22537
.30	.74082	**.70**	.49659	**1.10**	.33287	**1.50**	.22313
.31	.73345	.71	.49164	1.11	.32956	1.51	.22091
.32	.72615	.72	.48675	1.12	.32628	1.52	.21871
.33	.71892	.73	.48191	1.13	.32303	1.53	.21654
.34	.71177	.74	.47711	1.14	.31982	1.54	.21438
.35	.70469	.75	.47237	1.15	.31664	1.55	.21225
.36	.69768	.76	.46767	1.16	.31349	1.56	.21014
.37	.69073	.77	.46301	1.17	.31037	1.57	.20805
.38	.68386	.78	.45841	1.18	.30728	1.58	.20598
.39	.67706	.79	.45384	1.19	.30422	1.59	.20393

Table XIII The Exponential Function   **A.29**

TABLE XIII
The Exponential Function (continued)

x	e^{-x}	x	e^{-x}	x	e^{-x}	x	e^{-x}
1.60	.20190	**2.00**	.13534	**2.40**	.09072	**2.80**	.06081
1.61	.19989	2.01	.13399	2.41	.08982	2.81	.06020
1.62	.19790	2.02	.13266	2.42	.08892	2.82	.05961
1.63	.19593	2.03	.13134	2.43	.08804	2.83	.05901
1.64	.19398	2.04	.13003	2.44	.08716	2.84	.05843
1.65	.19205	2.05	.12873	2.45	.08629	2.85	.05784
1.66	.19014	2.06	.12745	2.46	.08543	2.86	.05727
1.67	.18825	2.07	.12619	2.47	.08458	2.87	.05670
1.68	.18637	2.08	.12493	2.48	.08374	2.88	.05613
1.69	.18452	2.09	.12369	2.49	.08291	2.89	.05558
1.70	.18268	**2.10**	.12246	**2.50**	.08208	**2.90**	.05502
1.71	.18087	2.11	.12124	2.51	.08127	2.91	.05448
1.72	.17907	2.12	.12003	2.52	.08046	2.92	.05393
1.73	.17728	2.13	.11884	2.53	.07966	2.93	.05340
1.74	.17552	2.14	.11765	2.54	.07887	2.94	.05287
1.75	.17377	2.15	.11648	2.55	.07808	2.95	.05234
1.76	.17204	2.16	.11533	2.56	.07730	2.96	.05182
1.77	.17033	2.17	.11418	2.57	.07654	2.97	.05130
1.78	.16864	2.18	.11304	2.58	.07577	2.98	.05079
1.79	.16696	2.19	.11192	2.59	.07502	2.99	.05029
1.80	.16530	**2.20**	.11080	**2.60**	.07427	**3.00**	.04979
1.81	.16365	2.21	.10970	2.61	.07353	3.01	.04929
1.82	.16203	2.22	.10861	2.62	.07280	3.02	.04880
1.83	.16041	2.23	.10753	2.63	.07208	3.03	.04832
1.84	.15882	2.24	.10646	2.64	.07136	3.04	.04783
1.85	.15724	2.25	.10540	2.65	.07065	3.05	.04736
1.86	.15567	2.26	.10435	2.66	.06995	3.06	.04689
1.87	.15412	2.27	.10331	2.67	.06925	3.07	.04642
1.88	.15259	2.28	.10228	2.68	.06856	3.08	.04596
1.89	.15107	2.29	.10127	2.69	.06788	3.09	.04550
1.90	.14957	**2.30**	.10026	**2.70**	.06721	**3.10**	.04505
1.91	.14808	2.31	.09926	2.71	.06654	3.11	.04460
1.92	.14661	2.32	.09827	2.72	.06587	3.12	.04416
1.93	.14515	2.33	.09730	2.73	.06522	3.13	.04372
1.94	.14370	2.34	.09633	2.74	.06457	3.14	.04328
1.95	.14227	2.35	.09537	2.75	.06393	3.15	.04285
1.96	.14086	2.36	.09442	2.76	.06329	3.16	.04243
1.97	.13946	2.37	.09348	2.77	.06266	3.17	.04200
1.98	.13807	2.38	.09255	2.78	.06204	3.18	.04159
1.99	.13670	2.39	.09163	2.79	.06142	3.19	.04117

TABLE XIII
The Exponential Function (continued)

x	e^{-x}	x	e^{-x}	x	e^{-x}	x	e^{-x}
3.20	.04076	**3.60**	.02732	**4.00**	.01832	**4.40**	.01228
3.21	.04036	3.61	.02705	4.01	.01813	4.41	.01216
3.22	.03996	3.62	.02678	4.02	.01795	4.42	.01203
3.23	.03956	3.63	.02652	4.03	.01777	4.43	.01191
3.24	.03916	3.64	.02625	4.04	.01760	4.44	.01180
3.25	.03877	3.65	.02599	4.05	.01742	4.45	.01168
3.26	.03839	3.66	.02573	4.06	.01725	4.46	.01156
3.27	.03801	3.67	.02548	4.07	.01708	4.47	.01145
3.28	.03763	3.68	.02522	4.08	.01691	4.48	.01133
3.29	.03725	3.69	.02497	4.09	.01674	4.49	.01122
3.30	.03688	**3.70**	.02472	**4.10**	.01657	**4.50**	.01111
3.31	.03652	3.71	.02448	4.11	.01641	4.51	.01100
3.32	.03615	3.72	.02423	4.12	.01624	4.52	.01089
3.33	.03579	3.73	.02399	4.13	.01608	4.53	.01078
3.34	.03544	3.74	.02375	4.14	.01592	4.54	.01067
3.35	.03508	3.75	.02352	4.15	.01576	4.55	.01057
3.36	.03474	3.76	.02328	4.16	.01561	4.56	.01046
3.37	.03439	3.77	.02305	4.17	.01545	4.57	.01036
3.38	.03405	3.78	.02282	4.18	.01530	4.58	.01025
3.39	.03371	3.79	.02260	4.19	.01515	4.59	.01015
3.40	.03337	**3.80**	.02237	**4.20**	.01500	**4.60**	.01005
3.41	.03304	3.81	.02215	4.21	.01485	4.61	.00995
3.42	.03271	3.82	.02193	4.22	.01470	4.62	.00985
3.43	.03239	3.83	.02171	4.23	.01455	4.63	.00975
3.44	.03206	3.84	.02149	4.24	.01441	4.64	.00966
3.45	.03175	3.85	.02128	4.25	.01426	4.65	.00956
3.46	.03143	3.86	.02107	4.26	.01412	4.66	.00947
3.47	.03112	3.87	.02086	4.27	.01398	4.67	.00937
3.48	.03081	3.88	.02065	4.28	.01384	4.68	.00928
3.49	.03050	3.89	.02045	4.29	.01370	4.69	.00919
3.50	.03020	**3.90**	.02024	**4.30**	.01357	**4.70**	.00910
3.51	.02990	3.91	.02004	4.31	.01343	4.71	.00900
3.52	.02960	3.92	.01984	4.32	.01330	4.72	.00892
3.53	.02930	3.93	.01964	4.33	.01317	4.73	.00883
3.54	.02901	3.94	.01945	4.34	.01304	4.74	.00874
3.55	.02872	3.95	.01925	4.35	.01291	4.75	.00865
3.56	.02844	3.96	.01906	4.36	.01278	4.76	.00857
3.57	.02816	3.97	.01887	4.37	.01265	4.77	.00848
3.58	.02788	3.98	.01869	4.38	.01253	4.78	.00840
3.59	.02760	3.99	.01850	4.39	.01240	4.79	.00831

Table XIII The Exponential Function   **A.31**

TABLE XIII
The Exponential Function (continued)

x	e^{-x}	x	e^{-x}	x	e^{-x}
4.80	.00823	**6.00**	.00248	**9.00**	.00012
4.81	.00815	6.10	.00244	9.10	.00011
4.82	.00807	6.20	.00203	9.20	.00010
4.83	.00799	6.30	.00184	9.30	.00009
4.84	.00791	6.40	.00166	9.40	.00008
4.85	.00783	6.50	.00150	9.50	.00007
4.86	.00775	6.60	.00136	9.60	.00007
4.87	.00767	6.70	.00123	9.70	.00006
4.88	.00760	6.80	.00111	9.80	.00006
4.89	.00752	6.90	.00101	9.90	.00005
4.90	.00745	**7.00**	.00091	**10.00**	.00005
4.91	.00737	7.10	.00083	10.10	.00004
4.92	.00730	7.20	.00075	10.20	.00004
4.93	.00723	7.30	.00068	10.30	.00003
4.94	.00715	7.40	.00061	10.40	.00003
4.95	.00708	7.50	.00055	10.50	.00003
4.96	.00701	7.60	.00050	10.60	.0002
4.97	.00694	7.70	.00045	10.70	.00002
4.98	.00687	7.80	.00041	10.80	.00002
4.99	.00681	7.90	.00037	10.90	.00002
5.00	.00674	**8.00**	.00034	**11.00**	.00002
5.10	.00610	8.10	.00030	11.10	.00002
5.20	.00552	8.20	.00027	11.20	.00001
5.30	.00499	8.30	.00025	11.30	.00001
5.40	.00452	8.40	.00022	11.40	.00001
5.50	.00409	8.50	.00020	11.50	.00001
5.60	.00370	8.60	.00018	11.60	.00001
5.70	.00335	8.70	.00017	11.70	.00001
5.80	.00303	8.80	.00015	11.80	.00001
5.90	.00274	8.90	.00014	11.90	.00001

Answers
to Selected
Exercises

CHAPTER 1

Section 1.5 (pages 12–14)

1. Choice (d) **2.** No. Both factors may not be related.

3. Not necessarily **4.** Not necessarily **5.** Choice (b)

6. Do not necessarily agree **7.** No **8.** Both **9.** No

10. Not necessarily. Other factors have to be considered. **11.** No

Mastery Tests: Form A (pages 15–17)

1. Choice (d)

2. The first paragraph detailing the number of pints of blood donated last week.

3. The second paragraph **4.** Choice (d) **5.** No

6. Medical records indicate that this is probably true.

7. Descriptive statistics **8.** Descriptive statistics

9. Inferential statistics **10.** No

Mastery Tests: Form B (pages 17–19)

1. No

2. The tobacco industry claims that other factors (hitherfore undetected), may be causing cancer.

3. No. They only received a 5% increase on what they were currently earning.

4. No **5.** Not necessarily **6.** Not necessarily

7. The first paragraph **8.** The second paragraph **9.** Yes

10. Not necessarily. More research is needed.

CHAPTER 2

Section 2.2 (pages (32–36)

1.

Ages (years)	Tally	Frequency
18–23	𝍸 𝍸 𝍸 ||	17
24–29	𝍸 𝍸 𝍸 |	16
30–35	𝍸 𝍸 𝍸 |||	18
36–41	𝍸 𝍸	10
42–47	𝍸 𝍸	10
48–54	𝍸 |||	8
55–60	||||	4
61–66	𝍸	5
67–73	𝍸 𝍸 |	11
74–79	|	1
		100

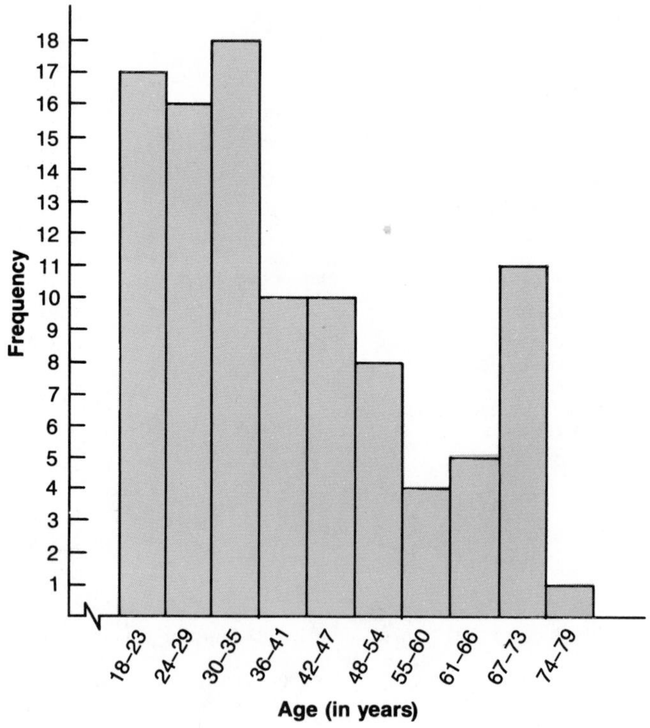

3.

Score	Tally	Frequency
7.7–14.5	|||	3
14.6–21.4	⊮ ⊮	10
21.5–28.3	||	2
28.4–35.2	⊮		6
35.3–42.1	⊮ ||	7
42.2–49.0	|||	3
49.1–55.9	⊮	5
56.0–62.8	⊮	5
62.9–69.7	||	2
69.8–76.6	⊮ ||	7
		50

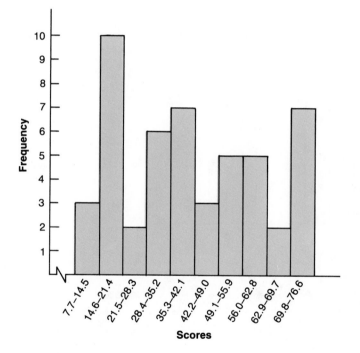

5.

Price of House	Tally	Frequency
$120,000–135,499	卌	5
135,500–150,999	卌	5
151,000–166,499	|||	3
166,500–181,999	卌 |||	8
182,000–197,499	|||	3
197,500–212,999	卌	5
213,000–228,499	||||	4
228,500–243,999		0
244,000–259,499	||||	4
259,500–275,000	卌	<u>5</u>
		42

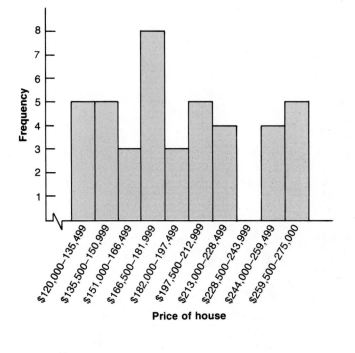

7. a. The frequency distribution using 10 classes.

Number of Licenses	Tally	Frequency
50–76	卌 |||	8
77–103	卌 |||	8
104–130	|||	3
131–157	卌 |||	8
158–184	||||	4
185–211	卌 |||	8
212–238	||	2
239–265	||	2
266–292	卌		6
293–319	卌 卌		11
		60

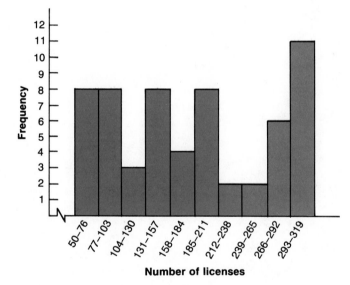

b. The frequency distribution using 5 classes.

Number of Licenses	Tally	Frequency
50–103	⌗ ⌗ ⌗ |	16
104–157	⌗ ⌗ |	11
158–211	⌗ ⌗ ||	12
212–265	||||	4
266–319	⌗ ⌗ ⌗ ||	17
		60

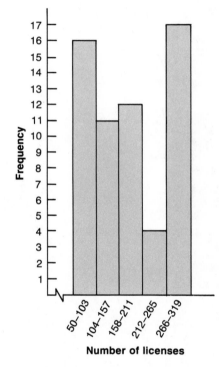

c. The frequency distribution using 15 classes.

Number of Licenses	Tally	Frequency
50–67	⅂⅂⅂⅂	5
68–85	⅂⅂⅂⅂	5
86–103	⅂⅂⅂⅂ |	6
104–121		0
122–139	⅂⅂⅂⅂	5
140–157	⅂⅂⅂⅂ |	6
158–175	|||	3
176–193	⅂⅂⅂⅂ |	6
194–211	|||	3
212–229	|	1
230–247	|	1
248–265	||	2
266–283	||||	4
284–301	⅂⅂⅂⅂ ⅂⅂⅂⅂ ||	12
302–319	|	1
		60

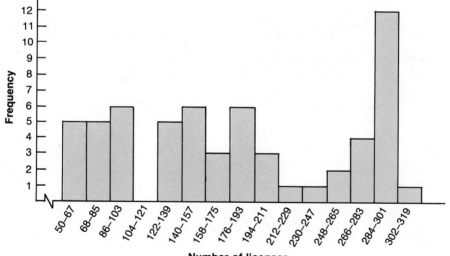

9. a.

Number of Products Completed	Tally	Frequency
10–12	|	1
13–15	||	2
16–18	|||||	5
19–21	|||||	5
22–24		0
25–27	|	1
28–30	||||	4
31–33		0
34–36	|	1
37–39	|	1
		20

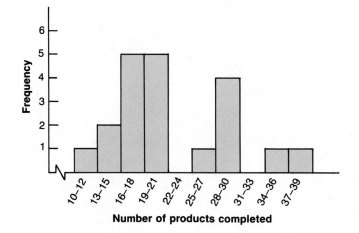

11. To avoid misinterpreting the data.

13. a. 26%   **b.** 19%   **c.** 55%   **d.** 11%   **e.** 89%

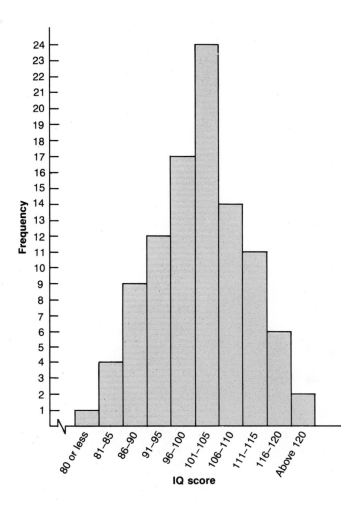

15. a.

Number of Stocks	Tally	Frequency
1–6	ⅢⅢ |||	8
7–12	ⅢⅢ	5
13–18	ⅢⅢ ⅢⅢ |||	13
19–24	ⅢⅢ ⅢⅢ ⅢⅢ ||||	19
25–30	ⅢⅢ	5
		50

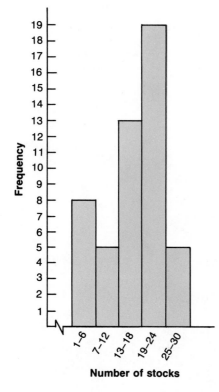

b.

Number of Stocks	Tally	Frequency
1–2	||	2
3–4	|||	3
5–6	|||	3
7–8	|||	3
9–10	||	2
11–12		0
13–14	|||	3
15–16	ǁǁǁ ||	7
17–18	|||	3
19–20	||	2
21–22	ǁǁǁ ||	7
23–24	ǁǁǁ ǁǁǁ	10
25–26		0
27–28	||	2
29–30	|||	3
		50

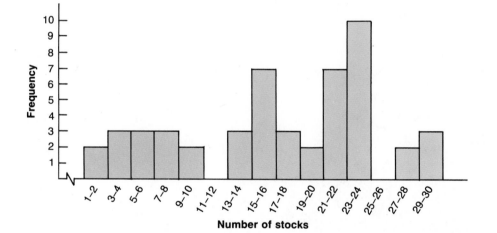

Section 2.3 (pages 56–63)

1.

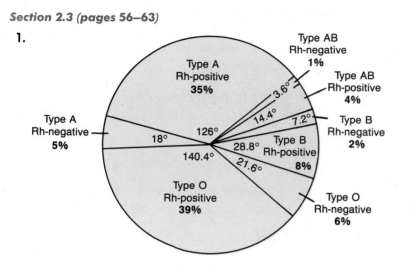

3. a. 1792.8 **b.** 475.2 **c.** 574.56 **d.** 1477.44

5.

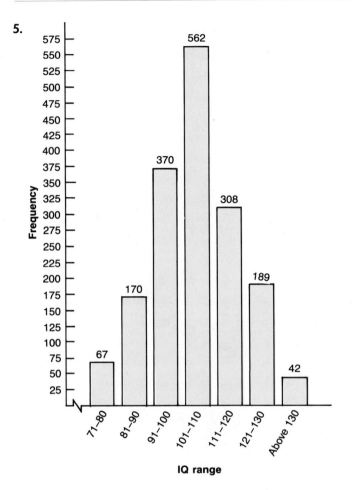

6. a. Humber Bridge 1400; Verrazano Narrows Bridge 1300; Golden Gate Bridge 1300; Mackinac Straits Bridge 1100; Bosporus Bridge 1100; George Washington Bridge 1000.

   b. 150 meters   **c.** 175 meters

7. a. 1975   **b.** Between 1983 and 1984

   c. From a low of 30¢ in 1975 to a high of $1.60 in 1984.

9.

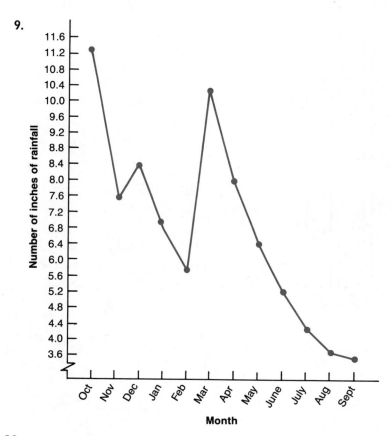

10. a. 140   **b.** 665

11.

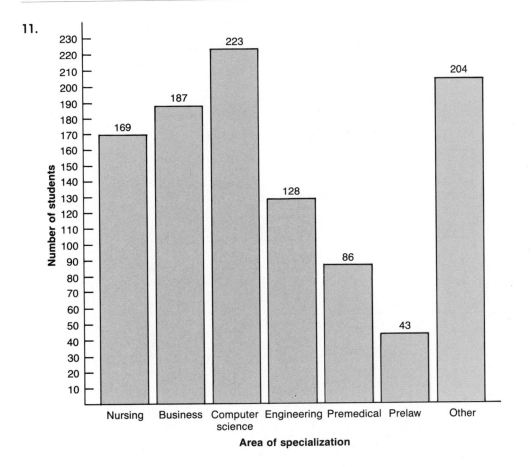

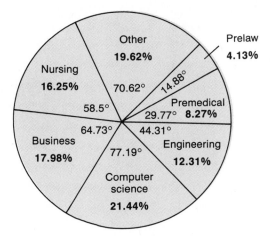

13. a. China 850 million; India 625 million; Soviet Union 250 million; United States 225 million

    b. 600 million

15. a. Likely   **b.** Likely   **c.** Not likely   **d.** Not likely   **e.** Not likely

Section 2.4 (pages 66–68)

1.

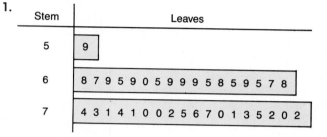

Stem	Leaves
5	9
6	8 7 9 5 9 0 5 9 9 9 5 8 5 9 5 7 8
7	4 3 1 4 1 0 0 2 5 6 7 0 1 3 5 2 0 2

3.

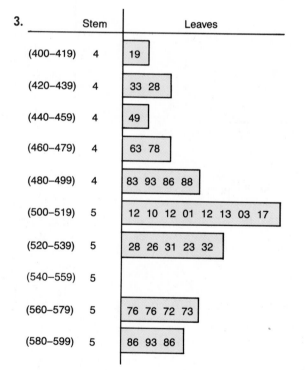

	Stem	Leaves
(400–419)	4	19
(420–439)	4	33 28
(440–459)	4	49
(460–479)	4	63 78
(480–499)	4	83 93 86 88
(500–519)	5	12 10 12 01 12 13 03 17
(520–539)	5	28 26 31 23 32
(540–559)	5	
(560–579)	5	76 76 72 73
(580–599)	5	86 93 86

5.

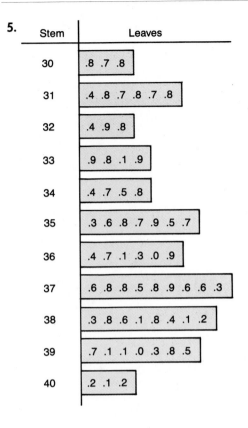

Stem	Leaves
30	.8 .7 .8
31	.4 .8 .7 .8 .7 .8
32	.4 .9 .8
33	.9 .8 .1 .9
34	.4 .7 .5 .8
35	.3 .6 .8 .7 .9 .5 .7
36	.4 .7 .1 .3 .0 .9
37	.6 .8 .8 .5 .8 .9 .6 .6 .3
38	.3 .8 .6 .1 .8 .4 .1 .2
39	.7 .1 .1 .0 .3 .8 .5
40	.2 .1 .2

Section 2.5 (page 71)

1. Both graphs are statistically correct, but because different spacings are used on the vertical scale, the graphs appear different.

2. When you double the diameter of a swimming pool, its area becomes four times as great.

Section 2.6 (pages 75–76)

1. 129.36

3. Price index for 1982 is 91.49. This indicates that in 1982 prices were 8.51% less when compared to 1985.

5.

Year	Cost Index	Comment
1981	100.00	
1982	114.58	This indicates an increase of 14.58% when compared to 1981.
1983	123.96	This indicates an increase of 23.96% when compared to 1981.
1984	145.83	This indicates an increase of 45.83% when compared to 1981.
1985	181.67	This indicates an increase of 81.67% when compared to 1981.

7.

Year	Cost Index	Comment
1981	100.00	
1982	106.88	This indicates an increase of 6.88% when compared to 1981.
1983	107.81	This indicates an increase of 7.81% when compared to 1981.
1984	110.00	This indicates an increase of 10% when compared to 1981.
1985	111.88	This indicates an increase of 11.88% when compared to 1981.

9. This means a decrease of 9% when compared to January 1982.

10. Decrease

Mastery Tests: Form A (pages 80–82)

1. 15

2.

Interval	Frequency	Relative Frequency
91–100	5	$\frac{5}{16}$ or 0.3125
81–90	6	$\frac{6}{16}$ or 0.375
71–80	2	$\frac{2}{16}$ or 0.125
61–70	2	$\frac{2}{16}$ or 0.125
51–60	1	$\frac{1}{16}$ or 0.0625

3.

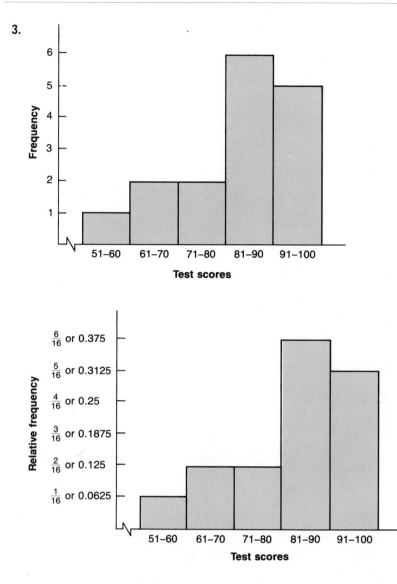

4.

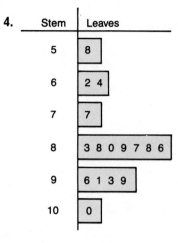

Stem	Leaves
5	8
6	2 4
7	7
8	3 8 0 9 7 8 6
9	6 1 3 9
10	0

5. Choice (a)   **6.** Choice (c)   **7.** Choice (c)

8.

Number of Calls	Tally	Frequency
16–19 |卌 |||	8
20–23	|||	3
24–27	|||	3
28–31	||||	4
32–35	||||	4
36–39	||||	4
40–43	||	2
44–47	||	2
48–51		0
52–56			1
		31

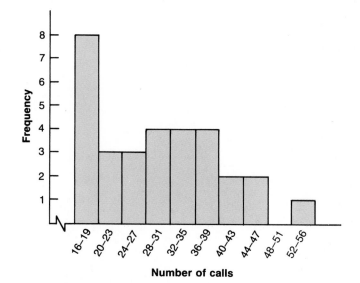

9.

Year	Number of Alarms Index
1980	100
1981	115.48
1982	133.33
1983	125.00
1984	97.62
1985	84.52

10.

Year	Cost Index
1980	81.63
1981	89.80
1982	85.71
1983	87.76
1984	93.88
1985	100.00

Mastery Tests: Form B (pages 82–85)

1.

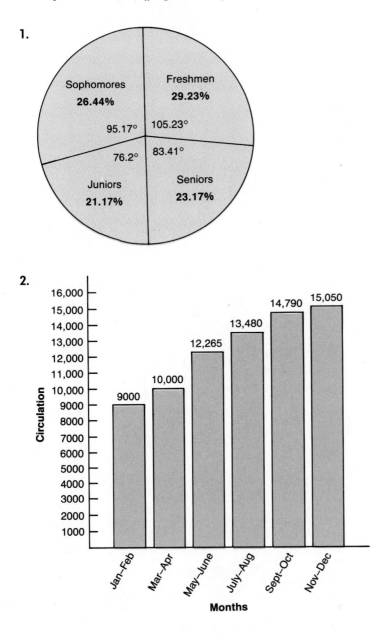

2.

3.

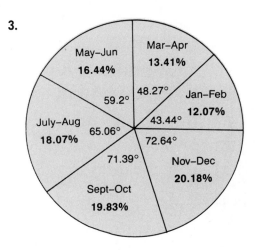

4. Technically, both are correct. However, the vertical scales are very misleading.

5. The area of the larger circle should be twice as large as the area of the smaller circle.

6.

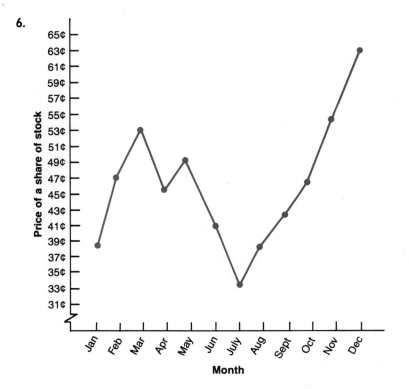

7.

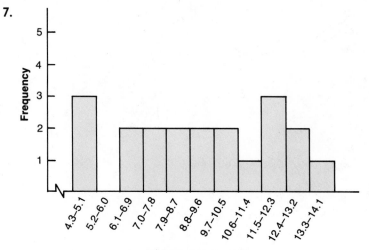

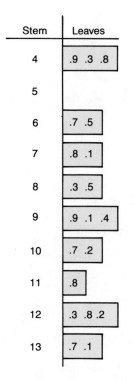

8.

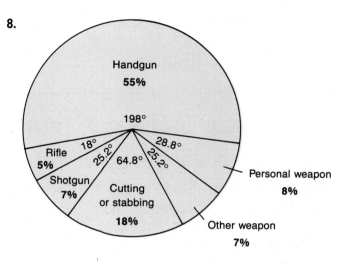

9. The number of low-income families is not the same as the number of high-income families.

10. a. Likely   **b.** Likely   **c.** Likely

CHAPTER 3

Section 3.2 (page 94)

1. a. $\sum_{i=1}^{7} y_i$ **b.** $\sum_{i=1}^{7} y_i^2$ **c.** $\sum_{i=1}^{7} y_i f_i$ **d.** $\sum_{i=1}^{7} 17 y_i$

e. $\sum_{i=1}^{n} (2y_i + x_i)$

3. a. 68   **b.** 4624   **c.** 1156   **d.** 73   **e.** 1297   **f.** 151   **g.** 5485

5. a. 105   **b.** 1120

Section 3.3 (pages 106–109)

1. Mean = 16.5; median = 15.5; mode 17

3. Mean = $19,477.78; median = $18,800; mode = $18,600

5. Mean = 185; median = 191.5; mode = 167 and 195

7. 3.23   **9.** $35.85

11. Mean = 16.6; We can't tell exactly what the mode equals. We can only say that it is between 16 and 20 cars; median = 16.66 (rounded off).

13. a. Manufacturer A is using the mean; Manufacturer B is using the mode; Manufacturer C is using the median.

b. It depends upon the individual.

15. Harmonic mean $= \dfrac{16}{9}$ or 1.78; Geometric mean $= 2$

Section 3.5 *(pages 118–121)*

1. a. Range $= 88$ pounds

b. Sample variance $= 622.24$; Sample standard deviation $= 24.945$

3. Sample mean $= 10.9$; Sample variance $= 6.767$; Sample standard deviation $= 2.601$; Average deviation $= 2.1$.

5. Population variance $= 221.438$; Population standard deviation $= 14.881$; Average deviation $= 12.25$

7. New sample mean $= 40.2$; New sample variance $= 182.7$; New sample standard deviation $= 13.517$; New average deviation $= 11.04$; The mean, standard deviation and average deviation are three times as great. The other measures are nine times as great.

9. Population variance $= 63.856$; Population standard deviation $= 7.991$

11. a. Population mean $= \$1.127$; Population standard deviation $= 0.088$

b. Population mean $= \$1.127$; Population standard deviation $= 0.088$

c. Both methods yield the same results.

13. The standard deviation for the middle-class neighborhoods is greater.

Section 3.6 *(page 124)*

1. When $k = 2$ then at least $\dfrac{3}{4}$ of the numbers are between $52.8 - 2\,(13.681)$ and $52.8 + 2\,(13.681)$, or between 25.438 and 80.162. Also when $k = 3$ then at least $\dfrac{8}{9}$ of the numbers will be between $52.8 - 3\,(13.681)$ and $52.8 + 3\,(13.681)$, or between 11.757 and 93.843.

3. 69.81 percent. $k = 1.82$

Section 3.8 *(pages 136–140)*

1. Michele $= $ 60th percentile; Ricardo $= $ 85th percentile

3. 43rd percentile

5. a.

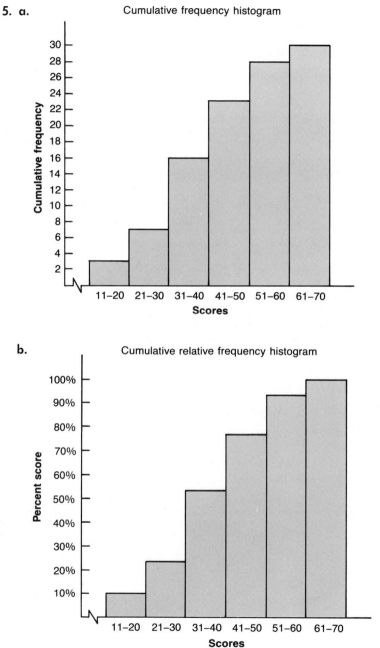

Cumulative frequency histogram

b.

Cumulative relative frequency histogram

7. a. 151.27   **b.** 115.91   **c.** 79.12

8. a. Anthony, Cathy, Joe, Cesar, Aida, Greta

   b. Cesar, Aida, Greta   **c.** Anthony, Cathy

10. Abellard, since his z-score is higher.

11. a. 140   **b.** 69%   **c.** 124.3

13. Pete = 118.622 minutes; Willie = 94.238 minutes; Tom = 104.754 minutes

14. He may not necessarily be accepted. We don't know what percentile rank an 89 corresponds to.

15. a. 15.87%   **b.** 15.87%   **c.** 97.73%

Mastery Tests: Form A (pages 144–145)

1. Choice (d)   **2.** Choice (e)   **3.** Choice (d)   **4.** Choice (c)

5. Choice (b)   **6.** Choice (c)   **7.** Choice (c)   **8.** Choice (a)

9. Choice (b)   **10.** Choice (d)

Mastery Tests: Form B (pages 145–148)

1. 69.9869   **2.** Choice (c)   **3.** Choice (a)   **4.** Choice (b)

5. Choice (d)   **6.** Choice (b)   **7.** Choice (a)   **8.** Choice (a)

9. Choice (c)

11. a. Lake C in New York and Lake P in New Jersey

   b. Lake F in New York and Lake O in New Jersey

13. Sample mean = 10; Sample standard deviation = 2.898

CHAPTER 4

Section 4.2 (pages 164–167)

1. a. $\dfrac{26}{200}$ or $\dfrac{13}{100}$ **b.** $\dfrac{21}{200}$ **c.** $\dfrac{10}{200}$ or $\dfrac{1}{20}$ **d.** $\dfrac{117}{200}$

3. $\dfrac{1}{4}$

5. $\dfrac{1}{720}$ **7.** $\dfrac{6}{76}$ or $\dfrac{3}{38}$

9. a. $\dfrac{60}{300}$ or $\dfrac{1}{5}$ **b.** $\dfrac{144}{300}$ or $\dfrac{12}{25}$ **c.** 0

11. a. $\dfrac{98}{228}$ or $\dfrac{49}{114}$

13. $\dfrac{1}{9}$ **14.** 0 **15.** $\dfrac{9}{39}$ or $\dfrac{3}{13}$

16. $\dfrac{6}{39}$ or $\dfrac{2}{13}$ **17.** $\dfrac{1}{6}$

Section 4.3 (pages 172–174)

1. 960 **3.** 20 **5.** 120

7. 132,600 **9.** 5040

11.

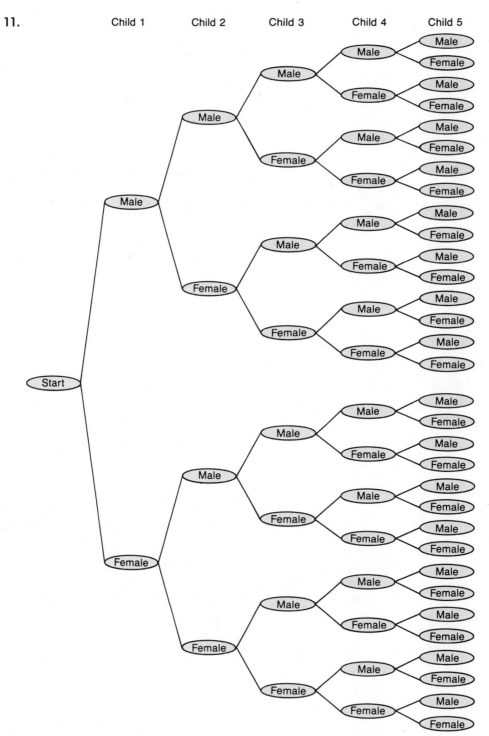

12. a. 5040   **b.** 3600   **c.** $\dfrac{1440}{5040}$ or $\dfrac{18}{63}$ **13.** 4536

15. 6

17. 18   **18.** $\dfrac{1}{120}$

Section 4.4 (pages 181–183)

1. a. 5040   **b.** 720   **c.** 2   **d.** 1   **e.** $\dfrac{1}{3}$ **f.** 6   **g.** 35   **h.** 20   **i.** 5040

j. 120   **k.** 30   **l.** 6   **m.** 5040   **n.** 2   **o.** 1   **p.** 1

3. $_{12}P_3 = 1320$

5. 40,320   **7. a.** $9! = 362,800$ **b.** $\dfrac{1}{9}$

9. $_{10}P_6 = 151,200$ **10.** $_6P_6 = 720$

11. $_{10}P_4 = 5040$ **13.** 720

15. 19,958,400

17. $720 \times 64 = 46,080$ **19.** 120 assuming "position" because of the smoke matters

Section 4.5 (pages 193–195)

1. a. 7   **b.** 15   **c.** 56   **d.** 1   **e.** 5   **f.** 84   **g.** 21

h. 70   **i.** 0 since it's impossible   **j.** 1

3. 84   **5.** 2,018,016

7. 133,784,560

9. 56,056

11. a. 36/190   **b.** 28/190   **c.** 6/190

15. a. 462   **b.** 462

   c. Since once we have determined the number of different ways in which the airlines can be selected, we already know how many ways the other airlines cannot be selected.

16. a. 816   **b.** 306   **c.** 18   **17.** $\dfrac{18}{1140}$ or $\dfrac{3}{190}$

Section 4.6 *(pages 200–201)*

1. $4.55   **3.** $42,750

5. Account I—$1035.78; Account II—986.46; Account III—888.54; Account IV—752.25; Account V—638.48

7. 7:2   **9.** 3:1

11. 7:1

Mastery Tests: Form A *(pages 203–205)*

1. Choice (a)   **2.** Choice (d)   **3.** Choice (b)   **4.** Choice (a)

5. Choice (b)   **6.** Choice (c)   **7.** Choice (a)   **8.** Choice (a)

9. Choice (b)   **10.** Choice (a)

11. a. 73,815/230,300 or 0.3205   **b.** 221,445/230,300 or 0.9616

   c. 495/230,300 or 0.0021

Mastery Tests: Form B *(pages 205–207)*

1. Choice (d)   **2.** Choice (c)   **3.** Choice (d)   **4.** Choice (b)

5. 77:23   **6. a.** 3:2   **b.** $3.40   **7.** 6/8 or 3/4

8. 1/5

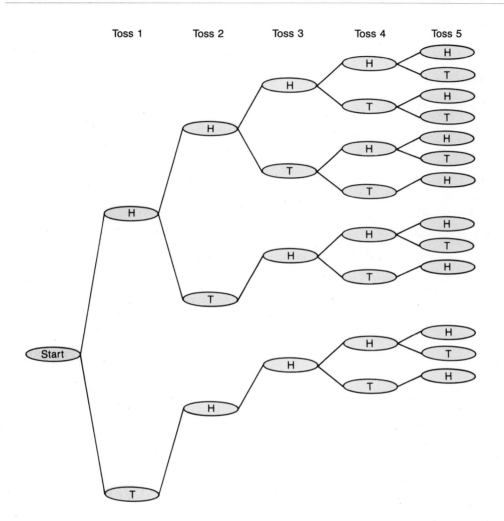

10. a. No. There are only 1000 codes possible.

   b. Yes. There are 17,576 codes possible.

11. The scheme using three letters followed by two numbers.

12. $-\$1161.11$ **13.** $1:5$ **14.** $\dfrac{1}{11,881,376}$ **15.** 210 (assuming order does not count)

16. 16,170   **17.** 210   **18.** 6720

CHAPTER 5

Section 5.2 (pages 219–222)

1. a. Not mutually exclusive   **b.** Not mutually exclusive

   c. Mutually exclusive   **d.** Mutually exclusive

   e. Not mutually exclusive

3. 0.86   **5.** 0.38

7. 0.83   **9.** $\dfrac{9}{14}$ **11.** $\dfrac{53}{60}$

13. $\dfrac{61}{360}$

Section 5.3 (pages 228–231)

1. $\dfrac{11}{68}$

3. a. $\dfrac{306}{1200}$ or $\dfrac{51}{200}$ **b.** $\dfrac{306}{364}$ or $\dfrac{153}{182}$ **c.** $\dfrac{306}{684}$ or $\dfrac{17}{38}$

5. $\dfrac{26}{58}$ or $\dfrac{13}{29}$ **7.** $\dfrac{11}{73}$

9. $\dfrac{69}{84}$ or $\dfrac{23}{28}$ **11.** $\dfrac{1}{5}$

13. a. $\dfrac{448}{1000}$ or $\dfrac{56}{125}$ **b.** $\dfrac{69}{200}$ **c.** $\dfrac{69}{448}$

Section 5.4 (pages 235–238)

1. 0.0456   **3.** 0.3732

5. 0.6603   **7.** 0.7790

9. a. $\dfrac{20}{9900}$ or $\dfrac{1}{495}$ **b.** $\dfrac{25}{10,000}$ or $\dfrac{1}{400}$

11. 0.1806   **13.** 0.2163   **15.** 0.0002

Section 5.5 (pages 246–247)

1. 0.2975   **3.** $\dfrac{3}{7}$ **5.** 0.3571

7. 0.1333   **9.** $\dfrac{8}{44}$ or 0.1818

Mastery Tests: Form A (pages 249–250)

1. a. $\dfrac{72}{310}$ or $\dfrac{36}{155}$ **b.** $\dfrac{126}{310}$ or $\dfrac{63}{155}$ **c.** $\dfrac{112}{310}$ or $\dfrac{56}{155}$

2. $\dfrac{1}{12}$ **3.** $\dfrac{6}{45}$ or $\dfrac{2}{15}$ **4.** $\dfrac{6}{14}$ or $\dfrac{3}{7}$ **5.** $\dfrac{56}{71}$

6. $\dfrac{43}{79}$ **7.** $\dfrac{1}{6}$ **8.** 0.6141   **9.** 0.7200   **10.** 0.8178

Mastery Tests: Form B (pages 250–253)

1. $\dfrac{3}{4}$ **2.** $\dfrac{20}{37}$ **3.** $\dfrac{23}{71}$ **4.** 0.1607   **5.** 0.4226

6. 0.5649   **7.** 0.9918   **8.** 0.9836   **9.** 0.0405   **10.** 0.02

CHAPTER 6

Section 6.2 (pages 263–265)

1. a. 1, 2, 3, . . . , 30   **b.** 0, 1, 2, . . .

c. 0, 1, 2, . . . , the capacity of the switchboard

d. 0, 1, 2, . . .

e. 0, 1, 2, . . . , the working population of the city

f. Any non-negative real number   **g.** Any non-negative real number

3. No. The sum of all the probabilities is more than 1.

5.

x	Prob (x)
0	$\frac{1}{8}$
1	$\frac{3}{8}$
2	$\frac{3}{8}$
3	$\frac{1}{8}$

7.

x	Prob (x)
0	$\frac{1}{16}$
1	$\frac{4}{16}$ or $\frac{1}{4}$
2	$\frac{6}{16}$ or $\frac{3}{8}$
3	$\frac{4}{16}$ or $\frac{1}{4}$
4	$\frac{1}{16}$

8.

x	Prob (x)	x	Prob (x)	x	Prob (x)
2	$\frac{1}{36}$	30	$\frac{2}{36}$	100	$\frac{1}{36}$
6	$\frac{2}{36}$	35	$\frac{2}{36}$	101	$\frac{2}{36}$
10	$\frac{1}{36}$	50	$\frac{1}{36}$	105	$\frac{2}{36}$
11	$\frac{2}{36}$	51	$\frac{2}{36}$	110	$\frac{2}{36}$
15	$\frac{2}{36}$	55	$\frac{2}{36}$	125	$\frac{2}{36}$
20	$\frac{1}{36}$	60	$\frac{2}{36}$	150	$\frac{2}{36}$
26	$\frac{2}{36}$	75	$\frac{2}{36}$	200	$\frac{1}{36}$

9.

x	Prob (x)
0	$\frac{9}{64}$
1	$\frac{30}{64}$ or $\frac{15}{32}$
2	$\frac{25}{64}$

Section 6.4 (pages 273–277)

1. Mean = 3.04; Variance = 2.8384; Standard deviation = 1.6848

3. $\mu = 2.85$; $\sigma^2 = 2.3675$; $\sigma = 1.5387$

5. $\mu = 8.36$

7. $\mu = 3.02$; $\sigma^2 = 2.6596$; $\sigma = 1.6308$

9. $\mu = 2.6667$; $\sigma = 1.4907$

11. **a.** $\mu = 2.97$; $\sigma = 1.2446$

 b. The interval $\mu \pm 2\sigma$ is 0.4808 to 5.4592. Using Chebyshev's Theorem with $k = 2$, then at least $\frac{3}{4}$ of the number of requests will lie within this interval. From the given information, Prob $(0.48 < x < 5.46)$ = Prob $(1 \leq x \leq 5)$ = 0.96.

Section 6.5 (pages 290–293)

1. 0.2945   3. 0.2097   5. 0.9619

7. 0.2616   9. 0.00086

11. 0.0055

13. **a.** 0.3108   **b.** 0.7033   **c.** 0.6075

15. **a.** 0.0156   **b.** 0.0156   **c.** 0.6562   **d.** 0.8907

Section 6.6 (pages 296–298)

1. 129; 8.575   3. 160; 9.798

5. 275.2; 6.207   7. 1800; 40.472   8. Yes

9. 255; 6.185

Section 6.7 (pages 300–301)

1. 0.3007

3. 0.1954   5. 0.7788

7. 0.5488

8. At least 8   9. 0.4232

Section 6.8 (pages 305–306)

1. 0.1758

3. **a.** 0.0128   **b.** 0.1282   **c.** 0.3590   **d.** 0.3590   **e.** 0.1282

5. 0.2847

7. 0.0166

9. 0.0959

Mastery Tests: Form A (pages 308–309)

1. Choice (a)   **2.** Choice (c)   **3.** Choice (d)   **4.** Choice (b)   **5.** Choice (a)

6. Choice (d)   **7.** Choice (d)   **8.** Choice (c)   **9.** Choice (a)   **10.** Choice (b)

Mastery Tests: Form B (pages 310–311)

1. 0.2545   **2.** 2.16   **3.** 1.0929   **4.** Approximately 1   **5.** 0.6201

6. 0.6873   **7.** 0.2341   **8.** 0.8647   **9.** At least 8   **10.** 0.4925

11.

$$\frac{\binom{12}{0}\binom{88}{10} + \binom{12}{1}\binom{88}{9} + \binom{12}{2}\binom{88}{8}}{\binom{100}{10}}$$

CHAPTER 7

Section 7.3 (pages 330–331)

1. a. 0.4664   **b.** 0.2357   **c.** 0.0239   **d.** 0.1949

e. 0.0068   **f.** 0.8917   **g.** 0.9361   **h.** 0.0687

3. a. -0.58 **b.** -0.95 **c.** -0.39 **d.** -1.64 or -1.65

5. a. 2.07   **b.** 2.18   **c.** -0.83 **d.** 2.68   **e.** 2.27   **f.** 2.93

7. a. 13.35th percentile   **b.** 92.51st percentile

c. 21.77th percentile   **d.** 97.72nd percentile

9. $z = -3$ becomes $x = 27$; $z = -2$ becomes $x = 34$; $z = -1$ becomes $x = 41$
$z = 0$ becomes $x = 48$; $z = +1$ becomes $x = 55$; $z = +2$ becomes $x = 62$
$z = +3$ becomes $x = 69$

11. 78.0728   **12.** $\mu = 41.6667$; $\sigma = 12.6667$

Section 7.4 (pages 335–337)

1. 0.1814   **3.** 0.1949   **5.** 3.25 years

7. 151.96   **9.** 4.1456

11. 276.4 minutes   **13.** 53.3%

Section 7.5 (pages 343–344)

1. a. 0.6368   **b.** 0.0107   **3.** 0.1075

5. a. 0.7486   **b.** 0.1155

7. a. 0.5832   **b.** 0.1184

9. a. 0.0724   **b.** Approximately 1

11. a. 0.8798   **b.** 0.0158   **c.** 0.1660

Mastery Tests: Form A (pages 347–348)

1. Choice (b)   **2.** Choice (c)   **3.** Choice (d)   **4.** Choice (c)

5. Choice (d)   **6.** Choice (c)   **7.** Choice (a)   **8.** Choice (b)

9. Choice (c)   **10.** Choice (e)

Mastery Tests: Form B (pages 348–349)

1. Choice (b)   **2.** 0.0113   **3.** 0.2119   **4.** 0.4972

5. 0.1492   **6.** 0.9188   **7. a.** 0.1271   **b.** 0.0328   **c.** 0.9772

8. a. 0.4052   **b.** 0.0336   **9.** 0.9066   **10.** 12,801.5

11. 159.2 minutes

CHAPTER 8

Section 8.2 (pages 357–358)

1. Those roofers whose numbers are 69, 19, 76, 92, and 41.

3. Those disk drives with the numbers 785, 612, 1887, 1745, 917, 1336, 1973, 1898, 402, 230, 163, 1488, 1059, 1637, 1810, 1391, 1630, 1988, 414, and 1451.

5. Those bonds with the numbers 7856, 6121, 9172, 4024, 2304, 1638, 4146, 6691, 256, 4542, 6115, and 9922.

7. Those customers whose credit card numbers are 36207, 34095, 32081, 57004, 60672, 15053, 48840, 60045, 12566, 17983, 31595, 20847, 08272, 26358, 85977, 53900, 65255, 85030, 64350, 46104, 22178, 06646, 06912, 41135, 67658, 14780, 12659, 66134, 64568, and 42607.

9. Those restaurants with the numbers 20, 166, 427, 79, 102, 332, 34, 81, 99, 143, 242, 73, 264, 129, 301, 73, 299, and 319.

Section 8.5 (pages 369–371)

1. a, b.

Years Selected	Sample Mean $\bar{x}$
6 and 10	8.0
6 and 4	5.0
6 and 7	6.5
6 and 9	7.5
6 and 8	7.0
10 and 4	7.0
10 and 7	8.5
10 and 9	9.5
10 and 8	9.0
4 and 7	5.5
4 and 9	6.5
4 and 8	6.0
7 and 9	8.0
7 and 8	7.5
9 and 8	8.5

c. 7.3333

d. 1.2472

3. a, b.

Numbers Selected	Sample Mean $\bar{x}$
5, 8, and 9	7.3333
5, 8, and 12	8.3333
5, 8, and 7	6.6667
5, 8, and 6	6.3333
8, 9, and 12	9.6667
8, 9, and 7	8.0000
8, 9, and 6	7.6667
9, 12, and 7	9.3333
9, 12, and 6	9.0000
5, 9, and 12	8.6667
5, 9, and 7	7.0000
5, 9, and 6	6.6667
5, 12, and 6	7.6667
5, 7, and 6	6.0000
9, 7, and 6	7.3333
12, 7, and 6	8.3333
8, 12, and 7	9.0000
8, 12, and 6	8.6667
8, 7, and 6	7.0000
5, 12, and 7	8.0000

c. $\mu_{\bar{x}} = 7.8333$

d. 1.0138

5. a. $\mu = 14$; $\sigma = 4.9396$

b.

Size 2		Size 3	
Numbers Selected	*Sample Mean*	*Numbers Selected*	*Sample Mean*
8 and 17	12.5	8, 17, and 12	12.3333
8 and 12	10.0	8, 17, and 11	12.0000
8 and 11	9.5	8, 17, and 22	15.6667
8 and 22	15.0	17, 12, and 11	13.3333
17 and 12	14.5	17, 12, and 22	17.0000
17 and 11	14.0	12, 11, and 22	15.0000
17 and 22	19.5	8, 12, and 11	10.3333
12 and 11	11.5	8, 12, and 22	14.0000
12 and 22	17.0	8, 11, and 22	13.6667
11 and 22	16.5	17, 11, and 22	16.6667

c. $\mu_{\bar{x}} = 14$, $\sigma_{\bar{x}} = 3.0249$; $\mu_{\bar{x}} = 14$; $\sigma_{\bar{x}} = 2.0166$

7. a. 1.2857 **b.** 0.9

Section 8.7 (pages 376–377)

1. 0.0094 **3.** 0.4911 **5.** 0.7487

7. Between 11.687 and 12.313 ounces

9. 0.8546

Mastery Tests: Form A (pages 379–380)

1. Those gardeners whose numbers are 19, 6, 21, 44, 11, and 27.

2, 3.

Numbers Selected	Sample Means
42 and 36	39.0
42 and 53	47.5
42 and 64	53.0
42 and 37	39.5
42 and 38	40.0
36 and 53	44.5
36 and 64	50.0
36 and 37	36.5
36 and 38	37.0
53 and 64	58.5
53 and 37	45.0
53 and 38	45.5
64 and 37	50.5
64 and 38	51.0
37 and 38	37.5

4. $\mu_{\bar{x}} = 45$ **5.** 6.4704 **6.** 0.0436

7. It becomes $\frac{1}{10}$ as much as it was before.   **8.** 0.0014

9. Those agents whose numbers are 20, 166, 79, 102, 34, 81, 99, and 143.   **10.** 0.2643

Mastery Tests: Form B (pages 380–382)

1. 0.9920   **2.** 0.0262   **3.** Between 15.944 and 16.256 inches

4. Approximately 0   **5.** Mean = 14.7; Standard deviation = 0.3833   **6.** 0.0336

7. 0.7487   **8. a.** 0.4013   **b.** 0.0401   **9.** 0.0016

10. No. The probability of this happening is approximately 0.

CHAPTER 9

Section 9.3 (pages 392–394)

1. Between 59.2948 and 62.7052

3. Between \$47,656.57 and \$49,343.43

5. Between \$1.2381 and \$1.2799

7. Between \$1.0259 and \$1.0921

9. Between 2.0446 and 2.1554

11. a. Between 13.4185 and 18.5815   **b.** Between 12.6019 and 19.3981

Section 9.4 (pages 348–400)

1. Between 43.1632 and 56.8368 hours

3. Between \$10.11 and \$12.89

5. Between 7085.3112 and 8914.6888 copies

7. Between \$974.04 and \$994.72

9. Between 40.1896 and 87.239

Section 9.6 (pages 404–405)

1. Between 9.8495 and 14.6517

3. Between 2.1165 and 3.9658

5. 3   **7.** 7   **9.** 29

Section 9.7 (pages 410–411)

1. 0.0019   **3.** 0.0721

5. Between 72.865% and 83.135%

7. 0.1611

9. Between 73.82% and 77.24%

Mastery Tests: Form A (pages 413–414)

1. 0.8002   **2.** Between 23.31% and 36.69%   **3.** 0.0655

4. Between 0.9813 and 1.3887 hours   **5.** 21

6. Between \$69.95 and \$118.60   **7.** 139   **8.** 0.3121

9. Between 15.119 and 19.881   **10.** Between 14.3654 and 20.6346

Mastery Tests: Form B (pages 414–416)

1. 0.0475   **2.** Between 60.1071 and 97.8929 pictures

3. Between 213.55 and 226.45 minutes   **4.** 139   **5.** 0.2578   **6.** 0.0346

7. Between \$4.44 and \$13.56   **8.** Between 6.7918 and 8.8082 hours

9. Between \$558.34 and \$611.66   **10.** 0.1492

CHAPTER 10

Section 10.4 (pages 435–436)

1. Yes ($z = 3.41$)   **3.** Yes ($z = 6.998$)

5. Yes ($z = 8.73$)

7. No ($z = 1.58$)   **9.** Yes ($z = 2.197$)

11. Not acceptable ($z = 8.94$)

Section 10.5 (pages 439–440)

1. Yes ($t = -6.67$)   **3.** Yes ($t = -3.12$)

5. No ($t = -1.32$)

7. Yes ($t = -5.16$)   **9.** No ($t = 1.67$)

Section 10.6 (pages 445–447)

1. Yes ($z = 3.03$)   **3.** Yes ($z = 3.10$)

5. Accept null hypothesis ($z = 1.50$)

7. Reject null hypothesis ($z = 6.18$)

9. Yes ($z = 35.89$)

Section 10.7 (pages 450–452)

1. Not significant ($t = 1.36$)

3. Not significant ($t = 2.07$)

5. Significant ($t = 3.7897$)

7. Yes, significant ($t = 3.086$)

9. Yes, significant ($t = 5.9453$)

Section 10.8 (pages 455–457)

1. No ($z = 1.49$) **3.** No ($z = -0.38$)

5. No ($z = 1.09$)

7. No ($z = -1.08$) **9.** Yes ($z = 2.61$)

11. Yes ($z = -2.57$)

Mastery Tests: Form A (pages 460–462)

1. No. Assuming we are using a 2-sided test ($z = 1.86$) **2.** Yes ($z = 2.465$) **3.** No ($z = 6.698$)

4. Yes ($z = -8.8956$) **5.** Yes ($z = -8.89$) **6.** Yes ($z = 5.547$)

7. Yes ($z = 5.9867$) **8.** Yes ($z = -4.045$) **9.** Yes ($z = 2.86$)

10. No ($t = 1.27$)

Mastery Tests: Form B (pages 462–464)

1. Yes ($t = 6.53$) **2.** No ($z = -2.26$) **3.** No ($t = 1.59$)

4. No ($t = 1.157$) **5.** Yes ($z = 3.82$) **6.** Yes ($z = -3.83$)

7. No ($z = 0.52$) **8.** No ($z = -1.57$) **9.** No ($t = 1.07$)

10. No ($z = 1.79$) **11.** Yes ($z = 4.63$)

CHAPTER 11

Section 11.3 (pages 477–482)

1. a. Zero correlation **b.** Positive correlation **c.** Positive correlation

d. Positive correlation **e.** Positive correlation **f.** Zero correlation

g. Positive correlation **h.** Zero correlation

3. a. Coefficient of correlation $= 0.957$ **b.** Yes

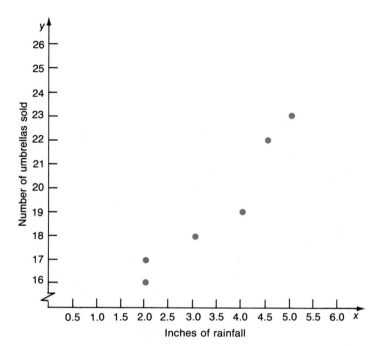

5. Coefficient of correlation $= 0.99$

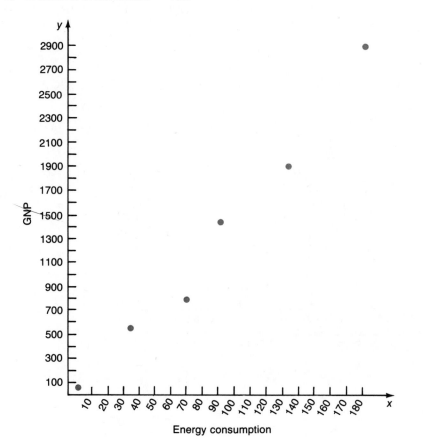

7. Coefficient of correlation = 0.639

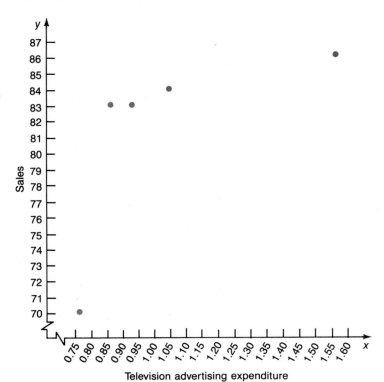

Television advertising expenditure

9. Coefficient of correlation $= 0.961$

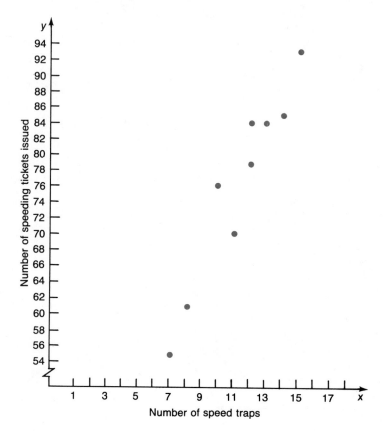

11. Coefficient of correlation = 0.992

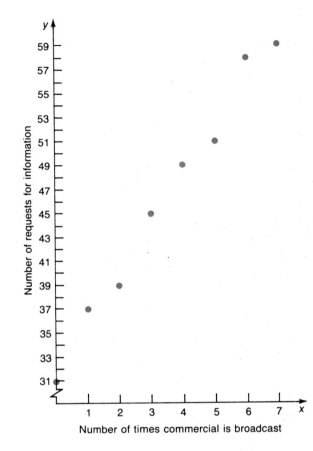

Number of requests for information

Number of times commercial is broadcast

Section 11.4 *(pages 483–484)*

1. Significant at the 5% level of significance

3. Significant at the 5% level of significance

5. Significant at the 5% level of significance

7. Significant at the 5% level of significance

9. Significant at the 5% level of significance

Section 11.6 *(pages 494–499)*

1. a. $\hat{y} = -9.61 + 0.729x$ **b.** 15.905 accidents

3. a. $\hat{y} = -6.89 + 0.833x$ **b.** 18.1 months

4. a. $\hat{y} = -28.35 + 1.384x$ **b.** 72.682 inches

5. a. $\hat{y} = 100.54 - 2.264x$ **b.** 73.372 complaints

6. a. $\hat{y} = 251128.4 - 9455.253x$ **b.** 128210.111

7. a. $\hat{y} = 6.32 + 2.704x$ **b.** 33.36

9. a. $\hat{y} = 39.82 - 0.739x$ **b.** 21.345

10. a. $\hat{y} = 49.27 + 2.246x$ **b.** 139.11

11. a. $\hat{y} = 0.09 + 11.89x$ **b.** 214.11

Section 11.7 (page 505)

1. 1.239

3. 0.443

4. 0.432

5. 2.228

6. 11943.854

7. 2.582

9. 2.961

10. 5.016

11. 8.919

Section 11.8 (page 511)

1. Between 12.1101 and 19.6999

3. Between 16.4823 and 19.7177

4. Between 71.3849 and 73.9791

5. Between 67.214 and 79.530

6. Between 89,627.9092 and 166,792.3128

7. Between 25.3861 and 41.3739

9. Between 12.4467 and 30.2433

10. Between 125.4876 and 152.7324

11. Between 188.1396 and 240.0804

Section 11.10 (pages 514–515)

1. a. $\hat{y} = 30.35024 + 0.3617x_1 + 0.5674x_2$ **b.** 82.37154

2. a. $\hat{y} = 10.020568 + 0.5913x_1 + 0.53702x_2$ **b.** 159.244

3. a. $\hat{y} = -4.6005 + 0.189x_1 + 0.005x_2$ **b.** 9.4405

Mastery Tests: Form A *(pages 520–522)*

1. Choice (b)

2. Choice (b)

3. Choice (a)

4. Choice (c)

5. 0.803

6. Between 6.8364 and 12.9236

7. Choice (b)

8. Choice (b)

9. No

10. Yes

Mastery Tests: Form B *(pages 522–525)*

1. $\hat{y} = 28.6475 - 3.8754x_1 + 0.9606x_2$

2. 23.576

3. **a.** -0.992 **b.** Yes

4. Between 46.5248 and 59.6052

5. **a.** 0.849   **b.** Yes

6. $\hat{y} = 11004.12 + 1642.59x$

7. 22.3

8. $\hat{y} = 10308.39 + 0.356x$

9. **a.** $\hat{y} = 60 + 0.5x$ **b.** Yes to some degree

10. **a.** $\hat{y} = 36 + 4.333x$ **b.** Yes   **c.** Approximately 9

CHAPTER 12

Section 12.2 *(pages 536–540)*

1. Reject null hypothesis ($\chi^2 = 9.0431$).

3. Reject null hypothesis ($\chi^2 = 9.124$).

5. Do not reject null hypothesis ($\chi^2 = 4.202$).

7. Do not reject null hypothesis ($\chi^2 = 5.419$).

9. Do not reject null hypothesis ($\chi^2 = 0.586$).

11. Do not reject null hypothesis ($\chi^2 = 0.7117$).

Section 12.3 (pages 544–546)

1. Do not reject null hypothesis ($\chi^2 = 9.99$).

3. Reject null hypothesis ($\chi^2 = 34.19$).

5. Do not reject null hypothesis ($\chi^2 = 6.54$).

7. Reject null hypothesis ($\chi^2 = 33.18$).

9. Reject null hypothesis ($\chi^2 = 37.43$).

11. Do not reject null hypothesis ($\chi^2 = 0.15$).

Section 12.4 (pages 550–554)

1. Do not reject null hypothesis ($\chi^2 = 2.474$).

3. Do not reject null hypothesis ($\chi^2 = 1.445$).

5. Reject null hypothesis ($\chi^2 = 59.41$).

7. Reject null hypothesis ($\chi^2 = 13.395$).

9. Do not reject null hypothesis ($\chi^2 = 2.296$).

Mastery Tests: Form A (pages 556–559)

1. Choice (b)

2. Choice (c)

3. Choice (b)

4. Choice (c)

5. 43.53

6. Reject null hypothesis ($\chi^2 = 43.53$).

7. Choice (a)

8. Do not reject null hypothesis ($\chi^2 = 0.954$).

9. Do not reject null hypothesis ($\chi^2 = 8.37$).

10. Reject null hypothesis ($\chi^2 = 35.64$).

Mastery Tests: Form B (pages 559–562)

1. Reject null hypothesis ($\chi^2 = 103.72$).

2. Do not reject null hypothesis ($\chi^2 = 2.348$).

3. Reject null hypothesis ($\chi^2 = 26.07$).

4. Reject null hypothesis ($\chi^2 = 101.7056$).

5. Reject null hypothesis ($\chi^2 = 9.1993$).

6. Do not reject null hypothesis ($\chi^2 = 5.04$ if we do not combine cells. However, since 30% of the cells have expected values which are less than 5, some of the cells should be combined).

7. Do not reject null hypothesis ($\chi^2 = 0.25$).

8. Do not reject null hypothesis ($\chi^2 = 1.675$).

9. Reject null hypothesis ($\chi^2 = 138.018$).

10. Reject null hypothesis ($\chi^2 = 54.947$).

CHAPTER 13

Section 13.4 (pages 585–590)

1.

Source	SS	DF	MS	F-ratio
Brands	24.33	3	8.11	0.59
Error	109.33	8	13.67	
Total	133.67	11		

Do not reject null hypothesis.

2.

Source	SS	DF	MS	F-ratio
Sections	1565.76	4	391.44	12.17
Error	643.2	20	32.16	
Total	2208.96	24		

Reject null hypothesis.

3.

Source	SS	DF	MS	F-ratio
Locations	10062	3	3354	5.07
Error	7945	12	662.08	
Total	18007	15		

Reject null hypothesis.

4.

Source	SS	DF	MS	F-ratio
Locations	2.19	3	0.73	0.03
Error	314.25	12	26.19	
Total	316.44	15		

Do not reject null hypothesis.

5.

Source	SS	DF	MS	F-ratio
Brands	829.73	2	414.87	10.57
Error	471.2	12	39.27	
Total	1300.93	14		

Reject null hypothesis.

6.

Source	SS	DF	MS	F-ratio
Pills	94.15	3	31.39	2.06
Error	243.6	16	15.23	
Total	337.75	19		

Do not reject null hypothesis.

7.

Source	SS	DF	MS	F-ratio
Vitamins	1.26	3	0.42	0.9
Error	5.59	12	0.47	
Total	6.85	15		

Do not reject null hypothesis.

8.

Source	SS	DF	MS	F-ratio
Employees	452.96	4	113.24	1.76
Error	1352.93	21	64.43	
Total	1805.88	25		

Do not reject null hypothesis.

9.

Source	SS	DF	MS	F-ratio
Movies	4	4	22.125	1.116
Conditions	65.35	3	21.125	1.0988
Error	237.9	12	19.825	
Total	391.75	19		

Do not reject null hypotheses in both cases.

10.

Source	SS	DF	MS	F-ratio
Techniques	880.8	3	293.6	7.19
Regions	156.3	4	39.075	0.96
Error	489.7	12	40.808	
Total	1526.8	19		

Reject null hypothesis concerning technique used.

Do not reject null hypothesis concerning regions.

11.

Source	SS	DF	MS	F-ratio
States	3.5	3	1.167	1.00
Crimes	3134	3	1044.67	895.18
Error	10.5	9	1.167	
Total	3148	15		

Do not reject null hypothesis concerning states.

Reject null hypothesis concerning types of crime.

Mastery Tests: Form A (pages 592–593)

1. 151.04		**7.** 37.76
2. 1106.4		**8.** 55.32
3. 1257.44		**9.** 0.68
4. 4		**10.** Choice (b)
5. 20		**11.** Choice (a)
6. 24		**12.** Choice (c)

Mastery Tests: Form B (pages 594–597)

1.

Source	SS	DF	MS	F-ratio
Drugs	358.6	3	119.33	7.05
Error	271.2	16	16.95	
Total	629.8	19		

Reject null hypothesis.

2.

Source	SS	DF	MS	F-ratio
Seasons	3.96	3	1.32	0.629
Quantities	1.31	2	0.655	0.312
Error	12.59	6	2.098	
Total	17.86	11		

Do not reject null hypotheses in both cases.

3.

Source	SS	DF	MS	F-ratio
Stores	2738.5	4	684.625	10.76
Wrappings	1157.75	3	385.917	6.065
Error	763.5	12	63.625	
Total	4659.75	19		

Reject null hypotheses in both cases.

4.

Source	SS	DF	MS	F-ratio
Diets	403.75	3	134.58	1.99
Error	1083.2	16	67.7	
Total	1486.95	19		

Do not reject null hypothesis.

5.

Source	SS	DF	MS	F-ratio
Hospitals	54.7	2	27.35	0.74
Error	1111.3	30	37.04	
Total	1166	32		

Do not reject null hypothesis.

6.

Source	SS	DF	MS	F-ratio
Beaches	45.73	2	22.87	0.63
Error	437.6	12	36.47	
Total	483.33	14		

Do not reject null hypothesis.

7.

Source	SS	DF	MS	F-ratio
Brands	1008.53	2	504.27	1.59
Error	3800.4	12	316.7	
Total	4808.93	14		

Do not reject null hypothesis.

8.

Source	SS	DF	MS	F-ratio
Examiners	90	4	22.5	0.49
Error	914	20	45.7	
Total	1004	24		

Do not reject null hypothesis.

9.

Source	SS	DF	MS	F-ratio
Rations	201.5	2	100.75	12.64
Weights	56.92	3	18.97	2.38
Error	47.83	6	7.97	
Total	306.25	11		

Reject null hypothesis concerning rations.

Do not reject null hypothesis concerning weights.

10.

Source	SS	DF	MS	F-ratio
Methods	176.69	3	58.9	1.02
Error	691.75	12	57.65	
Total	868.44	15		

Do not reject null hypothesis.

CHAPTER 14

Section 14.3 (pages 610–614)

1. Do not reject null hypothesis.

3. Do not reject null hypothesis.

5. Reject null hypothesis.

7. Do not reject null hypothesis.

9. Reject null hypothesis.

Section 14.4 (pages 618–620)

1. Reject null hypothesis.

3. Reject null hypothesis.

Section 14.5 (pages 626–628)

1. Reject null hypothesis ($z = 3.76$).

3. Do not reject null hypothesis ($z = -1.95959$). Decision is very close. Actually, more data should be collected.

5. Do not reject null hypothesis ($z = 0.14$).

Section 14.7 (pages 635–636)

1. a. Do not reject null hypothesis ($R = 0.1$).   **b.** Reject null hypothesis ($R = -0.9$).

c. Do not reject null hypothesis ($R = -0.1$).

3. Random

5. Random

Section 14.8 (page 641)

1. Do not reject null hypothesis ($\chi^2 = 1.5951$).

2. Do not reject null hypothesis ($\chi^2 = 1.1822$).

Mastery Tests: Form A (pages 645–648)

1. Reject null hypothesis.

2. Do not reject null hypothesis.

3. Reject null hypothesis.

4. Random

5. Do not reject null hypothesis.

6. $\dfrac{12}{n(n+1)} \sum_{i=1}^{r} \dfrac{R_i^2}{n_i} - 3(n+1)$

7. Choice (d)

8. Each of the tied observations is assigned the mean of the ranks that they occupy.

9. Choice (c)

10. Choice (d)

Mastery Tests: Form B (pages 649–651)

1. Do not reject null hypothesis.

2. Reject null hypothesis.

3. Do not reject null hypothesis.

4. Random

5. Random

6. Random

7. Do not reject null hypothesis.

8. Do not reject null hypothesis.

9. Do not reject null hypothesis.

10. Do not reject null hypothesis.

11. Reject null hypothesis.

INDEX

FREQUENTLY USED FORMULAS

Relative frequency $\dfrac{f_i}{n}$

Mean $\dfrac{\Sigma x}{n} = \dfrac{x_1 + x_2 + \cdots + x_n}{n}$

Weighted mean $\bar{x}_w = \dfrac{\Sigma xw}{\Sigma w}$

Variance $\sigma^2 = \dfrac{\Sigma(x - \mu)^2}{n}$ or $\dfrac{\Sigma x^2}{n} - \dfrac{(\Sigma x)^2}{n^2}$

Standard deviation $\sigma = \sqrt{\dfrac{\Sigma(x - \mu)^2}{n}}$

Average deviation $\dfrac{\Sigma|x - \mu|}{n}$

Sample standard deviation $\sqrt{\dfrac{\Sigma(x - \bar{x})^2}{n - 1}}$

Percentile rank of X $\dfrac{B + \frac{1}{2}E}{n} \cdot 100$

z-score $z = \dfrac{x - \mu}{\sigma}$ **Original score** $x = \mu + z\sigma$

Probability, p $p = \dfrac{f}{n}$

$_nP_r = \dfrac{n!}{(n - r)!}$ $_nP_n = n!$ $_nC_r = \dfrac{n!}{r!(n - r)!}$

Number of permutations with repetitions $\dfrac{n!}{p!q!r! \cdots}$

Mathematical expectation $m_1p_1 + m_2p_2 + m_3p_3 + \cdots$

Addition rule (for mutually exclusive events) $p(A \text{ or } B) = p(A) + p(B)$

Addition rule (general case) $p(A \text{ or } B) = p(A) + p(B) - p(A \text{ and } B)$

Complement of event A $p(A') = 1 - p(A)$

Conditional probability formula $p(A|B) = \dfrac{p(A \text{ and } B)}{p(B)}$

Multiplication rule $p(A \text{ and } B) = p(A|B) \cdot p(B)$

Multiplication rule (for independent events) $p(A \text{ and } B) = p(A) \cdot p(B)$

Bayes' rule $\dfrac{p(B|A_n)p(A_n)}{p(B|A_1)p(A_1) + p(B|A_2)p(A_2) + \cdots + p(B|A_n)p(A_n)}$

Mean of a probability distribution $\mu = \Sigma xp(x)$

ABOUT THE AUTHOR

Historian, public servant, and author, DANIEL J. BOORSTIN, who was Librarian of Congress Emeritus, directed the Library from 1975 to 1987. He had previously been director of the National Museum of History and Technology, and senior historian of the Smithsonian Institution in Washington, D.C. Before that he was the Preston and Sterling Morton Distinguished Service Professor of History at the University of Chicago, where he taught for twenty-five years.

Born in Atlanta, Georgia, and raised in Tulsa, Oklahoma, Boorstin graduated with highest honors from Harvard College and received his doctorate from Yale University. As a Rhodes Scholar at Balliol College, Oxford, England, he won a coveted double first in two degrees in law and was admitted as a barrister-at-law of the Inner Temple, London. He was also a member of the Massachusetts bar. He has been visiting professor at the University of Rome, the University of Geneva, the University of Kyoto in Japan, and the University of Puerto Rico. In Paris he was the first incumbent of a chair in American history at the Sorbonne, and at Cambridge University, England, he was Pitt Professor of American History and Institutions and Fellow of Trinity College. Boorstin lectured widely in the United States and all over the world. He received numerous honorary degrees and was decorated by the governments of France, Belgium, Portugal, and Japan.

The Discoverers, Boorstin's history of man's search to know the world and himself, was published in 1983. A Book-of-the-Month Club Main Selection, *The Discoverers* was on the *New York Times* best-seller list for half a year and won the Watson Davis Prize of the History of Science Society. This and his other books have been translated into more than twenty languages.

Boorstin's many books include *The Americans: The Colonial Experience* (1958), which won the Bancroft Prize; *The Americans: The National Experience* (1965), which won the Parkman Prize; and *The Americans: The Democratic Experience* (1973), which won the Pulitzer Prize for History and the Dexter Prize and was a Book-of-the-Month Club Main Selection. Among his other books are *The Mysterious Science of the Law* (1941), *The Lost World of Thomas Jefferson* (1948), *The Genius of American Politics* (1953), *The Image* (1962), and *The Republic of Technology* (1978). For young people he has written the *Landmark History of the American People.* His textbooks for high schools, *A History of the United States* (1980), written with Brooks M. Kelley, has been widely adopted. He edited *An American Primer* (1996) and the thirty-volume series *The Chicago History of American Civilization,* among other works. He died in 2004.

HIDDEN HISTORY
Exploring Our Secret Past

A collection of 24 incisive essays that examine rhythms, patterns, and institutions of everyday American life—from intimate portraits of legendary figures to expansive discussions of historical phenomena.

"Highly representative of his awesome scope ... eminently readable and provocative."   —*Washington Post Book World*

History/0-679-72223-8/$12.00 (Can. $16.00)

THE IMAGE
A Guide to Pseudo-Events in America
With an Afterword by George F. Will

In this analysis of America's inundation by illusion, Boorstin introduces the concept of "pseudo-events"—events such as press conferences and presidential debates, which are staged solely in order to be reported— and redefines *celebrity* as "a person who is known for his well-knownness." The result is an essential resource for anyone who wants to distinguish the manifold deceptions of our culture from its few enduring truths.

"A very informative and entertaining and chastising book."   —*Harper's*

History/0-679-74180-1/$12.00 (Can. $15.00)

- wifi
- gym m
- adv.
- mae m adxp op
- firm M
- wireless prop
- serv Tab
- staca chor.

Hypocrisy - skrmps
Invectives > immorality

(coconut oil [w] Lavender drop) - deoderant!

577 BF epitaph.

cucumber + pineapple
⇒ Juice - defer.